The Collected Papers of Hans Arnold Heilbronn

CANADIAN MATHEMATICAL SOCIETY SERIES OF MONOGRAPHS AND ADVANCED TEXTS

Monographies et Études de la Société Mathématique du Canada

Frank H. Clarke *Optimization and Nonsmooth Analysis*

Erwin Klein and Anthony C. Thompson *Theory of Correspondences: Including Applications to Mathematical Economics*

I. Gohberg, P. Lancaster, and L. Rodman *Invariant Subspaces of Matrices with Applications*

Jonathan Borwein and Peter Borwein *PI and the AGM—A Study in Analytic Number Theory and Computational Complexity*

Subhashis Nag *The Complex Analytic Theory of Teichmüller Spaces*

Manfred Kracht and Erwin Kreyszig *Methods of Complex Analysis in Partial Differential Equations with Applications*

Ernst J. Kani and Robert A. Smith *The Collected Papers of Hans Arnold Heilbronn*

The Collected Papers of Hans Arnold Heilbronn

Edited by

ERNST J. KANI
Queen's University, Kingston

and

the late ROBERT A. SMITH
University of Toronto

A Wiley-Interscience Publication

JOHN WILEY & SONS

New York • Chichester • Brisbane • Toronto • Singapore

Library of Congress Cataloging in Publication Data:

Heilbronn, Hans Arnold, 1908–1975.
The collected papers of Hans Arnold Heilbronn.

(Canadian Mathematical Society series of monographs
and advanted texts = Monographies et études de la
Société mathématique du Canada)
"A Wiley-Interscience publication."
1. Numbers, Theory of—Collected works.
2. Mathematical analysis—Collected works.
3. Heilbronn, Hans Arnold, 1980–1975—Bibliography.
4. Mathematicians—Germany—Biography. I. Kani,
Ernst J. II. Smith, Robert A. (Robert Arnold), 1937–1983.
III. Title. IV. Series: Canadian Mathematical
Society series of monographs and advanced texts.
QA241.H372515 1988 510 86-28999
ISBN 0-471-81655-8

Printed in the United States of America

10 9 8 7 6 5 4 3 2 1

HANS ARNOLD HEILBRONN
8 October 1908–28 April 1975

Contributors

Bryan J. Birch, Mathematical Institute, 24-29 St. Giles, Oxford, England

John H. H. Chalk, Department of Mathematics, University of Toronto, Ontario, Canada

Sarvadaman Chowla, School of Mathematics, Institute for Advanced Study, Princeton, New Jersey, USA

Max Deuring, Göttingen, Fed. Rep. Germany (Deceased)

Patrick Du Val, 7 Legion Court, Cambridge, England

Richard Foote, Department of Mathematics, University of Vermont, Burlington, Vermont, USA

Albrecht Fröhlich, Imperial College of Science and Technology, Department of Mathematics, Huxley Building, Queen's Gate, London, England

Walter K. Hayman, Department of Mathematics, Imperial College of Science and Technology, Queen's Gate, London, England

Donald J. Lewis, Department of Mathematics, The University of Michigan, Ann Arbor, Michigan, USA

Tressilian Nicholas, Trinity College, Cambridge, England

Robert A. Rankin, Department of Mathematics, University of Glasgow, Glasgow, Scotland

Peter Scherk, Department of Mathematics, University of Toronto, Toronto, Ontario, Canada (Deceased)

John C. Shepherdson, School of Mathematics, University of Bristol, University Walk, Bristol, England

Harold Stark, Department of Mathematics, Massachusetts Institute of Technology, Cambridge, Massachusetts, USA

Olga Taussky (Mrs. Todd), Alfred P. Sloan Laboratory of Mathematics and Physics, California Institute of Technology, Pasadena, California, USA

Robert C. Vaughan, Department of Mathematics, Imperial College of Science and Technology, Queen's Gate, London, England

Preface

Hans Heilbronn is probably best known to the mathematical world for his work on class numbers of imaginary quadratic fields; his outstanding contributions in this area include the proof of the *Gauss conjecture* on the finiteness of such fields with a given class number and the verification that there is "at most one more" (than actually exists) imaginary quadratic field of class number one. Nevertheless, his other contributions to number theory and analysis, especially to Waring's problem, L-functions, sieve problems, etc., are by no means of lesser importance and have had a substantial influence on the development of the subjects. Although some of his results have by now been superseded, many of the methods developed therein have not, and have opened up the way for a vast amount of research throughout the past thirty to forty years. For this and other reasons it seems appropriate to make the papers of Heilbronn readily available to the public.

At the insistence of the publisher, I have translated Heilbronn's articles written in German into English. In doing so, I have tried to preserve some of Heilbronn's (and Landau's) rather unique style in the translations, sometimes at the cost of smoothness of language. However, the translations are not meant to replace but rather to help the reader with the originals.

The project of collecting and editing these papers for publication was initiated by the late Professor Robert A. Smith, but his sudden and untimely death in March 1983 prevented him from completing this task. After his death, I was asked by Professor J. H. H. Chalk, Mrs. Smith, and Mrs. Heilbronn to continue the project since I had been a good friend of Bob Smith as well as Heilbronn's last (master) student just prior to his retirement in 1974.

In continuing this project I have been aided by many people, to all of whom I am very grateful. First and foremost I would like to thank the many contributors of recollections and commentaries for their excellent work in compiling such interesting contributions. (For example, in W. K. Hayman's commentary, a problem left open by Heilbronn is settled.) Some of the contributors, notably J. H. H. Chalk, A. Fröhlich, and R. C. Vaughan, have also helped me considerably with their advice on several points, as did P. Roquette. Furthermore, I am indebted to O. Taussky for providing me with the pictures of Heilbronn reproduced on page 36 of this volume, and to

Mrs. K. Smith and Mrs. A. Davenport for their substantial "behind the scenes" help in this project. I would also like to express my appreciation to E. O. Kreyszig for his offer to publish the volume in the series *CMS Monographs and Advanced Texts*, for his advice concerning several technical matters, and for his encouragement of the whole project. Finally, I would like to thank my wife Consuelo for her help with the typing and the rather tedious job of preparing the manuscript for publication.

ERNST J. KANI

Heidelberg
December 1984

Contents

HANS ARNOLD HEILBRONN

Curriculum Vitae 3

Bibliography of Heilbronn 5

Hans Arnold Heilbronn 9
J. W. S. Cassels and A. Fröhlich
Biographical Memoirs of Fellows of The Royal Society **22** (1976), 119–135

My Personal Recollections of Hans Heilbronn 27
Olga Taussky

Heilbronn at Bristol 39
John C. Shepherdson

Hans A. Heilbronn: Toronto Days, 1964–1974 45
J. H. H. Chalk

Brief Recollections of Hans Heilbronn 47
I. By M. Deuring
II. By P. Scherk
III. By T. C. Nicholas
IV. By P. Du Val
V. By R. A. Rankin
VI. By D. J. Lewis

PUBLICATIONS BY HANS ARNOLD HEILBRONN

1. Über die Verteilung der Primzahlen in Polynomen 57
Math. Annal. **104** (1931), 794–799

1t. On the Distribution of Prime Numbers in Polynomials 63
Translation of [1]

2. Über den Primzahlsatz von Herrn Hoheisel **70**
Math. Z. **36** (1933), 394–423

2t. On the Prime Number Theorem of Hoheisel **100**
Translation of [2]

3. (With E. Landau) Bemerkungen zur vorstehenden Arbeit von Herrn Bochner **143**
Math. Z. **37** (1933), 10–16

3t. (With E. Landau) Remarks on the Above Article of Bochner **150**
Translation of [3]

4. (With E. Landau) Ein Satz Über Potenzreihen **160**
Math. Z. **37** (1933), 17

4t. (With E. Landau) A Theorem on Power Series **162**
Translation of [4]

5. (With E. Landau) Anwendungen der N. Wienerschen Methode **164**
Math. Z. **37** (1933), 18–21

5t. (With E. Landau) Applications of the Method of N. Wiener **168**
Translation of [5]

6. Zu dem Integralsatz von Cauchy **173**
Math. Z. **37** (1933), 37–38

6t. On the Integral Theorem of Cauchy **175**
Translation of [6]

7. On the Class-Number in Imaginary Quadratic Fields **177**
Quart. J. Math. **5** (1934), 150–160

8. (With E. H. Linfoot) On the Imaginary Quadratic Corpora of Class-Number One **188**
Quart. J. Math. **5** (1934), 293–301

9. (With E. Landau and P. Scherk) Alle Grossen Ganzen Zahlen Lassen Sich als Summe von Höchstens 71 Primzahlen Darstellen **197**
Casopis Pest. Mat. Fys. **65** (1936), 117–141

9t. (With E. Landau and P. Scherk) All Large Integers can be Represented as a Sum of at Most 71 Prime Numbers **222**
Translation of [9]

10. Über das Waringsche Problem **252**
Acta Arith. **1** (1936), 212–221

10t. On Waring's Problem **262**
Translation of [10]

11. (With H. Davenport) On the Zeros of Certain Dirichlet Series I **272**
J. Lond. Math. Soc. **11** (1936), 181–185

12. (With H. Davenport) On the Zeros of Certain Dirichlet Series II **277**
J. Lond. Math. Soc. **11** (1936), 307–312

13. (With H. Davenport) On Waring's Problem for Fourth Powers **283**
Proc. Lond. Math. Soc. **41** (1936), 143–150

14. (With H. Davenport) On an Exponential Sum **291**
Proc. Lond. Math. Soc. **41** (1936), 449–453

15. (With H. Davenport) On Waring's Problem: Two Cubes and One Square **296**
Proc. Lond. Math. Soc. **43** (1937), 73–104

16. (With H. Davenport) Note on a Result in the Additive Theory of Numbers **328**
Proc. Lond. Math. Soc. **43** (1937), 142–151

17. On Real Characters **338**
Acta Arith. **2** (1937), 212–213

18. On an Inequality in the Elementary Theory of Numbers **340**
Proc. Camb. Phil. Soc. **33** (1937), 207–209

19. On Dirichlet Series which Satisfy a Certain Functional Equation **343**
Quart. J. Math. **9** (1938), 194–195

20. On Euclid's Algorithm in Real Quadratic Fields **345**
Proc. Camb. Phil. Soc. **34** (1938), 521–526

21. (With G. H. Hardy) Edmund Landau **351**
J. London Math. Soc. **13** (1938), 302–310

22. (With H. Davenport) On Indefinite Quadratic Forms in Five Variables **360**
J. Lond. Math. Soc. **21** (1946), 185–193

23. (With H. Davenport) On the Minimum of a Bilinear Form **369**
Quart. J. Math. **18** (1947), 107–121

24. (With H. Davenport) Asymmetric Inequalities for Non-homogeneous Linear Forms **384**
J. Lond. Math. Soc. **22** (1947), 53–61

25. On the Distribution of the Sequence $n^2\theta$(mod 1) **393**
Quart. J. Math. **19** (1948), 249–256

26. On Discrete Harmonic Functions **401**
Proc. Camb. Phil. Soc. **45** (1949), 194–206

27. On Euclid's Algorithm in Cubic Self-Conjugate Fields **414**
Proc. Camb. Phil. Soc. **46** (1950), 377–382

28. On Euclid's Algorithm in Cyclic Fields **420**
Can. J. Math. **3** (1951), 257–268

29. On the Averages of Some Arithmetical Functions of Two Variables **432**
Mathematika **5** (1958), 1–7

30. On the Representation of Homotopic Classes by Regular Functions **439**
Bull. Acad. Pol. Sci. **6** (1958), 181–184

31. Review of *Mathematische Werke* by Erick Hecke **443**
Math. Gazette **45** (1961), 77–78

32. Old Theorems and New Methods in Class-Field Theory **444**
J. London Math. Soc. **37** (1962), 6–9

33. (With L. Howarth) Prof. Jacques Hadamard **448**
Nature **200** (1963), 937–938

34. On the Representation of a Rational as a Sum of Four Squares by Means of Regular Functions **450**
J. Lond. Math. Soc. **39** (1964), 72–76

35. (With P. Erdös) On the Addition of Residue Classes mod *p* **455**
Acta Arith. **9** (1964), 149–159

36a. Introduction to Papers on Combinatory Analysis and Sums of Squares **466**
36b. Introduction to Papers on Waring's Problem **468**
36c. Introduction to Papers on Goldbach's Problem **473**
Collected Papers of G. H. Hardy (edited), Vol. I, Oxford at the Clarendon Press, 1966, pp. 263–264, 377–381, 533–534

37. Zeta-Functions and L-Functions **475**
Algebraic Number Theory (J. W. S. Cassels and A. Fröhlich, eds.), Academic Press, London, 1967, pp. 204–230

38. (With S. Bachmuth and H. Y. Mochizuki) Burnside Metabelian Groups **502**
Proc. Roy. Soc. Lond. **A307** (1968), 235–250

39. On the Average Length of a Class of Finite Continued Fractions **518**
Abhandlungen aus Zahlentheorie und Analysis zur Erinnerung an Edmund Landau (P. Turán, ed.), VEB Deutscher Verlag der Wissenschaften, Berlin, 1969, pp. 89–96

40. (With P. Scherk) Sums of Complexes in Torsion-Free Abelian Groups **526**
Can. Math. Bull. **12** (1969), 479–480

41. (With H. Davenport) On the Density of Discriminants of Cubic Fields **528**
Bull. Lond. Math. Soc. **1** (1969), 345–348

42. (With H. Davenport) On the Density of Discriminants of Cubic Fields II **532**
Proc. Roy. Soc. Lond. **A322** (1971), 405–420

43. On the 2-Class Group of Cubic Fields **548**
Studies in Pure Mathematics (L. Mirsky, ed.) Academic Press, London–New York, 1971, pp. 117–119

44. On Real Simple Zeros of Dedekind ζ-Functions **551**
Proc. 1972 Number Theory Conference, Univ. Colorado, Boulder, 1972, pp. 108–110

45. On Real Zeros of Dedekind ζ-Functions **554**
Can. J. Math. **25** (1973), 870–873

COMMENTARIES

The Papers on Prime Number Problems and Applications of Brun's Sieve ([1], [2], [9], [18]) **561**
R. C. Vaughan

The Papers on Analysis ([3]–[6], [26], [30]) **567**
W. K. Hayman

The Papers on Class Numbers of Imaginary Quadratic Fields ([7], [8], [45]) **571**
H. Stark

The Papers on Waring's Problem ([10], [13]–[16], [22], [25], [34]) **576**
B. J. Birch and R. C. Vaughan

The Papers on Euclid's Algorithm ([20], [27], [28]) **586**
H. Stark

The Papers on Zeta- and L-Functions ([11], [12]) **588**
H. Stark

The Papers on the Geometry of Numbers ([23], [24]) **589**
J. H. H. Chalk

The Papers on Discriminants of Cubic Fields ([41], [42]) **592**
A. Fröhlich

Miscellaneous Papers ([17], [19], [29], [35], [38], [39], [43]) **598**
B. J. Birch, S. Chowla, R. Foote, A. Fröhlich, and
R. C. Vaughan

Acknowledgments **605**

Hans Arnold Heilbronn

degree n, $|\mathrm{disc}(k)| = 3^{\frac{1}{2}n}$ But, as is proved in

H. Hasse, Zahlentheorie, Akademie V., 1948, p. 452

$$|\mathrm{disc}(k)| \geq \left(\frac{\pi}{4}\right)^n \left(\frac{n^n}{n!}\right)^2$$

$$|\mathrm{disc.}\, k|^{\frac{1}{n}} \geq \frac{\pi e^2}{4}(1+o(1))$$ for large n by Stirling's Formula. Actually, one can verify that for $n \geq 120$

$$3^{\frac{1}{2}} < \frac{\pi}{4}\, n^2\, n!^{-\frac{2}{n}}$$

Now we may assume that k is normal over $Q(\sqrt{-3})$ and that $\mathrm{Gal}(k/Q\sqrt{-3})$ is simple. As the class-number of $Q(\sqrt{-3})$, this Galois group is not abelian, hence its order is at least 60, i.e. $n \geq 120$.

The same argument will work, mutatis mutandis, for $Q(\sqrt{-1})$ and $Q(\sqrt{5})$, and for the cubic field of disc. -23. But it seems hopeless to-day to obtain an ∞ty of such fields.

Best wishes

yours sincerely

H. Heilbronn

Extract from a letter of Heilbronn of 18 December 1973 to one of the editors (E.K.).

Curriculum Vitae

Lebenslauf.*

Ich, Hans Heilbronn, wurde am 8. 10. 1908 in Berlin als Sohn des Kaufmanns Alfred Heilbronn geboren. Ich besitze die preußische Staatsangehörigkeit. Von Ostern 1914 bis Ostern 1926 besuchte ich das Heinrich von Kleist-Realgymnasium, Berlin, das ich mit dem Zeugnis der Reife verließ. Ich studierte im S.-S. 1926 Medizin in Berlin, alsdann Mathematik und Naturwissenschaften im W.-S. 26/27 und S.-S. 27 in Berlin, im W.-S. 27/28 und S.-S. 28 in Freiburg i. Brsg., vom W.-S. 28/29 bis zum S.-S. 30 in Göttingen. Seit dem Herbst 1930 bin ich als Assistent im Mathematischen Institut der Universität Göttingen beschäftigt. Ich danke allen meinen Lehrern für das große Interesse, das sie mir jederzeit entgegengebracht haben. Mein besonderer Dank gilt Herrn Professor Landau für die wertvollen Anregungen und Ratschläge, die er mir zuteil werden ließ.

[I, Hans Heilbronn, the son of the merchant Alfred Heilbronn, was born on 8 October 1908 in Berlin. I am Prussian by nationality. From Easter 1914 to Easter 1926 I attended the Heinrich Kleist Realgymnasium (Secondary school), Berlin, which I left with the "Zeugnis der Reife" (lit. proof of maturity, i.e. graduation diploma). In the summer semester (S-S.) 1926 I studied medicine in Berlin, then mathematics and the natural sciences in the winter semester (W-S.) 1926/27 and S-S. 27 in Berlin, in W-S. 27/28 and S-S. 28 in Freiburg (im Breisgau), and from W-S. 28/29 to S-S. 30 in Göttingen. Since the autumn of 1930 I have been employed as Assistant at the Mathematical Institute of the University of Göttingen. I thank all my teachers for the great interest which they have shown towards me at all times. My particular thanks are due to Professor Landau, for the valuable suggestions and advice which he bestowed on me.]

*From Heilbronn's dissertation (1931).

Summary

8 Oct. 1908	Born in Berlin
1926–1930	Studied Medicine, Mathematics & Sciences in Berlin, Freiburg and Göttingen
1930–1933	Assistant at Universität Göttingen
16 Dec. 1931	(Oral) Ph.D. exam
1934–1935	At University of Bristol (supported by Academic Assistance Council)
May 1935–1940	At University of Cambridge with Bevan Fellowship
1940	Interned at Isle of Man; later released
1940–1945	Served with British army
1945	At University College, London
1946–1949	Reader at University of Bristol
1949–1963	Professor & Head of Department at Bristol
April 1963	Resigned position at Bristol
March 1964	Married Dorothy Greaves
1964–1975	Professor at University of Toronto, Canada
28 April 1975	Died in Toronto

Bibliography of Heilbronn

[1] Über die Verteilung der Primzahlen in Polynomen.
Mathematische Annalen **104** (1931), 794–799.

[2] Über den Primzahlsatz von Herrn Hoheisel.
Mathematische Zeitschrift **36** (1933), 394–423.

[3] (With E. Landau) Bemerkungen zur vorstehenden Arbeit von Herrn Bochner.
Mathematische Zeitschrift **37** (1933), 10–16.

[4] (With E. Landau) Ein Satz über Potenzreihen.
Mathematische Zeitschrift **37** (1933), 17.

[5] (With E. Landau) Anwendungen der N. Wienerschen Methode.
Mathematische Zeitschrift **37** (1933), 18–21.

[6] Zu dem Integralsatz von Cauchy.
Mathematische Zeitschrift **37** (1933), 37–38.

[7] On the class-number in imaginary quadratic fields.
The Quarterly Journal of Mathematics, Oxford Series **5** (1934), 150–160.

[8] (With E. H. Linfoot) On the imaginary quadratic corpora of class-number one.
The Quarterly Journal of Mathematics, Oxford Series **5** (1934), 293–301.

[9] (With E. Landau and P. Scherk) Alle grossen ganzen Zahlen lassen sich als Summe von höchstens 71 Primzahlen darstellen.
Casopis pro pestování matematiky a fysiky **65** (1936), 117–141.

[10] Über das Waringsche Problem.
Acta Arithmetica **1** (1936), 212–221.

[11] (With H. Davenport) On the zeros of certain Dirichlet series. I.
Journal of the London Mathematical Society **11** (1936), 181–185.

[12] (With H. Davenport) On the zeros of certain Dirichlet series. II.
Journal of the London Mathematical Society **11** (1936), 307–312.

[13] (With H. Davenport) On Waring's problem for fourth powers.
Proceedings of the London Mathematical Society **41** (1936), 143–150.

[14] (With H. Davenport) On an exponential sum.
Proceedings of the London Mathematical Society **41** (1936), 449–453.

[15] (With H. Davenport) On Waring's problem: Two cubes and one square.
Proceedings of the London Mathematical Society **43** (1937), 73–104.

[16] (With H. Davenport) Note on a result in the additive theory of numbers.
Proceedings of the London Mathematical Society **43** (1937), 142–151.

[17] On real characters.
Acta Arithmetica **2** (1937), 212–213.

[18] On an inequality in the elementary theory of numbers.
Proceedings of the Cambridge Philosophical Society **33** (1937), 207–209.

[19] On Dirichlet series which satisfy a certain functional equation.
The Quarterly Journal of Mathematics, Oxford Series **9** (1938), 194–195.

[20] On Euclid's algorithm in real quadratic fields.
Proceedings of the Cambridge Philosophical Society **34** (1938), 521–526.

[21] (With G. H. Hardy) Edmund Landau.
Journal of the London Mathematical Society **13** (1938), 302–310.

[22] (With H. Davenport) On indefinite quadratic forms in five variables.
Journal of the London Mathematical Society **21** (1946), 185–193.

[23] (With H. Davenport) On the minimum of a bilinear form.
The Quarterly Journal of Mathematics, Oxford Series **18** (1947), 107–121.

[24] (With H. Davenport) Asymmetric inequalities for non-homogeneous linear forms.
Journal of the London Mathematical Society **22** (1947), 53–61.

[25] On the distribution of the sequence $n^2\theta \pmod 1$.
The Quarterly Journal of Mathematics, Oxford Series **19** (1948), 249–256.

[26] On discrete harmonic functions.
Proceedings of the Cambridge Philosophical Society **45** (1949), 194–206.

[27] On Euclid's algorithm in cubic self-conjugate fields.
Proceedings of the Cambridge Philosophical Society **46** (1950), 377–382.

[28] On Euclid's algorithm in cyclic fields.
Canadian Journal of Mathematics **3** (1951), 257–268.

[29] On the averages of some arithmetical functions of two variables.
Mathematika **5** (1958), 1–7.

[30] On the representation of homotopic classes by regular functions.
Bulletin de l'Académie Polonaise des Sciences **6** (1958), 181–184.

[31] Review of *Mathematische Werke* by Erick Hecke.
The Mathematical Gazette **45** (1961), 77–78.

[32] Old theorems and new methods in class-field theory.
Journal of the London Mathematical Society **37** (1962), 6–9.

[33] (With L. Howarth) Prof. Jacques Hadamard
Nature **200** (1963), 937–938.

[34] On the representation of a rational as a sum of four squares by means of regular functions.
Journal of the London Mathematical Society **39** (1964), 72–76.

[35] (With P. Erdös) On the addition of residue classes mod *p*.
Acta Arithmetica **9** (1964), 149–159.

[36] a. Introduction to papers on combinatorial analysis and sums of squares.
b. Introduction to papers on Waring's problem.
c. Introduction to papers on Goldbach's problem.
Collected Papers of G. H. Hardy (edited by a committee appointed by the London Mathematical Society), Vol. I, Oxford at the Clarendon Press, 1966, pp. 263–264, 377–381, 533–534.

[37] Zeta functions and L-functions.
Algebraic Number Theory (J. W. S. Cassels and A. Fröhlich, eds.), Academic Press, London, 1967, pp. 204–230.

[38] (With S. Bachmuth and H. Y. Mochizuki) Burnside metabelian groups.
Proceedings of the Royal Society, London **A 307** (1968), 235–250.

[39] On the average length of a class of finite continued fractions.
Abhandlungen aus Zahlentheorie und Analysis zur Erinnerung an Edmund Landau (edited by P. Turán), VEB Deutscher Verlag der Wissenschaften, Berlin, 1969, pp. 89–96.

[40] (With P. Scherk) Sums on complexes in torsion-free abelian groups.
Canadian Mathematical Bulletin **12** (1969), 479–480.

[41] (With H. Davenport) On the density of discriminants of cubic fields.
Bulletin of the London Mathematical Society **1** (1969), 345–348.

[42] (With H. Davenport) On the density of discriminants of cubic fields II.
Proceedings of the Royal Society, London **A 322** (1971), 405–420.

[43] On the 2-class group of cubic fields.
Studies in Pure Mathematics (L. Mirsky, ed.), Academic Press, London-New York, 1971, pp. 117–119.

[44] On real simple zeros of Dedekind ζ-functions.
Proceedings of the 1972 Number Theory Conference, University of Colorado, Boulder, 1972, pp. 108–110.

[45] On real zeros of Dedekind ζ-functions.
Canadian Journal of Mathematics **25** (1973), 870–873.

Hans Arnold Heilbronn

8 October 1908–28 April 1975
Elected F.R.S. 1971

J. W. S. Cassels, F.R.S. and A. Fröhlich, F.R.S.

HANS ARNOLD HEILBRONN was born in Berlin on 8 October 1908. No doubt, his home and upbringing was one typical for the cultured German-Jewish middle class, which had in those days been thoroughly assimilated into German life. In many small ways Heilbronn's habits, his directness, his correct manner—in the true sense of the word—his strong, genuine sense of propriety, his apparent stiffness, as well as his accent, bore witness to his German background

From 1914 to 1926 the boy attended the Realgymnasium Berlin-Schmargenhof, a school comparable to an English grammar school, the prefix 'Real' indicating emphasis on the sciences and on modern languages, rather than the classics. In 1926 he entered university, reading mathematics, physics and chemistry, but evidently his interests veered more and more towards mathematics. As customary in Germany, the young student moved around, first attending his home university in Berlin, then going on to Freiburg and ending up in Göttingen, at that time the undisputed centre of German mathematics. One would suppose that he fitted well into the accepted pattern of student life in the Germany of those days. It was only the rebels who abhorred duelling—Heilbronn carried a duelling scar from his Göttingen days throughout his life.

In 1930 the budding mathematician became assistant to E. Landau, the leader of a flourishing school of analytic number theory in Göttingen. In 1933 Heilbronn was awarded the D.Phil., the contents of the doctoral dissertation being paper 2 in our publications list, p. 134. By the end of 1933 he had published six papers, some jointly with Landau, and he had already made a mathematical name for himself, as was to become evident in terms of practical assistance after Hitler's rise to power.

By all accounts Landau was a rather formidable man and a demanding task master. Heilbronn seems, however, to have managed very well. Landau came to have a very high opinion of him and, judging by letters he wrote on his behalf in 1933, considered him his star pupil. On the other hand, those years with Landau were formative ones for Heilbronn. Throughout his life his principal research interest remained with number theory, largely analytic. In the shorter term, Landau also clearly exerted an influence on Heilbronn's mathematical style—witness his early papers.

This may be the right place to record an anecdote which, many years ago, Heilbronn told one of us (A. F.), who was then his research student. Heilbronn had been rather negative in reply to a conjecture put forward by A. F., but then, just as A. F. was leaving the room, he added: 'You must never take too much notice of pessimistic comments from your supervisor, or from any other mathematician, however great.' (Advice for which A. F. has always been grateful—he incidentally did prove his conjecture.) Heilbronn then recalled that he, when a student of Landau, had once raised the question of improving the estimates for trigonometric sums along the lines on which the Russian mathematician Vinogradov soon afterwards succeeded—only to be greeted with the answer from Landau that this was hopeless. Heilbronn added: 'I might not have got anywhere, but at that time I never even tried.'

In 1933, after the Nazis had come to power, Heilbronn, like Jewish scientists all over Germany, lost his position. In November he came to Cambridge, supporting himself for the time being. 'There is no possibility for me to continue scientific work in Germany', he writes in a letter to the Academic Assistance Council (about which more later), and 'I have means for half a year'. In mid-December Heilbronn received an invitation from H. R. Hassé, Head of the Department of Mathematics in the University of Bristol, offering hospitality and financial support, and on 16 January 1934 he came to Bristol, taking up residence at a students' hall of residence, Wills Hall, 'where the food is excellent (if you don't compare it with Trinity of course)', as he writes in a letter to H. Davenport.

These few sentences do scant justice to the efforts which had gone into finding some place for Heilbronn to go to. It is fascinating to take a look behind the scenes. The pivot of all the activity on behalf of refugee scientists was the Academic Assistance Council (A.A.C.). This is not the place to record all the good it did. One has to remember that Heilbronn was just one of many.

The first mention of Heilbronn's name in the records of the A.A.C. is in a letter from Harold Davenport, dated 31 August 1933. Davenport had just returned from a visit to Germany, having spent May–July in Göttingen and having, as he explains, come into contact with most mathematicians there. He singles out Heilbronn as specially deserving support, being 'one of the most promising young German mathematicians'. (Incidentally, this confirms that the two knew each other by then—whether this was their first meeting is not known.) In October 1933 there follows a letter of support from Hardy. In the meantime Mordell had been active in Manchester and the Vice-Chancellor writes from there offering accommodation for Heilbronn, but asking for financial support from the A.A.C. Not unexpectedly the A.A.C., notoriously short of funds, had to say no. Similar regretful negative replies went to Hardy, and to Heilbronn himself. Then, on 8 November, Hassé from Bristol says: 'I am considering the possibility of inviting a young German Jewish mathematician to take a temporary position.' He mentions Heilbronn, following a recommendation by Harold Bohr, and adds that he has asked the Jewish community in Bristol to help raise £200 p.a. 'Can the A.A.C. help?' Reply on 10 November: 'Our annual grant is £182'—plus a conditional offer of considering a small supplement if most of the £182 can be raised locally. On 29 November another letter from Hassé: 'The Jewish community will give £75 for three years'—so he needs another £107. No reply recorded. Then on 11 December, Hassé once more with better news: 'I now have promises for a further £65 p.a. mainly from university colleagues', with hopes for more. It is not clear where the balance came from, but in any case Heilbronn got his invitation.

Heilbronn's stay in Bristol extended over nearly a year and a half. During this period he sprang into mathematical prominence with his proof of the Gauss conjecture on class numbers of imaginary quadratic fields (see paper 7). The scholarship in Bristol was, of course, only a temporary one, and one sees the ever-helpful Hassé raising the question of a more permanent position for the refugee, although, as he explains, he would like to keep him in Bristol for good. It seems that at the end of 1934 Mordell came forward with the offer of a two-year scholarship in Manchester, and Heilbronn did indeed spend some time in Mordell's department. In May 1935, however, he was awarded the Bevan Fellowship at Trinity and some time afterwards he moved to Cambridge. It is fairly clear that Hardy was the prime mover behind the award. There are several letters of Hardy's in the A.A.C. files singling out Heilbronn for special support, and this to be given 'here' (i.e. in Cambridge) 'or in Oxford, not Canada or Australia'. Hardy continued to keep a benevolent eye on Heilbronn over many years, in particular again intervening on his behalf in 1940, when the need arose.

Heilbronn soon put down roots in Cambridge. He brought his parents and his sister over from Germany. Their house in Cambridge remained his true home base for many years, even after the war when he had taken up a chair in Bristol. Patrick Duval, in a recent letter, probably best sums up Heilbronn's personality and life pattern in those days.

> 'I met him when he first came to Cambridge, some years before the war: probably we were introduced by Davenport, but I'm not sure of that. He was I think rather shy, and not easy to get to know quickly. My mother lived in Cambridge then, and he went to tea with her once while I was abroad (it must have been 1935 I suppose); she wrote to me quoting Dr Johnson, "His intellect is as exalted as his stature; I am half afraid of him, yet he is no less amiable than formidable." This seems to me to give a very good impression of him in those days. After he got his parents and sister out of Germany, and settled them in a house in Chesterton Rd, we became very friendly; I was often at their house, and they used to come to ours. His mother was very musical, collected a circle of refugee musicians, and used often to give musical parties; she

> herself was an excellent pianist. I was at many of these parties; they rather bored Hans, who had no interest in music, but he was present at many of them, as a very correct and attentive host.'

Heilbronn's complete lack of gift or sympathy for music was well known among his friends, although for a short while in 1951 there seems to have been an attempt to redeem himself when he writes to Harold Davenport

> 'My latest vice is playing simple tunes on a children's flute. It made my family very depressed at times, but the dog tolerated it stoically.'

The Cambridge period also saw the beginning of the close collaboration with Harold Davenport. Altogether ten papers appeared under their joint names, the first in 1936 and the last in 1971 after Davenport's death. Being geographically separated most of the time after 1937, this collaboration was partly carried on by numerous letters and postcards. Heilbronn and Davenport became lifelong friends and their correspondence covers not only mathematics, but personal news, university politics, the British government, administrative problems and much else besides. Anne Davenport recalls:

> 'For many years, starting before the war and going on through the war when possible and after, till at least 1954—I think till 1957 or '58—Heilbronn and Harold went on a week's walking holiday—usually in the Easter vacation. They used to go off by car till they found a suitable place and they then spent the days walking or maybe rowing. Once they discovered Hartland, Devon, they often returned there. The walks got shorter with the years. All kinds of things were discussed—I think a detective story was planned—but there were also long companionable silences.'

Heilbronn and Davenport had many interests in common and showed a similar outlook in many spheres. Both were staunchly conservative. Later on, in the late fifties and early sixties, they were to launch jointly an energetic campaign against indiscriminate, rapid university expansion, being worried by the dangers of falling standards and the difficulties of recruiting sufficient new academic staff in mathematics of the right quality.

To return to events in chronological order, after the beginning of the war Heilbronn threw himself with characteristic vigour into the task of organizing Trinity's A.R.P. Fire Service, and he even succeeded, in order to facilitate fire fighting, in getting a door opened between Trinity and St John's. To quote the junior bursar of Trinity in a letter written on his behalf in 1940, as Heilbronn was legally a foreigner this was 'the only effective expression of his loyalty to his country which circumstances allowed him'.

In 1940 when the Bevan Fellowship expired, the College Council at Trinity resolved to continue Heilbronn's salary and rights. Other events, however, intervened. Heilbronn had actually applied for British citizenship in April 1939, but because of the war his naturalization was to be delayed until October 1946. Thus, when the crisis of 1940 came, he, like many other staunchly anti-Nazi refugees, found himself interned as an 'enemy alien'. This he deeply resented. Like many other 'friendly enemy aliens' he had to wait many months for his freedom. Again the ubiquitous Academic Assistance Council acted on his behalf, as on that of many other refugee scientists. We mentioned Hardy's intervention earlier. On his release he enlisted in the Pioneer Corps and was subsequently transferred to the Royal Corps of Signals and in 1943 to Military Intelligence.

Heilbronn served with the British army until autumn 1945. In the autumn of that year he took up a post at University College London where Davenport had become professor. Not surprisingly there is no record of creative work by Heilbronn during his army service. He did, however, find his way back into active research very quickly. His first postwar paper (paper 21) jointly with Davenport was submitted in March 1946.

In 1946 Heilbronn returned to Bristol, first as reader and from 1949 as professor and head of department. He made his home in Bristol at the Hawthorn Hotel, just across from the Royal Fort where the Department of Mathematics was then housed. During the vacations—at least in the first years—he tended to return to the house in Chesterton Road, Cambridge. Patrick Duval's description of Heilbronn as he was in the early years in Bristol cannot be bettered.

> 'During the two years I was there [in Bristol], he became a very close family friend. He was still rather stiff and formal in external manner. I remember the shock it gave our colleagues when my wife and I both automatically called him Hans; no one had dreamed of such a familiarity; but by the time we left the whole Department was following our example; and I believe basically he was glad of this, and I'm sure the whole atmosphere became much easier as a result. I have a feeling he wanted to be popular, but found it uphill work. He delighted in hospitality, and used to give good dinner parties.
>
> 'He was very keen on all kinds of sport, especially rowing and tennis. In interviewing students for admission he always asked about their sporting interests, and was unfavourably impressed if they had none.'

Duval then refers to the duelling scar which we have already mentioned and continues:

> 'The whole time that he was professor in Bristol, he was also coach of the University Boat Club, and worked really hard at this. After I went to London, when rowing students had to take exams during Henley week, he got me to arrange for them to sit under my invigilation in U.C. Several years running, after we had left Bristol, he entertained my wife (also a tennis fan) on a very hospitable scale for the week of the Bristol tennis tournament.'

Heilbronn's fondness of children is stressed in a letter from Anne Davenport, and she also recalls his formality of manner.

> 'When at University College London [in 1945] Hans stayed regularly at least one night a week with us. It was still "Heilbronn" and "Davenport". I got tired to asking "Will you have . . . Dr Heilbronn" and when it became Professor Heilbronn I went on strike. I announced that I was

going to call him Hans—he was rather taken aback and asked what he was to do. Only then did the two men get onto first name terms.'

During his tenure of the chair of pure mathematics, Heilbronn built up an excellent department in Bristol. He seems to have had his ear to the ground; for example, using his long stays in Cambridge to get to know of the brightest graduates and using his mastery of committees to secure Bristol lectureships for them. Among the appointments for which he was responsible were those of J. C. Shepherdson (at least indirectly), D. A. Burgess, C. Davis, C. Hooley, J. M. Marstrand and E. R. Reifenberg.

During this Bristol period Heilbronn exerted a major influence on mathematicians of the younger generation, which goes well beyond what has become apparent in the printed word. Some measure of his success is the number of former colleagues or students who now hold senior positions.

J. C. Shepherdson, who was to succeed Heilbronn to the chair of pure mathematics and who has made his name in mathematical logic, actually wrote his first paper on a number theoretic problem, suggested by Heilbronn (Shepherdson 1947). This was subsequently superseded by results of H. B. Mann.

One of us, as his research student in the late forties, was introduced by Heilbronn into algebraic number theory, in those days a topic which distinctly lacked popularity in Great Britain. Heilbronn's suggestion of a thesis subject was based on his idea that one could gain information on the ideal class group by viewing it as a module over the Galois group. This approach has nowadays found many adherents, but in those days it was rather novel. In this instance it led in particular to the determination of all absolutely Abelian fields whose degree was a power of prime l and whose class number was not divisible by l (Fröhlich 1954a, 1955), and to a criterion for the genus of a cyclic field of degree l^r to have a class number prime to l (Fröhlich 1954b). Although Heilbronn himself never published anything on this topic of ideal class groups, it interested him deeply and he kept on coming back to it. He suggested problems in this area again to his last Ph.D. student in Toronto, T. Callahan, who has since published papers on this subject (Callahan 1974a, b, 1976).

To return to the fifties and to Bristol, many other members of the department, notably the number theoreticians, were stimulated by Heilbronn and helped by his active interest in their work. He was always accessible. His manner was most direct and his intellectual honesty did not allow him to pass over in politeness whatever he felt was below standard. But although some may have found him a bit frightening at first acquaintance, he was a kind and modest man, who entirely lacked pomposity or aggressiveness.

In the late fifties developments in universities began to worry Heilbronn, and he became increasingly upset by what, in his eyes, were wrong decisions, both nationally and locally. As already noted earlier, he was entirely out of sympathy with academic policy and government planning in the Robbins period, and he spent a great deal of time and energy in trying to convince policy makers and colleagues of the dangers he saw ahead. It must have been a very frustrating time for him. In Bristol University he had always been a vigorous champion of his department. He had fought many battles, won some and lost some. Late in 1962 he began to feel that he had been let down, and with characteristic uncompromising directness he submitted his resignation in April 1963. At that stage he had no definite idea where he would go, but this did not seem to worry him at all. He left the university in March 1964.

Heilbronn's decision was deeply regretted by all his departmental colleagues. But the effect of his resignation on himself was wholly to the good. He behaved like a man suddenly released from captivity. Whereas during the last few years before, it had become increasingly difficult to have a mathematical conversation with him, in the face of his preoccupation with policy problems and administrative irritations, he suddenly wanted to talk again and listen again, and he was once more full of ideas.

The year 1964 also saw a new departure in Heilbronn's personal life. In March he married Mrs Dorothy Greaves, who had been widowed some years earlier and whom he had met in Cambridge. They both shared in particular a passion for bridge. This had indeed become Heilbronn's principal relaxation and one was told that he was an expert at the game.

After leaving Bristol Heilbronn, with his wife, first spent some time at the California Institute of Technology at Pasadena, following an invitation from Olga Taussky-Todd, an old friend and colleague. Later that year the couple settled down in Toronto, where he had accepted the offer of a chair at the university. Heilbronn had spent some time 24 years earlier in Canada, then as visiting professor at the University of British Columbia, Vancouver. He now found it quite easy to settle down permanently in Canada and he eventually took out Canadian citizenship in 1970. The couple left their mark, both among their friends at bridge and among their friends in the university, and the many mathematical visitors who came through and experienced their generous hospitality.

Mathematically Toronto gained a great deal from Heilbronn's presence. He found some number theoreticians in Toronto, notably John Chalk whom he had first met in 1945 at University College. With great vigour and drive he built up an active research school, supervising a number of Ph.Ds, and running seminars and study groups at which he took the major part in presenting new developments in his own way. According to the participants, the lecture courses he gave on various classical topics or recent advances in algebraic number theory contained new ideas as regards approach and presentation, but none of this was ever published.

Heilbronn also took an active part in the wider mathematical life of Canada. As the delegate of the Canadian Mathematical Congress to the International Congress of Mathematicians in Nice (1970) he officially presented the invitation for the next International Congress to meet in Vancouver in 1974. He was an active and conscientious member of the committee that planned the academic programme for Vancouver. Only ill health prevented him from taking a more active part in the running of the 1974 Congress itself.

Heilbronn's robust health took its first setback in the fifties, when he developed a peptic ulcer. When, after some time, he was categorically forbidden wine, he distributed his excellent cellar among his friends. In November 1973 he had a heart attack. Although he did recover, he nevertheless lost some of his driving energy, and his physical activities had to be curtailed. He continued, however, to play an active rôle in the life of the department in Toronto, and was scarcely absent from his office for even a day. He had become intensely interested in the history of the transcontinental railroad system in Canada, and in summer 1964 he and one of us, together with our wives, travelled across the continent to Vancouver by rail, Heilbronn at each stage explaining some interesting local aspect or anecdote connected with the original construction. On 28 April 1975 Heilbronn died during an operation to implant a pacemaker.

Heilbronn was a member of the London Mathematical Society, serving on its council for many years. Together with Davenport he oversaw the investments of the Society over a considerable period. He was President of the Society from 1959 to 1961. He was elected a Fellow of the Royal Society in 1951. He was also a Fellow of the Cambridge Philosophical Society and a member of the American Mathematical Society. Once in Canada he took an active part in the business of the Canadian Mathematical Congress, as already noted above, in connection with the International Congress of Mathematicians, and he was editor of the *Canadian Journal of Mathematics* 1967–69. In 1967 he was elected to the Royal Society of Canada and during 1971–73 he was a member of its council.

Heilbronn's mathematical work

(*a*) *Göttingen period* (papers 1–6)

During this period Heilbronn was a pupil of Landau and latterly his assistant.

In his first paper (1) he shows that the number $P(\xi)$ of primes represented by an integral polynomial $f(x)$ for integral x, $0<x<\xi$, satisfies the estimate $P(\xi) = O(\xi/\ln \xi)$. Previously only Nagell's weaker $P(\xi) = o(\xi)$ was known. For this he used a formulation of Brun's sieve which had recently been given by Landau and doubtless the problem was suggested by him. Subsequent workers have obtained more information about the constant implied by the O (see Halberstam-Richert 1974, especially §§5, 6), but presumably the O cannot be replaced by o.

The second paper (2) is Heilbronn's doctoral dissertation. In 1930 Hoheisel had proved the existence of a $\theta < 1$ such that for all large x there is a prime p such that $x<p<x+x^{\theta}$; and indeed that the number of such p is asymptotically $x^{\theta}/\ln x$. Heilbronn's method is simpler than Hoheisel's and he applies it also to primes in arithmetic progressions and to estimates of the sum of the Möbius function. The value for θ obtained by Hoheisel, namely

$1-(3300)^{-1}+\varepsilon$, is very near 1 and Heilbronn's is distinctly better: $1-(250)^{-1}+\varepsilon$. Subsequently improved values of θ were obtained by a number of workers, the main milestones being $\frac{5}{8}+\varepsilon$ (Ingham 1937), $\frac{3}{5}+\varepsilon$ (Montgomery 1971, ch. 14, where there is an historical discussion) and $\frac{7}{12}+\varepsilon$ (Huxley 1972a, b, ch. 28), the best to date: for primes in arithmetic progressions, see also Gallagher (1972).

The three papers written together with Landau (3, 4, 5) deal with applications and improved proofs of Tauberian theorems, in particular what is now known as the Landau–Ikehara theorem (for the modern context see H. R. Pitt 1958, especially §6.1). The only other paper from the Göttingen period is a simple proof of a strong form of Cauchy's theorem (6).

(*b*) *The class-number of imaginary quadratic fields* (papers 7, 8, 16, 19)

Soon after his arrival in Britain, Heilbronn made a considerable stir in the mathematical world by proving a long-standing conjecture of Gauss. This was that

$$h(d)\to\infty \tag{i}$$

as $d\to-\infty$ where $h(d)$ is the class-number of the quadratic number-field of discriminant* d. Already before 1918 Hecke had shown that (i) would follow from an extended Riemann hypothesis. In 1933 Deuring had proved the unexpected result that the *falsity* of the (ordinary) Riemann hypothesis implies that $h(d)\geqslant 2$ for all large enough d and in a manuscript which was available to Heilbronn but only published subsequently Mordell (1934) had extended Deuring's argument to show that the falsity of the Riemann hypothesis implies (i). What Heilbronn does in 7 is to show more generally that the falsity of the generalized Riemann hypothesis used by Hecke implies (i). It follows (*tertium non datur*) that (i) is true unconditionally. Quite shortly afterwards Siegel (1935) gave the explicit estimate

$$\ln h(d)\sim\tfrac{1}{2}\ln|d| \quad (d\to-\infty)$$

together with a corresponding theorem for $d>0$ (i.e. for real quadratic fields) involving the regulator with simple proof. An appropriate generalization to fields of any degree was obtained by Brauer (1950).

Numerical evidence strongly suggested that there are at most nine negative values of d for which $h(d)=1$, namely

$$-3,\quad -4,\quad -7,\quad -8,\quad -11,\quad -19,\quad -43,\quad -67,\quad -163. \tag{ii}$$

At almost the same time as 7 Heilbronn published a joint paper, 8, with E. H. Linfoot (who subsequently had a distinguished career as an astronomer) showing by methods similar to those of 7 that there could be at most one further d for which $h(d)=1$. The proof of Heilbronn and Linfoot consisted in deducing a contradiction from the hypothesis that two negative values of d with $h(d)=1$ exist other than those given by (ii) and so gave no possibility of deciding whether or not one extra $d<0$ with $h(d)=1$ exists. The

* So $d<0$ means that the field is imaginary. More precisely, Gauss had made the conjecture in terms of quadratic forms.

little-known German mathematician Heegner claimed to have disproved the existence of the additional d (Heegner 1952) but the contemporary mathematical consensus was that the paper was wrong-headed and the problem remained until practically simultaneously Stark (1967*a*) showed that the additional d did not exist and Baker (1966) gave a general theorem which could be applied to this and to many other problems. A subsequent examination of Heegner's paper showed that it gave, in fact, a viable approach (Birch 1968, Deuring 1968, Siegel 1968, Stark 1975b). For later developments see A. Baker (1975, especially ch. 5). For an ineffective generalization of the Heilbronn–Linfoot paper see Tatuzawa (1951).

The two papers 7 and 8 just discussed were both, apparently, written in the brief period after Heilbronn left Germany when he was at Bristol. The later and minor paper 19 supplies a neat proof of a formula which plays a key rôle in the generalization by Siegel (1935) of 7.

Paper 16 comes into the same circle of ideas. It neatly disproves a conjecture of Chowla about iterated character sums which, by the result of Hecke mentioned above, would have given good estimates for the class-number of imaginary quadratic fields. (For later related work see Selberg & Chowla 1967 and Bateman *et al.* 1975.)

(*c*) *Analytic number theory* (papers 9–15, 18, 21, 40)

In 1934 and 1935 the Russian mathematician I. M. Vinogradov published a number of difficult and obscure papers giving vastly superior estimates in Waring's problem to those which had been obtained by Hardy and Littlewood. In 9, which appeared shortly after Heilbronn had taken up his fellowships at Trinity, he simplified Vinogradov's method substantially and somewhat improved the estimates generally. This was followed shortly by four papers with Davenport on specific problems of the Waring type (13 is a technical lemma; 12, on sums of fourth powers, was subsequently superseded by Davenport 1939; 14 is on sums of two cubes and a square, cf. Roth 1949; and 15 is on the sum of a prime and a kth power).

Schnirelmann had shown that there is some constant c such that every sufficiently large integer n is the sum of at most c prime numbers and in 18 Heilbronn, Landau & Scherk show that $c = 71$ is an admissible value. This was a very substantial improvement over existing estimates but was rapidly superseded by Vinogradov (1937) who showed that every sufficiently large odd integer is the sum of at most three primes ('Goldbach's theorem', see also Vinogradov 1954, 1971).

The papers 10 and 11 both written in collaboration with Davenport go in a rather different direction. Let $Q(x,y) = ax^2+bxy+cy^2$ be a definite quadratic form with integral coefficients a, b, c and discriminant $-d = 4ac-b^2 > 0$. The Epstein zeta-function is defined by

$$\zeta(s, Q) = \sum_{\substack{(x,y)\neq(0,0)\\ x,y \text{ integral}}} Q(x,y)^{-s}.$$

When the class-number* $h(-d) = 1$ then $\zeta(s, Q)$ is indeed a Dedekind zeta-function with an Euler product; in particular it can have no zeros in $\Re s > 1$ (and a 'generalized Riemann Hypothesis' conjectures that the non-trivial zeros are all on $\Re s = \frac{1}{2}$). Potter and Titchmarsh had shown that, whatever the value of $h(-d)$, there are always infinitely many zeros on $s = \frac{1}{2}$. In these two papers it is shown that if $h(-d) \neq 1$ there are always infinitely many zeros in $\Re s > 1$ and so any presumed Riemann hypothesis fails spectacularly (for later information on the zeros of Epstein's zeta-functions see Stark 1967b). In 10 Davenport and Heilbronn also consider the function $\zeta(s, a) = \sum_{n=0}^{\infty} (n+a)^{-s}$ defined for any real $a > 0$. For $a = 1$ this is Riemann's zeta-function and so has no zeros in $\Re s > 1$. They show that if $a \neq 1$ is rational or if a is transcendental then again there are infinitely many zeros in $\Re s > 1$. (Their method was extended by Cassels (1961) to cover the remaining case: a algebraic.)

All the papers mentioned so far in this section date from Heilbronn's Trinity period. Paper 21 was written with Davenport after the war and is rather different in nature. Let $\lambda_1, \ldots, \lambda_5$ be non-zero real numbers. Then it is shown that $\sum_{j=1}^{5} \lambda_j x_j^2$ takes arbitrarily small values for integral values of the variables (not all zero). When the ratios λ_i/λ_j are all rational this is a special case of a classical theorem of Meyer so they can assume that, say, λ_1/λ_2 is irrational. This is dealt with by an elegantly simple modification of the Waring problem techniques. The result was the beginning of a large programme by Davenport and others on small values taken by quadratic forms (and, later, by forms of higher degree); but in this Heilbronn took no part.

At the very end of his life Heilbronn returned to the zeros of ζ-functions in paper 40. If $\zeta_C(s)$ is the zeta-function of the algebraic field C then it is classical that the residue κ at $s = 1$ satisfies $\kappa^{-1} = O(\ln|d|)$, where d is the absolute discriminant, unless there is a real zero s_0 with $\frac{1}{2} < s_0 < 1$ (which would, of course, contradict the generalized Riemann hypothesis). In certain cases L. Goldstein and J. Sunley had obtained effective estimates for κ by essentially algebraic techniques. In 40 using techniques of Brauer and the analytic theory of the Artin L-function Heilbronn shows quite generally that the problem of the existence of such an s_0 reduces to the corresponding problem for quadratic number fields. Following up this approach, Stark (1974, 1975a) obtains the striking consequence that the Brauer–Siegel theorem about the magnitude of the class numbers of algebraic number fields can be made effective in a large number of cases.

(*d*) *Euclidean algorithm*

In 20, written during the Trinity period, Heilbronn showed that there are only finitely many real quadratic fields with a euclidean algorithm. Partial results had been obtained by a number of mathematicians: in particular Erdös and Chao Ko had disposed of the case when the discriminant is prime and Heilbronn's proof is based on theirs. The estimates were such that it was

* We have written $h(-d)$ for the $h(d)$ of the papers so as to preserve conformity with the notation used earlier.

hopeless to attempt to enumerate all the euclidean real quadratic fields, or at least it was not attempted.

After the war Davenport invented a new approach based on ideas of the Geometry of Numbers which gave a much better estimate of the discriminants of euclidean real fields and so enabled them all to be determined. Davenport's geometrical approach also showed the finiteness of the number of euclidean fields of two other types: non-totally real cubic fields and totally complex quartic fields (i.e. whenever the group of units has rank 1). (See Davenport 1950a, b, 1951 or Cassels 1952 for a slightly different treatment.) It appears that these are the only types of field that can be dealt with by geometric methods. At about the same time Heilbronn returned to the problem and showed (26, 27) that his method applied to cyclic cubic fields (which, being totally real, are not covered by Davenport's results) and, more generally, for a wide class of cyclic fields. (For numerical results see Smith 1969.) This approach has not been carried further and it apparently remains an open question in general whether a given type of algebraic number field contains infinitely many fields with a euclidean algorithm. (It should perhaps be remarked that the term 'euclidean algorithm' here refers always to that using the norm function. The general type of algorithm discussed recently by Samuel and others raises entirely different problems.)

(*e*) *Geometry of numbers* (papers 22, 23)

Both of these were written with Davenport. The first is rather special and discusses the minimum of $|(\alpha x+\beta y)(\gamma z+\delta t)|$ for given real $\alpha,\beta,\gamma,\delta$ and for integral x,y,z,t satisfying $xt-yz=\pm 1$. It is shown that the first two minima are 'isolated' and correspond to the first two minima of the Markoff chain but that then the spectrum has a limit point. The methods are more or less standard.

The result of 23 is more interesting, though the techniques are equally standard. Let ξ,η be linear forms in variables u,v with determinant $\Delta\neq 0$ and let λ,μ be real numbers. Then it is shown that there are integral values of the variables such that

$$\begin{gathered}\xi+\lambda\geqslant 0,\quad \eta+\mu\geqslant 0,\\ (\xi+\lambda)(\eta+\mu)<|\Delta|.\end{gathered}\tag{iii}$$

It is shown that this statement becomes false if $|\Delta|$ in (iii) is replaced by $\kappa|\Delta|$ with any $\kappa<1$ and there is an examination of various special cases. This theorem was generalized to n dimensions by Chalk (1947) and Macbeath (1952) gave a further generalization.

(*f*) *Miscellaneous*

Paper 17 proves an inequality which is, apparently, motivated by problems of additive number theory and 24 discusses a generalization of harmonic functions to real-valued functions defined on the integral two-dimensional lattice.

In 25 it is shown that for every $\eta > 0$ there is a $C(\eta) > 0$ with the following property: for every real θ and for every integer $N \geqslant 1$ there exist integers n, g such that

$$1 \leqslant n \leqslant N; \quad |n^2\theta - g| \leqslant C(\eta) N^{-\frac{1}{2}+\eta}.$$

A weaker form with $\frac{2}{5}$ instead of $\frac{1}{2}$ in the exponent had been obtained already in 1927 by Vinogradov, and Heilbronn's proof is similar in general shape but depends on a more delicate analysis of the remainder term. Heilbronn raises the question whether the $\frac{1}{2}$ in the exponent is best-possible but no progress has been made here. The paper did, however, spark off a series of generalizations in other directions (Davenport 1967, several papers by Danicic, e.g. 1959, 1967, and Cook, e.g. 1972, 1975, Liu 1974).

Paper 28 arose from work of Fröhlich (1954b) on cyclic fields whose degree is a power of a prime l. He obtained necessary and sufficient conditions for the class number of the principal genus to be prime to l. These conditions are in terms of the mutual congruence behaviour of the discriminant prime divisors and in 28 Heilbronn deals with the density of, say, pairs of primes with given mutual congruence behaviour. The actual consequences for cyclic fields have never been stated explicitly.

Paper 29 answers a question posed by Kuratowski by showing that if $f(x)$ is a complex-valued function defined on an open set A of the complex plane then there is a regular function $F(x)$ and a continuous function $g(x)$, both defined on A, such that $f(x) = F(x)\,e^{g(x)}$ and there is some discussion of the problem of when every homotopy class of maps from one set A of the complex plane to another such set contains a map given by a regular function.

Paper 31 is similarly a response to a problem. It is well known that every rational $x > 0$ can be put in the shape

$$x = y_1^2 + y_2^2 + y_3^2 + y_4^2,$$

where $y_1, \ldots, y_4$ are rational. Heilbronn shows by a rather elaborate argument that the y_j can be chosen in such a way that they are continuous functions of x: indeed there are integral functions $f_j(x)$ such that $\sum (f_j(x))^2 = 1$ and $y_j = x^{\frac{1}{2}} f_j(x)$ is rational whenever x is rational. There is a rather weaker theorem for sums of higher powers.

Paper 32 with Erdös discusses the number F of solutions of a congruence

$$e_1 a_1 + \ldots + e_k a_k \equiv N \pmod{p},$$

where $a_1, \ldots, a_k, N$ are given and $e_1, \ldots, e_k$ take only the values 0, 1. It is shown that $F > 0$ provided that $k > 3(6p)^{\frac{1}{2}}$ and that $F \sim p^{-1} 2^k$ provided that k is large compared with $p^{\frac{2}{3}}$. (For the latest state of the problem, see Olson 1975.)

Paper 36 is concerned with the length $n(a) = n_N(a)$ of the continued fraction for a/N where the integer $N > 1$ is given and a is prime to N. It is shown that the average value of $n(a)$ is asymptotic to $12\pi^{-2}(\ln 2)(\ln N)$. The error term was subsequently improved by Porter (1975). The detailed argument of 36 was used by Manin (1972) for the computation of his 'modular symbols'. See also Yao & Knuth (1975).

Although 38 was published only in 1969 the work which it records had been done much earlier. Davenport and Heilbronn had obtained estimates for the numbers of cubic fields (for the two cases: totally real and otherwise) whose discriminants were below a given bound but, as they could not obtain results as precise as they had hoped, the work was put on one side. When Davenport was on the point of death Heilbronn must have felt that the results should be recorded and wrote 38. This impelled him to consider the problem again and the more precise results of 39, although published over the joint names, were not in fact discovered until after Davenport's death. There is one intriguing result for the non-cyclic cubic fields which might possibly hint at a quite general phenomenon. Except for the finite number of primes dividing the discriminant a rational prime p can have one of three possible decomposition patterns in such a field k. It is shown that (roughly speaking) the relative densities of the fields k in which a given prime p has each of these three behaviours are the same as the densities of the sets of primes p when the field k is kept fixed.

The other papers call for no special comment.

We are most grateful to Mrs Anne Davenport for her help, and in particular for the loan of the collection of letters which Harold Davenport had received from Heilbronn, and to Miss Esther Simpson, O.B.E., secretary of the Society for the Protection of Science and Learning Limited, formerly the Academic Assistance Council, for information and for her kind permission to have access to the file of Hans Heilbronn in the Society's archives. We also wish to acknowledge the assistance of the Department of Western MSS at the Bodleian Library, where these archives are now housed. We received helpful comments and information from Mrs D. Heilbronn, Dr T. Callahan, Professor J. H. Chalk, Professor H. S. M. Coxeter, Dr P. Du Val and Professor J. C. Shepherdson, and we wish to express our thanks to all of them.

References

Baker, A. 1966 Linear forms in the logarithms of algebraic numbers. *Mathematika* **13**, 204–210.

Baker, A. 1975 *Transcendental number theory*. Cambridge University Press.

Bateman, P. T., Purdy, G. B. & Wagstaff, S. S., Jr, 1975 Some numerical results on Fekete polynomials. *Math. Comp.* **29**, 7–23.

Birch, B. J. 1968 Diophantine analysis and modular functions. *Proc. Bombay Colloquium on Algebraic Geometry* (Tata Institute), pp. 35–42.

Brauer, R. 1950 On the zeta-functions of algebraic number fields II. *Am. J. Math.* **72**, 739–746.

Callahan, T. H. 1974a The three-class groups of non-Galois cubic fields—I. *Mathematika* **21**, 72–89.

Callahan, T. H. 1974b The three-class groups of non-Galois cubic fields—II. *Mathematika* **21**, 168–188.

Callahan, T. H. 1976 Dihedral extensions of order $2p$ with class number divisible by p. *Can. J. Math.* (To appear.)

Cassels, J. W. S. 1952 The inhomogeneous minimum of binary quadratic, ternary cubic and quaternary quartic forms. *Proc. Camb. Phil. Soc.* **48**, 72–86 (Addendum: 519–520).

Cassels, J. W. S. 1961 Footnote to a note of Davenport and Heilbronn. *J. Lond. math. Soc.* **36**, 177–184.

Chalk, J. H. H. 1947 On the positive values of linear forms I, II. *Q. Jl Math.* **18**, 215–227; **19**, 67–80.

Cook, R. J. 1972 The fractional parts of an additive form. *Proc. Camb. Phil. Soc.* **72**, 209–212.

Cook, R. J. 1975 On the fractional parts of a set of points III. *J. Lond. math. Soc.* **9**, 490–494.

Danicic, I. 1959 On the fractional parts of θx^2 and ϕx^2. *J. Lond. math. Soc.* **34**, 353–357.

Danicic, I. 1967 The distribution (mod 1) of pairs of quadratic forms with integer variables. *J. Lond. math. Soc.* **42**, 618–623.

Davenport, H. 1939 On Waring's problem for fourth powers. *Ann. Math.* **40**, 731–747.

Davenport, H. 1950a Euclid's algorithm in cubic fields of negative discriminant. *Acta math., Stockh.* **84**, 159–179.

Davenport, H. 1950b Euclid's algorithm in certain quartic fields. *Trans. Am. math. Soc.* **68**, 508–532.

Davenport, H. 1951 Indefinite binary quadratic forms and Euclid's algorithm in real quadratic fields. *Proc. Lond. math. Soc.* **53**, 65–82.

Davenport, H. 1967 On a theorem of Heilbronn. *Q. Jl Math.* **18**, 339–344.

Deuring, M. 1968 Imaginäre quadratische Zahlkörper mit der Klassenzahl Eins. *Invent. Math.* **5**, 169–179.

Fröhlich, A. 1954a On the absolute class group of Abelian fields. *J. Lond. math. Soc.* **29**, 211–217.

Fröhlich, A. 1954b The generalization of a theorem of L. Rédei's. *Q. Jl Math.* (2), **5**, 130–140.

Fröhlich, A. 1955 On the absolute class group of Abelian fields II. *J. Lond. math. Soc.* **30**, 72–80.

Gallagher, P. X. 1972 Primes in progressions to prime-power modulus. *Invent. Mat.* **16**, 191–201.

Halberstam, H. & Richert, H.-E. 1974 *Sieve methods.* London: Academic Press.

Heegner, K. 1952 Diophantische Analysis und Modulfunktionen. *Math. Z.* **56**, 227–253.

Huxley, M. N. 1972a On the difference between consecutive primes. *Invent. Mat.* **15**, 164–170.

Huxley, M. N. 1972b *The distribution of prime numbers.* Oxford University Press.

Ingham, A. E. 1937 On the difference between two consecutive primes. *Q. Jl Math.* **8**, 255–266.

Liu, M. C. 1974 Simultaneous approximation of two additive forms. *Proc. Camb. Phil. Soc.* **75**, 77–82.

Macbeath, A. M. 1952 A theorem on non-homogeneous lattices. *Ann. Math.* **56**, 269–293.

Manin, Ju. I. 1972 Parabolic points and the zeta-functions of modular curves (in Russian). *Izv. Akad. Nauk SSSR* (*ser. mat.*) **36**, 19–66.

Montgomery, H. L. 1971 *Topics in multiplicative number theory.* (Springer Lecture Notes in Mathematics no. 227.)

Mordell, L. J. 1934 On the Riemann and imaginary quadratic fields with given class number. *J. Lond. math. Soc.* **9**, 289–298.

Olsen, J. E. 1975 Sums of sets of group elements. *Acta arith.* **28**, 147–156.

Pitt, H. R. 1958 *Tauberian theorems.* Tata Institute and Oxford University Press. See especially §6.1.

Porter, J. W. 1975 On a theorem of Heilbronn. *Arithmetika* **22**, 20–28.
Roth, K. F. 1949 Proof that almost all positive integers are sums of a square, a positive cube and a fourth power. *J. Lond. math. Soc.* **25**, 4–13.
Selberg, A. & Chowla, S. 1967 On Epstein's zeta-function. *J. reine angew. Math.* **227**, 86–110.
Shepherdson, J. C. 1947 On the addition of elements of a sequence. *J. Lond. Math. Soc.* **22**, 85–88.
Siegel, C. L. 1935 Über die Classenzahl quadratischer Zahlkörper. *Acta arith.* **1**, 83–86.
Siegel, C. L. 1968 Zum Beweise des Starkschen Satzes. *Invent. Math.* **5**, 180–191.
Smith, J. R. 1969 On Euclid's algorithm in some cyclic cubic fields. *J. Lond. math. Soc.* **44**, 577–582.
Stark, H. M. 1967a A complete determination of the complex quadratic fields of class number one. *Mich. math. J.* **14**, 1–27.
Stark, H. M. 1967b On zeros of Epstein's zeta function. *Mathematika* **14**, 47–55.
Stark, H. M. 1974 Some effective cases of the Brauer–Siegel theorem. *Invent. Math.* **23**, 135–152.
Stark, H. M. 1975a The analytic theory of algebraic numbers. *Bull. Am. math. Soc.* **81**, 961–972.
Stark, H. M. 1975b Class-number of complex quadratic fields. *Modular functions of one variable I* (Antwerp summer school), pp. 153–175. (Springer Lecture Notes in Mathematics no. 320.)
Tatuzawa, B. 1951 On a theorem of Siegel. *Jap. J. Math.* **21**, 163–178.
Vinogradov, I. M. 1937 Representation of an odd integer as the sum of three primes. *Dokl. Akad. Nauk SSR (Comptes Rendus)* **15**, 169–172.
Vinogradov, 1954 *The method of trigonometric sums in the theory of numbers*. Interscience.
Vinogradov, I. M. 1971 *The method of trigonometric sums in the theory of numbers* (in Russian). Moscow: Nauka.
Yao, A. C. & Knuth D. E. 1975 Analysis of the subtractive algorithm for greatest common divisors. *Proc. nat Acad. Sci. U.S.A.* **72**, 4720–4722.

List of publications by H. A. Heilbronn

(1) 1931 Über die Verteilung der Primzahlen in Polynomen. *Math. Annal.* **104**, 794–799.
(2) 1933 Über den Primzahlsatz von Herrn Hoheisel. *Math. Z.* **36**, 394–423.
(3) (With E. Landau) Bemerkungen zur vorstehenden Arbeit von Herrn Bochner. *Math. Z.* **37**, 10–16.
(4) (With E. Landau) Ein Satz über Potenzreihen. *Math. Z.* **37**, 17.
(5) (With E. Landau) Anwendungen der N. Wienerschen Methode. *Math. Z.* **37**, 18–21.
(6) Zu dem Integralsatz von Cauchy. *Math. Z.* **37**, 37–38.
(7) 1934 On the class-number in imaginary quadratic fields. *Q. Jl Math.* **5**, 150–160.
(8) (With E. H. Linfoot) On the imaginary quadratic corpora of class-number one. *Q. Jl Math.* **5**, 293–301.
(9) 1936 Über das Waringsche Problem. *Acta arith.* **1**, 212–221.
(10) (With H. Davenport) On the zeros of certain Dirichlet series, 9. *J. Lond. math. Soc.* **11**, 181–185.
(11) (With H. Davenport) On the zeros of certain Dirichlet series II. *J. Lond. math. Soc.* **11**, 307–312.
(12) (With H. Davenport) On Waring's problem for fourth powers. *Proc. Lond. math. Soc.* **41**, 143–150.
(13) (With H. Davenport) On an exponential sum. *Proc. Lond. math. Soc.* **41**, 449–453.
(14) 1937 (With H. Davenport) On Waring's problem: Two cubes and one square. *Proc. Lond. math. Soc.* **43**, 73–104.

(15) 1937 (With H. DAVENPORT) Note on a result in the additive theory of numbers. *Proc. Lond. math. Soc.* **43**, 142–151.

(16) On real characters. *Acta Arith.* **2**, 212–213.

(17) On an inequality in the elementary theory of numbers. *Proc. Camb. Phil. Soc.* **38**, 207–209.

(18) 1936 (With E. LANDAU & P. SCHERK) Alle grossen Zahlen lassen sich als Summe von höchstens 71 Primzahlen darstellen. *Časopis Pěst. Mat. Fys.* **65**, 117–141.

(19) 1938 On Dirichlet series which satisfy a certain functional equation. *Q. Jl Math.* **9**, 194–195.

(20) On Euclid's algorithm in real quadratic fields. *Proc. Camb. Phil. Soc.* **34**, 521–526.

(21) 1946 (With H. DAVENPORT) On indefinite quadratic forms in five variables. *J. Lond. math. Soc.* **21**, 185–193.

(22) 1947 (With H. DAVENPORT) On the minimum of a bilinear form. *Q. Jl Math.* **18**, 107–121.

(23) (With H. DAVENPORT) Asymmetric inequalities for non-homogeneous linear forms. *J. Lond. math. Soc.* **22**, 53–61.

(24) 1949 On discrete harmonic functions. *Proc. Camb. Phil. Soc.* **45**, 194–206.

(25) 1948 On the distribution of the sequence $n^2\theta$ (mod 1). *Q. Jl Math.* **19**, 249–256.

(26) 1950 On Euclid's algorithm in cubic self-conjugate fields. *Proc. Camb. Phil. Soc.* **46**, 377–382.

(27) 1951 On Euclid's algorithm in cyclic fields. *Can. J. Math.* **3**, 257–268.

(28) 1958 On the averages of some arithmetical functions of two variables. *Mathematika* **5**, 1–7.

(29) On the representation of homotopy classes by regular functions. *Bull. Acad. pol. Sci.* **6**, 181–184.

(30) 1962 Old theorems and new methods in class-field theory (Presidential Address). *J. Lond. math. Soc.* **37**, 6–9.

(31) 1964 On the representation of a rational number by four squares by means of regular functions. *J. Lond. math. Soc.* **39**, 72–76.

(32) (With P. ERDÖS) On the addition of residue classes mod p, *Acta arith.* **9**, 149–159.

(33) 1966 *Collected papers of G. H. Hardy*, vol. 1, pt 1, Additive number theory (edited). Oxford University Press.

(34) 1967 *Algebraic number theory* (ed. J. W. S. Cassels & A. Fröhlich), Ch. VIII, Zeta functions and L-functions. Academic Press.

(35) 1968 (With S. BACHMUTH & H. Y. MOKIZUKI) Burnside metabelian groups. *Proc. R. Soc. Lond.* A **307**, 235–250.

(36) 1969 On the average length of a class of finite continued fractions, *Abhandlungen aus Zahlentheorie und Analysis zur Erinnerung an Edmund Landau** (ed. P. Turán), pp. 87–96. Berlin: VEB Deutscher Verlag der Wissenschaften.

(37) (With P. SCHERK) Sums of complexes in torsion-free abelian groups. *Can. Math. Bull.* **12**, 479–480.

(38) (With H. DAVENPORT) On the density of discriminants of cubic fields. *Bull. Lond. Math. Soc.* **1**, 345–348.

(39) 1971 (With H. DAVENPORT) On the density of discriminants of cubic fields. II. *Proc. R. Soc. Lond.* A **322**, 405–420.

(40) 1973 On real zeros of Dedekind ζ-functions. *Can. J. Math.* **25**, 870–873.

(41) 1971 On the 2-class group of cubic fields. *Studies in pure mathematics, papers presented to Richard Rado*, pp. 117–119. London: Academic Press.

(42) 1938 (With G. H. HARDY) E. Landau (obituary). *J. Lond. math. Soc.* **13**, 310–318.

* Also published under the title *Number theory and analysis* by the Plenum Press, New York.

My Personal Recollections of Hans Heilbronn

Olga Taussky

Every mathematician who knew Heilbronn as a mathematician or has heard about him as a mathematician is well aware of the fact that he was a giant among mathematicians. Otherwise little is known about him, and superficial observation of his appearance, manners, and lecture style could easily lead to wrong conclusions; for example, there was something military in his looks and speech. Added to this is the fact that as a student he had joined a fraternity whose practices in those days included duelling between its members. Because of their pride in this, victims of facial cuts encouraged the attending physician to retain these cuts (called *Schmisse*) instead of healing them. However, H. could be one of the most peace-loving, kindest, and most fatherly persons I have known. Furthermore, with his tall, lean appearance (at least before the gourmet cooking at Trinity College, Cambridge, England made him put on weight), his manners of great strength at lectures and otherwise, he appeared to be a man of solid health, physical and emotional. However, this too turned out incorrect. He drifted into severe stomach problems, partially of nervous origin, earlier in life than one would have expected. He had to undergo major surgery for this. He also developed heart problems, which led to his death. We all have our breaking points, of course, and he certainly had a load to carry in view of the political situation which hit our generation in Europe.

H.'s main line was analysis. He told me in Cambridge in 1935 that a theorem only seemed attractive to him if it contained an "O" or "o." However, I do not think that he was too serious about this. Although both of us considered number theory our main mathematical attraction, the fact that I am an algebraist prevented us from becoming true colleagues. However, we had a good many contacts in our work.

I met him in Göttingen in 1931–1932; later in Cambridge, England, in 1935–37; later at Caltech; and sporadically at congresses, like the international congresses at Oslo, Harvard, Amsterdam, and Vancouver. At Oslo I saw him only rarely. I was busy talking topological algebra to van Dantzig. But I do remember that he told me that he was deeply impressed being present at a technical conversation between C. L. Siegel and Hecke, both leaders in a subject so meaningful to him. He was chairman of my lecture at Amsterdam. At Vancouver he was already a very sick man. But we also met at a number of specialized conferences, like the British Association meeting at Cambridge in 1949. There he sent a note to me after my lecture: "Don't become a suffixist." My lecture concerned my work on the n-dimensional Laplace differential equation.

Later meetings where we met were Brighton in England, Ann Arbor, Boulder, and Stony Brook. We always had mathematical conversations. In particular he let me tell him about my ideas on integral matrices and encouraged me in this direction.

I will now describe my contacts with H. at the various places at which we spent common longer intervals in time.

1. *Göttingen.* Here Heilbronn was still a student but at the same time he was an assistant to Landau. It was very rare for a student to be appointed an official assistant. But Heilbronn had impressed Landau greatly by discovering a statement in the introduction of one of Landau's books which when interpreted with some mischief can be interpreted as wrong. As assistant to Landau, Heilbronn had to work very hard. I do not remember that he ever complained about it. I suppose it was in connection with Landau's four-hour course, but may not have been. He certainly had to help prepare the mathematical riddles which accompanied the big dinner party to which Landau invited some of the junior members of the department. I recall that the seating arrangement had to be computed by the male guests, who had to find their female partner by a nontrivial computation. The females did not have to work on this. There was also a mathematical crossword puzzle.

I did not take part in all the mathematical games. But there was one game one could not avoid. In this game you were defeated and immediately eliminated if a participant asked you a question which you could not answer correctly and three other participants could. Mahler defeated me at once with a question about prime numbers to which the correct answer would have been "Look it up in a table".

Landau knew that number theory was my subject, so when he arranged for a series of small lectures about certain number theoreticians he chose me for Mertens since I was educated in Austria. Since I was suffering from a slight attack of flu at that time, I sent my write-up to Heilbronn to read out for me. He told me later that Landau was satisfied, but he (H.) had omitted a paragraph in which I referred to the fact that Mertens called a sequence convergent only if the ε's could be obtained effectively—hence he obeyed the demands of the intuitionists.

Another contact with Heilbronn was via Emmy Noether's class field seminar. She had trained herself recently in this subject, with the hope of digging into this still unexplained subject with her abstract tools. She had gathered quite a large attendance together, and everybody had to lecture. Heilbronn chose, of course, something analytical and spent the night before his lecture studying and preparing. Although there is no question about the fact that his presentation was accurate, he himself felt that he had not reached the degree of insight into the material he demanded of himself. He praised my lecture as the only one of real competency. This may have even been the case, but I had written a thesis in class field theory while the others had had to pick up their topic for the purpose of the seminar. My lecture concerned the extension of Hilbert's theorem 94 to relative cyclic fields of prime power degree.

Heilbronn's attitude towards Emmy was one of great respect. One of his closest student colleagues was her thesis student at that time. To some degree he could have been called a "Noether boy." In my article "My personal recollections of Emmy Noether" in the 1982 Dekker book on Emmy, I tell how the Noether boys played a big, but flattering, joke on her. I think H. had a leading role in that joke. However, when I once referred to the much discussed fact that she had not been given a chair in Göttingen, he said that in spite of everything he could not imagine her experiencing the challenge of giving the calculus course. Actually, from what I know of Emmy, she was quite pleased with the position she had and had no grudge about it. However, Heilbronn added that he himself would not like to be pupil of a woman. I reminded him of this later at Cambridge. He had forgotten about it and felt truly ashamed about it, saying that he had not realized that he had still been so immature at that time.

He was in contact with Emmy's favourite student Deuring, and this led him later to a mathematical achievement that determined his whole future and put him into the class of top mathematicians. For he proved a conjecture of Gauss: there exist only a finite number of imaginary quadratic fields with given ideal class number h (stated by Gauss, of course, in the language of quadratic forms). He based his proof on an achievement by Max Deuring: If the Riemann hypothesis is wrong, then the Gauss conjecture is true for $h = 1$. This he generalized to: If a certain *generalized* Riemann hypothesis (GRH) is false, then the Gauss conjecture is true for arbitrary h. Since E. Hecke had earlier proved that the same conclusion holds if GRH is true, this settled the Gauss conjecture. While this proof is totally noneffective, Heilbronn was able to (partially) overcome this shortcoming in the case $h = 1$, for he achieved, in collaboration with Linfoot, the result that there are either nine or ten imaginary quadratic fields with class number 1. Harold Stark and Alan Baker settled the "or" simultaneously by showing that there are only nine, but it was found that Kurt Heegner had already been able to show this earlier. Baker and Stark were able to find all fields with class number two. But quite recently (as yet unpublished[1]) the problem of obtaining effective bounds on the discriminants of imaginary quadratic fields with a given class number has been settled by the work of Dorian Goldfeld, who in 1976 reduced the problem to the existence of a suitable Birch–Swinnerton–Dyer L-function with a triple zero at the center point of the critical strip, and finally by Benedict Gross and Don Zagier, who,

[1] *Added in proof*. The Gross–Zagier article has recently appeared in *Inventiones Mathematicae* **84** (1986), 225–320. See also the survey articles by J. Oesterle, *Seminaire Bourbaki* 1983/84, Exp. 631, and by D. Goldfeld, *Bulletin of the American Mathematical Society* (*New Series*) **13** (1985), 23–37. (Ed.)

as a by-product of their study of heights of so-called Heegner points on (Jacobians of) modular curves, were able to actually exhibit such an *L*-function.

But it all started with Heilbronn and his use of Deuring's work. I am indebted to Carlos Julio Moreno for telling me about the term "Deuring–Heilbronn phenomenon" introduced by Yu. V. Linnik in his work on the least prime in an arithmetic progression [*Rec. Math.* (*Matematičeskii Sbornik*) *Novaja Serija* **15** (57) (1944) 347–368]. Deuring's paper was published in *Mathematische Zeitschrift* **37** (1933), 405–415 and Heilbronn's in *Quarterly Journal of Mathematics Oxford* **5** (1934). Heilbronn himself had moved to Great Britain by the time of his achievement.

Let me then return to Göttingen. In due course Heilbronn completed his thesis during 1932, and Courant was one of his examiners at his oral examination. I think Courant gave him quite a grilling but praised his performance very strongly afterwards. Courant was the director of the mathematics institute. He himself had a gathering of brilliant young men working with him, of whom he was very proud. I do not think that Courant had any frictions with Landau, but he and Emmy were not on the best terms. I suppose they were too strong minded, both of them. Heilbronn was not close to the "Courant group"—at least I do not think so. First of all, the Schmisse were not liked; they seemed to characterize him as a person different from what he really was. Also, the devotion of the Courant group to partial differential equations did not agree with Heilbronn's deep devotion to number theory.

2. *Cambridge*. After Göttingen I had no connection with Heilbronn until the fall of 1935 when I came to Cambridge, England, where I had the good fortune of having been awarded a research fellowship at Girton College. The academic years 1932–1934 I had spent back in Vienna with small assistantships and 1934–1935 at Bryn Mawr College, already partially supported by Girton College. Heilbronn welcomed me very warmly. He had left Germany and had been at Bristol University, where he achieved his success regarding the Gauss conjecture.

A great award was given to Heilbronn in the form of a five-year fellowship at Trinity College, Cambridge. This was Newton's college and at the time I am describing it housed G. H. Hardy, J. E. Littlewood, and a number of young, highly promising mathematicians. So Heilbronn was not to be mathematically lonely. Apart from the British mathematicians mentioned, he had the company in Cambridge of a group of excellent German young men—B. H. Neumann, K. Hirsch, and R. Rado, all offsprings of I. Schur's famous algebra school. They were not attached to a college and were a good bit more anxious about their academic future than Heilbronn needed to be. But later the war came and everything changed for everybody. I seemed to be the only one brought up in Austria in this group of German language colleagues and, strangely, they made me feel, to some degree, an outsider. However, I was immediately incorporated into their weekly algebra meeting.

Heilbronn's life at Trinity was, of course, a great deal more comfortable than that of Neumann, Hirsch, and Rado. Of course, at that time he had already accomplished more than they. But he was concerned about them all. Actually, all four of them came from Berlin. In order to cheer up the German-speaking group, he dedicated a corner of his living room to the housing of a German lending library to which people were asked to donate books they could spare. At the end of the academic year, the books were auctioned in a party given by Heilbronn in his rooms, so people could buy books which they had liked. I remember that when a book by a German professor of history was auctioned there, I put myself out a little since he seemed a rather depressed person. Suddenly, he got up and bowed towards me saying, "Do you realize you are buying this book for a price exceeding what it costs when it is new?."

Heilbronn at this time was heavily engaged in work with Davenport, an analyst who combined interest in number theory with interest in the German language and had worked with Mordell and with Hasse in Germany. He was the only one of the young British mathematicians who joined the weekly algebra gatherings. Neumann, Hirsch, and Rado were closely attached to Philip Hall, a group theory expert and a Burnside fan who had already made a name for himself with a lengthy article on p-groups and soon afterwards generalized Sylow subgroups to subgroups whose order is prime to their index, now called Hall groups. He continued with important contributions and is regarded as one of the leaders in group theory. My work in class field theory, starting with my thesis, was in fact concerned with p-groups, coming from E. Artin's proof of the principal ideal theorem. However, I was unable to persuade the actual group theoreticians, including P. Hall, to come to the aid of these still unsolved problems.

Heilbronn was very much interested in my work at all times and had a great trust in me from the Göttingen days onwards. Although he was not uninterested in joining up with me to some extent in my Cambridge days, it did not materialize. He even dangled a very interesting problem before me. It was actually a problem combining Galois theory, group theory, and algebraic number theory. I have no idea what made him interested in this question; it seems far from his line of interest. Later he handed it over to his most successful pupil, A. Fröhlich, and, as I had expected, let him work on it by himself. Much of Fröhlich's work, to this day, shows a sparkling of this influence. Fröhlich's thesis was published in *Quarterly Journal of Mathematics Oxford* (2) **1** (1950), 270–283; **3** (1952), 98–106.

I myself realized at once that this problem needed full-time devotion. But I had returned from my year at Bryn Mawr, which happens to be not too far from Princeton, and I had accompanied Emmy Noether occasionally on her weekly trips to that place, as much as my financial situation permitted. Already during my last year in Vienna the paper by Pontryagin on continuous fields had fallen into my hands. It was the first item outside of class field theory that had fascinated me.

And it was a good thing that this happened. The problems to which my thesis adviser had introduced me are still unsolved, called "hopeless" by Artin, although they nevertheless once in a while have yielded interesting outputs. Even Heilbronn became interested in them during his last years, and I will return to this. At this moment I want to describe what made me flee from Heilbronn's problem—maybe not a wise decision. I had become interested in topological algebra, its links with quaternions, and Princeton was a hotbed at this time regarding this subject. I had produced some results and had given myself further tasks. Even H. Hopf in Zürich, a most powerful scholar and teacher, had heard about them from a visitor from Princeton and wrote me a long letter to Cambridge, asking me to show it around there. Almost nobody there cared about it actually, and I was again almost entirely on my own. Heilbronn was interested in a passive way, and Neumann checked whatever I achieved and gave moral support and some help. I did not achieve a great deal in a subject which needs teamwork by experts, which I did not have. But I achieved a little. When Hardy asked me to lecture in his "conversation class," Heilbronn, my kind friend, advised me not to lecture on my class field theory but on my topological algebra. This was excellent advice, because it enabled me to explain the background easily. Hardy considered my lecture a great success and allowed me consequently to use his name in job applications.

This was not the only time in my Cambridge days that Heilbronn assisted me with advice and support; on two occasions at least he became positively protective of me. It was when others treated me with contempt and tried to exclude me from activities for which I was competent and which were meaningful to my work and even later my career. In one of these cases he himself was approached to take on the activity. But he declined, saying that I was more suitable. He himself did not tell me about his action. But he might have harmed himself; those days were times of severe unemployment.

While in Cambridge, he tried to be one of the Cambridge set. He took people like his mother or Mrs. Pell Wheeler, the head of the Bryn Mawr College Mathematics Department, together with myself on a punt ride. He was interested in rowing and the regatta at Henley. As far as mathematics goes, he was greatly helpful in making up scholarship entrance examinations.

I recall two incidents when H. and I discussed work of mine.

(a) During my last year in Vienna I was an assistant of Hans Hahn, and in this capacity I became supervisor to one of his last Ph.D. students. The subject was in functional analysis and concerned sequences which were monotonically decreasing of order α, α between 0 and 1. Earlier theoretical work of Hahn, based on ideas of Helly, was used, and a relation with Cesàro means was established. I was quite proud of my ability to cope with a subject so alien to my training while I was busy with a number of other commitments. But, unfortunately, an analyst with the stature of Heilbronn was not very impressed. Hahn had hoped that I would issue this thesis as part of a survey article, but I never got to it. Later I heard that someone else had obtained the same results and was apparently not too proud to publish them. This comes when you associate with giants like H.

(b) In the fall of 1936 he checked two small manuscripts on class field theory for me before submitting them to the London Mathematical Society. When he had finished his task, he said: "This is quite nice, but it will not make you world famous." I agreed with him. However, one of them plays a role in the class field tower problem and is cited in several group theory books and occasionally simply included. The other one was generalized by John H. Smith.

In the fall of 1937 I started my first regular teaching at the University of London. I visited Cambridge occasionally and ran into Heilbronn from time to time at meetings in London; for example I attended E. Cartan's Rouse Ball lecture in 1938 at Cambridge. Heilbronn even came to lecture at my college in London. But the war clouds became darker and lower. Heilbronn was interned as a enemy alien, but he obtained a release by voluntarily joining a special army unit. When the war was over, his old friend Davenport, who held a chair at University College, London, gave him temporary employment until he was awarded a chair at Bristol. This is where Fröhlich became his thesis student, a fact that made Heilbronn extremely proud and happy. He stayed at Bristol for a number of years but purchased a house for his sister in Cambridge and stayed there during vacations.

He also came down to visit our flat in London and gave us a beautiful wedding present: a cut glass decanter (with a silver top). Later, during his visit at Pasadena he gave us a pressure cooker, but we never managed to use it successfully.

I had the one and only one argument with him during the International Congress at Amsterdam. (I suppose every function must have a singularity.) During a bus tour I sat with Heilbronn (instead of my husband). I drifted into saying something I had wanted to say for a long time but was too scared to say —criticizing him for not doing any research. He replied that he was now a dean and did I want him to be a bad dean? May all deans all over the world forgive me!—at least those who have not solved Gauss or equivalent conjectures—I replied "yes." At that he yelled at me, and soon afterwards I got up and joined my husband. But this is not the end of the story. Later we visited Cambridge, and Mrs. Mordell, who drove us around, suggested that we visit H. He welcomed us warmly. And further, he showed me the manuscript of his paper [34], "On the representation of a rational number by four squares," a very fine paper, and I, of course, am particularly interested in sums of squares. Recently I told this story to Fröhlich and he thanked me warmly for my courage—he himself had been concerned about a man like H. spending years of his life without visible research success. The paper is published in the *Journal of the London Mathematical Society* **39** (1964), 72–76 and shows that Heilbronn had not lost his old strength.

What Heilbronn proves is a refinement of Lagrange's classical theorem that every positive rational x can be expressed as $y_1^2 + y_2^2 + y_3^2 + y_4^2$, where the y_i are rational numbers. He shows that the y_i can be obtained from analytical functions. He says that he was asked whether this generalization holds, and later he says "[his] attention was drawn to such questions." He does not say from whom the stimulus came. And it seems that the person or chain of persons who approached him did not communicate or even know where it all came from. I am greatly indebted to G. Kreisel for informing me about the background of this and giving me the privilege of filling in a piece of history. The local person, that is, the colleague in Bristol, was J. C. Shepherdson, and he had it from a letter (a copy of which still exists) by Kreisel himself, who had raised the question in a *Mathematical Reviews* Reviewer's Note of a book by R. L. Goodstein, *Mathematical Reviews* **24 A** (1962), 1821. Heilbronn's result thus emerges as the first special case of a question of interest to logicians. The interest of logicians in sums of squares is well known. A more general question concerning polynomials was solved by C. N. Delzell [see "A continuous, constructive solution to Hilbert's 17th's problem," *Inventiones Mathematicae* **76** (1984), 365–384.]

Although H. had not published anything for a long time (until the appearance of the above article), he had nevertheless suggested many problems to his colleagues and friends. One such problem, which concerns the analogue of Waring's problem for holomorphic functions is discussed by W. K. Hayman in his recollections of Heilbronn (cf. p. 566 of this volume). Another problem which he posed was to estimate the area of the smallest triangle obtained by selecting three out of N points lying in a disc of unit area, and this was considered by P. Erdös, K. F. Roth, W. Schmidt, and, more recently, by J. Komlos, J. Pint, and E. Szemerédi in their article "A lower bound on Heilbronn's problem," which appeared in the *Journal of the London Mathematical Society* (2) **25** (1982), 13–24. In this article they disprove what has been called Heilbronn's conjecture; however, P. Erdös has informed me that he has the impression that H. denied having made a conjecture there.

Although analytic number theory was his main interest, Heilbronn also had considerable interest in computational and numerical mathematics. I remember that he talked on a computational problem when I got him invited to Westfield College, University of London, to address the student's mathematics club, and that H. P. F. Swinnerton-Dyer used a result of H. in tabulating the class numbers and fundamental units for cubic cyclic fields. (See my article in *Proceedings of Symposia in Applied Mathematics*, Vol. VI, American Mathematical Society, 1956, p. 193.) Furthermore, his paper [26] "On discrete harmonic functions" may be considered a contribution to numerical mathematics, as well as his solution of a matrix problem raised by John Todd and quoted by him in his paper "The problem of error in digital computation," in *Error in Digital Computation*, Vol. I, (L. B. Ball, Ed.), Wiley, 1965, p. 32.

3. *Caltech.* One day the news reached me that Heilbronn had resigned at Bristol and was living for the time being at Cambridge. In those days the

departments still had plenty of funds, and I immediately consulted our chairman, Professor Bohnenblust, as to whether we could invite Heilbronn for at least a temporary stay. This was granted, and Heilbronn immediately replied that he was interested. This visit came off, and to the surprise of all of us he announced that he would be joined soon afterwards by his wife, Dorothy, a friend and bridge partner of many years—he had not been married previously. In the meantime, he had also been offered permanent employment by the University of Toronto, an offer which he accepted.

Dorothy enjoyed herself very much in Pasadena, but H. found the heat very troublesome. He did some research while at Caltech and lectured on the Cauchy-Davenport theorem. Heilbronn's invitation to Caltech led later to an invitation to Santa Barbara, where his lectures in number theory in the summer of 1967 were very helpful. Notes of these lectures were taken.

In 1965 we all met again at the Brighton number theory conference, run by Fröhlich, which led to the very much appreciated book by Cassels and Fröhlich. There Heilbronn gave one of the major lectures, "Zeta functions and L-functions," prepared for publication by Burgess and Halberstam. Twice during that lecture he became rather nervous, perhaps because of questions interrupting him. Later, people criticized him for trying to be more modern than he was trained to be in his presentation. I reassured him after his lecture, and he appreciated this.

Both of us visited the number theory meeting at Stony Brook, and something happened that gave me tremendous satisfaction. H. became very much interested in the topic of my lecture there: my recent work on Hilbert's Theorem 94 and its application to quadratic fields with ideal 3-class groups of exponent 3. (I was criticized for not using cohomological tools which were applicable there.) The thesis of H.'s student Tom Callahan was an outgrowth of this—H. invited me to be external thesis examiner. (I had also been Fröhlich's external thesis examiner in 1950!) Similarly, the thesis of C. Chang, written under the guidance of R. A. Smith, was connected with this topic.

Both these theses are attached to the joint work of Scholz and myself [*Journal für die Reine und Angewandte Mathematik* **171** (1934), 19–40] on imaginary quadratic fields with 3-class groups of type (3, 3) and to a paper in *Monatshefte für Mathematik und Physik* **40** (1933), 323–350, written by Scholz while the joint paper was in progress. Callahan's thesis was published in two parts in *Mathematika* **21** (1974), 72–87, 168–188, and Chang's thesis, Toronto 1977, is included in a paper by Chang and Foote, *Canadian Journal of Mathematics* **32** (1980), 1229–1243. The connection with our work is as follows.

With the class group as assumed, the class field has as Galois group with respect to the field in question the abelian group of type (3, 3). There are then four subfields of the class field each of relative degree 6 with respect to **Q** and the normal closure of a non-normal cubic extension of **Q**. In the second of Callahan's papers a geometric method is used. The main result of the first paper is an inequality for the ranks of the three fields involved: the quadratic

Left to right: O. Taussky, H. Heilbronn, Mrs. & Mr. L. J. Mordell.

(Pictures taken between 1935 and 1939.)

R. Rado, P. Erdös, H. Heilbronn.

field, the non-normal cubic field, and the field of order 6. For a field of p-class group of type (p, p), there are exactly $p + 1$ cyclic unramified extensions. By Hilbert's Theorem 94, at least one class of order p capitulates in such a field, that is, becomes principal. What this class is is still not known in general, although the case $p > 2$ formed the subject of my own thesis, written under Furtwängler, published in *Journal für die Reine und Angewandte Mathematik* **168** (1932), 193–210. The same problem forms the subject of Chang's thesis and the paper of Chang and Foote. In addition, I introduced a division of the capitulating classes into two types in a paper published in 1970 in *Journal für die Reine und Angewandte Mathematik* **171**. This was the subject I lectured on at Stony Brook in Heilbronn's presence and which apparently stimulated him.

At Toronto, H. gave several advanced courses; the last of these was a course on class field theory in 1973–1974. Notes of this course were written up by S. Pierce and T. Callahan (with a little help from R. A. Smith and considerable assistance from H.).

Scientific life at Toronto agreed with H. very much. However, his health declined. The last time I saw him was at the Vancouver Congress in 1974.

It seems to me that I am coming back to the start of this article at this point. For H. was a "tough guy" and wanted to be considered as one. John Todd remembers that he offered him a drink and H. asked for straight gin, not a martini or even pink gin. But there was also some element of weakness or ill health which prevented him from working too hard. He had the talent, but not the strength, to achieve even more or to write a book on his favorite subject which would be of everlasting usefulness. But the mathematical world is not likely to forget what he did achieve.

Heilbronn in Bristol

John C. Shepherdson

Bristol was the first university to find financial support for Heilbronn when he came to England after losing his position in Germany in 1933. As described in detail in the obituary written by Cassels and Fröhlich, he was here for about a year and a half in 1934–1935. He was extremely grateful for the support he received at that time from Bristol and subsequently from Cambridge, and this manifested itself in a quite extraordinary loyalty to Britain and to Bristol and Cambridge in particular.

I first met him here when he returned as a Reader in 1946. I had just been appointed as a very young and inexperienced Assistant Lecturer. Heilbronn was not on the committee which interviewed me because he had not been appointed himself, but I think he may have had a hand in my appointment. I had met him once at Trinity, playing bridge with him and Hermann Bondi and Maurice Pryce. And it was certainly consistent with the standard method he adopted later for filling posts, of using his inside knowledge of the Cambridge scene to select promising young men.

Although H. R. Hassé was still head of the department, he was due to retire in 3 years. I think everyone expected that Heilbronn would be his successor; from the start he was not only the most powerful and distinguished mathematician in the department but also the most dominant and influential. In fact the department was not very active in research at that time and several of the older members who had been very promising young men in the 1920s had done very little recently. There was certainly less pressure than there is now, a feeling that lecturing was what one was paid for and research something one did mainly for one's own satisfaction. They were more leisurely days and no one would feel guilty about going off to play golf in the afternoon. And although lecturing loads were much higher than now the total teaching load seemed much less. Many of the lectures were based on textbooks and at a very elementary level; we had an intermediate level which was not much above A level. We didn't give out duplicated lecture notes, exercise sheets, and solution sheets, and I think there were no tutorials until Heilbronn introduced them after the Cambridge model. Administration was almost unnecessary; with a staff of less than ten there was no need for departmental meetings, and with a handful of students in the final year (literally! there were four in 1951) and other classes of fifteen or so (apart from the elementary service courses), there was no need for elaborate student records. Since most of us, including Heilbronn, submitted papers to journals in handwriting, a half-time secretary was quite adequate for the whole department.

I always found Heilbronn rather frightening. He was tall and powerfully built, with a rather severe looking face made even more intimidating by his dueling scars. He had a forceful personality and would express his views strongly in a loud voice which brooked no rebuttal. And, of course, one knew that on a mathematical point he was virtually certain to be right. In fact he was extremely concerned about the welfare of his staff and very generous. But he did like to get his own way, and when he succeeded to the Henry Overton Wills Chair of Mathematics in 1949 he ran the department like a benevolent despot. His standard reply to a suggestion he disagreed with was: "Well, when you are running your own department you can do it that way." As it turned out, he was wrong; when we did become heads of departments the greater democracy we had desired had arrived, and having done what our seniors wanted when we were young we found ourselves doing what our juniors wanted when we were older. Faced with the stultifying effect of the innumerable committees of today, some of us look back through rose-tinted spectacles to Heilbronn's reign when he could make all decisions quickly between his in-tray and his wastepaper basket and all we had to do was grumble occasionally. He certainly did his best to protect us from administrative work and give us as much time as possible to do mathematics. He was also very good at fighting for the department, at getting new posts and promotions. The present log jam for promotion to "Reader and Senior Lecturer" did not exist then, but it must have taken all his force of personality and eminence as a mathematician to get Readerships for two of us before we were 30, and with only half a dozen published papers. On the other hand he was not prepared to push for ancillary resources he thought unimportant. He was a very staunch conservative and very much opposed to any waste of public money. He felt it would be immoral to press for more than the minimum necessary secretarial assistance or stationery or even books for the library. Several of us kept pressing him to get more money for the library at a time when the library grant was £75 a year. He used to spend more than that on his personal library but took the view that we could borrow freely from it and that the department library should have only really essential books, particularly the ones needed by the students.

He was very broadminded about the mathematical activity of the department and quietly encouraged us all to work on whatever we wanted to. When I arrived in Bristol I thought I wanted to work on theoretical atomic physics, but I soon decided that to do that I would have to learn a lot more physics. So when H.H. (as we all used to refer to him) suggested a little problem in elementary number theory to me I tried it. Luckily he gave me a reference where I could find out what the problem was, because I hadn't the courage to disclose the depths of my ignorance by telling him when he was describing it that I didn't know what a residue class was. I succeeded in solving this problem and asked H.H. what I should read if I wanted to work in number theory. He told me I should start with the five volumes of Landau's *Zahlentheorie*. I found this prospect too daunting and drifted into set theory and mathematical logic. H.H. made no attempt to influence me and continued to take an interest in my work. He made no attempt to emphasize number theory unduly, his objective in filling posts being to get the best mathematician, preferably young, that he could find. He succeeded in attracting some very good pure mathematicians: In number theory: D. A. Burgess, C. S. Davis, C. Hooley, in geometric measure theory: E. R. Reifenberg, J. M. Marstrand; in geometry: P. Du Val; in topology: H. B. Griffiths. And he also helped Leslie Howarth, who was appointed to a Chair of Applied Mathematics in 1949, to build up one of the strongest schools of fluid mechanics in the world. But we were not successful in persuading him to appoint a lecturer in Statistics. He regarded that as something any competent analyst could do if he put his mind to it—though the implication was that perhaps he wouldn't want to. He introduced regular weekly or fortnightly seminars and made it very clear we were all expected to attend. We felt it necessary to offer a very good excuse if we did not attend. The speakers did not always realise they were speaking to a general audience. I think number theorists in particular felt they had to talk at a high level in the presence of someone as eminent as Heilbronn. I remember listening to Carl Ludwig Siegel for 90 minutes nonstop, and another number theorist who lost me after 90 seconds when he used some term I didn't know the meaning of. But H.H. would sweeten the pill by inviting, at his own expense, three or four or us to dinner with the speaker at Hort's, one of the best and oldest established restaurants in the town. He was very well known there and was treated with deference by the head waiter; there was no chance there of the barbarism which he complained of at another restaurant—of having cold wine brought to room temperature by immersing the bottle in hot water.

From the time he arrived in 1946 until just before he left in 1964 H.H. lived in a small bare room in the Hawthorns Hotel, just across the road from the mathematics department. At least he lived there during term time; during vacations he went back to Cambridge to live in the house he had bought for his sister. He would go back to the Hawthorns for lunch and, unless there was a meeting he had to attend, not reappear until 5.00 p.m. We assumed he was having a siesta. Then he would work in the department until dinner time and usually go back there after dinner. He dressed conventionally but rather carelessly. His main relaxation was bridge, which he played with members of the department and also at the Bristol Bridge Club. I don't think he participated in any sport except walking, but he was a staunch supporter of the university boat club. He coached them, gave financial support, and went to see them compete. He was also very keen on watching tennis and used to attend every day of the Redland Tennis Tournament in Bristol which in those days used to attract star players. With characteristic generosity he would invite his colleagues and friends and their wives to accompany him. He was also very generous with wedding presents and gave us the largest and best quality pressure cooker obtainable. This was one of his favourite gifts; I think he hoped it would revolutionise cooking the way the wheel and axle had revolutionised transport. Unlike Olga Taussky, we found this the most useful of our wedding presents; we used to take it with us to sterilise babies' bottles, and after 27 years of constant use it is, I hope, still going strong.

H.H. appeared to be in robust health, but in fact in his last few years here he had a duodenal ulcer and continued lecturing even when in great pain. Mathematics and Bristol university were his consuming interests, and he was unselfish enough to devote much time and effort (perhaps, as Olga Taussky has suggested, too much time) to work for the mathematical and university community. He was proud to be president of the London Mathematical Society and willing to take his turn as Dean of the Faculty of Science, and he worked hard at both jobs. I think one of the reasons he left Bristol in 1964 was that he felt he had failed in developing mathematics here as much as he had hoped for. He believed that after having been moved from the Physics Department to the Engineering Department he had eventually succeeded in getting adopted a plan for a Mathematics Institute or at least a self-contained building for the Mathematics Department containing lecture rooms and library as well as offices. He felt he had been let down when this fell through and the new plan split the department into a new building with offices and two lecture rooms, with the other existing lecture rooms and library in the Engineering building. Actually, I think his desire to see mathematics accorded its proper place may have clouded his judgement on the right way to spend public money, for although it would have been very much more convenient to have one self-contained building, the university was not then using its lecture rooms to capacity. But he was also disturbed by the Robbins proposals for a great expansion of the universities. Many of us expressed views in agreement with him against the creation of small new universities but he had little support for his more general view that the rapid expansion of existing universities was a mistake, that "more" means "worse." It did not, in the event, lead to the dramatic fall in standards he had feared. On the other hand, I think the motivation of students has changed; one can no longer assume that someone who studies mathematics at a university is keenly interested in mathematics but only that they are fairly good at it and think it offers reasonable prospects of a job. And with hindsight I think many would agree that the *rapidity* of the expansion of the universities in the 1960s was a mistake which may have contributed to the student unrest at the end of the decade, may possibly have ultimately caused the recent cuts, and has undoubtedly led to many talented young people today being denied academic posts.

H.H.'s announcement in 1963 that he planned to resign, without any definite alternative job to go to, came as a complete surprise to us, and we tried unsuccessfully to talk him out of it—though we knew that there was virtually no chance of him changing his mind on anything once it was made up. When he eventually left for North America and got married at the same time, we were amazed, for we had always thought of him as a confirmed bachelor and deeply attached to his adopted country. And it was somewhat ironic in that the university expansion which he feared had already taken place over there. Only when he had gone did we realise the full extent of our loss.

Hans A. Heilbronn: Toronto Days, 1964–1975

J. H. H. Chalk

A glance at his list of publications shows that Hans Heilbronn maintained a continuing interest in number theory throughout his mathematical career. He liked to claim that he was also a classical complex analyst, and both these aspects are reflected in his published articles. His early work in analytic number theory began in Germany, where he was Landau's assistant in the early 1930s, and it continued to flourish in Cambridge, where he formed a deep friendship with Harold Davenport. This led to a series of joint articles (cf. [11]–[16]), and further joint articles followed when they worked together in Manchester (1938) and London (1946). He left London in 1946 to take up an appointment in Bristol as Reader in Mathematics (later Professor), but, like many number theorists in England at that time, he made occasional visits to London to attend Davenport's seminar. In fact, it was Davenport's progress on problems concerning Euclid's algorithm that inspired Hans to settle the case for cyclic fields (cf. [28]). For those few of us in Canada working in number theory in the late 1950s, it was important to keep abreast of recent developments, and visits to London (and Cambridge) for the latest news was almost mandatory, if only to ensure that one was not working on a problem already solved. At a sherry party given by Professor and Mrs. Mordell at their home in Cambridge in the summer of 1963, I first heard of Hans's resignation from the Chair of Mathematics at Bristol. With support from our Chairman (Professor De Lury, himself a former number theorist), a letter was sent to Hans to invite him to join our department. His reply was characteristically short and to the effect that the conditions of the appointment were quite acceptable and that he required a month to decide. Shortly afterwards another note arrived to inform us that he was planning to marry. Then, following our letter of congratulations, he replied, accepting the post. On arrival in Toronto for the fall term of 1964, he settled with his wife Dorothy in a spacious apartment within easy reach of the Toronto campus.

It was a good choice, as we were soon to discover, for he was a splendid host, and in the course of the 11 years he was with us a succession of distinguished visitors were welcomed there. The main room contained, to my initial surprise, a rather large colour television set, not something I would have expected from our early acquaintance in 1946. It turned out that his interest in sports, particulary rowing, now included the viewing of hockey games.

After the first year, Hans started his graduate lectures, and these attracted several colleagues as well as graduate students, in particular the late R. A. (Bob) Smith and myself. Although he listened politely to questions on classical elementary number theory, it became clear that he was much more interested in arithmetical theory in a wider context. At first he concentrated on the theory of algebraic function fields with a view to giving his version of Weil's theorem on the Riemann hypothesis. This gave one an overall picture of the subject matter, and the following year he developed it more systematically with details, following the pattern in Eichler's book *On Algebraic Numbers and Functions*. With this preparation, the lectures in subsequent years were devoted to class field theory with applications to both algebraic function fields and algebraic number theory.

On the appearance of Weil's book *Basic Number Theory*, Hans adopted its language and notation and presented it in his own inimitable style—strict economy of words and symbols and rather large steps requiring detailed work on the part of the audience. Fortunately, Bob Smith was adept at handling these, and one of the highlights for me was a joint publication by Bob and myself on an alternative proof, using function fields, of Bombieri's theorem on exponential sums "along a curve." This was really a suggestion from Hans, and we merely carried out the programme and extended it a bit. Here class field theory made a brief but essential appearance at one stage of the argument. Other instances where a known fact on class fields was used to complete the solution of a number theory problem can be found in Hans's own work (cf. [2], [29]). Hans himself was never keen to write, even lecturing from a bare minimum of notes, and often none at all. However, his lectures on class fields were written up, with considerable attention to detail and with certain modifications, by our colleagues T. Callahan and S. Pierce but have regrettably remained unpublished. Subsequently, Bob Smith prepared a manuscript for a textbook introducing valuation theory with a view to applications in algebraic number theory and algebraic function fields. Although this is far from complete, there is ample evidence that much of it was inspired by Hans's lectures. Hans always enjoyed discussions of unsolved problems and was a source for ideas, and even if these did not always lie within the grasp of the questioner there was often the compensation of finding that a certain approach could be ruled out or that a similar but easier way was worth trying. Sometimes, he would take on a particular problem himself (e.g., [26], [39]) and even publish it, but he always gave careful acknowledgment of the source.

Finally, on a more personal note, I would like to add my own tribute to Hans. From a brief acquaintance in London (1946), where his presence and personality permitted few variations of opinion, to subsequent occasional meetings at conferences where, with the passing years, more avuncular aspects of his nature prevailed, a deep friendship evolved that I would not have thought possible in those early years.

Brief Recollections of Heilbronn

I. By M. Deuring *

Ich traf Hans Heilbronn zuerst im Jahre 1926, als ich mit dem Studium der Mathematik begann. Die wissenschaftliche Arbeit war damals, ganz inoffiziell, in verschiedene Gruppen aufgeteilt, die den einzelnen Professoren zugeordnet waren. Zu Landaus, der Zahlentheorie gewidmeten Gruppe, gehörte Hans Heilbronn. Auch der damals schon schwerkranke Hilbert hatte noch einige Schüler, von denen insbesondere Arnold Schmidt und G. Gentzen zu nennen sind. Um R. Courant scharten sich die Adepten der reellen Analysis; ich nenne da H. Lewy, K. Friedrichs und F. Rellich. Ich selbst hatte mich sehr bald den Algebraikern um E. Noether und, etwas später, B. L. van der Waerden angeschlossen. Landau hielt damals eine vielsemestrige Reihe von Vorlesungen über Zahlentheorie, die ihren Niederschlag in dem bekannten Werk "*Vorlesungen über Zahlentheorie*" gefunden haben. Bei der Abfassung und bei der Drucklegung dieser Vorlesungen war Heilbronn ein willkommener und eifriger Mitarbeiter. Beiden, Landau sowohl wie Heilbronn, kam dabei zustatten, daß sie pflichtbewußte Preussen waren. Ich erinnere mich noch sehr gut an ein Semester, in dem ich als Hilfsassistent für die Übungen zur Differential- und Integralrechnung von Landau tätig war. Jeder von uns Assistenten nahm einen Teil der schriftlich ausgeführten Übungsarbeiten mit nach Hause, sah sie durch und korrigierte sie. Am Abend, bevor die Arbeiten zurückgegeben wurden, erschienen wir mit unseren korrigierten Heften bei Landau in seinem Hause. Dann ging Landau daran, jede Übungsarbeit im Einzelnen sorgfältig durchzusehen. Er korrigierte hier und da sehr genau etwas, was der Assistent übersehen oder irrtümlicherweise beanstandet hatte. Landau war streng aber gerecht mit den Studenten und Heilbronn unterstützte ihn dabei. Das dauerte solange wie auch immer, manchmal bis 11 Uhr abends. Und es geschah mindestens einmal, daß Landau nach Schluß der Arbeit sagte: Herr Heilbronn, ich habe da heute einen Brief von Hardy (oder von wem auch immer) erhalten, den müssen wir jetzt noch zusammen studieren. Die ungeheuere Arbeitskraft Landaus war wirklich beeindruckend und ich glaube, daß dadurch Heilbronn eine ziemliche Last aufgebürdet wurde.

Ich verlor Heilbronn dann bald aus dem Auge, weil ich im Jahre 1931 nach Leipzig zu van der Waerden ging. 1933 mußte Heilbronn, ebenso wie Landau,

*Deceased December 20, 1984.

die Tätigkeit in Göttingen aufgeben, und ging, wie bekannt, nach England. Damals hatte ich gerade eine kleine Arbeit geschrieben über die Klassenzahlen imaginärer quadratischer Zahlkörper, über die Vermutung von Gauß, daß die Klassenzahl mit der Diskriminante ins Unendliche wächst, die zu beweisen mir aber nicht gelang. Heilbronn hat dann, schon in England, meinen Ansatz verbessert und ergänzt, mit dem Ergebnis, daß es ihm gelang, die Gauß'sche Vermutung zu bestätigen, wenn auch nur in einer ineffektiven Weise, wie man heute sagt. Obwohl ich etwas enttäuscht war, daß mir nicht eine solche gute Idee gekommen war, war ich doch ganz froh, daß gerade er das Glück hatte, die Lösung zu finden, denn ich dachte mir gleich, daß das seiner Karriere in England nützen müsse—und dies war es ja auch in der Tat der Fall gewesen.

Ich habe dann Heilbronn nur einmal wieder getroffen, nämlich 1958 auf dem Internationalen Mathematikerkongreß in Edinburgh.

[I first met Hans Heilbronn in the year 1926, when I began my studies of mathematics. The scientific work was then, quite unofficially, divided into several groups which were assigned to the individual professors. Hans Heilbronn belonged to Landau's group, which was dedicated to number theory. Even Hilbert, who was at that time already quite ill, still had some students; of these one should name in particular Arnold Schmidt and G. Gentzen. The adepts of real analysis flocked around R. Courant; here I mention H. Lewy, K. Friedrichs, and F. Rellich. I myself very soon joined the algebraists around E. Noether and, somewhat later, B. L. van der Waerden. At that time Landau gave a series of lectures on number theory which lasted several semesters; these had their repercussions in the well-known work "*Vorlesungen über Zahlentheorie*." In the composition and proofreading of these Vorlesungen, Heilbronn was a welcome and eager collaborator. Both Landau and Heilbronn had to their advantage the fact that they were duty-conscious Prussians. I still remember very well one semester in which I was a tutor for Landau's calculus course. Each of the tutors took home a portion of the written exercises and checked and graded them. The evening prior to returning the exercises we appeared with our corrected notebooks at Landau's house. Then Landau engaged himself in looking through every exercise in detail. He very precisely corrected here and there something which the assistant had overlooked or wrongly criticised. Landau was very strict but just to the students, and Heilbronn supported him in this respect. This lasted as long as was necessary, sometimes until 11 p.m. And it happened at least once that at the end of the work Landau said: "Mr. Heilbronn, today I received a letter from Hardy [or from whomsoever] which we must study together now." Landau's enormous energy was really impressive, and I believe that through it quite a burden was imposed upon Heilbronn.

I soon lost sight of Heilbronn, because in 1931 I joined van der Waerden in Leipzig. In 1933 Heilbronn, like Landau, had to resign from his position in Göttingen and went to England, as is known. At that time I had just written a short article on the class numbers of imaginary quadratic fields, on the conjecture of Gauss, which, however, I did not succeed in proving. Then Heilbronn, already in England, improved and completed my attempt [*Ansatz*], with the result that he succeeded in verifying the Gauss conjecture, albeit only in an ineffective manner, as one says today. Even though I was somewhat disappointed that such a good idea had not come to me, I was nevertheless quite glad that it was Heilbronn who had the luck to find the solution, for I thought immediately that this would be useful for his career in England—and this was indeed the case.

I met Heilbronn only once after this, in 1958 at the International Mathematical Congress in Edinburgh.]

Note. H. Wefelscheid pointed out to the editor that the first statement above seems to conflict with Heilbronn's curriculum vitae (cf. p. 3), and M. Kneser suggested that because of this they probably did not meet until the spring of 1929 when Deuring returned from Rome. Moreover, Heilbronn could not have helped Landau with the "Vorlesungen," since they were already published in 1927 (and he is not mentioned in Landau's preface).

II. By P. Scherk*

...I never asked Hans about his fencing. But there was clear evidence on his head, "*Schmisse*" (saber cuts), that he had been a member of a dueling student association (*schlagende Verbindung*) and that he had taken part in these light saber (*Schläger*) duels (*Mensuren*). You probably know that every member of such an association had to fight a certain number of these duels under very precisely defined rules.

I am not sure whether Hans got his Ph.D. in Berlin or in Göttingen. At any rate he already had his Ph.D. when I first met him in Göttingen in the winter of 1931–1932. At that time he was Landau's assistant. That winter Landau lectured (four times a week!) on the ζ-function, and the following summer on Dirichlet series; I believe the two cooperated in several papers at the time. Landau was a marvellous technician and a murderous slave driver; you could not imagine a better teacher. At the time Hans had written on an application of Brun's sieve method: If $F(x)$ is a polynomial in $\mathbf{Z}[X]$ whose coefficients are relatively prime, then the sequence $F(n)$ contains infinitely many prime numbers. That paper impressed me very much.

Our paper on the Goldbach conjecture started out with a faulty paper by Romanov which Landau tore to pieces. Then Hans Heilbronn got R.'s results in a correct fashion and improved on them. (He was already in Cambridge.)

*Extract from a letter of June 6, 1975 to J. H. H. Chalk. P. Scherk passed away on June 6, 1985.

Landau rewrote and improved H.H.'s paper and results. Landau was in Berlin, and I was his assistant. He expected me to work through the whole long paper in three days. I did, and several common versions followed which led to " ≤ 71 primes." The great paper by Vinogradov appeared less than a year later.

That year Landau wrote his Cambridge Tract. It contains Hans's big paper on quadratic forms or rather on their class numbers. Landau considered that paper one of the great achievements of the century!

I think the Torontonians as a group were never conscious of Hans's world class calibre...

III. By T. C. Nicholas*

I have looked up the minutes of the Trinity College Council for information about Hans Heilbronn to see if they would jog my memories, which at my present age of 95 have become unreliable....

As regards the Bevan Fellowship, Professor A. A. Bevan, who lived in the rooms I now inhabit, died in 1934 and left a legacy of £10,000 to Trinity College, the use of which was to be decided by a committee consisting of all the Fellows of the College who were professors in the University of Cambridge on the day of his death. This body of about a dozen Fellows decided that the income should be used for an experimental period of 10 years to provide a research fellowship to be called the Bevan Fellowship, with a higher stipend than the normal research fellowship, and that after 10 years it should be decided whether to continue the fellowship or to use the money for College building.

It was the period when the persecution of Jews had begun in Germany and many of distinction were leaving that country. The first Bevan Fellow elected in October 1934 was a very distinguished German classical scholar Professor Fraenkel, but he was very soon elected to a professorship at Oxford and Hans Heilbronn was elected in his place in May 1935 for 5 years and took up residence in Cambridge with his mother.

I remember him as being large in stature and rather brusque in voice and manner with conspicuous scars on his cheeks from sword cuts—the very opposite to my impression of a fugitive! Our ways did not often meet while he was a fellow. Professor Duff remembers that he was a very active member of the Fire Party constituted when war with Germany broke out in September 1939 to protect Trinity College in the event of air attack.

When the invasion of Holland took place in the spring of 1940, he was

*Extract from a letter of February 1984 to Mrs. Davenport. Mr. Tressilian Nicholas was Senior Bursar at Trinity College from 1929 to 1956.

interned in the Isle of Man with a number of other Cambridge residents but was later released on volunteering to serve in the British Army in October 1940. I recall meeting him some years later and being surprised when he told me he was serving in the Guards Armoured Division. A council minute records that during his internment payment of his stipend as Bevan Fellow was to be made to Mrs. Heilbronn, and his Fellowship must have expired in 1940.

No further appointment to the Bevan Fellowship was made during the war, and the income was added to the capital. At the expiration of the 10 years it was decided to use the bequest in part payment of the cost of a new College hostel for undergraduates in Green Street, which is called the Bevan Hostel.

IV. By P. Du Val*

I cannot recall... the exact date when he came to England; I would guess about 1932–1933; and I am fairly sure that he was elected to a short-term research fellowship in Trinity (my own College, and also Coxeter's and Davenport's), probably at Littlewood's instance, and while I was still a Fellow, as I think it can only have been through conversation in Hall that I quickly got to know him fairly well. As soon as he could, he got his parents and sister out of Germany and settled them in a house in Cambridge (Chesterton Road) where I was a frequent visitor. His mother was an excellent pianist; they had many musical friends and gave very good musical parties, at which Hans, though quite unmusical himself (rather rare in a mathematician), was an assiduous and charming host.

His manner was very correct and formal, and he could be rather intimidating on first acquaintance. I remember well that when I joined his staff in Bristol, all our colleagues called him Professor Heilbronn, very deferentially, but my wife and I, who already knew him well, called him Hans from the start, and by the time we left all the department were doing the same.

On one occasion, when I was not in England (it must have been about 1934–1935), he came to tea with my mother, who lived in Cambridge, and she wrote to me, quoting (I think) Dr. Johnson: "His intellect is an exalted as his stature; I am half afraid of him, yet he is no less amiable than formidable."

In the general hysteria of 1940 he was interned, with so many other anti-Nazi refugees (including Segré and Mahler), in the Isle of Man; and on being released joined the Pioneer Corps as a private; but I lost touch with him at about this time, as I went to Turkey and was there during the rest of the war and till 1949 and then for three years in America.

*Extract from a letter of January 30, 1984 to J. H. H. Chalk.

As well as tennis, he was an enthusiast for many sports, especially rowing and sailing, but here again I remember this best in his Bristol period. Shepherdson will perhaps have remarked that as well as being professor of mathematics, he was also coach of the University Boat Club and used to arrange for good oarsmen who had exams in Cowes week to sit for them separately in London at times which did not interfere with their commitments on the river.

V. By R. A. Rankin*

I was a Part III student in 1936–1937 and attended a course on algebraic functions which Harold [Davenport] gave with the avowed purpose of learning the subject himself. Heilbronn was present, and my impression is that they had consulted closely together about the content of the course. I also attended, between then and 1939, at least two different courses on number theory given by H.H. He was an excellent lecturer in the German style: i.e., he did not believe in using notes and so occasionally got tied up when things became complicated. He was supposed to act, together with Hardy, for my Ph.D. oral, but this took place in June 1940, when he was interned in the Isle of Man, so that Littlewood had to act instead. Before that he helped Hardy run his weekly "conversation class" and was very helpful to me when I was a research student; I remember that he drew my attention to an early paper of Landau which enabled me to improve the main result of my thesis.

He retained his knowledge of Latin beyond his school days, and when the LMS required a Latin address (to Royalty or the Privy Council?) he was able to compose it.

He had a reputation as a misogynist, which I am sure was undeserved, considering that he got married later on. When I went to Watson's chair in Birmingham in 1951 he gave me his considered advice, which consisted mostly of telling me how to deal with women students who might come to see me. This was to have a bag of caramels in my desk to offer them. He assured me that it was not possible to cry while sucking a caramel! I never had any occasion to try this out.

VI. By D. J. Lewis

I first met Hans Heilbronn in the fall of 1959. Prior to meeting him I had heard tales of his austere and formidable character and his not caring to

*Extract from a letter of March 1984 to Mrs. Davenport.

humor fools. I had a problem to which I thought he might have the answer, and so with great trepidation I approached him, quite fearful that he might squash me as one would a pesky mosquito. To my surprise and pleasure I found he was quite willing to spend some time discussing my problem and that his reputation for an encyclopedic knowledge of number theory was well warranted. During the next few years I saw him from time to time, usually in social context—either when dining at Trinity College or when we were both guests at the Davenport home. It was clear that he and Harold Davenport were fond friends who sought and enjoyed each other's company. Although their walking holidays in the west of England no longer occurred, in Davenport's company Heilbronn thought and talked best while pacing.

The early 1960s were not the happiest of times for Heilbronn. He was not particularly active as a research mathematician and chafed at the responsibilities of being head of department at Bristol University. Indeed, it appeared that he channeled his creative and competitive energies into bridge tournaments and coaching the rowing team of the University. He was extremely uncomfortable with the changes occurring in the English university system. He felt that the opening of many new universities would spread the limited mathematical talent too thin and would lead to weak and unsupportive environments for mathematicians outside of Cambridge and London.

As president of the London Mathematical Society he sought to stem the tide, unmindful and uncomprehending the social forces at work demanding more university places. It would be interesting to hear his comments in 1984 as the present government closes some London colleges and plans to reduce the size of other British universities.

The move to Toronto and his marriage to Dorothy produced a rebirth in Heilbronn. He was free of administrative duties and accepted that in the new world one would not strive to educate only the elite and so would have many universities of varying competence. His interest in mathematics returned, perhaps in part because he now had the challenge of having a significant number of young people around him who needed direction and viewed him as their mentor. It was a fun time to visit Toronto, for the number theory seminar was very alive. During this time Heilbronn was an active participant in conferences and research institutes, especially the 1969 Research Institute at Stony Brook and the 1973 Institute at Ann Arbor for recent doctorates in number theory. The young people found him a font of valuable information and he eagerly and enthusiastically discussed their work. He worked diligently to bring the 1974 International Congress to Canada. During this interval he had a genuine zest for life, and I still remember the excitement with which he described his and Dorothy's trip to the Alaskan glaciers. His death cut short the impact he was beginning to have on number theory in mid-America.

Publications of Hans Arnold Heilbronn

1.

Über die Verteilung der Primzahlen in Polynomen

Mathematische Annalen 104 (1931), 794–799
[Commentary on p. 561]

Hauptsatz. *Es sei $f(x)$ ein ganzwertiges Polynom*[1] *vom Grade $n \geqq 2$. $P(\xi, f) = P(\xi)$ bezeichne für $\xi \geqq 2$ die Anzahl der natürlichen Zahlen $x \leqq \xi$, für die $f(x) = p$ Primzahl ist. Dann ist*

$$P(\xi) \leqq \alpha_1 \frac{\xi}{\log \xi}. \tag{1}$$

Hier und in der Folge bezeichnen lateinische Buchstaben ganze, rationale Zahlen, p Primzahlen, griechische Buchstaben reelle Zahlen, $\alpha_1, \ldots, \alpha_{10}$ positive Konstanten, die nur von den Koeffizienten von $f(x)$ abhängen.

Der Beweis des Hauptsatzes erfolgt mit Hilfe der elementaren Siebmethode des Eratosthenes, die Herr Brun[2] mit großem Erfolg in die moderne Zahlentheorie eingeführt hat und die von den Herren Schnirelmann[3] und Landau[4] in eine Fassung gebracht worden ist, die sich ohne weiteres zum Beweise des obigen Hauptsatzes benutzen läßt. Die Siebmethode ist zwar schon mehrfach auf das Problem der Verteilung der Primzahlen in Polynomen angewandt worden[5], doch ist das einzige bisher in der Richtung von (1) vorliegende Ergebnis das von Herrn Nagell[6] bewiesene Resultat:

$$\lim_{\xi = \infty} \frac{P(\xi)}{\xi} = 0. \tag{2}$$

---

[1]) Ein Polynom heißt ganzwertig, wenn es für ganze, rationale Argumente ganze, rationale Werte annimmt.

[2]) Viggo Brun, Le crible d'Eratosthène et le théorème de Goldbach, Comptes rendus hebdomadaires des séances de l'académie des sciences **168** (1919), S. 544–546.

[3]) L. Schnirelmann, Sur les propriétés additives des nombres, Iswestija Donskowo Polytechnitschewskowo Instituta (Nowotscherkask) **14** (1930), S. 3–28.

[4]) Edmund Landau, Die Goldbachsche Vermutung und der Schnirelmannsche Satz, Nachrichten von der Gesellschaft der Wissenschaften zu Göttingen, 1930, S. 255–276.

[5]) Hans Rademacher, Beiträge zur Viggo Brunschen Methode in der Zahlentheorie, Abhandlungen aus dem Mathematischen Seminar der Hamburgischen Universität **3** (1924), S. 12–30.

[6]) Trygve Nagell, Zur Arithmetik der Polynome, Abhandlungen aus dem Mathematischen Seminar der Hamburgischen Universität **1** (1922), S. 179–194.

Ob (1) scharf ist, weiß ich nicht, da man bisher für kein Polynom

$$(3) \qquad \lim_{\xi=\infty} P(\xi) = \infty$$

bewiesen hat.

Ohne Beschränkung der Allgemeinheit darf $f(x)$ erstens als irreduzibel, zweitens ohne feste Primteiler angenommen werden.

$v(p)$ bezeichne die Anzahl der inkongruenten Wurzeln der Kongruenz

$$(4) \qquad f(x) \equiv 0 \pmod{p}.$$

Dann ist

$$(5) \qquad 0 \leqq v(p) \leqq n,$$

und es gibt $v(p)$ Zahlen

$$(6) \qquad 0 \leqq w_{1,p} < \ldots < w_{v(p),p} < p,$$

so daß aus (4)

$$(7) \qquad x \equiv w_{u,p} \pmod{p} \text{ für ein } u \text{ mit } 1 \leqq u \leqq v(p)$$

folgt.

Ferner gibt es bekanntlich zu $f(x)$ ein b mit $b \mid n!$, so daß $bf(x)$ ein Polynom mit ganzen, rationalen Koeffizienten ist. Für $p > n$ ändert also $v(p)$ seinen Wert nicht, wenn man statt $f(x)$ das Polynom $bf(x)$ zugrunde legt. Hieraus ergibt sich, daß der folgende Hilfssatz, der für ganzzahlige Polynome von Herrn Rademacher [7]) als Korollar aus einem Satze von Herrn Nagell [8]) gefolgert worden ist, auch für ganzwertige Polynome gilt.

Hilfssatz 1. *Für $\eta > n$ ist*

$$(8) \qquad \frac{\alpha_2}{\log \eta} \leqq \prod_{n<p\leqq\eta} \left(1 - \frac{v(p)}{p}\right) \leqq \frac{\alpha_3}{\log \eta}.$$

Definition. *Es sei $d > 0$, $h \gtreqless 0$, $k \geqq 0$, $p_r \nmid d$ für $1 \leqq r \leqq k$, $\varrho = \varrho(\xi) > \alpha_4 \geqq 10n$;*

$$(9) \qquad 10n \leqq \alpha_4 < p_1 < \ldots < p_k \leqq \varrho.$$

Dann bezeichne $F_{h,\xi,f}(d; p_1, \ldots, p_k) = F(d; p_1, \ldots, p_k)$ ($F(d)$ für $k = 0$) die Anzahl der x mit

$$(10) \qquad 1 \leqq x \leqq \xi,$$

$$(11) \qquad x \equiv h \pmod{d},$$

$$(12) \qquad x \not\equiv w_{u,p_r} \pmod{p_r} \text{ für } 1 \leqq r \leqq k, \; 1 \leqq u \leqq v(p_r).$$

---

[7]) loc. cit., S. 27.

[8]) T. Nagell, L'intermédiaire des mathématiciens (2) 1 (1922), S. 129–131.

Hilfssatz 2.[9])

$$(13)\quad F(d; p_1, \ldots, p_k) = F(d) - \sum_{1 \leqq r_1 \leqq k} v(p_{r_1}) F(d p_{r_1}) + \sum_{1 \leqq r_2 < r_1 \leqq k} v(p_{r_1}) v(p_{r_2}) F(d p_{r_1} p_{r_2}; p_1, \ldots, p_{r_2 - 1}).$$

Beweis. Es ist für $k > 0$

$$(14)\quad F(d; p_1, \ldots, p_k) = F(d; p_1, \ldots, p_{k-1}) - v(p_k) F(d p_k; p_1, \ldots, p_{k-1}).$$

Denn $F(d; p_1, \ldots, p_k)$ ist die Differenz der Lösungszahl von

$$1 \leqq x \leqq \xi,\ x \equiv h \pmod d,\ x \not\equiv w_{u, p_r} \pmod{p_r} \text{ für } 1 \leqq r \leqq k-1,\ 1 \leqq u \leqq v(p_r)$$

minus der Lösungszahl von

$$1 \leqq x \leqq \xi,\ x \equiv h \pmod d,\ x \not\equiv w_{u, p_r} \pmod{p_r} \text{ für } 1 \leqq r \leqq k-1,\ 1 \leqq u \leqq v(p_r),$$
$$x \equiv w_{u, p_k} \pmod{p_k} \text{ für ein } u \text{ mit } 1 \leqq u \leqq v(p_k).$$

Wendet man (14) sukzessive auf $F(d; p_1, \ldots, p_{k-1})$, $F(d; p_1, \ldots, p_{k-2})$, $\ldots$, $F(d; p_1)$ statt auf $F(d; p_1, \ldots, p_k)$ an, so erhält man

$$(15)\quad F(d; p_1, \ldots, p_k) = F(d) - \sum_{1 \leqq r_1 \leqq k} v(p_{r_1}) F(d p_{r_1}; p_1, \ldots, p_{r_1 - 1}).$$

Wendet man (15) mit $r_1 - 1$ statt k auf die in (15) hinter dem Summenzeichen stehenden Glieder an, so folgt die Behauptung.

Hilfssatz 3. *Es sei*

$$(16)\quad t > 0,\ k = k_0 > k_1 > \ldots > k_t \geqq 0,$$

$$(17)\quad r_1 > r_2 > \ldots > r_{2t} > 0,\ r_j \leqq k_{\left[\frac{j-1}{2}\right]} \text{ für } 1 \leqq j \leqq 2t.$$

Dann ist

$$(18)\quad F(1; p_1, \ldots, p_k) \leqq F(1) + \sum_{i=1}^{2t-1} (-1)^i \sum_{r_1, \ldots, r_i} v(p_{r_1}) \ldots v(p_{r_i}) F(p_{r_1} \ldots p_{r_i}) + \sum_{r_1, \ldots, r_{2t}} v(p_{r_1}) \ldots v(p_{r_{2t}}) F(p_{r_1} \ldots p_{r_{2t}}; p_1, \ldots, p_{\mathrm{Min}(r_{2t}-1, k_t)}).$$

Beweis. Für $t = 1$ Hilfssatz 2.

Für $t > 1$ ist nach Hilfssatz 2

$$(19)\quad F(p_{r_1} \ldots p_{r_{2t}}; p_1, \ldots, p_{\mathrm{Min}(r_{2t}-1, k_t)}) = F(p_{r_1} \ldots p_{r_{2t}}) - \sum_{r_{2t+1}} v(p_{r_{2t+1}}) F(p_{r_1} \ldots p_{r_{2t+1}}) + \sum_{r_{2t+1}, r_{2t+2}} v(p_{r_{2t+1}}) v(p_{r_{2t+2}}) F(p_{r_1} \ldots p_{r_{2t+1}}; p_1, \ldots, p_{r_{2t+2}-1}).$$

Aus (19) folgt Hilfssatz 3 durch Schluß von t auf $t+1$.

---

[9]) Vgl. Landau, loc. cit. § 2, wo die entsprechenden Hilfssätze in größter Breite entwickelt werden.

Hilfssatz 4. Voraussetzung: (16), (17),

$$\tau = 1 + \sum_{i=1}^{2t} (-1)^i \sum_{r_1, \ldots, r_i} \frac{v(p_{r_1})}{p_{r_1}} \cdots \frac{v(p_{r_i})}{p_{r_i}}. \tag{20}$$

Behauptung. $F(1; p_1, \ldots, p_k) \leqq \xi\tau + n \prod_{m=0}^{t-1} (n k_m)^2$.

Beweis. Da $\left| F(d) - \frac{\xi}{d} \right| \leqq 1$, ist nach Hilfssatz 3

$$F(1; p_1, \ldots, p_k) \leqq F(1) + \sum_{i=1}^{2t} (-1)^i \sum_{r_1, \ldots, r_i} v(p_{r_1}) \ldots v(p_{r_i}) F(p_{r_1} \ldots p_{r_i}) \tag{21}$$

$$\leqq \xi + \sum_{i=1}^{2t} (-1)^i \sum_{r_1, \ldots, r_i} v(p_{r_1}) \ldots v(p_{r_i}) \frac{\xi}{p_{r_1} \ldots p_{r_i}} + \sum_{i=1}^{2t} \sum_{r_1, \ldots, r_i} v(p_{r_1}) \ldots v(p_{r_i})$$

$$\leqq \xi\tau + \sum_{i=1}^{2t} \sum_{r_1, \ldots, r_i} n^i \leqq \xi\tau + \sum_{i=1}^{2t} n^i \prod_{m=0}^{t-1} k_m^2 \leqq \xi\tau + n \prod_{m=0}^{t-1} (n k_m)^2.$$

Hilfssatz 5. Voraussetzung: (16), (17), (20),

$$L_m = \prod_{k_m < s \leqq k_{m-1}} \left(1 - \frac{v(p_s)}{p_s}\right) \geqq \frac{4}{5} \quad \textit{für} \quad 1 \leqq m \leqq t. \tag{22}$$

Behauptung: $\tau < 2 \prod_{m=1}^{t} L_m$.

Beweis. *Siehe Landau*, loc. cit. § 1.

Hilfssatz 6. *Das in* (9) *auftretende* $\alpha_4 \geqq 10n$ *sei so groß gewählt, daß, wenn* $\pi(\sigma)$ *für* $\sigma \geqq \alpha_4$ *die Anzahl der* σ *nicht übersteigenden Primzahlen bezeichnet,*

$$\pi(\sigma) \leqq \frac{\sigma}{n} \tag{23}$$

ist. Es sei

$$\alpha_5 = \frac{\alpha_2}{27\alpha_3} \leqq \frac{1}{27}, \tag{24}$$

$$\varrho = \xi^{\alpha_5}. \tag{25}$$

$p_1, \ldots, p_k$ *durchlaufen alle durch* (9) *zugelassenen Primzahlen, also ist* $k = k(\xi)$ *eindeutig bestimmt. Es sei*

$$\xi \geqq \alpha_6, \tag{26}$$

wo $\alpha_6 \geqq 2$ *passend wählbar ist, so daß stets*

$$k \geqq 1. \tag{27}$$

Dann können $t, k_0, \ldots, k_t$ *so gewählt werden, daß sie* (16), (22) *und*

$$\prod_{m=0}^{t-1} (n k_m)^2 \leqq \xi^{\frac{2}{3}} \tag{28}$$

genügen.

Beweis. $k_0 = k$. Da für $1 \leqq s \leqq k$ $1 - \frac{v(p_s)}{p_s} \geqq \frac{4}{5}$, gibt es ein kleinstes k_1 mit

$$L_1 = \prod_{k_1 < s \leqq k_0} \left(1 - \frac{v(p_s)}{p_s}\right) \geqq \frac{4}{5}, \quad \text{falls } k_1 > 0, \tag{29}$$

ein kleinstes k_2 mit

$$L_2 = \prod_{k_2 < s \leqq k_1} \left(1 - \frac{v(p_s)}{p_s}\right) \geqq \frac{4}{5}, \quad \text{falls } k_2 > 0, \tag{30}$$

ein kleinstes k_3 und so fort, wodurch $t, k_0, \ldots, k_t$ eindeutig bestimmt sind und (16) und (22) erfüllen.

Dann ist

$$\left(1 - \frac{v(p_{k_i})}{p_{k_i}}\right) L_i < \frac{4}{5} \quad \text{für } 1 \leqq i \leqq t-1, \tag{31}$$

also

$$L_i < \frac{4}{5}\left(1 - \frac{n}{10n}\right)^{-1} = \frac{4}{5} \cdot \frac{10}{9} = \frac{8}{9}; \tag{32}$$

da nach (23) für $0 \leqq m \leqq t-1$

$$n k_m \leqq p_{k_m}, \tag{33}$$

ist also nach Hilfssatz 1 für $0 \leqq m \leqq t-1$

$$\log(n k_m) \leqq \log p_{k_m} \leqq \alpha_3 \prod_{n < p \leqq p_{k_m}} \left(1 - \frac{v(p)}{p}\right)^{-1} \tag{34}$$

$$= \alpha_3 \prod_{n < p \leqq \varrho} \left(1 - \frac{v(p)}{p}\right)^{-1} \prod_{i=1}^{m} L_i \leqq \alpha_3 \frac{\log \varrho}{\alpha_2} \cdot \left(\frac{8}{9}\right)^m = \frac{\log \varrho}{27 \alpha_5} \left(\frac{8}{9}\right)^m,$$

$$\sum_{m=0}^{t-1} \log(n k_m) \leqq \frac{\log \varrho}{27 \alpha_5} \sum_{m=0}^{t-1} \left(\frac{8}{9}\right)^m < \frac{1}{27} \log \xi \sum_{m=0}^{\infty} \left(\frac{8}{9}\right)^m = \frac{1}{3} \log \xi, \tag{35}$$

$$\prod_{m=0}^{t-1} (n k_m)^2 \leqq \xi^{\frac{2}{3}}.$$

Beweis des Hauptsatzes. Es gibt ein $\alpha_7 \geqq \sqrt{\alpha_6}$, so daß

$$|f(x)| \geqq x \quad \text{für } x \geqq \alpha_7. \tag{36}$$

Ohne Beschränkung der Allgemeinheit sei $\xi \geqq \alpha_7^2$. Dann folgt aus $f(x) = p$, $\sqrt{\xi} < x \leqq \xi$ (in den Bezeichnungen des Hilfssatzes 6)

$$p = f(x) \geqq x > \sqrt{\xi} > \varrho, \tag{37}$$

$$p_s \nmid f(x) \quad \text{für } 1 \leqq s \leqq k, \tag{38}$$

also genügt x den Bedingungen (10), (11), (12) für $d=1$; d. h. die Anzahl der x mit $f(x)=p$, $\sqrt{\xi}<x\leqq\xi$ ist $\leqq F(1; p_1, \ldots, p_k)$, also gilt

$$
\begin{aligned}
(39) \quad P(\xi) &\leqq \sqrt{\xi} + F(1; p_1, \ldots, p_k) \\
&\leqq \sqrt{\xi} + \xi\tau + n \prod_{m=0}^{t-1} (n k_m)^2 && \text{(Hilfssatz 4)} \\
&\leqq \sqrt{\xi} + \xi \cdot 2 \prod_{\alpha_4 < p \leqq \varrho} \left(1 - \frac{v(p)}{p}\right) + n \xi^{\frac{2}{3}} && \text{(Hilfssatz 5 und 6)} \\
&\leqq \alpha_8 \frac{\xi}{\log \xi} + \xi \alpha_9 \prod_{n < p \leqq \varrho} \left(1 - \frac{v(p)}{p}\right) + \alpha_{10} \frac{\xi}{\log \xi} \\
&\leqq (\alpha_8 + \alpha_{10}) \frac{\xi}{\log \xi} + \xi \alpha_9 \alpha_3 \frac{1}{\log \varrho} \leqq \alpha_1 \frac{\xi}{\log \xi} && \text{(Hilfssatz 1).}
\end{aligned}
$$

(Eingegangen am 16. 12. 1930.)

[Göttingen]

1t.

On the Distribution of Prime Numbers in Polynomials

Translation of
Mathematische Annalen 104 (1931), 794–799.
[Commentary on p. 561]

***Main Theorem**. Let $f(x)$ be an integral-valued polynomial*[1] *of degree $n \geqq 2$. For $\xi \geqq 2$ let $P(\xi, f) = P(\xi)$ denote the number of natural numbers $x \leqq \xi$, for which $f(x) = p$ is a prime number. Then*

$$P(\xi) \leqq \alpha_1 \frac{\xi}{\log \xi}. \tag{1}$$

Here and in what follows, Latin letters denote rational integers; p, prime numbers; Greek letters, real numbers; and $\alpha_1, \ldots, \alpha_{10}$, positive constants which depend only on the coefficients of $f(x)$.

The proof of the main theorem is carried out with the help of the elementary sieve method of Eratosthenes which was introduced into modern number theory by Brun[2] with great success and which Schnirelmann[3] and Landau[4] put into a form which can readily be used for the proof of the above main theorem. Although the sieve method has already been applied several times to the problem of the distribution of prime numbers,[5] until now the only existing result in the direction of (1) is the result proved by Nagell[6]:

$$\lim_{\xi=\infty} \frac{P(\xi)}{\xi} = 0. \tag{2}$$

[1]A polynomial is called integral-valued if it assumes rational integral values at rational integral arguments.

[2]Viggo Brun, Le crible d'Eratosthène et le théorème de Goldbach, *Comptes rendus hebdomadaires des séances de l'académie des sciences* **168** (1919), 544–546.

[3]L. Schnirelmann, Sur les propriétés additives des nombres, *Iswestija Donskowo Polytechnitschewskowo Instituta (Nowotscherkask)* **14** (1930), 3–28.

[4]Edmund Landau, Die Goldbachsche Vermutung und der Schnirelmannsche Satz, *Nachrichten von der Gesellschaft der Wissenschaften zu Göttingen*, 1930, 255–276.

[5]Hans Rademacher, Beiträge zur Viggo Brunschen Methode in der Zahlentheorie, *Abhandlungen aus dem Mathematischen Seminar der Hamburgischen Universität* **3** (1924), 12–30.

[6]Trygve Nagell, Zur Arithmetik der Polynome, *Abhandlungen aus dem Mathematischen Seminar der Hamburgischen Universität* **1** (1922), 179–194.

I do not know whether (1) is sharp, since for no polynomial has it been proven that

$$\lim_{\xi=\infty} P(\xi) = \infty. \tag{3}$$

Without loss of generality one can assume that $f(x)$ is first irreducible and second without fixed prime divisors.

Let $v(p)$ denote the number of incongruent solutions of the congruence

$$f(x) \equiv 0 (\mathrm{mod}\ p). \tag{4}$$

Then

$$0 \leqq v(p) \leqq n, \tag{5}$$

and there exist $v(p)$ numbers

$$0 \leqq w_{1,p} < \cdots < w_{v(p),p} < p, \tag{6}$$

such that from (4)

$$x \equiv w_{u,p} (\mathrm{mod}\ p) \quad \text{for some } u \text{ with } 1 \leqq u \leqq v(p) \tag{7}$$

follows.

Moreover, it is known that for $f(x)$ there exists a b with $b|n!$ such that $bf(x)$ is a polynomial with rational integral coefficients. Thus, for $p > n$, $v(p)$ remains unchanged if one replaces $f(x)$ by $bf(x)$. From this it follows that the following lemma, which, for integral polynomials, was deduced by Rademacher[7] as a corollary of a theorem of Nagell,[8] is also valid for integral-valued polynomials.

Lemma 1. *For $\eta > n$ we have*

$$\frac{\alpha_3}{\log \eta} \leqq \prod_{n<p\leqq\eta} \left(1 - \frac{v(p)}{p}\right) \leqq \frac{\alpha_3}{\log \eta}. \tag{8}$$

Definition. Let $d > 0$, $h \gtreqless 0$, $k > 0$, $p_r \nmid d$ for $1 \leqq r \leqq k$, $p = p(\xi) > \alpha_4 \geqq 10n$;

$$10n \leqq \alpha_4 < p_1 < \cdots < p_k \leqq \varrho. \tag{9}$$

[7] loc. cit., p. 27.

[8] T. Nagell, *L'intermédiaire des mathématiciens* (2) **1** (1922), 129–131.

Then $F_{h,\xi,f}(d; p_1,\ldots, p_k) = F(d; p_1,\ldots, p_k)$ [or $F(d)$, if $k = 0$] denotes the number of x with

$$1 \leqq x \leqq \xi, \tag{10}$$

$$x \equiv h(\operatorname{mod} d), \tag{11}$$

$$x \not\equiv w_{u,p_r}(\operatorname{mod} p_r) \quad \text{for } 1 \leqq r \leqq k, 1 \leqq u \leqq v(p_r). \tag{12}$$

Lemma 2.[9]

$$(13)\quad F(d; p_1,\ldots, p_k) = F(d) - \sum_{1\leqq r_1\leqq k} v(p_{r_1})F(dp_{r_1})$$

$$+ \sum_{1\leqq r_2<r_1\leqq k} v(p_{r_1})v(p_{r_2})F(dp_{r_1}p_{r_2}; p_1,\ldots, p_{r_2-1}).$$

Proof. For $k > 0$ we have

$$(14)\quad F(d; p_1,\ldots, p_k) = F(d; p_1,\ldots, p_{k-1})$$

$$-v(p_k)F(dp_k; p_1,\ldots, p_{k-1}).$$

In fact, $F(d; p_1,\ldots, p_k)$ is the difference of the number of solutions of

$$1 \leqq x \leqq \xi,\ x \pm h(\operatorname{mod} d),\ x \not\equiv w_{u,p_r}(\operatorname{mod} p_r)$$

$$\text{for } 1 \leqq r \leqq k-1, 1 \leqq u \leqq v(p_r)$$

minus the number of solutions of

$$1 \leqq x \leqq \xi,\ x \equiv h(\operatorname{mod} d),\ x \not\equiv w_{u,p_r}(\operatorname{mod} p_r)$$

$$\text{for } 1 \leqq r \leqq k-1, 1 \leqq u \leqq v(p_r),$$

$$x \equiv w_{u,p_k}(\operatorname{mod} p_k) \quad \text{for some } u \text{ with } 1 \leqq u \leqq v(p_k).$$

Applying (14) successively to $F(d; p_1,\ldots, p_{n-1})$, $F(d; p_1,\ldots, p_{k-2}),\ldots,$ $F(d; p_1)$ in place of $F(d; p_1,\ldots, p_r)$ yields

$$(15)\quad F(d; p_1,\ldots, p_k) = F(d) - \sum_{1\leqq r_1\leqq k} v(p_{r_1})F(dp_{r_1}; p_1,\ldots, p_{r_1-1}).$$

[9]Cf. Landau, loc. cit. §2, where the corresponding lemmas are developed at great length.

Applying (15) with $r_1 - 1$ in place of k to the terms following the summation sign in (15) proves the assertion.

Lemma 3. *Let*

$$t > 0,\ k = k_0 > k_1 > \cdots > k_t \geqq 0, \tag{16}$$

$$r_1 > r_2 > \cdots > r_{2t} > 0,\ r_j \leqq k_{\left[\frac{j-1}{2}\right]} \quad \text{for } 1 \leqq j \leqq 2t. \tag{17}$$

Then

$$\begin{aligned} (18)\quad & F(1; p_1, \ldots, p_k) \\ & \leqq F(1) + \sum_{i=1}^{2t-1} (-1)^i \sum_{r_1, \ldots, r_i} v(p_{r_1}) \ldots v(p_{r_i}) F(p_{r_1} \cdots p_{r_i}) \\ & \quad + \sum_{r_1, \ldots, r_{2t}} v(p_{r_1}) \ldots v(p_{r_{2t}}) F(p_{r_1} \cdots p_{r_{2t}}; p_1, \ldots, p_{\mathrm{Min}(r_{2t}-1, k_t)}). \end{aligned}$$

Proof. For $t = 1$, Lemma 2.
For $t > 1$, we have by Lemma 2

$$\begin{aligned} (19)\quad & F(p_{r_1} \cdots p_{r_{2t}}; p_1, \ldots, p_{\mathrm{Min}(r_{2t}-1, k_t)}) \\ & = F(p_{r_1} \cdots p_{r_{2t}}) - \sum_{r_{2t+1}} v(p_{r_{2t+1}}) F(p_{r_1} \cdots p_{r_{2t+1}}) \\ & \quad + \sum_{r_{2t+1}, r_{2t+2}} v(p_{r_{2t+1}}) v(p_{r_{2t+2}}) F(p_{r_1} \cdots p_{r_{2t+1}}; p_1, \ldots, p_{r_{2t+2}-1}). \end{aligned}$$

Lemma 3 follows from (19) by induction on t.

Lemma 4. *Hypothesis*: (16), (17),

$$\tau = 1 + \sum_{i=1}^{2t} (-1)^i \sum_{r_1, \ldots, r_i} \frac{v(p_{r_1})}{p_{r_1}} \cdots \frac{v(p_{r_i})}{p_{r_i}}. \tag{20}$$

Claim: $F(1; p_1, \ldots, p_k) \leqq \xi\tau + n \prod_{m=0}^{t-1} (nk_m)^2.$

Proof. Since $\left| F(d) - \frac{\xi}{d} \right| \leqq 1$, we have by Lemma 3

(21) $F(1; p_1, \ldots, p_k)$

$$\leqq F(1) + \sum_{i=1}^{2t} (-1)^i \sum_{r_1, \ldots, r_i} v(p_{r_1}) \ldots v(p_{r_i}) F(p_{r_1} \cdots p_{r_i})$$

$$\leqq \xi + \sum_{i=1}^{2t} (-1)^i \sum_{r_1, \ldots, r_i} v(p_{r_1}) \ldots v(p_{r_i}) \frac{\xi}{p_{r_1} \cdots p_{r_i}}$$

$$+ \sum_{i=1}^{2t} \sum_{r_1, \ldots, r_i} v(p_{r_1}) \ldots v(p_{r_i})$$

$$\leqq \xi\tau + \sum_{i=1}^{2t} \sum_{r_1, \ldots, r_i} n^i \leqq \xi\tau + \sum_{i=1}^{2t} n^i \prod_{m=0}^{t-1} k_m^2 \leqq \xi\tau + n \prod_{m=0}^{t-1} (nk_m)^2.$$

Lemma 5. *Hypothesis*: (16), (17), (20),

$$L_m = \prod_{k_m < s \leqq k_{m-1}} \left(1 - \frac{v(p_s)}{p_s}\right) \geqq \frac{4}{5} \quad \textit{for } 1 \leqq m \leqq t. \tag{22}$$

Claim: $\tau < 2 \prod_{m=1}^{t} L_m$.

Proof. See Landau, loc. cit, §1.

Lemma 6. *Suppose that the constant $\alpha_4 \geqq 10n$ which appears in* (9) *has been chosen so large that for $\sigma > \alpha_4$ we have*

$$\pi(\alpha) \leqq \frac{\sigma}{n} \tag{23}$$

where $\pi(\sigma)$ denotes the number of prime numbers which do not exceed σ. Let

$$\alpha_5 = \frac{\alpha_2}{27\alpha_3} \leqq \frac{1}{27}, \tag{24}$$

$$\varrho = \xi^{\alpha_5}. \tag{25}$$

$p_1, \ldots, p_k$ run through all the prime numbers permitted by (9); *in particular, $k = k(\xi)$ is uniquely determined. Let*

$$\xi \geqq \alpha_6, \tag{26}$$

where α_6 is to be chosen such that we always have

(27) $$k \geqq 1.$$

Then $t, k_0, \ldots, k_t$ can be chosen such that they satisfy (16), (22); *and*

(28) $$\prod_{m=0}^{t-1} (nk_m)^2 \leqq \xi^{\frac{2}{3}}.$$

Proof. $k_0 = k$. Since $1 - \dfrac{v(p_s)}{p_s} \geqq \dfrac{4}{5}$ if $1 < s < k$, there exists a smallest k_1 with

(29) $$L_1 = \prod_{k_1 < s \leqq k_0} \left(1 - \frac{v(p_s)}{p_s}\right) \geqq \frac{4}{5}, \quad \textit{if } k_1 > 0,$$

a smallest k_2 with

(30) $$L_2 = \prod_{k_2 < s \leqq k_1} \left(1 - \frac{v(p_s)}{p_s}\right) \geqq \frac{4}{5}, \quad \textit{if } k_2 > 0,$$

a smallest k_3, and so forth, whereby $t, k_0, \ldots, k_t$ are uniquely determined and satisfy (16) and (22).

Then we have

(31) $$\left(1 - \frac{v(p_{k_i})}{p_{k_i}}\right) L_i < \frac{4}{5} \quad \textit{for } 1 \leqq i \leqq t-1,$$

and hence

(32) $$L_i < \frac{4}{5}\left(1 - \frac{n}{10n}\right)^{-1} = \frac{4}{5} \cdot \frac{10}{9} = \frac{8}{9};$$

since by (23) we have for $0 \leqq m \leqq t-1$

(33) $$nk_m \leqq p_{k_m},$$

and therefore by Lemma 1 for $0 \leqq m \leqq t-1$

(34) $$\log(nk_m) \leqq \log p_{k_m} \leqq \alpha_3 \prod_{n < p \leqq p_{k_m}} \left(1 - \frac{v(p)}{p}\right)^{-1}$$

$$= \alpha_3 \prod_{n < p \leqq \varrho} \left(1 - \frac{v(p)}{p}\right)^{-1} \prod_{i=1}^{m} L_i$$

$$\leqq \alpha_3 \frac{\log \varrho}{\alpha_2} \cdot \left(\frac{8}{9}\right)^m = \frac{\log \varrho}{27\alpha_5}\left(\frac{8}{9}\right)^m,$$

$$\sum_{m=0}^{t-1} \log(nk_m) \leqq \frac{\log \varrho}{27\alpha_5} \sum_{m=0}^{t-1} \left(\frac{8}{9}\right)^m < \frac{1}{27} \log \xi \sum_{m=0}^{\infty} \left(\frac{8}{9}\right)^m \tag{35}$$

$$= \frac{1}{3} \log \xi,$$

$$\prod_{m=0}^{t-1} (nk_m)^2 \leqq \xi^{\frac{2}{3}}.$$

Proof of the Main Theorem. There exists an $\alpha_7 \geqq \sqrt{\alpha_6}$ such that

$$|f(x)| \geqq x \text{ for } x \geqq \alpha_7. \tag{36}$$

Without loss of generality suppose $\xi \geqq \alpha_7^2$. Then it follows from $f(x) = p$, $\sqrt{\xi} < x \leqq \xi$ (using the notations of Lemma 6) that

$$p = f(x) \geqq x > \sqrt{\xi} > \varrho, \tag{37}$$

$$p_s \nmid f(x) \quad \text{for } 1 \leqq s \leqq k, \tag{38}$$

and so x satisfies the conditions (10), (11), (12) for $d = 1$; i.e., the number of x with $f(x) = p$, $\sqrt{\xi} < x \leqq \xi$ is $\leqq F(1; p_1, \ldots, p_k)$ and hence

$$P(\xi) \leqq \sqrt{\xi} + F(1; p_1, \ldots, p_k) \tag{39}$$

$$\leqq \sqrt{\xi} + \xi\tau + n \prod_{m=0}^{t-1} (nk_m)^2 \qquad \text{(Lemma 4)}$$

$$\leqq \sqrt{\xi} + \xi \cdot 2 \prod_{\alpha_4 < p \leqq \varrho} \left(1 - \frac{v(p)}{p}\right) + n\xi^{\frac{2}{3}} \qquad \text{(Lemmas 5 and 6)}$$

$$\leqq \alpha_8 \frac{\xi}{\log \xi} + \xi\alpha_9 \prod_{n < p \leqq \varrho} \left(1 - \frac{v(p)}{p}\right) + \alpha_{10} \frac{\xi}{\log \xi}$$

$$\leqq (\alpha_8 + \alpha_{10}) \frac{\xi}{\log \xi} + \xi\alpha_9\alpha_3 \frac{1}{\log \varrho} \leqq \alpha_1 \frac{\xi}{\log \xi} \qquad \text{(Lemma 1)}.$$

(Received 16 December 1930)

2.

Über den Primzahlsatz von Herrn Hoheisel[1]

Mathematische Zeitschrift 36 (1933), 394–423
[Commentary on p. 561]

Inhaltsübersicht.

	Seite
Inhaltsübersicht	394
Einleitung	394
Bezeichnungen	397
§ 1. Asymptotische Entwicklung von $L(s, \chi)$ und $L'(s, \chi)$	398
§ 2. Der Littlewoodsche Satz über die Nullstellen von $L(s, \chi)$	402
§ 3. Hilfssätze über die Teilerzahlen	413
§ 4. Asymptotische Entwicklung von $\frac{1}{L(s, \chi)}$ und $\frac{-L'(s, \chi)}{L(s, \chi)}$	416
§ 5. Beweis der Hauptsätze	418

Einleitung.

Herr Hoheisel[2]) bewies kürzlich den interessanten

Satz 1. *Es sei $\pi(x)$ für $x \geqq 2$ die Anzahl der x nicht übersteigenden Primzahlen. Dann gibt es ein positives*[3]) $a_1 < 1$ *derart, daß*

$$\lim_{x=\infty} \{\pi(x + x^{a_1}) - \pi(x)\} \frac{\log x}{x^{a_1}} = 1. \tag{1}$$

Für diesen Satz gebe ich in dieser Arbeit einen neuen Beweis, der sich von dem Beweis von Herrn Hoheisel in der Hauptsache dadurch unterscheidet, daß ich die Riemannsche Zetafunktion nur in der Nähe der Geraden $\sigma = 1$ betrachte. Über ihre Nullstellen benutze ich daher nur

---

[1]) Diese Arbeit hat der mathematisch-naturwissenschaftlichen Fakultät der Universität Göttingen als Inauguraldissertation vorgelegen.

[2]) Guido Hoheisel, Primzahlprobleme in der Analysis, Sitzungsberichte der Preußischen Akademie der Wissenschaften, physik.-math. Klasse, 1930, S. 580–588.

[3]) $a_1, a_2, \ldots$ bezeichnen im folgenden positive, absolute Konstanten.

den Satz von Herrn Littlewood[4]):

$$(2) \qquad \zeta(s) \neq 0 \quad \text{für} \quad t \geqq a_2, \quad \sigma \geqq 1 - a_3 \frac{\log\log t}{\log t},$$

den auch Herr Hoheisel zum Beweise von Satz 1 verwendet. Im Gegensatz zu Herrn Hoheisel benutze ich nicht:

1. Die Riemannsche Funktionalgleichung von $\zeta(s)$.
2. Die Hadamardsche Produktdarstellung von $\zeta(s)$.
3. Den Carlsonschen Satz über die Nullstellen von $\zeta(s)$, den Herr Hoheisel noch verschärft hat.

Durch diese Vereinfachungen wird mein Beweis unabhängig von der Tatsache, daß die Funktion $\frac{\zeta'}{\zeta}(s)$, auf deren Studium das Problem der Verteilung der Primzahlen sich leicht zurückführen läßt, die logarithmische Ableitung der meromorphen Funktion $\zeta(s)$ ist. Infolgedessen ist meine Beweismethode auch auf andere Dirichletsche Reihen anwendbar, z. B. auf die Funktion $\frac{1}{\zeta(s)}$, und ich erhalte als Analogon zu Satz 1 den

Satz 2. *Es sei $\mu(n)$ die Möbiussche Funktion; dann ist*

$$(3) \qquad \lim_{x=\infty} x^{-a_1} \sum_{x<n\leqq x+x^{a_1}} \mu(n) = 0.$$

Für die Konstante a_1 ergibt sich als hinreichende Bedingung[5])

$$(4) \qquad 1 - \frac{1}{4 + \frac{5}{a_3}} < a_1 < 1,$$

wo a_3 die Konstante aus dem obengenannten Littlewoodschen Satze ist.

Da es keine große Mehrarbeit erfordert, beweise ich statt der Sätze 1 und 2 sogleich die entsprechenden Sätze für die arithmetische Progression:

Satz 3. *Es seien k und l natürliche, teilerfremde Zahlen. $\pi(x; k, l)$ bezeichne für $x \geqq 2$ die Anzahl der x nicht übersteigenden Primzahlen $p \equiv l \pmod{k}$. Dann gibt es ein $a_4 < 1$, so daß*

$$(5) \qquad \lim_{x=\infty} \{\pi(x + x^{a_4}; k, l) - \pi(x; k, l)\} \frac{\log x}{x^{a_4}} = \frac{1}{\varphi(k)}.$$

[4]) J. E. Littlewood, Researches in the Theory of the Riemann ζ-Function, Proceedings of the London Mathematical Society, Ser. II, 20 (1922), p. XXV. Vgl. auch Edmund Landau, Vorlesungen über Zahlentheorie 2 (1927), Satz 397.

[5]) Herr Hoheisel erhält die (weniger leistende) Bedingung

$$1 - \frac{1}{4 + \frac{6}{a_3}} < a_1 < 1.$$

Satz 4. *Es seien k und l natürliche, teilerfremde Zahlen. Dann ist*

$$\lim_{x=\infty} x^{-a_1} \sum_{\substack{x<n\leqq x+x^{a_1}\\ n\equiv l \pmod{k}}} \mu(n) = 0. \tag{6}$$

Zum Beweis der Sätze 3 und 4 brauche ich natürlich den Littlewoodschen Satz in der allgemeineren Form[6]):

$$L(s,\chi) \neq 0 \quad \text{für} \quad t \geqq b_1, \quad \sigma \geqq 1 - a_5 \frac{\log\log t}{\log t}. \tag{7}$$

Hier ist $L(s,\chi)$ die zu einem Charakter $\chi \pmod{k}$ gehörige L-Reihe: b_1 bezeichnet eine positive, nur von k abhängige Konstante. Als hinreichende Bedingung für a_4 finde ich (analog zu (4)):

$$1 - \frac{1}{4+\frac{5}{a_5}} < a_4 < 1. \tag{8}$$

Ich werde im folgenden zeigen, daß diese Landausche Verallgemeinerung des Littlewoodschen Satzes mit $a_5 = \frac{1}{49{,}13}$ richtig ist. Da nun

$$1 - \frac{1}{4+\frac{5}{\frac{1}{49{,}13}}} = 1 - \frac{1}{4+245{,}65} < 1 - \frac{1}{250} < 1 \tag{9}$$

ist, folgt aus (8), daß

$$a_1 = a_4 = 1 - \frac{1}{250} \tag{10}$$

Konstanten sind, für die die Sätze 1 bis 4 richtig sind. Damit ergibt sich als Korollar zu Satz 3 der

Satz 5. *Zu jeder natürlichen Zahl k gibt es ein $x_k \geqq 2$, so daß, wenn l eine natürliche, zu k teilerfremde Zahl ist, für $x \geqq x_k$ zwischen x^{250} (exkl.) und $(x+1)^{250}$ (exkl.) stets eine Primzahl $p \equiv l \pmod{k}$ liegt.*

Beweis. Es ist für $x \geqq 2$

$$(x+1)^{250} > x^{250} + x^{249}, \tag{11}$$

also nach Satz 3

$$\begin{aligned} \pi((x+1)^{250}; k,l) - \pi(x^{250}; k,l) &\geqq \pi(x^{250}+x^{249}; k,l) - \pi(x^{250}; k,l) \\ &= \pi\left(x^{250} + (x^{250})^{1-\frac{1}{250}}; k,l\right) - \pi(x^{250}; k,l) \sim \frac{1}{\varphi(k)} \frac{(x^{250})^{1-\frac{1}{250}}}{\log x^{250}}, \end{aligned} \tag{12}$$

wo die rechte Seite mit wachsendem x über alle Grenzen wächst.

---

[6]) Siehe Landau, loc. cit.

In § 1 wird eine von den Herren Hardy und Littlewood[7]) für $\zeta(s)$ bewiesene Entwicklung auf die Dirichletschen L-Reihen und ihre ersten Ableitungen übertragen.

In § 2 beweise ich den oben zitierten Littlewood-Landauschen Satz von neuem, um ein möglichst großes a_5 zu gewinnen, und leite dann daraus bekannte Abschätzungen für $\frac{L'}{L}(s, \chi)$ und $\frac{1}{L(s, \chi)}$ ab.

In § 3 folgen einige elementare Ungleichungen über die Teilerzahlen, um eine Integralabschätzung zu beweisen, die in ähnlicher Form auch bei Herrn Hoheisel zum Beweise des Carlsonschen Satzes auftritt. Hierbei bediene ich mich einer elementaren zahlentheoretischen Methode von Herrn Ingham[8]).

In § 4 tritt der entscheidende Hilfssatz auf. Er liefert eine Darstellung von $\frac{1}{L(s, \chi)}$ bzw. $-\frac{L'}{L}(s, \chi)$ am Rande des kritischen Streifens als endliche Dirichletsche Reihe + einem „kleinen" Fehlerglied + einem Glied, das zwar nicht „klein", wohl aber „im Mittel klein" ist, was genügt, um in § 5 die Sätze 3 und 4 zu beweisen.

Bezeichnungen.

Dauernd seien zwei natürliche, teilerfremde Zahlen k und l gegeben.

$\chi(n)$ bezeichne einen Dirichletschen Charakter mod k.

$a_1, \ldots, a_{13}$ bezeichnen absolute, positive Konstanten.

$b_1, \ldots, b_6$ bezeichnen positive Konstanten, die nur von k abhängen.

O- und o-Abschätzungen gelten *bei festem k gleichmäßig in allen übrigen Variablen*[9]).

Komplexe Integrale werden stets über einen geraden Weg erstreckt, wenn nicht ausdrücklich etwas anderes gesagt wird.

$d_r(n)$ bezeichne die Anzahl der Zerlegungen der natürlichen Zahl n in r positive ganzzahlige Faktoren, wobei die Reihenfolge der Faktoren berücksichtigt wird (tritt nur für $r = 2, 3, 4$ auf); für $d_2(n)$ wird kurz $d(n)$ geschrieben.

---

[7]) G. H. Hardy and J. E. Littlewood, The zeros of Riemann's Zeta-Function on the critical line, Math. Zeitschr. **10** (1921), S. 283–317; der erwähnte Spezialfall ist ungefähr das Lemma 2 dieser Arbeit.

[8]) A. E. Ingham, Some asymptotic formulae in the theory of numbers, The Journal of The London Mathematical Society **2** (1927), p. 202–208.

[9]) $f = O(g)$ bedeute, $|f| < bg$ für alle in der betreffenden Voraussetzung zugelassenen Werte der Argumente, wo b nur von k abhängt. $f(z; z_1, \ldots, z_n) = o(g(z; z_1, \ldots, z_n))$ für $z \to \infty$ bedeutet, für jedes $\varepsilon > 0$ gibt es ein $\omega = \omega(\varepsilon, k) > 0$, das nur von ε und k abhängt, so daß für alle in der Voraussetzung zugelassenen Werte $z_1, \ldots, z_n$ für $z \geqq \omega$ stets $|f(z; z_1, \ldots, z_n)| < \varepsilon g(z; z_1, \ldots, z_n)$.

$\Lambda(n)$, $\mu(n)$, $\varphi(n)$ haben als Funktionen der natürlichen Zahl n die übliche Bedeutung: Es ist

$$\Lambda(n)=\begin{cases}\log p, \text{ wenn } n=p^{\alpha}, \ p \text{ Primzahl}, \ \alpha \geqq 1 \text{ ganz},\\ 0 \text{ sonst},\end{cases}$$

$$\mu(n)=\begin{cases}1 \text{ für } n=1,\\ (-1)^{r}, \text{ wenn } n \text{ Produkt von } r \text{ verschiedenen Primzahlen ist},\\ 0, \text{ wenn } n \text{ durch das Quadrat einer Primzahl teilbar ist},\end{cases}$$

$\varphi(n)$ die Anzahl der zu n teilerfremden Restklassen mod n.

Stets ist $s=\sigma+it$, $\sigma=\Re s$;

$$\zeta(s)=\sum_{n=1}^{\infty} n^{-s} \quad \text{für} \quad \sigma>1; \tag{13}$$

$$\zeta(s,w)=\sum_{n=0}^{\infty}(n+w)^{-s} \quad \text{für} \quad \sigma>1,\ 0<w\leqq 1; \tag{14}$$

$$L(s,\chi)=\sum_{n=1}^{\infty}\chi(n)n^{-s} \quad \text{für} \quad \sigma>1. \tag{15}$$

Ich erinnere an die für $\sigma>1$ gültigen Identitäten:

$$L(s,\chi)=k^{-s}\sum_{q=1}^{k}\chi(q)\zeta\left(s,\frac{q}{k}\right); \tag{16}$$

$$\frac{-L'(s,\chi)}{L(s,\chi)}=\sum_{n=1}^{\infty}\chi(n)\Lambda(n)n^{-s}; \tag{17}$$

$$\frac{1}{L(s,\chi)}=\sum_{n=1}^{\infty}\chi(n)\mu(n)n^{-s}. \tag{18}$$

§ 1.

Asymptotische Entwicklung von $L(s,\chi)$ und $L'(s,\chi)$.

Hilfssatz 1. *Für* $2\leqq t\leqq \pi T$, $\frac{1}{2}\leqq\sigma\leqq 3$, $0<w\leqq 1$, *ganzes* $m\neq 0$ *ist*

$$\int_{T}^{\infty}(u+w)^{-s}e^{2\pi imu}\,du=-\frac{1}{2\pi im}e^{2\pi imT}(T+w)^{-s}+O\left(\frac{tT^{-1-\sigma}}{m^2}\right). \tag{19}$$

Beweis. Durch partielle Integration erhält man

$$\int_{T}^{\infty}(u+w)^{-s}e^{2\pi imu}\,du \tag{20}$$

$$=-\frac{1}{2\pi im}e^{2\pi imT}(T+w)^{-s}+\frac{s}{2\pi im}\int_{T}^{\infty}(u+w)^{-s-1}e^{2\pi imu}\,du.$$

Für $u \geqq T$ strebt der Ausdruck

$$(21)\qquad \frac{(u+w)^{-\sigma-1}}{-\frac{t}{u+w}+2\pi m} = \frac{1}{(u+w)^{\sigma}(2\pi m(u+w)-t)}$$

mit wachsendem u monoton gegen 0. Für $u = T$ ist der absolute Betrag von (21) gleich

$$(22)\qquad \frac{(T+w)^{-\sigma-1}}{\left|-\frac{t}{T+w}+2\pi m\right|} \leqq \frac{T^{-\sigma-1}}{2\pi|m|-\pi} \leqq \frac{T^{-\sigma-1}}{\pi|m|}.$$

Also ist nach dem zweiten Mittelwertsatz

$$(23)\qquad \int_T^\infty (u+w)^{-s-1} e^{2\pi i m u}\,du = \int_T^\infty (u+w)^{-\sigma-1} e^{i(-t\log(u+w)+2\pi m u)}\,du$$

$$= \int_T^\infty (u+w)^{-\sigma-1} \frac{1}{-\frac{t}{u+w}+2\pi m} e^{i(-t\log(u+w)+2\pi m u)}\,d(-t\log(u+w)+2\pi m u)$$

$$= O\left(\frac{T^{-1-\sigma}}{|m|}\right) \text{ nach } (22).$$

Aus (20) und (23) folgt die Behauptung.

Hilfssatz 2. *Für* $2 < t < \pi T$, $\frac{1}{2} < \sigma < 3$, $0 < w \leqq 1$, *ganzes* T *ist*

$$(24)\qquad g(s) = g(s; T, w) = \sum_{m=1}^{\infty} \int_T^\infty (u+w)^{-s}(e^{2\pi i m u} + e^{-2\pi i m u})\,du \text{ \textit{konvergent},}$$

$$(25)\qquad g(s) \text{ \textit{regulär},}$$

$$(26)\qquad g(s) = O\left(\frac{T^{1-\sigma}}{t}\right).$$

Beweis. Für ganzes $m > 0$ gilt, da auch T ganz ist,

$$(27)\qquad \frac{-1}{2\pi i m} e^{2\pi i m T} + \frac{-1}{2\pi i(-m)} e^{2\pi i(-m)T} = 0.$$

Daher ist nach Hilfssatz 1

$$(28)\qquad \int_T^\infty (u+w)^{-s}(e^{2\pi i m u} + e^{-2\pi i m u})\,du = O\left(\frac{tT^{-1-\sigma}}{m^2}\right),$$

woraus sofort (24) folgt.

Da nach (28) die Reihe gleichmäßig in s konvergiert, genügt es zum Beweise von (25), zu zeigen, daß jedes der Integrale in s regulär ist, was durch (20) in Evidenz gesetzt wird. Damit ist (25) bewiesen.

Nach (28) ist

$$(29)\qquad g(s) = \sum_{m=1}^{\infty} O\left(\frac{tT^{-1-\sigma}}{m^2}\right) = O(tT^{-1-\sigma}) = O\left(\frac{T^{1-\sigma}}{t}\right).$$

Hilfssatz 3. *Unter den Voraussetzungen von Hilfssatz 2 ist*

$$(30)\qquad \zeta(s, w) \text{ regulär.}$$

$$(31)\qquad \zeta(s, w) = \sum_{n=0}^{T-1} (n+w)^{-s} + \frac{1}{2}(T+w)^{-s} + g(s; T, w) - \frac{(T+w)^{1-s}}{1-s},$$

$$(32)\qquad \zeta(s, w) = \sum_{n=0}^{T-1} (n+w)^{-s} + O\left(\frac{T^{1-\sigma}}{t}\right).$$

Beweis. Es sei zunächst $\sigma > 1$. Dann ist nach der Dirichletschen Summenformel für ganzes $V > T$

$$(33)\qquad \sum_{n=T}^{V} (n+w)^{-s}$$

$$= \frac{1}{2}\{(T+w)^{-s} + (V+w)^{-s}\} + \lim_{N=\infty} \sum_{m=-N}^{N} \int_T^V (u+w)^{-s} e^{2\pi i m u}\, du$$

$$= \frac{1}{2}\{(T+w)^{-s} + (V+w)^{-s}\} + \int_T^V (u+w)^{-s}\, du$$

$$+ \sum_{m=1}^{\infty} \int_T^V (u+w)^{-s} (e^{2\pi i m u} + e^{-2\pi i m u})\, du$$

$$= \frac{1}{2}\{(T+w)^{-s} + (V+w)^{-s}\} + \frac{(V+w)^{1-s}}{1-s} - \frac{(T+w)^{1-s}}{1-s}$$

$$+ g(s; T, w) - g(s; V, w),$$

wenn $g(s; T, w)$ und $g(s; V, w)$ die Bedeutung aus Hilfssatz 2 haben. Nach (26) ist

$$(34)\qquad \lim_{V=\infty} g(s; V, w) = 0;$$

also folgt aus (33), wenn V gegen ∞ geht:

$$(35)\qquad \sum_{n=T}^{\infty} (n+w)^{-s} = \frac{1}{2}(T+w)^{-s} - \frac{(T+w)^{1-s}}{1-s} + g(s; T, w).$$

Aus (35) und (14) folgt sofort (31) für $\sigma > 1$. Da die in (31) auf der rechten Seite auftretenden Funktionen für alle s der Voraussetzung regulär sind nach (25), gelten (31) und (30) für alle unsere s. (32) aber folgt sofort aus (31) und (26).

Hilfssatz 4. *Für* $2 < t < \pi T$, $\frac{1}{2} < \sigma < 3$ *ist*

$$(36)\qquad L(s, \chi) \text{ regulär,}$$

$$(37)\qquad L(s, \chi) = \sum_{1 \leq r \leq kT} \chi(r) r^{-s} + O\left(\frac{T^{1-\sigma}}{t}\right).$$

Beweis. Ohne Beschränkung der Allgemeinheit sei T **ganz.** Sonst betrachte man $[T]+1$ statt T. Das geht wegen

$$(38) \qquad \left| \sum_{kT < r \leqq k([T]+1)} \chi(r) r^{-s} \right| \leqq k(kT)^{-\sigma} = O\left(\frac{T^{1-\sigma}}{t}\right).$$

Aus (16) folgt zunächst (36) nach (30). Ferner folgt aus (16) und (32)

$$(39) \qquad L(s,\chi) = k^{-s} \sum_{q=1}^{k} \chi(q) \left\{ \sum_{n=0}^{T-1} \left(n + \frac{q}{k}\right)^{-s} + O\left(\frac{T^{1-\sigma}}{t}\right) \right\}$$

$$= \sum_{r=1}^{kT} \chi(r) r^{-s} + O\left(\frac{T^{1-\sigma}}{t}\right).$$

Hilfssatz 5. *Für* $3 \leqq t \leqq T$, $\frac{3}{4} \leqq \sigma \leqq 2$ *ist*

$$(40) \qquad -L'(s,\chi) = \sum_{1 \leqq r \leqq kT} \chi(r) \log r \, r^{-s} + O\left(\log T \frac{T^{1-\sigma}}{t}\right).$$

Beweis. Setzt man

$$(41) \qquad h(s) = L(s,\chi) - \sum_{1 \leqq r \leqq kT} \chi(r) r^{-s},$$

so ist $h(s)$ nach Hilfssatz 4 für alle s der Voraussetzung von Hilfssatz 4 regulär, und es genügt zum Beweise von (40),

$$(42) \qquad h'(s) = O\left(\log T \frac{T^{1-\sigma}}{t}\right)$$

für alle s der Voraussetzung von Hilfssatz 5 zu zeigen. Es sei nunmehr $s_0 = \sigma_0 + i t_0$ mit

$$(43) \qquad \frac{3}{4} \leqq \sigma_0 \leqq 2, \quad 3 \leqq t_0 \leqq T$$

beliebig, aber fest gewählt. Dann genügen alle s mit

$$(44) \qquad |s - s_0| \leqq \frac{1}{4 \log T} \left(< \frac{1}{4} \right)$$

den Voraussetzungen von Hilfssatz 4; es ist also für jedes s, das auf dem Kreise (44) liegt,

$$(45) \qquad h(s) = O\left(\frac{T^{1-\sigma}}{t}\right) = O\left(\frac{T^{1-\sigma_0 + \frac{1}{4 \log T}}}{t_0}\right) = O\left(\frac{T^{1-\sigma_0}}{t_0}\right).$$

Also ist nach Cauchy:

$$(46) \qquad |h'(s_0)| \leqq 4 \log T \, O\left(\frac{T^{1-\sigma_0}}{t_0}\right).$$

Damit ist (42), also auch (40) bewiesen.

§ 2.

Der Littlewoodsche Satz über die Nullstellen von $L(s, \chi)$.

Hilfssatz 6. *Für* $\xi > 0,\ 0 < \sigma < 1$ *ist*

$$\frac{1-\sigma}{\xi} - \frac{2^{-[\xi]}}{\xi} + \frac{2^{-[\xi]}}{[\xi]+2} < \log 2 \,\frac{1-\sigma}{\log \frac{1}{1-\sigma}}. \tag{47}$$

Beweis. Für reelles η ist

$$2^{\eta} - \eta > 0 \tag{48}$$

und

$$\eta(2^{\eta} - 1) \geqq 0. \tag{49}$$

Also ist

$$\xi(2^{\eta} - \eta) + \eta(2^{\eta} - 1) > 0, \tag{50}$$

$$2^{\eta}(\xi + \eta) - \eta(\xi + 1) > 0. \tag{51}$$

Also ist für $\eta > -\xi$

$$\frac{2^{\eta}}{\xi(\xi+1)} - \frac{\eta}{\xi(\xi+\eta)} > 0, \tag{52}$$

$$2^{\eta}\left(\frac{1}{\xi} - \frac{1}{\xi+1}\right) - \frac{1}{\xi} + \frac{1}{\xi+\eta} > 0. \tag{53}$$

Setzt man

$$\eta = \frac{\log \frac{1}{1-\sigma}}{\log 2} - \xi, \tag{54}$$

so folgt aus (53)

$$\frac{1}{1-\sigma} 2^{-\xi}\left(\frac{1}{\xi} - \frac{1}{\xi+1}\right) - \frac{1}{\xi} + \frac{\log 2}{\log \frac{1}{1-\sigma}} > 0, \tag{55}$$

$$2^{-[\xi]}\left(\frac{1}{\xi} - \frac{1}{[\xi]+2}\right) - \frac{1-\sigma}{\xi} + (1-\sigma)\frac{\log 2}{\log \frac{1}{1-\sigma}} > 0. \tag{56}$$

Das gibt die Behauptung.

Hilfssatz 7. *Für* $\xi \geqq \frac{3}{2}$ *ist*

$$-\frac{1}{\xi} + \frac{1}{[\xi]+2} \geqq -\frac{1}{[\xi]+2}. \tag{57}$$

Beweis. Für $\xi \geqq 2$ ist

$$2\xi \geqq \xi + 2 \geqq [\xi] + 2, \tag{58}$$

für $\frac{3}{2} \leqq \xi < 2$ ist

$$2\xi \geqq 3 = [\xi] + 2, \tag{59}$$

also für $\xi \geqq \frac{3}{2}$

$$2\xi \geqq [\xi] + 2, \quad \frac{1}{\xi} \leqq \frac{2}{[\xi]+2}, \tag{60}$$

woraus die Behauptung folgt.

Hilfssatz 8. *Für $3 \leqq t$, $0 < w \leqq 1$, $0 < \sigma < 1$, $r > 0$, $R = 2^{r-1}$, $1 \leqq N \leqq N' < 2N$, $N \leqq t$ ist, wenn r, N, N' ganz sind,*

$$\left| \sum_{n=N}^{N'} (n+w)^{-s} \right| \leqq a_6 N^{1-\sigma} \left(N^{-\frac{1}{R}} t^{\frac{1}{R(r+1)}} + t^{-\frac{1}{R(r+1)}} \right) \log t. \tag{61}$$

Beweis. Siehe Landau, Vorlesungen über Zahlentheorie, Band 2, Satz 389[10]).

Hilfssatz 9[11]). *Für $\frac{3}{4} \leqq \sigma < 1$, $t \geqq 3$, $0 < w \leqq 1$ ist*

$$\zeta(s, w) - w^{-s} = O\left(t^{\log 2 \frac{1-\sigma}{\log \frac{1}{1-\sigma}}} \log^2 t \right). \tag{62}$$

Beweis. Nach Hilfssatz 3 mit $T = [t] + 1$ ist

$$\zeta(s, w) = \sum_{n=0}^{[t]} (n+w)^{-s} + O(1). \tag{63}$$

Ich zerteile die Summe folgendermaßen:

$$\sum_{n=0}^{[t]} = \sum_{n=0}^{0} + \sum_{n=1}^{1} + \sum_{n=2}^{3} + \sum_{n=4}^{7} + \cdots + \sum_{n=2^{\left[\frac{\log t}{\log 2}\right]}}^{[t]}. \tag{64}$$

Das geht, denn es ist

$$2^{\left[\frac{\log t}{\log 2}\right]} \leqq 2^{\frac{\log t}{\log 2}} = t < 2^{\left[\frac{\log t}{\log 2}\right]+1}, \tag{65}$$

$$2^{\left[\frac{\log t}{\log 2}\right]} \leqq [t] < 2^{\left[\frac{\log t}{\log 2}\right]+1}. \tag{66}$$

Die erste Summe in (64) ist w^{-s}, die zweite ist $O(1)$. Die übrigen genügen nach (66) den Voraussetzungen von Hilfssatz 8, und ihre Anzahl ist gleich $\left[\frac{\log t}{\log 2}\right] = O(\log t)$. Es genügt also, für jede dieser Summen zu zeigen, daß sie

$$O\left(t^{\log 2 \frac{1-\sigma}{\log \frac{1}{1-\sigma}}} \log t \right) \tag{67}$$

---

10) Dort wird der Satz nur für $t > 3$ bewiesen. Aber daß er auch für $t \geqq 3$ richtig ist, folgt aus Stetigkeitsgründen.

11) Dieser Satz geht auf die Herren Hardy und Littlewood zurück. Er findet sich in etwas schwächerer Form bei Landau, Vorlesungen über Zahlentheorie, Bd. 2, Satz 394. Eine schärfere Abschätzung steht bei van der Corput-Koksma, Sur l'ordre de grandeur de la fonction $\zeta(s)$ de Riemann dans la bande critique, Annales de la faculté des sciences de l'université de Toulouse, 3. sér. 22 (1930), p. 1—39.

ist. Wenn, wie in Hilfssatz 8, N und N' erstes und letztes Glied der Summe bezeichnen, so ist

$$(68)\qquad 2 \leqq N \leqq N' \leqq t, \qquad N' < 2N.$$

Erster Fall. $t^{\frac{2}{3}} < N \leqq t$. Dann ist nach Hilfssatz 8 mit $r = 1$, $R = 1$

$$(69)\qquad \left|\sum_{n=N}^{N'} (n+w)^{-s}\right| \leqq a_6 \left(N^{-\sigma} t^{\frac{1}{2}} + N^{1-\sigma} t^{-\frac{1}{2}}\right) \log t$$

$$\leqq a_6 \left(t^{-\frac{2}{3}\cdot\frac{3}{4}} t^{\frac{1}{2}} + t^{\frac{1}{4}} t^{-\frac{1}{2}}\right) \log t \leqq 2 a_6 \log t.$$

Zweiter Fall. $2 \leqq N \leqq t^{\frac{2}{3}}$. Wir setzen $\xi = \frac{\log t}{\log N} \left(\geqq \frac{3}{2}\right)$ und wenden Hilfssatz 8 an mit $r = [\xi] + 1$, $R = 2^{[\xi]}$. Dann gilt, da $N = t^{\frac{1}{\xi}}$,

$$(70)\qquad \left|\sum_{n=N}^{N'} (n+w)^{-s}\right| \leqq a_6 t^{\frac{1-\sigma}{\xi}} \left(t^{-\frac{2^{-[\xi]}}{\xi} + \frac{2^{-[\xi]}}{[\xi]+2}} + t^{-\frac{2^{-[\xi]}}{[\xi]+2}}\right) \log t$$

$$\leqq 2 a_6 t^{\frac{1-\sigma}{\xi}} t^{-\frac{2^{-[\xi]}}{\xi} + \frac{2^{-[\xi]}}{[\xi]+2}} \log t \text{ (nach Hilfssatz 7)}$$

$$\leqq 2 a_6 t^{\log 2 \frac{1-\sigma}{\log \frac{1}{1-\sigma}}} \log t \text{ (nach Hilfssatz 6)}.$$

Damit ist Hilfssatz 9 bewiesen.

Hilfssatz 10. *Für* $\frac{3}{4} \leqq \sigma < 1$, $t \geqq 3$ *ist*

$$(71)\qquad L(s, \chi) = O\left(t^{\log 2 \frac{1-\sigma}{\log \frac{1}{1-\sigma}}} \log^2 t\right).$$

Beweis. Es ist nach (16) und Hilfssatz 9

$$(72)\qquad L(s, \chi) = k^{-s} \sum_{q=1}^{k} \chi(q) \left(\zeta\left(s, \frac{q}{k}\right) - \left(\frac{q}{k}\right)^{-s}\right) + \sum_{q=1}^{k} \chi(q) q^{-s}$$

$$= O\left(t^{\log 2 \frac{1-\sigma}{\log \frac{1}{1-\sigma}}} \log^2 t\right) + O(1).$$

Hilfssatz 11. *Es sei* $f(x)$ *regulär für* $|x| \leqq 1$, $f(0) = 1$, $M \geqq 0$,

$$(73)\qquad |f(x)| \leqq e^M \text{ für } |x| \leqq 1.$$

Dann gilt für $|x| < 1$, *wenn* ξ *alle (in ihrer Vielfachheit gezählten*[12]) *Wurzeln von* $f(x)$ *mit* $|\xi| < 1$ *durchläuft*[13]),

$$(74)\qquad \left|\frac{f'(x)}{f(x)} - \sum_{\xi} \left(\frac{1}{x-\xi} + \frac{\bar{\xi}}{1-\bar{\xi}x}\right)\right| \leqq \frac{2M}{(1-|x|)^2}.$$

---

[12]) Auch im folgenden werden natürlich Wurzeln stets in entsprechender Vielfachheit gezählt.

[13]) Wenn es kein solches ξ gibt, ist der Satz auch richtig, in dem eine leere Summe 0, ein leeres Produkt 1 bedeutet. Entsprechendes gilt auch für die folgenden Hilfssätze.

Beweis. Bekanntlich ist, da $|\xi| < 1$,

$$(75)\qquad \left|\frac{x-\xi}{1-\bar{\xi}x}\right| = 1 \text{ für } |x| = 1.$$

Setzt man

$$(76)\qquad g(x) = f(x)\prod_{\xi}\frac{1-\bar{\xi}x}{x-\xi},$$

so ist

$$(77)\qquad g(x) \text{ regulär für } |x| \leqq 1,$$

$$(78)\qquad g(x) \neq 0 \text{ für } |x| < 1,$$

$$(79)\qquad |g(x)| = |f(x)| \text{ für } |x| = 1 \text{ nach } (75),$$

$$(80)\qquad |g(x)| \leqq e^{M} \text{ für } |x| = 1,$$

$$(81)\qquad |g(x)| \leqq e^{M} \text{ für } |x| \leqq 1,$$

$$(82)\qquad g(0) = f(0)\prod_{\xi}\frac{1}{-\xi},$$

$$(83)\qquad |g(0)| \geqq 1.$$

Nach (77) und (78) gibt es ein $h(x)$ mit

$$(84)\qquad g(x) = e^{h(x)},$$

so daß

$$(85)\qquad h(x) \text{ regulär für } |x| < 1,$$

$$(86)\qquad \Re(h(x)) \leqq M \text{ für } |x| < 1 \text{ nach } (81),$$

$$(87)\qquad \Re(h(0)) \geqq 0 \qquad \text{nach } (83).$$

Also gilt für $|x| < 1$

$$(88)\qquad |h'(x)| \leqq \frac{2(M - \Re(h(0)))}{(1-|x|)^2} \leqq \frac{2M}{(1-|x|)^2}.$$

Aus (76) und (84) folgt für $|x| < 1$

$$(89)\qquad h'(x) = \frac{g'(x)}{g(x)} = \frac{f'(x)}{f(x)} - \sum_{\xi}\left(\frac{1}{x-\xi} + \frac{\bar{\xi}}{1-\bar{\xi}x}\right).$$

Aus (89) und (88) folgt die Behauptung.

Hilfssatz 12. *Es sei* $a_7 = \frac{2}{31}$, $t_1 \geqq 3$, $\sigma_1 = 1 + a_7\frac{\log\log t_1}{\log t_1}$, $r = \frac{(\log\log t_1)^3}{\log t_1}$, $1 \leqq m \leqq 4$, *m ganz*, $s_m = \sigma_1 + imt_1$. *ϱ durchlaufe alle Nullstellen von $L(s,\chi)$ im Innern des Kreises:*

$$(90)\qquad |s - s_m| \leqq r.$$

Dann gilt für alle s des Kreisinnern bei $t_1 \to \infty$

$$(91)\qquad \left|\frac{L'}{L}(s,\chi) - \sum_{\varrho}\left(\frac{1}{s-\varrho} + \frac{\overline{\varrho - s_m}}{r^2 - (\overline{\varrho - s_m})(s - s_m)}\right)\right| \leqq \frac{2\log 2\, r^2 \log t_1\,(1+o(1))}{\log\log t_1\,(r - |s - s_m|)^2};$$

und speziell für $s = s_m$ *bei* $t_1 \to \infty$

$$(92)\qquad \left|\frac{L'}{L}(s_m, \chi) - \sum_{\varrho}\left(\frac{1}{s_m - \varrho} + \frac{\overline{\varrho - s_m}}{r^2}\right)\right| \leqq \frac{2\log 2 \log t_1 (1 + o(1))}{\log\log t_1}.$$

Beweis. Es sei t_1 sofort so groß, daß der Kreis (90) ganz zu dem Bereich

$$(93)\qquad \sigma \geqq \frac{3}{4}, \qquad t \geqq 3$$

gehört und $r < 1$ ist. Aus (90) folgt, wenn $\sigma < 1$,

$$(94)\qquad \sigma \geqq \sigma_1 - r > 1 - r, \qquad 1 - \sigma < r,$$

$$(95)\qquad \frac{1-\sigma}{\log \frac{1}{1-\sigma}} < \frac{r}{\log \frac{1}{r}}.$$ [14]

Daher ist nach Hilfssatz 10 für alle s aus (90)

$$(96)\qquad L(s, \chi) = O\left(t^{\log 2 \frac{r}{\log \frac{1}{r}}} \log^2 t\right);$$

denn für $\sigma \geqq 1$ ist (96) nach (37) trivial. Aus (96) folgt

$$(97)\qquad L(s, \chi) = O\left(e^{\log t \frac{(\log\log t_1)^3 \log 2}{\log t_1 (\log\log t_1 - 3 \log\log\log t_1)} + 2\log\log t}\right).$$

Wegen

$$(98)\qquad \frac{\log t}{\log t_1} = 1 + o(1),$$

$$(99)\qquad \frac{\log\log t_1}{\log\log t_1 - 3\log\log\log t_1} = 1 + o(1),$$

$$(100)\qquad 2\log\log t = o((\log\log t_1)^2)$$

folgt aus (97)

$$(101)\qquad |L(s, \chi)| \leqq e^{\log 2 (\log\log t_1)^2 (1 + o(1))}.$$

Ferner gilt wegen

$$(102)\qquad \left|\frac{1}{L(s_m, \chi)}\right| \leqq \zeta(\sigma_1) = O\left(\frac{1}{\sigma_1 - 1}\right) = O\left(\frac{\log t_1}{\log\log t_1}\right) \leqq e^{o((\log\log t_1)^2)}$$

für alle s aus (90)

$$(103)\qquad \left|\frac{L(s, \chi)}{L(s_m, \chi)}\right| \leqq e^{\log 2 (\log\log t_1)^2 (1 + o(1))}.$$

[14]) Die Funktion $\dfrac{u}{\log \frac{1}{u}}$ steigt monoton für $0 < u < 1$.

Wir setzen

$$x = \frac{s - s_m}{r}, \qquad \xi = \frac{\varrho - s_m}{r}, \tag{104}$$

$$f(x) = f\left(\frac{s - s_m}{r}\right) = \frac{L(s, \chi)}{L(s_m, \chi)}, \tag{105}$$

so daß ξ alle Wurzeln von $f(x)$ im Kreise $|x| < 1$ durchläuft.

Da $f(0) = 1$ ist, ist Hilfssatz 11 auf $f(x)$ mit

$$M = \log 2 (\log\log t_1)^2 (1 + o(1)) \tag{106}$$

anwendbar, und aus (74) folgt unter Berücksichtigung von (104), (105), (106) und

$$f'(x) = r\frac{L'(s, \chi)}{L(s_m, \chi)} \tag{107}$$

die Ungleichung

$$\left| r\frac{L'(s, \chi)}{L(s, \chi)} - \sum_{\varrho}\left(\frac{r}{s - \varrho} - \frac{\dfrac{\overline{\varrho - s_m}}{r}}{1 - \dfrac{\overline{\varrho - s_m}}{r}\cdot\dfrac{s - s_m}{r}}\right)\right| \tag{108}$$

$$\leqq 2 \log 2 (\log\log t_1)^2 (1 - o(1)) \frac{1}{\left(1 - \dfrac{|s - s_m|}{r}\right)^2}$$

$$= 2 \log 2\, r \frac{\log t_1}{\log\log t_1} (1 - o(1)) \frac{r^2}{(r - |s - s_m|)^2}.$$

Dividiert man (108) durch r, so folgt (91).

Hilfssatz 13. *Unter den Voraussetzungen von Hilfssatz* 12 *gilt*

$$\Re\left(\frac{1}{s_m - \varrho} + \frac{\overline{\varrho - s_m}}{r^2}\right) \geqq 0 \quad \textit{für jedes } \varrho,\text{[15]} \tag{109}$$

$$-\Re\left(\frac{L'}{L}(s_m, \chi)\right) \leqq 2 \log 2 \frac{\log t_1}{\log\log t_1}(1 + o(1)), \tag{110}$$

$$-\Re\left(\frac{L'}{L}(s_m, \chi)\right) \leqq 2 \log 2 \frac{\log t_1}{\log\log t_1}(1 + o(1)) - \Re\left(\frac{1}{s_m - \varrho}\right) \quad \textit{für jedes } \varrho. \tag{111}$$

Beweis. Bekanntlich ist $L(s, \chi) \neq 0$ für $\sigma > 1$, also

$$\Re(s_m - \varrho) > 0, \quad \Re\left(\frac{1}{s_m - \varrho}\right) > 0. \tag{112}$$

Es ist

$$\frac{1}{s_m - \varrho} + \frac{\overline{\varrho - s_m}}{r^2} = \frac{1}{s_m - \varrho}\left(1 - \frac{|\varrho - s_m|^2}{r^2}\right). \tag{113}$$

Aus (112), (113) und

$$|\varrho - s_m| < r \tag{114}$$

[15]) Wenn es kein ϱ mit $|\varrho - s_m| < r$ gibt, enthalten (109) und (111) keine Aussage.

folgt (109). Aus (92) folgt

$$(115) \qquad -\Re\left(\frac{L'}{L}(s_m, \chi)\right) + \Re\left(\sum_{\varrho}\left(\frac{1}{s_m - \varrho} + \frac{\overline{\varrho - s_m}}{r^2}\right)\right) \leqq 2 \log 2 \frac{\log t_1}{\log\log t_1}(1 + o(1)).$$

Aus (115) und (109), angewandt auf alle ϱ, folgt (110).

Aus (115) und (109), angewandt auf alle ϱ mit einer Ausnahme, folgt für dieses ausgenommene ϱ

$$(116) \qquad -\Re\left(\frac{L'}{L}(s_m, \chi)\right) \leqq 2 \log 2 \frac{\log t_1}{\log\log t_1}(1 + o(1)) - \Re\left(\frac{1}{s_m - \varrho}\right) - \Re\left(\frac{\overline{\varrho - s_m}}{r^2}\right).$$

Es ist nach (114)

$$(117) \qquad -\Re\left(\frac{\overline{\varrho - s_m}}{r^2}\right) \leqq \frac{1}{r} = o\left(\frac{\log t_1}{\log\log t_1}\right).$$

Aus (116) und (117) folgt (111).

Hilfssatz 14. *Für reelles φ ist*

$$(118) \quad P(\varphi) = 47 + 80 \cos\varphi + 49 \cos(2\varphi) + 20 \cos(3\varphi) + 4 \cos(4\varphi) \geqq 0.$$

Beweis. Setzt man $\cos\varphi = \eta$, so ist

$$(119) \quad \cos(2\varphi) = 2\eta^2 - 1, \; \cos(3\varphi) = 4\eta^3 - 3\eta, \; \cos(4\varphi) = 8\eta^4 - 8\eta^2 + 1,$$

$$\begin{aligned}(120) \quad P(\varphi) &= 47 + 80\eta + 98\eta^2 - 49 + 80\eta^3 - 60\eta + 32\eta^4 - 32\eta^2 + 4 \\ &= 2 + 20\eta + 66\eta^2 + 80\eta^3 + 32\eta^4 \\ &= 2(1+\eta)(1 + 9\eta + 24\eta^2 + 16\eta^3) \\ &= 2(1+\eta)^2(1 + 8\eta + 16\eta^2) = 2(1+\eta)^2(1+4\eta)^2 \geqq 0.\end{aligned}$$

Hilfssatz 15. *Für $\sigma > 1$ und reelles t ist*

$$\begin{aligned}(121) \quad -\Re\Big\{47 \frac{L'}{L}(\sigma, \chi^0) &+ 80 \frac{L'}{L}(\sigma + it, \chi) + 49 \frac{L'}{L}(\sigma + 2it, \chi^2) \\ &+ 20 \frac{L'}{L}(\sigma + 3it, \chi^3) + 4 \frac{L'}{L}(\sigma + 4it, \chi^4)\Big\} \geqq 0.\end{aligned}$$

Beweis. Für ganzes $m \geqq 0$ ist

$$(122) \qquad -\frac{L'}{L}(\sigma + imt, \chi^m) = \sum_{\substack{n=1 \\ (n,k)=1}}^{\infty} \Lambda(n) n^{-\sigma} (\chi(n) n^{-it})^m.$$

Wählt man für die in (122) auftretenden n die $\varphi_n \leqq 0$ so, daß

$$(123) \qquad \cos\varphi_n = \Re(\chi(n) n^{-it}),$$

so ist

$$(124) \qquad \cos(m\varphi_n) = \Re\left((\chi(n) n^{-it})^m\right).$$

Also ist die linke Seite der Behauptung gleich

$$(125)\qquad \sum_{\substack{n=1\\(n,k)=1}}^{\infty} \Lambda(n)\, n^{-\sigma} P(\varphi_n) \geqq 0.$$

Hilfssatz 16. *Für* $3 \leqq b_2 \leqq t$, $a_5 = \frac{1}{49{,}13}$, $\sigma \geqq 1 - a_5 \frac{\log\log t}{\log t}$ *ist*

$$(126)\qquad L(s,\chi) \neq 0.$$

Beweis (indirekt). Angenommen, es gäbe unendlich viele Wurzeln $\varrho = \beta + i\gamma$, $\beta = \Re(\varrho)$, von $L(s,\chi)$ mit

$$(127)\qquad \gamma \geqq 3, \quad 1 - a_5 \frac{\log\log\gamma}{\log\gamma} \leqq \beta \leqq 1.$$

Dann setze ich

$$(128)\qquad \sigma_1 = 1 + a_7 \frac{\log\log\gamma}{\log\gamma}$$

und wende Hilfssatz 13 auf $L(s,\chi^m)$ für $1 \leqq m \leqq 4$ an mit

$$(129)\qquad t_1 = \gamma, \quad s_m = \sigma_1 + i m \gamma.$$

Es gilt also für $\gamma \to \infty$ wegen $0 < s_1 - \varrho = O\left(\frac{\log\log\gamma}{\log\gamma}\right) = o\left(\frac{(\log\log\gamma)^3}{\log\gamma}\right)$

$$(130)\qquad -\Re\left(\frac{L'}{L}(\sigma_1 + i m\gamma, \chi^m)\right) \leqq 2\log 2 \frac{\log\gamma}{\log\log\gamma}(1 + o(1)) \text{ für } m = 4, 3, 2 \text{ nach (110)},$$

$$(131)\qquad -\Re\left(\frac{L'}{L}(\sigma_1 + i\gamma, \chi)\right) \leqq 2\log 2 \frac{\log\gamma}{\log\log\gamma}(1 + o(1)) - \frac{1}{\sigma_1 - \beta} \text{ nach (111)},$$

$$(132)\qquad -\Re\left(\frac{L'}{L}(\sigma_1, \chi^0)\right) = \frac{1}{\sigma_1 - 1}(1 + o(1)) = \frac{1}{a_7}\frac{\log\gamma}{\log\log\gamma}(1 + o(1));$$

denn $-\frac{L'}{L}(s,\chi^0)$ hat in $s = 1$ einen einfachen Pol mit dem Residuum 1. Multipliziert man die fünf Ungleichungen (nämlich (130) für $m = 4, 3, 2$, (131) und (132)) mit 4, 20, 49, 80, 47 und addiert man sie alsdann, so folgt nach Hilfssatz 15

$$(133)\qquad 0 \leqq \frac{\log\gamma}{\log\log\gamma}(1 + o(1))\left\{2\log 2(4 + 20 + 49 + 80) + \frac{47}{a_7}\right\} - \frac{80}{\sigma_1 - \beta}.$$

Aus (127) und (128) folgt

$$(134)\qquad -\frac{80}{\sigma_1 - \beta} \leqq -\frac{80}{a_5 + a_7}\frac{\log\gamma}{\log\log\gamma}.$$

Aus (133) und (134) folgt

$$(135)\qquad 0 \leqq \frac{\log\gamma}{\log\log\gamma}(1 + o(1))\left(306\log 2 + \frac{47}{a_7} - \frac{80}{a_5 + a_7}\right).$$

Dividiert man (135) durch $\frac{\log\gamma}{\log\log\gamma}$ und läßt man γ gegen ∞ gehen (das geht, denn wir hatten ja angenommen, daß es unendlich viele γ gibt, die (127), also auch (135) genügen), so erhält man

$$(136)\quad 0 \leqq 306 \log 2 + \frac{47}{a_7} - \frac{80}{a_5 + a_7} = 306 \log 2 + \frac{47 \cdot 31}{2} - \frac{80}{\frac{1}{49{,}13} + \frac{2}{31}}$$

$$< 306 \cdot 0{,}694 + 728{,}5 - \frac{80 \cdot 49{,}13 \cdot 31}{129{,}26}$$

$$= 940{,}864 - \frac{121\,842{,}4}{129{,}26} < 941 - \frac{121\,842{,}4}{129{,}3}$$

$$= \frac{121\,671{,}3 - 121\,842{,}4}{129{,}3} < 0.$$

Durch diesen Widerspruch ist unser Satz bewiesen.

Hilfssatz 17. *Es sei*

$$a_8 = \frac{1}{49{,}15} (< a_5), \quad t \geqq b_3 \geqq b_2 \geqq 3, \quad \sigma \geqq 1 - a_8 \frac{\log \log t}{\log t}.$$

Dann ist

$$(137)\quad \frac{L'}{L}(s, \chi) = O\left(\frac{\log t}{\log \log t}\right).$$

Beweis. Es sei

$$(138)\quad \sigma_1 = 1 + a_7 \frac{\log \log t}{\log t}, \qquad s_1 = \sigma_1 + it.$$

Ist $\sigma \geqq \sigma_1$, so ist (137) trivial wegen

$$(139)\quad \left|\frac{L'}{L}(s, \chi)\right| \leqq -\frac{\zeta'}{\zeta}(\sigma) \leqq -\frac{\zeta'}{\zeta}(\sigma_1) = \frac{1}{\sigma_1 - 1} + O(1) = O\left(\frac{\log t}{\log \log t}\right).$$

Also sei alsbald

$$(140)\quad \sigma < \sigma_1.$$

Wir wenden Hilfssatz 12 an mit $t_1 = t$, $m = 1$.

Dann folgt aus (92) und (139)

$$(141)\quad \sum_{\varrho} \left(\frac{1}{s_1 - \varrho} + \frac{\overline{\varrho - s_1}}{r^2}\right) = O\left(\frac{\log t}{\log \log t}\right).$$

Nunmehr wählen wir $b_3 \geqq b_2$ so groß, daß für $t \geqq b_3$ erstens

$$(142)\quad 0 < s_1 - s \leqq \frac{1}{2} r.$$

Das geht wegen

$$(143)\quad s_1 - s = O\left(\frac{\log \log t}{\log t}\right) = o(r) \text{ für } t \to \infty.$$

Zweitens sei b_3 so groß gewählt, daß bei $t \geqq b_3$ für alle in (141) auftretenden ϱ gilt

$$(144)\quad \beta = \Re(\varrho) \leqq 1 - \frac{a_5 + a_8}{2} \frac{\log \log t}{\log t}.$$

Das geht nach Hilfssatz 16, da $\frac{a_5+a_8}{2} < a_5$.[16]) Dann ist nach Voraussetzung und nach (144)

$$(145) \qquad |s-\varrho| \geqq \Re(s-\varrho)$$

$$= \sigma - \beta \geqq \left(1 - a_8 \frac{\log\log t}{\log t}\right) - \left(1 - \frac{a_5+a_8}{2}\frac{\log\log t}{\log t}\right) = \frac{a_5-a_8}{2}\frac{\log\log t}{\log t}.$$

Ferner ist

$$(146) \qquad 0 < \sigma_1 - \sigma = s_1 - s \leqq (a_7 + a_8)\frac{\log\log t}{\log t}.$$

Aus (145) und (146) folgt

$$(147) \qquad s_1 - s \leqq 2\frac{a_7+a_8}{a_5-a_8}|s-\varrho|,$$

$$(148) \quad |s_1-\varrho| \leqq (s_1-s) + |s-\varrho| \leqq \left(1+2\frac{a_7+a_8}{a_5-a_8}\right)|s-\varrho| = a_9|s-\varrho|.$$

Ferner folgt aus

$$(149) \qquad \beta < \sigma < \sigma_1$$

$$(150) \qquad s_1 - s < \Re(s_1-\varrho),$$

und auch

$$(151) \qquad |s-\varrho| < |s_1-\varrho| < r.$$

Aus (142) und (151) folgt

$$(152) \qquad \left|\frac{r^2-(s-\varrho)(\overline{\varrho-s_1})}{r^2(r^2-(s-s_1)(\overline{\varrho-s_1}))}\right| \leqq \frac{r^2+r\cdot r}{r^2\left(r^2-\frac{1}{2}r\cdot r\right)} = \frac{4}{r^2}.$$

Es ist

$$(153) \qquad \frac{1}{s-\varrho} + \frac{\overline{\varrho-s_1}}{r^2-(s-s_1)(\overline{\varrho-s_1})} = \frac{r^2-|\varrho-s_1|^2}{(s-\varrho)(r^2-(s-s_1)(\overline{\varrho-s_1}))},$$

$$(154) \qquad \frac{1}{s_1-\varrho} + \frac{\overline{\varrho-s_1}}{r^2} = \frac{r^2-|\varrho-s_1|^2}{(s_1-\varrho)r^2},$$

$$(155) \qquad \Re\left(\frac{1}{s_1-\varrho} + \frac{\overline{\varrho-s_1}}{r^2}\right) = \frac{\Re(s_1-\varrho)(r^2-|\varrho-s_1|^2)}{|s_1-\varrho|^2 r^2}.$$

Subtrahiert man (154) von (153), so ergibt sich für den absoluten Betrag der Differenz

---

[16]) Wegen $|\gamma - t| < r = o(1)$ gilt bei hinreichend großem b_3 für $t \geqq b_3$

$$\beta \leqq 1 - a_5\frac{\log\log\gamma}{\log\gamma} = 1 - a_5\frac{\log\log(t+o(1))}{\log(t+o(1))} \leqq 1 - \frac{a_5+a_8}{2}\frac{\log\log t}{\log t}.$$

$$(156)\quad \left|\frac{1}{s-\varrho}+\frac{\overline{\varrho-s_1}}{r^2-(s-s_1)(\overline{\varrho-s_1})}-\frac{1}{s_1-\varrho}-\frac{\overline{\varrho-s_1}}{r^2}\right|$$

$$=(r^2-|\varrho-s_1|^2)\left|\frac{(s_1-\varrho)r^2-(s-\varrho)r^2+(s-\varrho)(s-s_1)(\overline{\varrho-s_1})}{(s-\varrho)(r^2-(s-s_1)(\overline{\varrho-s_1}))(s_1-\varrho)r^2}\right|$$

$$=(s_1-s)\frac{r^2-|\varrho-s_1|^2}{|s-\varrho|\,|s_1-\varrho|}\left|\frac{r^2-(s-\varrho)(\overline{\varrho-s_1})}{r^2(r^2-(s-s_1)(\overline{\varrho-s_1}))}\right|$$

$$\leqq(s_1-s)\frac{r^2-|\varrho-s_1|^2}{|s-\varrho|\,|s_1-\varrho|}\frac{4}{r^2}\quad(\text{nach }(152))$$

$$\leqq\Re(s_1-\varrho)\,a_9\frac{r^2-|\varrho-s_1|^2}{|s_1-\varrho|^2}\frac{4}{r^2}\quad(\text{nach }(150)\text{ und }(148))$$

$$=4a_9\,\Re\left(\frac{1}{s_1-\varrho}+\frac{\overline{\varrho-s_1}}{r^2}\right)\text{ nach }(155).$$

Aus (156) folgt

$$(157)\quad \left|\sum_\varrho\left(\frac{1}{s-\varrho}+\frac{\overline{\varrho-s_1}}{r^2-(s-s_1)(\overline{\varrho-s_1})}\right)-\sum_\varrho\left(\frac{1}{s_1-\varrho}+\frac{\overline{\varrho-s_1}}{r^2}\right)\right|$$

$$\leqq 4a_9\,\Re\left(\sum_\varrho\left(\frac{1}{s_1-\varrho}+\frac{\overline{\varrho-s_1}}{r^2}\right)\right)\leqq 4a_9\left|\sum_\varrho\left(\frac{1}{s_1-\varrho}+\frac{\overline{\varrho-s_1}}{r^2}\right)\right|.$$

Aus (141) und (157) folgt

$$(158)\quad \sum_\varrho\left(\frac{1}{s-\varrho}+\frac{\overline{\varrho-s_1}}{r^2-(s-s_1)(\overline{\varrho-s_1})}\right)=O\left(\frac{\log t}{\log\log t}\right).$$

Aus (91), (142) und (158) folgt die Behauptung.

Hilfssatz 18. *Unter den Voraussetzungen von Hilfssatz 17 (d. h. auch bei demselben b_3) gilt*

$$(159)\quad \frac{1}{L(s,\chi)}=O\left(\frac{\log t}{\log\log t}\right).$$

Beweis. Erster Fall. Es sei $\sigma\geqq\sigma_0=1+\frac{\log\log t}{\log t}$. Dann ist (159) trivial wegen

$$(160)\quad \left|\frac{1}{L(s,\chi)}\right|\leqq\zeta(\sigma)\leqq\zeta(\sigma_0)=\frac{1}{\sigma_0-1}+O(1)=O\left(\frac{\log t}{\log\log t}\right).$$

Zweiter Fall. Es sei $\sigma<\sigma_0$. Dann ist

$$(161)\quad \log|L(s,\chi)|=\Re\log L(\sigma_0+it,\chi)+\Re\int_{\sigma_0+it}^{s}\frac{L'}{L}(u,\chi)\,du$$

$$\geqq-\log\zeta(\sigma_0)+\int_{\sigma_0}^{\sigma}O\left(\frac{\log t}{\log\log t}\right)du\quad(\text{nach Hilfssatz 17})$$

$$=\log(\sigma_0-1)+O(1)=-\log\log t+\log\log\log t+O(1),$$

$$(162)\quad \left|\frac{1}{L(s,\chi)}\right|=e^{-\log|L(s,\chi)|}\leqq e^{\log\log t-\log\log\log t+O(1)}=O\left(\frac{\log t}{\log\log t}\right).$$

§ 3.

Hilfssätze über die Teilerzahlen.

In den Hilfssätzen 19—21 dieses Paragraphen bezeichnen kleine lateinische Buchstaben natürliche Zahlen.

Hilfssatz 19. *Für* $z \geqq 2$, $2 \leqq r \leqq 4$ *ist*

$$\sum_{n \leqq z} \frac{d_r(n)}{n} = O(\log^r z). \tag{163}$$

Beweis. Es ist

$$\sum_{n \leqq z} \frac{d_r(n)}{n} = \sum_{n \leqq z} \sum_{u_1 \ldots u_r = n} \frac{1}{u_1 \ldots u_r} \leqq \sum_{\substack{u_1 \leqq z \\ \cdots\cdots \\ u_r \leqq z}} \frac{1}{u_1 \ldots u_r} = \Big(\sum_{u \leqq z} \frac{1}{u}\Big)^r = O(\log^r z). \tag{164}$$

Hilfssatz 20. *Für* $z \geqq 2$ *ist*

$$\sum_{n \leqq z} \frac{d^2(n)}{n} = O(\log^4 z). \tag{165}$$

Beweis. Aus $d(vw) \leqq d(v)d(w)$ folgt

$$d^2(n) = \sum_{u | n} d(n) \leqq \sum_{u | n} d(u) d\Big(\frac{n}{u}\Big) = d_4(n), \tag{166}$$

und Hilfssatz 19 mit $r = 4$ gibt die Behauptung.

Hilfssatz 21. *Für* $z \geqq 2$ *ist*

$$S = \sum_{n < m \leqq z} \frac{d(n)\, d(m)}{n m \log \frac{m}{n}} = O(\log^4 z). \tag{167}$$

Beweis. Es ist für $m > n$

$$\log \frac{m}{n} = -\log\Big(1 - \frac{m-n}{m}\Big) > \frac{m-n}{m}, \tag{168}$$

$$S < \sum_{n < m \leqq z} \frac{d(n)\, d(m)}{n(m-n)} = \sum_{n < m \leqq z} \frac{d(n)\, d(m)}{m(m-n)} + \sum_{n < m \leqq z} \frac{d(n)\, d(m)}{n m} \tag{169}$$

$$\leqq \sum_{n < m \leqq z} \frac{d(n)\, d(m)}{m(m-n)} + \Big(\sum_{n \leqq z} \frac{d(n)}{n}\Big)^2 = \sum_{n < m \leqq z} \frac{d(n)\, d(m)}{m(m-n)} + O(\log^4 z)$$

nach Hilfssatz 19 mit $r = 2$. Setzt man

$$m = ce, \quad n = fg, \quad m - n = r, \tag{170}$$

so folgt aus (169)

$$S \leqq O(\log^4 z) + \sum_{r \leqq z} \frac{1}{r} \sum_{\substack{c,e,f,g \\ ce \leqq z \\ ce - fg = r}} \frac{1}{ce}. \tag{171}$$

In dieser Summe treten c und e, sowie f und g symmetrisch auf: wir verkleinern die Summe also nicht, wenn wir sie mit 4 multiplizieren und zu den Summationsbedingungen noch die Ungleichungen

$$c \leqq e, \quad f \leqq g \tag{172}$$

hinzufügen. Aus (172) folgt überdies wegen

$$e^2 \geqq c\,e = r + fg > fg \geqq f^2 \tag{173}$$

noch die Summationsbedingung

$$e > f. \tag{174}$$

Aus (170) folgt

$$c\,e \equiv r \pmod f. \tag{175}$$

Wir können also die Ungleichung (171) ersetzen durch:

$$S \leqq O(\log^4 z) + 4 \sum_{r \leqq z} \frac{1}{r} \sum_{c \leqq z} \frac{1}{c} \sum_{f \leqq z} \sum_{\substack{e \\ f < e \leqq z \\ c\,e \equiv r \,(\mathrm{mod}\, f)}} \frac{1}{e}. \tag{176}$$

Wird $h = (c, f)$ gesetzt, so kann bei festen c, r, f (175) nur dann eine Lösung haben, wenn

$$h \mid r. \tag{177}$$

Es sei also (177) erfüllt und e_0 die kleinste Lösung von (175), die der Bedingung (174) genügt. Alle übrigen Lösungen von (175) mit $f < e \leqq z$ sind dann unter den Zahlen

$$e = e_0 + \frac{f}{h} u, \quad u \leqq z \tag{178}$$

enthalten. Aus (176) folgt daher unter Berücksichtigung von (177) und (178)

$$\begin{aligned} S &\leqq O(\log^4 z) + 4 \sum_{c \leqq z} \frac{1}{c} \sum_{f \leqq z} \sum_{\substack{r \leqq z \\ h \mid r}} \frac{1}{r} \left(\frac{1}{e_0} + \sum_{u \leqq z} \frac{1}{e_0 + \frac{f}{h} u} \right) \\ &\leqq O(\log^4 z) + 4 \sum_{c \leqq z} \frac{1}{c} \sum_{f \leqq z} \sum_{\substack{r \leqq z \\ h \mid r}} \frac{1}{r\,e_0} + 4 \sum_{c \leqq z} \sum_{f \leqq z} \sum_{\substack{r \leqq z \\ h \mid r}} \sum_{u \leqq z} \frac{h}{c\,r\,f\,u}. \end{aligned} \tag{179}$$

Wegen $e_0 > f$ ist

$$\sum_{c \leqq z} \frac{1}{c} \sum_{f \leqq z} \sum_{\substack{r \leqq z \\ h \mid r}} \frac{1}{r\,e_0} \leqq \sum_{c \leqq z} \frac{1}{c} \sum_{f \leqq z} \frac{1}{f} \sum_{r \leqq z} \frac{1}{r} = O(\log^3 z). \tag{180}$$

Zur Abschätzung der zweiten Summe in (179) setzen wir

$$c = h\,c', \quad f = h\,f', \quad r = h\,r'. \tag{181}$$

Dann ist

$$(182)\quad \sum_{\substack{c \leqq z}} \sum_{f \leqq z} \sum_{\substack{r \leqq z \\ h \mid r}} \sum_{u \leqq z} \frac{h}{c r f u} \leqq \sum_{h \leqq z} \sum_{c' \leqq z} \sum_{f' \leqq z} \sum_{r' \leqq z} \sum_{u \leqq z} \frac{h}{h c' h r' h f' u}$$

$$= \sum_{h \leqq z} \frac{1}{h^2} \left(\sum_{v \leqq z} \frac{1}{v} \right)^4 = O(\log^4 z).$$

Aus (179), (180) und (182) folgt die Behauptung.

Bezeichnung. Für den Rest dieser Arbeit sei für $0 \leqq t \leqq T$

$$(183)\quad A(s) = A(s; T, \chi) = 1 - \sum_{1 \leqq n \leqq kT} \chi(n) n^{-s} \sum_{1 \leqq n \leqq kT} \chi(n) \mu(n) n^{-s}.$$

Hilfssatz 22. *Für* $T \geqq 3$, $\frac{1}{2} \leqq \sigma \leqq 1$ *ist*

$$(184)\quad \int_0^T |A(s)|^2 \, dt = O(T^{4(1-\sigma)} \log^4 T).$$

Beweis. Bekanntlich ist für ganzes $v > 1$

$$(185)\quad \sum_{u \mid v} \chi\left(\frac{v}{u}\right) \chi(u) \mu(u) = \chi(v) \sum_{u \mid v} \mu(u) = 0,$$

also nach (183) bei passenden $e(n)$

$$(186)\quad A(s) = \sum_{kT < n \leqq (kT)^2} e(n) n^{-s},$$

wo

$$e(n) = e(n; T, \chi), \qquad |e(n)| \leqq d(n).$$

Also ist

$$(187)\quad \int_0^T |A(s)|^2 \, dt = \int_0^T \Big(\sum_{kT < m \leqq (kT)^2} e(m) m^{-\sigma - it} \sum_{kT < n \leqq (kT)^2} \overline{e(n)} n^{-\sigma + it} \Big) dt$$

$$= T \sum_{kT < m \leqq (kT)^2} |e(m)|^2 m^{-2\sigma} + \sum_{\substack{kT < m \leqq (kT)^2 \\ kT < n \leqq (kT)^2 \\ m \neq n}} e(m) \overline{e(n)} (mn)^{-\sigma} \int_0^T \left(\frac{n}{m}\right)^{it} dt$$

$$\leqq T \sum_{kT < m \leqq (kT)^2} \frac{d^2(m)}{m} m^{1-2\sigma} + \sum_{\substack{kT < m \leqq (kT)^2 \\ kT < n \leqq (kT)^2 \\ m \neq n}} \frac{d(m) d(n)}{mn} (mn)^{1-\sigma} \frac{2}{\left|\log \frac{m}{n}\right|}$$

$$\leqq T (kT)^{1-2\sigma} \sum_{kT < m \leqq (kT)^2} \frac{d^2(m)}{m} + 4 (kT)^{4(1-\sigma)} \sum_{kT < n < m \leqq (kT)^2} \frac{d(m) d(n)}{mn \log \frac{m}{n}}$$

$$\leqq k^{1-2\sigma} T^{2(1-\sigma)} \sum_{1 \leqq m \leqq (kT)^2} \frac{d^2(m)}{m} + 4 k^{4(1-\sigma)} T^{4(1-\sigma)} \sum_{1 \leqq n < m \leqq (kT)^2} \frac{d(m) d(n)}{mn \log \frac{m}{n}}$$

$$= O(T^{4(1-\sigma)} \log^4 T)$$

nach Hilfssatz 20 und 21.

§ 4.

Asymptotische Entwicklung von $\frac{1}{L(s,\chi)}$ und $\frac{-L'(s,\chi)}{L(s,\chi)}$.

In diesem Paragraphen sei stets

$$(3 \leqq b_2 \leqq)\, b_3 \leqq b_4 \leqq (\log T)^{a_8+2} \leqq t \leqq T, \tag{188}$$

$$\sigma = 1 - a_8 \frac{\log\log T}{\log T}. \tag{189}$$

Es sei $b_4 \geqq b_3$ so groß gewählt, daß für $b_4 \leqq \tau_1 \leqq \tau_2$

$$\frac{\log\log \tau_1}{\log \tau_1} \geqq \frac{\log\log \tau_2}{\log \tau_2} \tag{190}$$

ist, und aus (188) und (189) die zwei nachstehenden Ungleichungen folgen:

$$\frac{4}{5} \leqq \sigma < 1, \tag{191}$$

$$|\Sigma \chi(n)\, n^{-s}| \geqq \frac{1}{\log T}, \tag{192}$$

wo Σ, wie auch weiterhin in diesem Paragraphen, zur Abkürzung für $\sum\limits_{1 \leqq n \leqq kT}$ steht. Daß sich (190) und (191) durch hinreichend große Wahl von b_4 erreichen lassen, ist ohne weiteres klar. Aus (188), (189), (190) und (191) folgt, daß sich die Hilfssätze 4, 5, 17 und 18 auf alle unsere s anwenden lassen, was wir in diesem Paragraphen mehrfach benutzen werden. Nach Hilfssatz 4 und 18 ist

$$\begin{aligned} |\Sigma \chi(n)\, n^{-s}| &\geqq |L(s,\chi)| + O\left(\frac{T^{1-\sigma}}{t}\right) \geqq b_5 \frac{\log\log t}{\log t} + O\left(\frac{e^{\log T\, a_8 \frac{\log\log T}{\log T}}}{(\log T)^{a_8+2}}\right) \\ &\geqq b_5 \frac{\log\log T}{\log T} + O\left(\frac{1}{\log^2 T}\right) \geqq \frac{1}{\log T} \end{aligned} \tag{193}$$

bei passend großem b_4. Damit ist (192) bewiesen.

Hilfssatz 23. *Es ist*

$$\frac{1}{L(s,\chi)} = \sum_{1 \leqq n \leqq (kT)^3} c(n)\, n^{-s} + O\left(\log^2 T \frac{T^{1-\sigma}}{t}\right) + O(\log T)\, |A(s)|^2, \tag{194}$$

wo

$$c(n) = c(n;\, T, \chi), \qquad |c(n)| \leqq 3\, d_3(n). \tag{195}$$

Beweis. Es ist

$$\frac{1}{L(s,\chi)} = \frac{1}{\Sigma \chi(n)\, n^{-s}} + \frac{\Sigma \chi(n)\, n^{-s} - L(s,\chi)}{L(s,\chi)\, \Sigma \chi(n)\, n^{-s}}, \tag{196}$$

also nach Hilfssatz 4, 18 und (192)

$$(197)\qquad \frac{1}{L(s,\chi)} = \frac{1}{\sum \chi(n) n^{-s}} + O\left(\log^2 T \frac{T^{1-\sigma}}{t}\right)$$
$$= \frac{A^2(s)}{\sum \chi(n) n^{-s}} + 2 \sum \chi(n) \mu(n) n^{-s} - \sum \chi(n) n^{-s} \left(\sum \chi(n) \mu(n) n^{-s}\right)^2$$
$$+ O\left(\log^2 T \frac{T^{1-\sigma}}{t}\right),$$

woraus nach (192) die Behauptung folgt, wenn man

$$(198)\quad c(n) = -\sum_{\substack{uvw=n \\ 1\leqq u\leqq kT \\ 1\leqq v\leqq kT \\ 1\leqq w\leqq kT}} \chi(u)\chi(v)\mu(v)\chi(w)\mu(w) + \begin{cases} 2\chi(n)\mu(n) & \text{für } 1\leqq n\leqq kT, \\ 0 & \text{für } kT < n \leqq (kT)^3 \end{cases}$$

setzt.

Hilfssatz 24. *Es ist*

$$(199)\quad -\frac{L'}{L}(s,\chi) = \sum_{1\leqq n\leqq (kT)^4} g(n) n^{-s} + O\left(\log^2 T \frac{T^{1-\sigma}}{t}\right) + O(\log T)|A(s)|^2,$$

wo

$$(200)\qquad g(n) = g(n; T, \chi), \qquad |g(n)| \leqq 3(\log(kT))\, d_4(n).$$

Beweis. Es ist

$$(201)\quad -\frac{L'}{L}(s,\chi) = \frac{\sum \chi(n)\log n\, n^{-s}}{\sum \chi(n) n^{-s}} + \frac{L'}{L}(s,\chi)\frac{L(s,\chi) - \sum \chi(n) n^{-s}}{\sum \chi(n) n^{-s}}$$
$$- \frac{L'(s,\chi) + \sum \chi(n)\log n\, n^{-s}}{\sum \chi(n) n^{-s}}$$
$$= \frac{\sum \chi(n)\log n\, n^{-s}}{\sum \chi(n) n^{-s}} + O\left(\log^2 T \frac{T^{1-\sigma}}{t}\right)$$

nach (192) und den Hilfssätzen 4, 5 und 17. Da nach der in (193) ausgeführten Rechnung

$$(202)\qquad \log^2 T \frac{T^{1-\sigma}}{t} \leqq 1,$$

folgt aus (201) und Hilfssatz 17

$$(203)\qquad \frac{\sum \chi(n)\log n\, n^{-s}}{\sum \chi(n) n^{-s}} = O(\log T).$$

Ferner folgt aus (201)

$$(204)\quad -\frac{L'}{L}(s,\chi) = A^2(s)\frac{\sum \chi(n)\log n\, n^{-s}}{\sum \chi(n) n^{-s}} - 2\sum \chi(n)\log n\, n^{-s} \sum \chi(n)\mu(n) n^{-s}$$
$$- \sum \chi(n)\log n\, n^{-s} \sum \chi(n) n^{-s} \left(\sum \chi(n)\mu(n) n^{-s}\right)^2 + O\left(\log^2 T \frac{T^{1-\sigma}}{t}\right).$$

Aus (204) und (203) folgt die Behauptung, wenn man

$$(205)\quad g(n) = 2\sum_{\substack{uv=n \\ u\leqq kT \\ v\leqq kT}} \chi(u)\log u\, \chi(v)\mu(v) - \sum_{\substack{uvwq=n \\ u\leqq kT \\ v\leqq kT \\ w\leqq kT \\ q\leqq kT}} \chi(u)\log u\, \chi(v)\chi(w)\mu(w)\chi(q)\mu(q)$$

setzt.

§ 5.

Beweis der Hauptsätze.

Wir setzen

$$a_{10} = 249{,}8. \tag{206}$$

Dann ist

$$a_8(4 - a_{10}) + 5 = \frac{1}{49{,}15}(-245{,}8) + 5 = \frac{-0{,}05}{49{,}15} < 0. \tag{207}$$

Da bekanntlich $L(1 + it) \neq 0$, gibt es ein $b_6 < 1$, so daß

$$L(s, \chi) \neq 0 \quad \text{für} \quad 0 \leqq t \leqq b_4, \quad \sigma \geqq b_6. \tag{208}$$

Es sei $x > 0$ die in Satz 3 und 4 auftretende Variable. Wir wählen $T > 0$, so daß

$$T^{a_{10}} = x \tag{209}$$

ist, und setzen

$$\sigma_0 = 1 - a_8 \frac{\log \log T}{\log T}. \tag{210}$$

Da T mit x gegen ∞ wächst, können wir x sofort so groß annehmen, daß erstens

$$b_4 \leqq (\log T)^{a_8 + 2} \leqq T, \tag{211}$$

zweitens

$$\sigma_0 \geqq b_6, \tag{212}$$

drittens

$$T \leqq \frac{x}{(kT)^4} \tag{213}$$

gilt.

Dann ist wegen (211) § 4 anwendbar mit

$$\sigma = \sigma_0, \quad (\log T)^{a_8 + 2} \leqq t \leqq T. \tag{214}$$

Es ist also insbesondere für $b_4 \leqq t \leqq T$ nach (190)

$$a_5 \frac{\log \log t}{\log t} > a_8 \frac{\log \log t}{\log t} \geqq a_8 \frac{\log \log T}{\log T} = 1 - \sigma_0. \tag{215}$$

Aus (215), (190) und Hilfssatz 16 einerseits, (208) und (212) andererseits folgt

$$L(s, \chi) \neq 0 \quad \text{für} \quad \sigma \geqq \sigma_0, \quad 0 \leqq t \leqq T. \tag{216}$$

Unter diesen Annahmen gilt für $y \geqq x$ bei $x \to \infty$:

Hilfssatz 25. $\displaystyle\int_{\sigma_0}^{\sigma_0 + iT} y^{s-1} \frac{-L'}{L}(s, \chi)\, ds = o(1).$

Hilfssatz 26. $\int\limits_{\sigma_0}^{\sigma_0+iT} y^{s-1}\frac{1}{L(s,\chi)}\,ds = o(1).$

Beweis. Es ist

$$T^{1-\sigma_0} = T^{a_8\frac{\log\log T}{\log T}} = (\log T)^{a_8}, \tag{217}$$

$$y^{\sigma_0-1} \leqq x^{\sigma_0-1} = T^{a_{10}(\sigma_0-1)} = (\log T)^{-a_8 a_{10}}, \tag{218}$$

also nach (207)

$$y^{\sigma_0-1}(T^{1-\sigma_0})^4 \log^5 T \leqq (\log T)^{-a_8 a_{10}+4a_8+5} = o(1). \tag{219}$$

Wir setzen zur Abkürzung

$$v = (\log T)^{a_8+2}. \tag{220}$$

Erster Schritt $\left(\int\limits_{\sigma_0}^{\sigma_0+iv}\right)$. Es ist für $\sigma = \sigma_0$, $0 \leqq t \leqq v$

$$-\frac{L'}{L}(s,\chi) = O(\log T), \tag{221}$$

$$\frac{1}{L(s,\chi)} = O(\log T). \tag{222}$$

Denn für $t \geqq b_4$ folgt (221) bzw. (222) aus Hilfssatz 17 bzw. 18 und (215). Für $0 \leqq t \leqq b_4$ gilt sogar

$$-\frac{L'}{L}(s,\chi) = O(1) \quad \text{für} \quad \chi \neq \chi_0 \ (\chi_0 \text{ Hauptcharakter}), \tag{223}$$

$$\frac{1}{L(s,\chi)} = O(1); \tag{224}$$

denn diese Funktionen sind für $\sigma = 1$ regulär. Ferner gilt

$$-\frac{L'}{L}(s,\chi_0) = O\left(\frac{1}{1-\sigma_0}\right) = O(\log T), \tag{225}$$

da diese Funktion auf der Geraden $\sigma = 1$ als einzige Singularität einen Pol erster Ordnung in $s = 1$ besitzt. Damit sind (221) und (222) bewiesen. Also folgt

$$\left|\int\limits_{\sigma_0}^{\sigma_0+iv} y^{s-1}\frac{-L'(s,\chi)}{L(s,\chi)}\,ds\right| \leqq v\,y^{\sigma_0-1}O(\log T) \tag{226}$$

$$= O((\log T)^{a_8+2-a_8a_{10}+1}) \quad \text{(nach (218) und (220))}$$

$$= o(1) \quad \text{nach (207);}$$

und entsprechend

$$(227)\qquad \left|\int_{\sigma_0}^{\sigma_0+iv} y^{s-1}\frac{1}{L(s,\chi)}\,ds\right| = o(1).$$

Zweiter Schritt (Behandlung des ersten Fehlergliedes in den Hilfssätzen 23 und 24). Es ist

$$(228)\qquad \left|\int_{\sigma_0+iv}^{\sigma_0+iT} y^{s-1}\log^2 T\frac{T^{1-\sigma}}{t}\,ds\right| \leqq y^{\sigma_0-1}\log^2 T\cdot T^{1-\sigma_0}\int_1^T\frac{dt}{t}$$
$$= y^{\sigma_0-1}T^{1-\sigma_0}\log^3 T = o(1) \quad \text{nach } (219).$$

Dritter Schritt (Behandlung des zweiten Fehlergliedes in den Hilfssätzen 23 und 24). Es ist

$$(229)\qquad \left|\int_{\sigma_0+iv}^{\sigma_0+iT} y^{s-1}\log T\,|A(s)|^2\,ds\right| \leqq y^{\sigma_0-1}\log T\int_0^T |A(\sigma_0+it)|^2\,dt$$
$$= y^{\sigma_0-1}\log T\cdot O(T^{4(1-\sigma_0)}\log^4 T) \quad \text{(Hilfssatz 22)}$$
$$= o(1) \quad \text{nach } (219).$$

Vierter Schritt (Beweis von Hilfssatz 25). Nach Hilfssatz 24, (226), (228), (229) genügt es zu zeigen:

$$(230)\qquad \int_{\sigma_0+iv}^{\sigma_0+iT} y^{s-1}\sum_{1\leqq n\leqq (kT)^4} g(n)\,n^{-s}\,ds = o(1).$$

Es ist

$$(231)\qquad \int_{\sigma_0+iv}^{\sigma_0+iT} y^{s-1}\sum_{1\leqq n\leqq (kT)^4} g(n)\,n^{-s}\,ds$$
$$= i\sum_{1\leqq n\leqq (kT)^4}\frac{g(n)}{n}\left(\frac{y}{n}\right)^{\sigma_0-1}\int_v^T\left(\frac{y}{n}\right)^{it}dt$$
$$= O\left(\sum_{1\leqq n\leqq (kT)^4}\frac{d_4(n)}{n}\log(kT)\left(\frac{y}{n}\right)^{\sigma_0-1}\frac{1}{\log\frac{y}{n}}\right) \quad \text{(nach (200))}$$
$$= O\left(\log^4 T\log T\left(\frac{y}{(kT)^4}\right)^{\sigma_0-1}\frac{1}{\log T}\right) \quad \text{(nach Hilfssatz 19 und (213))}$$
$$= O\left(\left(\frac{y}{T^4}\right)^{\sigma_0-1}\log^4 T\right) = o(1) \quad \text{nach } (219).$$

Damit ist Hilfssatz 25 bewiesen.

Fünfter Schritt (Beweis von Hilfssatz 26). Nach Hilfssatz 23, (227), (228), (229) genügt es zu zeigen:

$$(232)\qquad \int_{\sigma_0+iv}^{\sigma_0+iT} y^{s-1}\sum_{1\leqq n\leqq (kT)^3} c(n)\,n^{-s}\,ds = o(1).$$

Es ist

$$(233)\qquad \int_{\sigma_0+iv}^{\sigma_0+iT} y^{s-1} \sum_{1\leqq n\leqq (kT)^3} c(n)\, n^{-s}\, ds$$

$$= i \sum_{1\leqq n\leqq (kT)^3} \frac{c(n)}{n}\left(\frac{y}{n}\right)^{\sigma_0-1} \int_v^T \left(\frac{y}{n}\right)^{it} dt$$

$$= O\left(\sum_{1\leqq n\leqq (kT)^3} \frac{d_3(n)}{n}\left(\frac{y}{n}\right)^{\sigma_0-1} \frac{1}{\log\frac{y}{n}}\right) \quad \text{(nach (195))}$$

$$= O\left(\log^3 T\left(\frac{y}{(kT)^3}\right)^{\sigma_0-1}\frac{1}{\log T}\right) \quad \text{(nach Hilfssatz 19 und (213))}$$

$$= O\left(\log^2 T\, y^{\sigma_0-1}\, T^{3(1-\sigma_0)}\right) = o(1) \quad \text{nach (219).}$$

Damit ist auch Hilfssatz 26 bewiesen.

Hilfssatz 27. *Es sei* $|h(n)| \leqq \log n$ *für* $n \geqq 3$, *also*

$$(234)\qquad H(s) = \sum_{n=1}^{\infty} h(n)\, n^{-s}$$

absolut konvergent für $\sigma > 1$. *Es sei ferner für* $1 < \sigma \leqq 2$

$$(235)\qquad \sum_{n=1}^{\infty} |h(n)|\, n^{-\sigma} \leqq a_{11}\frac{1}{\sigma-1}.$$

Dann ist für $z \geqq T \geqq 3,\ 1 < \eta < 2$

$$(236)\qquad \left|\frac{1}{2\pi i}\int_{\eta-iT}^{\eta+iT} \frac{z^s H(s)}{s}\, ds - \sum_{1\leqq n\leqq z} h(n)\right| \leqq a_{12}\frac{z^{\eta}\log^2 z}{(\eta-1)T}.$$

Beweis. Siehe Landau, Über einige Summen, die von den Nullstellen der Riemann'schen Zetafunktion abhängen, Acta Mathematica **35** (1912), S. 271–294, Hilfssatz 3 [17]).

Nunmehr setzen wir

$$(237)\qquad H_1(s) = \frac{1}{\varphi(k)}\sum_{\chi}\frac{1}{\chi(l)}\frac{-L'}{L}(s,\chi),$$

$$(238)\qquad H_2(s) = \frac{1}{\varphi(k)}\sum_{\chi}\frac{1}{\chi(l)}\frac{1}{L(s,\chi)},$$

indem wir über alle Charaktere χ mod k summieren. Die Funktionen $H_1(s)$ und $H_2(s)$ gestatten die für $\sigma > 1$ gültigen Entwicklungen:

---

[17]) Herr Landau beweist den Satz nur in der schwächeren Form, in der statt a_{12} nur eine von z, η und T unabhängige Konstante steht. Doch liefert sein Beweis auch genau den obigen Hilfssatz.

$$(239)\qquad H_1(s) = \sum_{\substack{n=1\\ n\equiv l \,(\mathrm{mod}\, k)}}^{\infty} \Lambda(n)\, n^{-s},$$

$$(240)\qquad H_2(s) = \sum_{\substack{n=1\\ n\equiv l \,(\mathrm{mod}\, k)}}^{\infty} \mu(n)\, n^{-s},$$

die sofort aus (237) und (238) einerseits, (17) und (18) andererseits folgen. Da nach (239) und (240) die Funktionen $H_1(s)$ und $H_2(s)$ den Voraussetzungen von Hilfssatz 27 genügen, ist für

$$(241)\qquad 1 < \eta = 1 + \frac{1}{\log x} < 2$$

$$(242)\qquad \left| \sum_{\substack{x < n \leqq x + x^{a_4}\\ n\equiv l \,(\mathrm{mod}\, k)}} \Lambda(n) - \frac{1}{2\pi i}\int_{\eta - iT}^{\eta + iT} \{(x + x^{a_4})^s - x^s\}\, H_1(s)\frac{ds}{s}\right|$$

$$\leqq \frac{a_{12}}{(\eta-1)T}\left((x + x^{a_4})^{1+\frac{1}{\log x}} \log^2(x + x^{a_4}) + x^{1+\frac{1}{\log x}}\log^2 x\right)$$

$$\leqq \frac{a_{13}}{T}\log x\; x \log^2 x = o(x^{a_4})$$

wegen

$$(243)\qquad \frac{x}{T} = x^{1-\frac{1}{a_{10}}},\quad 1 - \frac{1}{a_{10}} = 1 - \frac{1}{249{,}8} < 1 - \frac{1}{250} = a_4.$$

Entsprechend erhält man

$$(244)\qquad \left| \sum_{\substack{x < n \leqq x + x^{a_4}\\ n\equiv l \,(\mathrm{mod}\, k)}} \mu(n) - \frac{1}{2\pi i}\int_{\eta - iT}^{\eta + iT} \{(x + x^{a_4})^s - x^s\}\, H_2(s)\frac{ds}{s}\right| = o(x^{a_4}).$$

Auf dem abgeschlossenen Rechteck

$$(245)\qquad \sigma_0 \leqq \sigma \leqq \eta,\ |t| \leqq T$$

sind die Funktionen

$$(246)\qquad \{(x + x^{a_4})^s - x^s\}\, H_\nu(s)\frac{1}{s} \qquad (\nu = 1, 2)$$

regulär mit Ausnahme des Pols in $s = 1$ von höchstens erster Ordnung mit dem Residuum $(2-\nu)\frac{x^{a_4}}{\varphi(k)}$, da auf (245) überall $L(s, \chi) \neq 0$ nach (216). Um die in (242) und (244) auftretenden Integrale zu behandeln, wenden wir den Cauchyschen Integralsatz auf das Rechteck (245) an. Wir wollen zeigen

$$(247)\qquad \frac{1}{2\pi i}\int_{\eta - iT}^{\eta + iT} - (2-\nu)\frac{x^{a_4}}{\varphi(k)} = \frac{1}{2\pi i}\left(\int_{\eta - iT}^{\sigma_0 - iT} + \int_{\sigma_0 - iT}^{\sigma_0 + iT} + \int_{\sigma_0 + iT}^{\eta + iT}\right) = o(x^{a_4}),$$

wenn der Integrand die Funktion (246) ist. Da dieser für reelle s reell ist, genügt es,

$$\int_{\sigma_0}^{\sigma_0+iT} \{(x+x^{a_1})^s - x^s\} H_\nu(s) \frac{ds}{s} = o(x^{a_1}) \tag{248}$$

und

$$\int_{\sigma_0+iT}^{\eta+iT} \{(x+x^{a_1})^s - x^s\} H_\nu(s) \frac{ds}{s} = o(x^{a_1}) \tag{249}$$

zu zeigen. Für $t = T$, $\sigma_0 \leqq \sigma \leqq \eta$ ist nach Hilfssatz 17 bzw. 18

$$H_\nu(s) = O(\log T), \tag{250}$$

woraus sofort

$$\begin{aligned} \int_{\sigma_0+iT}^{\eta+iT} \{(x+x^{a_1})^s - x^s\} H_\nu(s) \frac{ds}{s} &= (\eta - \sigma_0)\, O(x)\, O(\log T) \frac{1}{T} \\ &= O\left(\frac{x \log T}{T}\right) = o(x^{a_1}) \text{ (nach (243))} \end{aligned} \tag{251}$$

folgt.

Ferner ist

$$\begin{aligned} \int_{\sigma_0}^{\sigma_0+iT} \{(x+x^{a_1})^s - x^s\} H_\nu(s) \frac{ds}{s} &= \int_x^{x+x^{a_1}} dy \int_{\sigma_0}^{\sigma_0+iT} y^{s-1} H_\nu(s)\, ds \\ &= \int_x^{x+x^{a_1}} o(1)\, dy = o(x^{a_1}) \text{ nach Hilfssatz 25 und 26.} \end{aligned} \tag{252}$$

Somit sind (248) und (249), also auch (247) bewiesen. Aus (247) mit $\nu = 2$ und (244) folgt sofort Satz 4. Aus (247) mit $\nu = 1$ und (242) folgt

$$\sum_{\substack{x < n \leqq x + x^{a_1} \\ n \equiv l \,(\mathrm{mod}\, k)}} \Lambda(n) = \frac{x^{a_1}}{\varphi(k)} + o(x^{a_1}). \tag{253}$$

Hieraus erhält man aber sofort Satz 3. Denn es ist

$$\pi(x+x^{a_1}; k, l) - \pi(x; k, l) \leqq \frac{1}{\log x} \sum_{\substack{x < n \leqq x + x^{a_1} \\ n \equiv l \,(\mathrm{mod}\, k)}} \Lambda(n) = \frac{1}{\varphi(k)} \frac{x^{a_1}}{\log x}(1 + o(1)), \tag{254}$$

$$\begin{aligned} \pi(x+x^{a_1}; k, l) - \pi(x; k, l) &\geqq \sum_{\substack{x < p^m \leqq x + x^{a_1} \\ p^m \equiv l \,(\mathrm{mod}\, k)}} 1 + O(\sqrt{x}) \\ &\geqq \frac{1}{\log(x + x^{a_1})} \sum_{\substack{x < n \leqq x + x^{a_1} \\ n \equiv l \,(\mathrm{mod}\, k)}} \Lambda(n) + O(\sqrt{x}) = \frac{1}{\varphi(k)} \frac{x^{a_1}}{\log x}(1 + o(1)). \end{aligned} \tag{255}$$

(Eingegangen am 28. Januar 1932.)

[Göttingen]

2t.

On the Prime Number Theorem of Hoheisel[1]

Translation of
Mathematische Zeitschrift 36 (1933), 394–423
[Commentary on p. 561]

Table of contents

	Page
Table of contents	100
Introduction	100
Notations	104
§1. Asymptotic expansions of $L(s,\chi)$ and $L'(s,\chi)$	105
§2. Littlewood's theorem on the zeros of $L(s,\chi)$	110
§3. Lemmas on divisor numbers	126
§4. Asymptotic expansions of $\dfrac{1}{L(s,\chi)}$ and $\dfrac{-L'(s,\chi)}{L(s,\chi)}$	131
§5. Proof of the main theorems	134

Introduction

Hoheisel[2] recently proved the interesting

Theorem 1. *For $x \geqq 2$ let $\pi(x)$ be the number of prime numbers which do not exceed x. Then there exists a positive*[3] $a_1 < 1$ *such that*

$$(1) \qquad \lim_{x=\infty} \{\pi(x + x^{a_1}) - \pi(x)\} \frac{\log x}{x^{a_1}} = 1.$$

For this theorem I shall give in this paper a new proof which differs from that of Hoheisel mainly in the fact that I study the Riemann zeta function

[1] This paper was presented to the mathematical-scientific faculty of the University of Göttingen [mathematisch-naturwissenschaftliche Fakultät der Universität Göttingen] as the inaugural dissertation.

[2] Guido Hoheisel, Primzahlprobleme in der Analysis, *Sitzungsberichte der Preußischen Akademie der Wissenschaften*, *physik-math. Klasse*, 1930, 580–588.

[3] In the following, $a_1, a_2, \ldots$ denote positive, absolute constants.

only near the line $\sigma = 1$. About its zeros I therefore use only the theorem of Littlewood[4]:

$$\zeta(s) \neq 0 \quad \text{for } t \geqq a_2, \quad \sigma \geqq 1 - a_3\frac{\log\log t}{\log t}, \tag{2}$$

which is also employed by Hoheisel in his proof of Theorem 1. Contrary to Hoheisel I do not need:

1. Riemann's functional equation of $\zeta(s)$.
2. Hadamard's product expansion of $\zeta(s)$.
3. Carlson's theorem on the zeros of $\zeta(s)$ which was sharpened by Hoheisel.

By means of these simplifications my proof becomes independent of the fact that the function $\dfrac{\zeta'(s)}{\zeta(s)}$, to which the study of the problem of the distribution of the prime numbers can be easily reduced, is the logarithmic derivative of the meromorphic function $\zeta(s)$. Consequently my method of proof is also applicable to other Dirichlet series, e.g., to the function $\dfrac{1}{\zeta(s)}$, and I obtain as an analogue to Theorem 1

Theorem 2. *If $\mu(n)$ denotes the Möbius function, then*

$$\lim_{x=\infty} x^{-a_1} \sum_{x<n\leqq x+x^{a_1}} \mu(n) = 0. \tag{3}$$

For the constant a_1 one obtains as a necessary condition[5]

$$1 - \frac{1}{4 + \dfrac{5}{a_3}} < a_1 < 1, \tag{4}$$

where a_3 is the constant occurring in Littlewood's theorem mentioned above.

[4] J. E. Littlewood, Researches in the Theory of the Riemann ζ-Function. *Proceedings of the London Mathematical Society*, *Ser.* II, **20** (1922), p. xxv. Cf. also Edmund Landau, *Vorlesungen über Zahlentheorie* **2** (1927), Satz 397.

[5] Hoheisel obtains the (less efficient) condition

$$1 - \frac{1}{4 + \dfrac{6}{a_3}} < a_1 < 1.$$

Since it does not require any more work, I shall immediately prove, in place of Theorems 1 and 2, the corresponding theorems for arithmetic progressions:

Theorem 3. *Let k and l be natural, relatively prime numbers. For $x \geqq 2$ let $\pi(x; k, l)$ denote the number of prime numbers $p \equiv l \pmod{k}$ which do not exceed x. Then there exists $a_4 < 1$ such that*

$$(5) \qquad \lim_{x=\infty} \{\pi(x + x^{a_4}; k, l) - \pi(x; k, l)\} \frac{\log x}{x^{a_4}} = \frac{1}{\varphi(k)}.$$

Theorem 4. *Let k and l be natural, relatively prime numbers. Then*

$$(6) \qquad \lim_{x=\infty} x^{-a_4} \sum_{\substack{x<n\leqq x+x^{a_4} \\ n\equiv l \pmod{k}}} \mu(n) = 0.$$

For the proof of Theorems 3 and 4 I naturally need the theorem of Littlewood in the more general form[6]:

$$(7) \qquad L(s, \chi) \neq 0 \quad \text{for } t \geqq b_1, \quad \sigma \geqq 1 - a_5 \frac{\log\log t}{\log t}.$$

Here, $L(s, \chi)$ is the L-series belonging to a character $\chi \pmod{k}$; b_1 denotes a positive constant which depends only on k. As a sufficient condition for a_4 I shall find [analogous to (4)]:

$$(8) \qquad 1 - \frac{1}{4 + \frac{5}{a_5}} < a_4 < 1.$$

Below I shall show that this generalization of Littlewood's theorem which is due to Landau holds with $a_5 = \frac{1}{49.13}$. Since

$$(9) \qquad 1 - \frac{1}{4 + \frac{5}{\frac{1}{49.13}}} = 1 - \frac{1}{4 + 245.65} < 1 - \frac{1}{250} < 1$$

it follows from (8) that

$$(10) \qquad a_1 = a_4 = 1 - \frac{1}{250}$$

[6]See Landau, loc. cit.

are constants for which Theorems 1 to 4 are valid. Therefore one obtains as a corollary to Theorem 3

Theorem 5. *For every natural number k there exists* $x_k \geqq 2$ *such that, if l is a natural number prime to k, for* $x \geqq x_k$ *there always exists a prime number* $p \equiv l \pmod{k}$ *which lies between* x^{250} *(exclusive) and* $(x+1)^{250}$ *(exclusive).*

Proof. For $x \geqq 2$ we have

$$(x+1)^{250} > x^{250} + x^{249}, \tag{11}$$

and therefore by Theorem 3

$$\begin{aligned}(12)\quad & \pi\left((x+1)^{250}; k, l\right) - \pi(x^{250}; k, l) \\ & \geqq \pi(x^{250} + x^{249}; k, l) - \pi(x^{250}; k, l) \\ & = \pi\left(x^{250} + (x^{250})^{1-\frac{1}{250}}; k, l\right) - \pi(x^{250}; k, l) \sim \frac{(x^{250})^{1-\frac{1}{250}}}{\varphi(k)\log x^{250}},\end{aligned}$$

where the right side increases with x over all bounds.

In §1 an expansion, proved for $\zeta(s)$ by Hardy-Littlewood, will be carried over to the Dirichlet L-series and their first derivatives.

In §2 I shall prove the above-cited theorem of Littlewood and Landau[7] anew in order to produce an a_5 which is as large as possible, and then derive from it known estimates for $\frac{L'}{L}(s, \chi)$ and $\frac{1}{L(s, \chi)}$.

In §3 some elementary inequalities on divisor numbers will be presented in order to prove an integral estimate which also appears in similar form in Hoheisel's proof of the theorem of Carlson. Hereby I shall make use of an elementary number-theoretic method of Ingham.[8]

In §4 the critical lemma appears. It gives a representation of $\frac{1}{L(s, \chi)}$ and $-\frac{L'}{L}(s, \chi)$ in the critical strip as a finite Dirichlet series + a "small" error term + a term which, although not "small," is nevertheless "small in the mean," which is enough in order to give

in §5 the proof of Theorems 3 and 4.

[7] G. H. Hardy and J. E. Littlewood, The zeros of Riemann's zeta-function on the critical line, *Math. Zeitschr.* **10** (1921), 283–317; the special case mentioned above is approximately Lemma 2 of that article.

[8] A. E. Ingham, Some asymptotic formulae in the theory of numbers, *The Journal of the London Mathematical Society* **2** (1927), 202–208.

Notations

We constantly assume that we are given two natural numbers k and l which are relatively prime.

Let $\chi(n)$ denote a Dirichlet character mod k.

$a_1, \dots, a_{13}$ denote absolute, positive constants.

$b_1, \dots, b_6$ denote positive constants which depend only on k.

O- and o-estimates are, *for k fixed*, valid *uniformly* in all the other variables.[9]

Complex integrals will always be taken over a straight path, if nothing to the contrary is stated explicitly.

Let $d_r(n)$ denote the number of decompositions of the natural number n into r positive integral factors, whereby the order of the factors is taken into account (appears only for $r = 2, 3, 4$); $d_2(n)$ will be abbreviated to $d(n)$.

$\Lambda(n), \mu(n), \varphi(n)$ have, as functions of the natural number n, their usual meaning:

$$\Lambda(n) = \begin{cases} \log p & \text{if } n = p^\alpha,\ p \text{ prime number, } \alpha \geqq 1 \text{ integral} \\ 0 & \text{otherwise} \end{cases}$$

$$\mu(n) = \begin{cases} 1 & \text{for } n = 1 \\ (-1)^r & \text{if } n \text{ is a product of } r \text{ distinct prime numbers} \\ 0 & \text{if } n \text{ is divisible by the square of a prime number} \end{cases}$$

$\varphi(n)$ is the number of residue classes mod n which are prime to n.

Throughout, $s = \sigma + it$, $\sigma = \Re s$

$$\zeta(s) = \sum_{n=1}^{\infty} n^{-s} \quad \text{for } \sigma > 1; \tag{13}$$

$$\zeta(s, w) = \sum_{n=0}^{\infty} (n + w)^{-s} \quad \text{for } \sigma > 1, 0 < w \leqq 1; \tag{14}$$

$$L(s, \chi) = \sum_{n=1}^{\infty} \chi(n) n^{-s} \quad \text{for } \sigma > 1. \tag{15}$$

I recall the following identities which are valid for $\sigma > 1$:

[9] $f = O(g)$ means $|f| < bg$ for all the arguments allowed by the hypothesis in question, where b depends only on k. $f(z; z_1, \dots, z_n) = o(g(z; z_1, \dots, z_n))$ for $z \to \infty$ means that for every $\varepsilon > 0$ there exists $\omega = \omega(\varepsilon, k) > 0$ which depends only on ε and k such that for all values $z_1, \dots, z_n$ permitted by the hypothesis and for $z \geqq \omega$ one always has $|f(z; z_1, \dots, z_n)| < \varepsilon g(z; z_1, \dots, z_n)$.

$$L(s,\chi) = k^{-s} \sum_{q=1}^{k} \chi(q)\zeta\left(s, \frac{q}{k}\right); \tag{16}$$

$$\frac{-L'(s,\chi)}{L(s,\chi)} = \sum_{n=1}^{\infty} \chi(n)\Lambda(n)n^{-s}; \tag{17}$$

$$\frac{1}{L(s,\chi)} = \sum_{n=1}^{\infty} \chi(n)\mu(n)n^{-s}. \tag{18}$$

§1. Asymptotic expansions of $L(s,\chi)$ and $L'(s,\chi)$

Lemma 1. *For $2 \leqq t \leqq \pi T$, $\frac{1}{2} \leqq \sigma \leqq 3$, $0 < w \leqq 1$ and integral $m \neq 0$ we have*

$$\int_T^{\infty} (u+w)^{-s} e^{2\pi imu}\, du = -\frac{1}{2\pi im} e^{2\pi imT}(T+w)^{-s} + O\left(\frac{tT^{-1-\sigma}}{m^2}\right). \tag{19}$$

Proof. By integration by parts one obtains

$$\int_T^{\infty} (u+w)^{-s} e^{2\pi imu}\, du \tag{20}$$

$$= -\frac{1}{2\pi im} e^{2\pi imT}(T+w)^{-s} + \frac{s}{2\pi im} \int_T^{\infty} (u+w)^{-s-1} e^{2\pi imu}\, du.$$

For $u \geqq T$ the expression

$$\frac{(u+w)^{-\sigma-1}}{-\dfrac{t}{u+w} + 2\pi m} = \frac{1}{(u+w)^{\sigma}(2\pi m(u+w) - t)} \tag{21}$$

tends with increasing u monotonically toward 0. For $u = T$ the absolute value of (21) is equal to

$$\frac{(T+w)^{-\sigma-1}}{\left|-\dfrac{t}{T+w} + 2\pi m\right|} \leqq \frac{T^{-\sigma-1}}{2\pi|m| - \pi} \leqq \frac{T^{-\sigma-1}}{\pi|m|}. \tag{22}$$

Therefore, by the second mean value theorem,

$$(23)\quad \int_T^\infty (u+w)^{-s-1} e^{2\pi i m u}\,du$$

$$= \int_T^\infty (u+w)^{-\sigma-1} e^{i(-t\log(u+w)+2\pi m u)}\,du$$

$$= \int_T^\infty (u+w)^{-\sigma-1} \frac{1}{-\dfrac{t}{u+w}+2\pi m} e^{i(-t\log(u+w)+2\pi m u)}$$

$$\cdot d(-t\log(u+w)+2\pi m u)$$

$$= O\left(\frac{T^{-1-\sigma}}{|m|}\right) \quad \text{by (22)}.$$

From (20) and (23) follows the assertion.

Lemma 2. *For $2 < t < \pi T$, $\frac{1}{2} < \sigma < 3$, $0 < w \leqq 1$ and integral T we have that*

$$(24)\quad g(s) = g(s; T, w)$$

$$= \sum_{m=1}^{\infty} \int_T^\infty (u+w)^{-s}\left(e^{2\pi i m u} + e^{-2\pi i m u}\right)du \qquad \textit{is convergent},$$

$$(25)\qquad g(s) \textit{ is regular},$$

$$(26)\qquad g(s) = O\left(\frac{T^{1-\sigma}}{t}\right).$$

Proof. For integral $m > 0$ we have, since T is also integral,

$$(27)\qquad \frac{-1}{2\pi i m} e^{2\pi i m T} + \frac{-1}{2\pi i(-m)} e^{2\pi i(-m)T} = 0.$$

Therefore, by Lemma 1 we have

$$\int_T^\infty (u+w)^{-s}(e^{2\pi imu}+e^{-2\pi imu})\,du = O\left(\frac{tT^{-1-\sigma}}{m^2}\right), \tag{28}$$

from which (24) follows immediately.

Since by (28) the series converges uniformly in s, it suffices to show that each of the integrals is regular in s, which is put into evidence by (20). Thus (25) is proved.

According to (28) we have

$$g(s) = \sum_{m=1}^{\infty} O\left(\frac{tT^{-1-\sigma}}{m^2}\right) = O(tT^{-1-\sigma}) = O\left(\frac{T^{1-\sigma}}{t}\right). \tag{29}$$

Lemma 3. *Under the assumptions of Lemma 2*

$$\zeta(s,w) \text{ is regular}, \tag{30}$$

$$\zeta(s,w) = \sum_{n=0}^{T-1}(n+w)^{-s} + \frac{1}{2}(T+w)^{-s} + g(s;T,w) - \frac{(T+w)^{1-s}}{1-s}, \tag{31}$$

$$\zeta(s,w) = \sum_{n=0}^{T-1}(n+w)^{-s} + O\left(\frac{T^{1-\sigma}}{t}\right). \tag{32}$$

Proof. For the time being let $\sigma > 1$. Then by Dirichlet's sum formula we have for integral $V > T$

$$\begin{aligned}\sum_{n=T}^{V}(n+w)^{-s} &= \frac{1}{2}\{(T+w)^{-s}+(V+w)^{-s}\} + \lim_{N=\infty}\sum_{m=-N}^{N}\cdot\int_T^V (u+w)^{-s}e^{2\pi imu}\,du \\ &= \frac{1}{2}\{(T+w)^{-s}+(V+w)^{-s}\} + \int_T^V (u+w)^{-s}\,du\end{aligned} \tag{33}$$

$$+\sum_{m=1}^{\infty}\int_{T}^{V}(u+w)^{-s}(e^{2\pi imu}+e^{-2\pi imu})\,du$$

$$=\frac{1}{2}\{(T+w)^{-s}+(V+w)^{-s}\}$$

$$+\frac{(V+w)^{1-s}}{1-s}-\frac{(T+w)^{1-s}}{1-s}$$

$$+g(s;T,w)-g(s;V,w),$$

if $g(s;T,w)$ and $g(s;V,w)$ have the same meaning as in Lemma 2. By (26) we have

$$\lim_{V=\infty} g(s;V,w)=0; \tag{34}$$

hence it follows from (33) that, as V goes toward ∞, we have

$$\sum_{n=T}^{\infty}(n+w)^{-s}=\frac{1}{2}(T+w)^{-s}-\frac{(T+w)^{1-s}}{1-s}+g(s;T,w). \tag{35}$$

From (35) and (14), (31) follows immediately for $\sigma > 1$. Since the functions appearing on the right-hand side of (31) are by (25) regular for all s satisfying the hypothesis, (31) and (30) are valid for all our s. On the other hand, (32) follows immediately from (31) and (26).

Lemma 4. *For* $2 < t < \pi T$, $\frac{1}{2} < \sigma < 3$,

$$L(s,\chi) \text{ is regular}, \tag{36}$$

$$L(s,\chi)=\sum_{1\leqq r\leqq kT}\chi(r)r^{-s}+O\left(\frac{T^{1-\sigma}}{t}\right). \tag{37}$$

Proof. Without loss of generality let T be integral. Otherwise consider $[T]+1$ instead of T. This is possible because

$$\left|\sum_{kT<r\leqq k([T]+1)}\chi(r)r^{-s}\right|\leqq k(kT)^{-\sigma}=O\left(\frac{T^{1-\sigma}}{t}\right). \tag{38}$$

To begin with, (36) follows from (16), in view of (30). Moreover, from (16) and (32) it follows that

$$(39)\qquad L(s,\chi)=k^{-s}\sum_{q=1}^{k}\chi(q)\left\{\sum_{n=0}^{T-1}\left(n+\frac{q}{k}\right)^{-s}+O\left(\frac{T^{1-\sigma}}{t}\right)\right\}$$

$$=\sum_{r=1}^{kT}\chi(r)r^{-s}+O\left(\frac{T^{1-\sigma}}{t}\right).$$

***Lemma* 5.** *For* $3\leqq t\leqq T$, $\frac{3}{4}\leqq\sigma\leqq 2$

$$(40)\qquad -L'(s,\chi)=\sum_{1\leqq r\leqq kT}\chi(r)\log(r)\,r^{-s}+O\left(\log T\frac{T^{1-\sigma}}{t}\right).$$

Proof. If one puts

$$(41)\qquad h(s)=L(s,\chi)-\sum_{1\leqq r\leqq kT}\chi(r)r^{-s},$$

then, by Lemma 4, $h(s)$ is regular for all s satisfying the hypothesis of Lemma 4, and so for the proof (40) it suffices to verify that

$$(42)\qquad h'(s)=O\left(\log T\frac{T^{1-\sigma}}{t}\right)$$

for all s satisfying the hypothesis of Lemma 5. At this stage let $s_0=\sigma_0+it_0$ with

$$(43)\qquad \frac{3}{4}\leqq\sigma_0\leqq 2,\ 3\leqq t_0\leqq T$$

be arbitrary but fixed. Then all s with

$$(44)\qquad |s-s_0|\leqq\frac{1}{4\log T}\left(<\frac{1}{4}\right)$$

satisfy the hypothesis of Lemma 4; therefore, for all s lying in the disc (44), we have

$$(45)\qquad h(s)=O\left(\frac{T^{1-\sigma}}{t}\right)=O\left(\frac{T^{1-\sigma_0+\frac{1}{4\log T}}}{t_0}\right)=O\left(\frac{T^{1-\sigma_0}}{t_0}\right).$$

Therefore, by Cauchy, we have

$$|h'(s_0)| \leqq 4 \log T O\left(\frac{T^{1-\sigma_0}}{t_0}\right). \tag{46}$$

Herewith (42), hence also (40) is proved.

§2. Littlewood's Theorem on the zeros of $L(s,\chi)$

Lemma 6. *For* $\xi > 0$, $0 < \sigma < 1$,

$$\frac{1-\sigma}{\xi} - \frac{2^{-[\xi]}}{\xi} + \frac{2^{-[\xi]}}{[\xi]+2} < \log 2 \frac{1-\sigma}{\log \frac{1}{1-\sigma}}. \tag{47}$$

Proof. For real η we have

$$2^{\eta} - \eta > 0 \tag{48}$$

$$\eta(2^{\eta} - 1) \geqq 0. \tag{49}$$

Thus

$$\xi(2^{\eta} - \eta) + \eta(2^{\eta} - 1) > 0, \tag{50}$$

$$2^{\eta}(\xi + \eta) - \eta(\xi + 1) > 0. \tag{51}$$

Therefore, for $\eta > -\xi$,

$$\frac{2^{\eta}}{\xi(\xi+1)} - \frac{\eta}{\xi(\xi+\eta)} > 0, \tag{52}$$

$$2^{\eta}\left(\frac{1}{\xi} - \frac{1}{\xi+1}\right) - \frac{1}{\xi} + \frac{1}{\xi+\eta} > 0. \tag{53}$$

If we put

$$\eta = \frac{\log \frac{1}{1-\sigma}}{\log 2} - \xi, \tag{54}$$

then it follows from (53) that

$$\frac{1}{1-\sigma}2^{-\xi}\left(\frac{1}{\xi}-\frac{1}{\xi+1}\right)-\frac{1}{\xi}+\frac{\log 2}{\log\frac{1}{1-\sigma}}>0, \tag{55}$$

$$2^{-[\xi]}\left(\frac{1}{\xi}-\frac{1}{[\xi]+2}\right)-\frac{1-\sigma}{\xi}+(1-\sigma)\frac{\log 2}{\log\frac{1}{1-\sigma}}>0. \tag{56}$$

This gives the assertion.

***Lemma* 7.** *For* $\xi \geqq \frac{3}{2}$,

$$-\frac{1}{\xi}+\frac{1}{[\xi]+2} \geqq -\frac{1}{[\xi]+2}. \tag{57}$$

Proof. For $\xi \geqq 2$

$$2\xi \geqq \xi + 2 \geqq [\xi] + 2, \tag{58}$$

for $\frac{3}{2} \leqq \xi < 2$,

$$2\xi \geqq 3 = [\xi] + 2, \tag{59}$$

and hence for $\xi \geqq \frac{3}{2}$,

$$2\xi \geqq [\xi] + 2, \qquad \frac{1}{\xi} \leqq \frac{2}{[\xi]+2}, \tag{60}$$

whence the assertion.

***Lemma* 8.** *For* $3 \leqq t$, $0 < w \leqq 1$, $0 < \sigma < 1$, $r > 0$, $R = 2^{r-1}$, $1 \leqq N \leqq N' < 2N$, $N \leqq t$, *we have, if* r, N, N' *are integral,*

$$\left|\sum_{n=N}^{N'}(n+w)^{-s}\right| \leqq a_6 N^{1-\sigma}\left(N^{-\frac{1}{R}}t^{\frac{1}{R(r+1)}}+t^{-\frac{1}{R(r+1)}}\right)\log t. \tag{61}$$

Proof. See Landau, *Vorlesungen über Zahlentheorie*, Vol. 2, Theorem 389.[10]

[10] There the theorem is proved only for $t > 3$. However, it follows by continuity that it is also true for $t \geqq 3$.

Lemma 9.[11] *For* $\frac{3}{4} \leqq \sigma < 1$, $t \geqq 3$, $0 < w \leqq 1$

$$\zeta(s, w) - w^{-s} = O\left(t^{\log 2 \frac{1-\sigma}{\log \frac{1}{1-\sigma}}} \log^2 t\right). \tag{62}$$

Proof. By Lemma 3 with $T = [t] + 1$ we have

$$\zeta(s, w) = \sum_{n=0}^{[t]} (n + w)^{-s} + O(1). \tag{63}$$

I partition the sum as follows:

$$\sum_{n=0}^{[t]} = \sum_{n=0}^{0} + \sum_{n=1}^{1} + \sum_{n=2}^{3} + \sum_{n=4}^{7} + \cdots + \sum_{n=2^{\left[\frac{\log t}{\log 2}\right]}}^{[t]} \tag{64}$$

This is possible because

$$2^{\left[\frac{\log t}{\log 2}\right]} \leqq 2^{\frac{\log t}{\log 2}} = t < 2^{\left[\frac{\log t}{\log 2}\right]+1}, \tag{65}$$

$$2^{\left[\frac{\log t}{\log 2}\right]} \leqq [t] < 2^{\left[\frac{\log t}{\log 2}\right]+1}. \tag{66}$$

The first sum in (64) is w^{-s}, the second one $O(1)$. By (66) the rest satisfy the hypotheses of Lemma 8, and there are $\left[\frac{\log t}{\log 2}\right] = O(\log t)$ such sums. It is therefore sufficient to show that each such sum is

$$O\left(t^{\log 2 \frac{1-\sigma}{\log \frac{1}{1-\sigma}}} \log t\right). \tag{67}$$

[11] This theorem is due to Hardy and Littlewood. A slightly weaker version may be found in Landau, *Vorlesung über Zahlentheorie*, vol. **2** Satz 394. A sharper estimate is presented by Corput-Koksma, Sur l'ordre de grandeur de la fonction $\zeta(s)$ de Riemann dans la bande, critique, *Annales de la faculté des sciences de l'université de Toulouse*, Ser. 3, **22** (1930), 1–39.

If, as in Lemma 8, N and N' denote the first and last member of the sum, then we have

$$(68) \qquad 2 \leqq N \leqq N' \leqq t, \qquad N' < 2N.$$

First Case. $t^{2/3} < N \leqq t$. Then by Lemma 8 with $r = 1$, $R = 1$ we obtain

$$(69) \qquad \left|\sum_{n=N}^{N'} (n+w)^{-s}\right| \leqq a_6\left(N^{-\sigma}t^{\frac{1}{2}} + N^{1-\sigma}t^{-\frac{1}{2}}\right)\log t$$

$$\leqq a_6\left(t^{-\frac{2}{3}\cdot\frac{3}{4}}t^{\frac{1}{2}} + t^{\frac{1}{4}}t^{-\frac{1}{2}}\right)\log t \leqq 2a_6 \log t.$$

Second Case. $2 \leqq N \leqq t^{2/3}$. We put $\xi = \dfrac{\log t}{\log N}\left(\geqq \dfrac{3}{2}\right)$ and apply Lemma 8 with $r = [\xi] + 1$, $R = 2^{[\xi]}$. Then we have, since $N = t^{1/\xi}$,

$$(70) \qquad \left|\sum_{n=N}^{N'} (n+w)^{-s}\right| \leqq a_6 t^{\frac{1-\sigma}{\xi}}\left(t^{\frac{-2-[\xi]}{\xi}+\frac{2-[\xi]}{[\xi]+2}} + t^{-\frac{2[\xi]}{[\xi]+2}}\right)\log t$$

$$\leqq 2a_6 t^{\frac{1-\sigma}{\xi}} t^{\frac{2-[\xi]}{\xi}+\frac{2-[\xi]}{[\xi]+2}} \log t \quad \text{(by Lemma 7)}$$

$$\leqq 2a_6 t^{\log 2 \frac{1-\sigma}{\log\frac{1}{1-\sigma}}} \log t \quad \text{(by Lemma 6)}.$$

This proves Lemma 9.

Lemma 10. *For* $\frac{3}{4} \leqq \sigma < 1$, $t \geqq 3$,

$$(71) \qquad L(s,\chi) = O\left(t^{\log 2 \frac{1-\sigma}{\log\frac{1}{1-\sigma}}} \log^2 t\right).$$

Proof. By (16) and Lemma 9 we have

$$(72) \qquad L(s,\chi) = k^{-s}\sum_{q=1}^{k} \chi(q)\left(\zeta\left(s,\frac{q}{k}\right) - \left(\frac{q}{k}\right)^{-s}\right) + \sum_{q=1}^{k} \chi(q) q^{-s}$$

$$= O\left(t^{\log 2 \frac{1-\sigma}{\log \frac{1}{1-\sigma}}} \log^2 t\right) + O(1).$$

Lemma 11. *Let $f(x)$ be regular for $|x| \leqq 1$, $f(0) = 1$, $M \geqq 0$,*

$$|f(x)| \leqq e^M \text{ for } |x| \leqq 1. \tag{73}$$

Then, if ξ runs over all roots (counted with their multiplicity[12]*) of $f(x)$ with $|\xi| < 1$,*[13] *we have for $|x| < 1$*

$$\left|\frac{f'(x)}{f(x)} - \sum_{\xi}\left(\frac{1}{x-\xi} + \frac{\bar{\xi}}{1-\bar{\xi}x}\right)\right| \leqq \frac{2M}{(1-|x|)^2}. \tag{74}$$

Proof. Since $|\xi| < 1$,

$$\left|\frac{x-\xi}{1-\bar{\xi}x}\right| = 1 \quad \text{for } |x| = 1, \tag{75}$$

as is well known. If one puts

$$g(x) = f(x)\prod_{\xi}\frac{1-\bar{\xi}x}{x-\xi}, \tag{76}$$

then

$$g(x) \text{ is regular} \quad \text{for } |x| \leqq 1, \tag{77}$$

$$g(x) \neq 0 \quad \text{for } |x| < 1, \tag{78}$$

$$|g(x)| = |f(x)| \text{ for } |x| = 1 \quad \text{by (75)}, \tag{79}$$

$$|g(x)| \leqq e^M \quad \text{for } |x| = 1, \tag{80}$$

$$|g(x)| \leqq e^M \quad \text{for } |x| \leqq 1, \tag{81}$$

$$g(0) = f(0)\prod_{\xi}\frac{1}{-\xi}, \tag{82}$$

[12]Also in what follows, roots are, of course, counted with their corresponding multiplicities.

[13]If there is no such ξ, the theorem is also true if an empty sum denotes 0 and an empty product denotes 1. The same applies also to the following lemmas.

(83) $$|g(0)| \geqq 1.$$

By (77) and (78) there exists an $h(x)$ with

(84) $$g(x) = e^{h(x)},$$

such that

(85) $$h(x) \text{ is regular} \quad \text{for } |x| < 1$$

(86) $$\Re(h(x)) \leqq M \quad \text{for } |x| < 1 \text{ by (81)},$$

(87) $$\Re(h(0)) \geqq 0 \qquad \text{by (83)}.$$

Therefore we have for $|x| < 1$

(88) $$|h'(x)| \leqq \frac{2(M - \Re(h(0)))}{(1 - |x|)^2} \leqq \frac{2M}{(1 - |x|)^2}.$$

From (76) and (84) it follows that for $|x| < 1$,

(89) $$h'(x) = \frac{g'(x)}{g(x)} = \frac{f'(x)}{f(x)} - \sum_{\xi} \left(\frac{1}{x - \xi} + \frac{\bar{\xi}}{1 - \bar{\xi}x} \right).$$

The assertion follows from (89) and (88).

Lemma 12. *Let* $a_7 = \frac{2}{31}$, $t_1 \geqq 3$, $\sigma_1 = 1 + a_7 \frac{\log\log t_1}{\log t_1}$, $r = \frac{(\log\log t_1)^3}{\log t_1}$, $1 \leqq m \leqq 4$, *for all* $s_m = \sigma_1 + imt_1$. *Let* ϱ *run over all roots of* $L(s, \chi)$ *in the interior of the disc*

(90) $$|s - s_m| \leqq r.$$

There for all s *in the interior of the disc we have, while* $t_1 \to \infty$,

(91)

$$\left| \frac{L'}{L}(s, \chi) - \sum_{\varrho} \left(\frac{1}{s - \varrho} + \frac{\overline{\varrho - s_m}}{r^2 - (\overline{\varrho - s_m})(s - s_m)} \right) \right| \leqq \frac{2 \log 2\, r^2 \log t_1 (1 + o(1))}{\log\log t_1 (r - |s - s_m|)^2};$$

and, in particular, for $s = s_m$ *while* $t_1 \to \infty$,

$$(92)\qquad \left|\frac{L'}{L}(s_m,\chi)-\sum_{\varrho}\left(\frac{1}{s_m+\varrho}+\frac{\overline{\varrho-s_m}}{r^2}\right)\right| \leqq \frac{2\log 2\log t_1(1+o(1))}{\log\log t_1}.$$

Proof. Let t_1 be immediately so large that the disc (90) is contained entirely in the region

$$(93)\qquad \sigma \geqq \frac{3}{4}, \qquad t \geqq 3$$

and such that $r < 1$. From (90) it follows that if $\sigma < 1$, then

$$(94)\qquad \sigma \geqq \sigma_1 - r > 1 - r, \qquad 1-\sigma < r,$$

$$(95)\qquad \frac{1-\sigma}{\log\frac{1}{1-\sigma}} < \frac{r}{\log\frac{1}{r}}.^{14}$$

Hence, by Lemma 10, we have for all s in (90)

$$(96)\qquad L(s,\chi) = O\left(t^{\log 2\frac{r}{\log\frac{1}{r}}}\log^2 t\right);$$

because for $\sigma \geqq 1$, (96) is trivial by (37). From (96) follows

$$(97)\qquad L(s,\chi) = O\left(e^{\log t\frac{(\log\log t_1)^3\log 2}{\log t_1(\log\log t_1 - 3\log\log\log t_1)}+2\log\log t}\right).$$

Since

$$(98)\qquad \frac{\log t}{\log t_1} = 1 + o(1),$$

$$(99)\qquad \frac{\log\log t_1}{\log\log t_1 - 3\log\log\log t_1} = 1 + o(1),$$

$$(100)\qquad 2\log\log t = o\left((\log\log t_1)^2\right),$$

[14] The function $\frac{u}{\log 1/u}$ increases monotonically for $0 < u < 1$.

it follows from (97) that

$$|L(s,\chi)| \leqq e^{\log 2(\log\log t_1)^2(1+o(1))}. \tag{101}$$

Moreover, since

$$\left|\frac{1}{L(s_m,\chi)}\right| \leqq \zeta(\sigma_1) = O\left(\frac{1}{\sigma_1 - 1}\right) = O\left(\frac{\log t_1}{\log\log t_1}\right) \leqq e^{o((\log\log t_1)^2)} \tag{102}$$

we have for all s in (90)

$$\left|\frac{L(s,\chi)}{L(s_m,\chi)}\right| \leqq e^{\log 2(\log\log t_1)^2(1+o(1))}. \tag{103}$$

We put

$$x = \frac{s - s_m}{r}, \qquad \xi = \frac{\varrho - s_m}{r}, \tag{104}$$

$$f(x) = f\left(\frac{s - s_m}{r}\right) = \frac{L(s,\chi)}{L(s_m,\chi)}, \tag{105}$$

so that ξ runs over all roots of $f(x)$ in the disc $|x| < 1$.
Since $f(0) = 1$, Lemma 11 is applicable to $f(x)$ with

$$M = \log 2(\log\log t_1)^2(1 + o(1)), \tag{106}$$

and from (74), taking (104), (105), (106), and

$$f'(x) = r\frac{L'(s,\chi)}{L(s_m,\chi)} \tag{107}$$

into consideration, the inequality

$$\left| r\frac{L'(s,\chi)}{L(s,\chi)} - \sum_{\varrho}\left(\frac{r}{s-\varrho} + \frac{\dfrac{\overline{\varrho - s_m}}{r}}{1 - \dfrac{\overline{\varrho - s_m}}{r}\cdot\dfrac{s - s_m}{r}}\right)\right| \tag{108}$$

$$\leqq 2\log 2(\log\log t_1)^2(1+o(1))\frac{1}{\left(1-\frac{|s-s_m|}{r}\right)^2}$$

$$= 2\log 2r\frac{\log t_1}{\log\log t_1}(1+o(1))\frac{r^2}{(r-|s-s_m|)^2}$$

follows. If one divides (108) by r, then (91) follows.

Lemma 13. *Under the hypothesis of Lemma* 12 *we have*

$$\Re\left(\frac{1}{s_m-\varrho}+\frac{\overline{\varrho-s_m}}{r^2}\right)\geqq 0 \quad \textit{for each } \varrho,\tag{109}$$ [15]

$$-\Re\left(\frac{L'}{L}(s_m,\chi)\right)\leqq 2\log 2\frac{\log t_1}{\log\log t_1}(1+o(1)),\tag{110}$$

(111)

$$-\Re\left(\frac{L'}{L}(s_m,\chi)\right)\leqq 2\log 2\frac{\log t_1}{\log\log t_1}(1+o(1))-\Re\left(\frac{1}{s_m-\varrho}\right)\quad \textit{for each } \varrho.$$

Proof. It is well known that $L(s,\chi)\neq 0$ for $\sigma>1$, and therefore

$$\Re(s_m-\varrho)>0,\qquad \Re\left(\frac{1}{s_m-\varrho}\right)>0.\tag{112}$$

We have

$$\frac{1}{s_m-\varrho}+\frac{\overline{\varrho-s_m}}{r^2}=\frac{1}{s_m-\varrho}\left(1-\frac{|\varrho-s_m|^2}{r^2}\right).\tag{113}$$

From (112), (113), and

$$|\varrho-s_m|<r,\tag{114}$$

(109) follows. From (92) follows

[15] If no ϱ with $|\varrho-s_m|<r$ exists, then (109) and (111) are vacuous.

(115)

$$-\Re\left(\frac{L'}{L}(s_m,\chi)\right)+\Re\left(\sum_{\varrho}\left(\frac{1}{s_m-\varrho}+\frac{\overline{\varrho-s_m}}{r^2}\right)\right)\leqq 2\log 2\frac{\log t_1}{\log\log t_1}(1+o(1)).$$

From (115) and (109), applied to all ϱ, follows (110).

From (115) and (109), applied to all ϱ with one exception, it follows for this exceptional ϱ that

(116)

$$-\Re\left(\frac{L'}{L}(s_m,\chi)\right)\leqq 2\log 2\frac{\log t_1}{\log\log t_1}(1+o(1))-\Re\left(\frac{1}{s_m-\varrho}\right)-\Re\left(\frac{\overline{\varrho-s_m}}{r^2}\right).$$

By (114) we have

$$-\Re\left(\frac{\overline{\varrho-s_m}}{r^2}\right)\leqq\frac{1}{r}=o\left(\frac{\log t_1}{\log\log t_1}\right). \tag{117}$$

From (116) and (117) we obtain (111).

Lemma 14. *For real φ we have*

$$P(\varphi)=47+80\cos\varphi+49\cos(2\varphi)+20\cos(3\varphi)+4\cos(4\varphi)\geqq 0. \tag{118}$$

Proof. Put $\cos\varphi=\eta$; then

$$\cos(2\varphi)=2\eta^2-1,\ \cos(3\varphi)=4\eta^3-3\eta,\ \cos(4\varphi)=8\eta^4-8\eta^2+1, \tag{119}$$

$$\begin{aligned}P(\varphi)&=47+80\eta+98\eta^2-49+80\eta^3-60\eta+32\eta^4-32\eta^2+4\\&=2+20\eta+66\eta^2+80\eta^3+32\eta^4\\&=2(1+\eta)(1+9\eta+24\eta^2+16\eta^3)\\&=2(1+\eta)^2(1+8\eta+16\eta^2)=2(1+\eta)^2(1+4\eta)^2\geqq 0.\end{aligned} \tag{120}$$

Lemma 15. *For $\sigma>1$ and real t,*

$$\begin{aligned}-\Re\Big\{&47\frac{L'}{L}(\sigma,\chi^0)+80\frac{L'}{L}(\sigma+it,\chi)+49\frac{L'}{L}(\sigma+2it,\chi^2)\\&+20\frac{L'}{L}(\sigma+3it,\chi^3)+4\frac{L'}{L}(\sigma+4it,\chi^4)\Big\}\geqq 0.\end{aligned} \tag{121}$$

Proof. For integral $m \geqq 0$,

$$- \frac{L'}{L}(\sigma + imt, \chi^m) = \sum_{\substack{n=1 \\ (n,k)=1}}^{\infty} \Lambda(n) n^{-\sigma} (\chi(n) n^{-it})^m. \tag{122}$$

If for the n appearing in (122) one chooses the $\varphi_n \geqq 0$ such that

$$\cos \varphi_n = \Re(\chi(n) n^{-it}), \tag{123}$$

then

$$\cos(m\varphi_n) = \Re((\chi(n) n^{-it})^m). \tag{124}$$

Therefore, the left-hand side of the assertion is equal to

$$\sum_{\substack{n=1 \\ (n,k)=1}}^{\infty} \Lambda(n) n^{-\sigma} P(\varphi_n) \geqq 0. \tag{125}$$

Lemma 16. *For* $3 \leqq b_2 \leqq t$, $a_5 = \dfrac{1}{49.13}$, $\sigma \geqq 1 - a_5 \dfrac{\log\log t}{\log t}$ *we have*

$$L(s, \chi) \neq 0. \tag{126}$$

Proof (Indirect). Suppose there exist infinitely many roots $\varrho = \beta + i\gamma$, $\beta = \Re(\varrho)$, of $L(s, \chi)$ with

$$\gamma \geqq 3, \quad 1 - a_5 \frac{\log\log\gamma}{\log\gamma} \leqq \beta \leqq 1. \tag{127}$$

Then I put

$$\sigma_1 = 1 + a_7 \frac{\log\log\gamma}{\log\gamma} \tag{128}$$

and apply Lemma 13 to $L(s, \chi^m)$ for $1 \leqq m \leqq 4$ with

$$t_1 = \gamma, \quad s_m = \sigma_1 + im\gamma. \tag{129}$$

Since $0 < s_1 - \varrho = O\left(\dfrac{\log\log\gamma}{\log\gamma}\right) = o\left(\dfrac{(\log\log\gamma)^3}{\log\gamma}\right)$, we therefore have for $\gamma \to \infty$,

$$-\Re\left(\frac{L'}{L}(\sigma_1 + im\gamma, \chi^m)\right) \leqq 2\log 2\frac{\log\gamma}{\log\log\gamma}(1 + o(1)) \tag{130}$$

for $m = 4, 3, 2$ by (110),

$$-\Re\left(\frac{L'}{L}(\sigma_1 + i\gamma, \chi)\right) \leqq 2\log 2\frac{\log\gamma}{\log\log\gamma}(1 + o(1)) - \frac{1}{\sigma_1 - \beta} \quad \text{by (111)}, \tag{131}$$

$$-\Re\left(\frac{L'}{L}(\sigma_1, \chi^0)\right) = \frac{1}{\sigma_1 - 1}(1 + o(1)) = \frac{1}{a_7}\frac{\log\gamma}{\log\log\gamma}(1 + o(1)); \tag{132}$$

because $-\frac{L'}{L}(s, \chi^0)$ has at $s = 1$ a simple pole with residue 1. If one multiplies the five inequalities [given by (130) for $m = 4, 3, 2$, (131), and (132)] by $4, 20, 49, 80, 47$ and adds them, then it follows from Lemma 15 that

$$0 \leqq \frac{\log\gamma}{\log\log\gamma}(1 + o(1)) \times \left\{2\log 2(4 + 20 + 49 + 80) + \frac{47}{a_7}\right\} - \frac{80}{\sigma_1 - \beta}. \tag{133}$$

From (127) and (128) it follows that

$$-\frac{80}{\sigma_1 - \beta} \leqq \frac{80}{a_5 + a_7}\frac{\log\gamma}{\log\log\gamma}. \tag{134}$$

From (133) and (134) it follows that

$$0 \leqq \frac{\log\gamma}{\log\log\gamma}(1 + o(1))\left(306\log 2 + \frac{47}{a_7} - \frac{80}{a_5 + a_7}\right). \tag{135}$$

If one divides (135) by $\frac{\log\gamma}{\log\log\gamma}$ and lets γ tend to ∞ [this is possible, since we have in fact assumed that there exist infinitely many γ which satisfy (127) and hence (135)], then one obtains

$$0 \leqq 306 \log 2 + \frac{47}{a_7} - \frac{80}{a_5 + a_7} \tag{136}$$

$$= 306 \log 2 + \frac{47 \cdot 31}{2} - \frac{80}{\frac{1}{49.13} + \frac{2}{31}}$$

$$< 306 \cdot 0.694 + 728.5 - \frac{80 \cdot 49.13 \cdot 31}{129.26}$$

$$= 940.864 - \frac{121\,842.4}{129.26} < 941 - \frac{121\,842.4}{129.3}$$

$$= \frac{121\,671.3 - 121\,842.4}{129.3} < 0.$$

This contradiction proves our lemma.

Lemma 17. *Let*

$$a_8 = \frac{1}{49.15} (< a_5), \quad t \geqq b_3 \geqq b_2 \geqq 3, \quad \sigma \geqq 1 - a_8 \frac{\log \log t}{\log t}.$$

Then

$$\frac{L'}{L}(s, \chi) = O\left(\frac{\log t}{\log \log t}\right). \tag{137}$$

Proof. Let

$$\sigma_1 = 1 + a_7 \frac{\log \log t}{\log t}, \quad s_1 = \sigma_1 + it. \tag{138}$$

If $\sigma \geqq \sigma_1$, then (137) is trivial because

$$\left| \frac{L'}{L}(s, \chi) \right| \leqq -\frac{\zeta'}{\zeta}(\sigma) \leqq -\frac{\zeta'}{\zeta}(\sigma_1) \tag{139}$$

$$= \frac{1}{\sigma_1 - 1} + O(1) = O\left(\frac{\log t}{\log \log t}\right).$$

Therefore, henceforth let

$$\sigma < \sigma_1. \tag{140}$$

We apply Lemma 12 with $t_1 = t$, $m = 1$.
Then from (92) and (139) it follows that

$$\sum_{\varrho}\left(\frac{1}{s_1 - \varrho} + \frac{\overline{\varrho - s_1}}{r^2}\right) = O\left(\frac{\log t}{\log\log t}\right). \tag{141}$$

Now we choose $b_3 \geqq b_2$ so large that for $t \geqq b_3$ we have, first of all,

$$0 < s_1 - s \leqq \frac{1}{2}r. \tag{142}$$

This is possible because

$$s_1 - s = O\left(\frac{\log\log t}{\log t}\right) = o(r) \quad \text{for } t \to \infty. \tag{143}$$

Second, choose b_3 so large that with $t \geqq b_3$ we have, for all ϱ appearing in (141), that

$$\beta = \Re(\varrho) \leqq 1 - \frac{a_5 + a_8}{2}\frac{\log\log t}{\log t}. \tag{144}$$

This is possible by Lemma 16, since $\dfrac{a_5 + a_8}{2} < a_5$.[16] Then we have by hypothesis and (144) that

$$\begin{aligned} |s - \varrho| &\geqq \Re(s - \varrho) \\ &= \sigma - \beta \geqq \left(1 - a_8\frac{\log\log t}{\log t}\right) - \left(1 - \frac{a_5 + a_8}{2}\frac{\log\log t}{\log t}\right) \\ &= \frac{a_5 - a_8}{2}\frac{\log\log t}{\log t}. \end{aligned} \tag{145}$$

Moreover,

[16] Because $|\gamma - t| < r = O(1)$, one has for sufficiently large b_3 and $t \geqq b_3$ that

$$\beta \leqq 1 - a_5\frac{\log\log\gamma}{\log\gamma} = 1 - a_5\frac{\log\log(t + o(1))}{\log(t + o(1))} \leqq 1 - \frac{a_5 + a_8}{2}\frac{\log\log t}{\log t}.$$

$$(146) \qquad 0 < \sigma_1 - \sigma = s_1 - s \leqq (a_7 + a_8)\frac{\log\log t}{\log t}.$$

From (145) and (146) follows

$$(147) \qquad s_1 - s \leqq 2\frac{a_7 + a_8}{a_5 - a_8}|s - \varrho|,$$

$$(148) \qquad |s_1 - \varrho| \leqq (s_1 - s) + |s - \varrho|$$

$$\leqq \left(1 + 2\frac{a_7 + a_8}{a_5 - a_8}\right)|s - \varrho| = a_9|s - \varrho|.$$

Moreover, it follows from

$$(149) \qquad \beta < \sigma < \sigma_1$$

that

$$(150) \qquad s_1 - s < \Re(s_1 - \varrho),$$

and also that

$$(151) \qquad |s - \varrho| < |s_1 - \varrho| < r.$$

From (142) and (151) we have

$$(152) \qquad \left|\frac{r^2 - (s - \varrho)(\overline{\varrho - s_1})}{r^2\left(r^2 - (s - s_1)(\overline{\varrho - s_1})\right)}\right| \leqq \frac{r^2 + r \cdot r}{r^2\left(r^2 - \frac{1}{2}r \cdot r\right)} = \frac{4}{r^2}.$$

We have

$$(153) \qquad \frac{1}{s - \varrho} + \frac{\overline{\varrho - s_1}}{r^2 - (s - s_1)(\overline{\varrho - s_1})} = \frac{r^2 - |\varrho - s_1|^2}{(s - \varrho)\left(r^2 - (s - s_1)(\overline{\varrho - s_1})\right)}$$

$$(154) \qquad \frac{1}{s_1 - \varrho} + \frac{\overline{\varrho - s_1}}{r^2} = \frac{r^2 - |\varrho - s_1|^2}{(s_1 - \varrho)r^2},$$

$$(155) \qquad \Re\left(\frac{1}{s_1 - \varrho} + \frac{\overline{\varrho - s_1}}{r^2}\right) = \frac{\Re(s_1 - \varrho)\left(r^2 - |\varrho - s_1|^2\right)}{|s_1 - \varrho|^2 r^2}.$$

Subtracting (154) from (153), one obtains for the absolute value of the difference

(156)

$$\left|\frac{1}{s-\varrho}+\frac{\overline{\varrho-s_1}}{r^2-(s-s_1)(\overline{\varrho-s_1})}-\frac{1}{s_1-\varrho}-\frac{\overline{\varrho-s_1}}{r^2}\right|$$

$$=\left(r^2-|\varrho-s_1|^2\right)\left|\frac{(s_1-\varrho)r^2-(s-\varrho)r^2+(s-\varrho)(s-s_1)(\overline{\varrho-s_1})}{(s-\varrho)\left(r^2-(s-s_1)(\overline{\varrho-s_1})\right)(s_1-\varrho)r^2}\right|$$

$$=(s_1-s)\frac{r^2-|\varrho-s_1|^2}{|s-\varrho|\,|s_1-\varrho|}\left|\frac{r^2-(s-\varrho)(\overline{\varrho-s_1})}{r^2\left(r^2-(s-s_1)(\overline{\varrho-s_1})\right)}\right|$$

$$\leqq (s_1-s)\frac{r^2-|\varrho-s_1|^2}{|s-\varrho|\,|s_1-\varrho|}\frac{4}{r^2} \qquad \text{(by (152))}$$

$$\leqq \Re(s_1-\varrho)a_9\frac{r^2-|\varrho-s_1|^2}{|s_1-\varrho|^2}\frac{4}{r^2} \qquad \text{(by (150) and (148))}$$

$$=4a_9\Re\left(\frac{1}{s_1-\varrho}+\frac{\overline{\varrho-s_1}}{r^2}\right) \qquad \text{(by (155)).}$$

From (156) follows

$$\text{(157)}\quad \left|\sum_{\varrho}\left(\frac{1}{s-\varrho}+\frac{\overline{\varrho-s_1}}{r^2-(s-s_1)(\overline{\varrho-s_1})}\right)-\sum_{\varrho}\left(\frac{1}{s_1-\varrho}+\frac{\overline{\varrho-s_1}}{r^2}\right)\right|$$

$$\leqq 4a_9\Re\left(\sum_{\varrho}\left(\frac{1}{s_1-\varrho}+\frac{\overline{\varrho-s_1}}{r^2}\right)\right)\leqq 4a_9\left|\sum_{\varrho}\left(\frac{1}{s_1-\varrho}+\frac{\overline{\varrho-s_1}}{r^2}\right)\right|.$$

From (141) and (157) follows

$$\text{(158)}\qquad \sum_{\varrho}\left(\frac{1}{s-\varrho}+\frac{\overline{\varrho-s_1}}{r^2-(s-s_1)(\overline{\varrho-s_1})}\right)=O\left(\frac{\log t}{\log\log t}\right).$$

The assertion follows from (91), (142), and (158).

Lemma 18. *Under the hypotheses of Lemma* 17 (*i.e., also with the same* b_3), *one has*

$$\frac{1}{L(s,\chi)} = O\left(\frac{\log t}{\log\log t}\right). \tag{159}$$

Proof. *First Case.* Let $\sigma \geqq \sigma_0 + 1 + \dfrac{\log\log t}{\log t}$. Then (159) is trivial because

$$\left|\frac{1}{L(s,\chi)}\right| \leqq \zeta(\sigma) \leqq \zeta(\sigma_0) = \frac{1}{\sigma_0 - 1} + O(1) = O\left(\frac{\log t}{\log\log t}\right). \tag{160}$$

Second Case. Let $\sigma < \sigma_0$. Then

$$\begin{aligned} \log|L(s,\chi)| &= \Re \log L(\sigma_0 + it, \chi) + \Re \int_{\sigma_0+it}^{s} \frac{L'}{L}(u,\chi)\,du \\ &\geqq -\log\zeta(\sigma_0) + \int_{\sigma_0}^{\sigma} O\left(\frac{\log t}{\log\log t}\right) du \quad \text{(by Lemma 17)} \\ &= \log(\sigma_0 - 1) + O(1) = -\log\log t + \log\log\log t + O(1), \end{aligned} \tag{161}$$

$$\left|\frac{1}{L(s,\chi)}\right| = e^{-\log|L(s,\chi)|} \leqq e^{\log\log t - \log\log\log t + O(1)} = O\left(\frac{\log t}{\log\log t}\right). \tag{162}$$

§3. Lemmas on divisor numbers

In Lemmas 19–21 of this section, lowercase Latin letters denote natural numbers.

Lemma 19. *For* $z \geqq 2$, $2 \leqq r \leqq 4$

$$\sum_{n \leqq z} \frac{d_r(n)}{n} = O(\log^r z). \tag{163}$$

Proof.

$$\sum_{n \leqq z} \frac{d_r(n)}{n} = \sum_{n \leqq z} \sum_{u_1 \ldots u_r = n} \frac{1}{u_1 \ldots u_r} \tag{164}$$

$$\leqq \sum_{\substack{u_1 \leqq z \\ u_r \leqq z}} \frac{1}{u_1 \ldots u_r} = \left(\sum_{u \leqq z} \frac{1}{u} \right)^r = O(\log^r z).$$

Lemma 20. *For $z \geqq 2$,*

$$\sum_{n \leqq z} \frac{d^2(n)}{n} = O(\log^4 z). \tag{165}$$

Proof. From $d(vw) \leqq d(v) \cdot d(w)$ follows

$$d^2(n) = \sum_{u|n} d(n) \leqq \sum_{u|n} d(u) d\left(\frac{n}{u}\right) = d_4(n), \tag{166}$$

and Lemma 19 with $r = 4$ yields the assertion.

Lemma 21. *For $z \geqq 2$*

$$S = \sum_{n<m \leqq z} \frac{d(n)d(m)}{nm \log \dfrac{m}{n}} = O(\log^4 z). \tag{167}$$

Proof. For $m > n$,

$$\log \frac{m}{n} = -\log\left(1 - \frac{m-n}{m}\right) > \frac{m-n}{m}, \tag{168}$$

$$\begin{aligned} S &< \sum_{n<m \leqq z} \frac{d(n)d(m)}{n(m-n)} \\ &= \sum_{n<m \leqq z} \frac{d(n)d(m)}{m(m-n)} + \sum_{n<m \leqq z} \frac{d(n)d(m)}{nm} \\ &\leqq \sum_{n<m \leqq z} \frac{d(n)d(m)}{m(m-n)} + \left(\sum_{n \leqq z} \frac{d(n)}{n} \right)^2 \\ &= \sum_{n<m \leqq z} \frac{d(n)d(m)}{m(m-n)} + O(\log^4 z) \end{aligned} \tag{169}$$

by Lemma 19 with $r = 2$. If one puts

$$m = ce, \qquad n = fg, \qquad m - n = r, \tag{170}$$

then it follows from (169) that

$$S \leqq O(\log^4 z) + \sum_{r \leqq z} \frac{1}{r} \sum_{\substack{c,e,f,g \\ ce \leqq z \\ ce - fg = r}} \frac{1}{ce}. \tag{171}$$

In this sum c and e appear symmetrically; similarly for f and g. Therefore, the sum is not diminished if we multiply it by 4 and add the inequalities

$$c \leqq e, \qquad f \leqq g \tag{172}$$

to the summation conditions. Because

$$e^2 \geqq ce = r + fg > fg \geqq f^2, \tag{173}$$

the summation condition

$$e > f \tag{174}$$

follows from (172).

From (170) follows

$$ce \equiv r \pmod{f}. \tag{175}$$

We can thus replace the inequality (171) by

$$S \leqq O(\log^4 z) + 4 \sum_{r \leqq z} \frac{1}{r} \sum_{c \leqq z} \frac{1}{c} \sum_{f \leqq z} \sum_{\substack{e \\ f < e \leqq z \\ ce \equiv r \pmod{f}}} \frac{1}{e}. \tag{176}$$

If we put $h = (c, f)$, then for fixed c, r, f, (175) can have a solution only if

$$h \mid r. \tag{177}$$

Therefore, suppose (177) holds and let e_0 be the smallest solution of (175) which satisfies the condition (174). Then all other solutions of (175) with $f < e \leqq z$ are contained in the numbers

$$e = e_0 + \frac{f}{h} u, \qquad u \leqq z. \tag{178}$$

From (176), taking (177) and (178) into account, it follows that

$$(179)\quad S \leqq O(\log^4 z) + 4 \sum_{c \leqq z} \frac{1}{c} \sum_{f \leqq z} \sum_{\substack{r \leqq z \\ h|r}} \frac{1}{r} \left(\frac{1}{e_0} + \sum_{u \leqq z} \frac{1}{e_0 + \frac{f}{h} u} \right)$$

$$\leqq O(\log^4 z) + 4 \sum_{c \leqq z} \frac{1}{c} \sum_{f \leqq z} \sum_{\substack{r \leqq z \\ h|r}} \frac{1}{r e_0} + 4 \sum_{c \leqq z} \sum_{f \leqq z} \sum_{\substack{r \leqq z \\ h|r}} \sum_{u \leqq z} \frac{h}{crfu}.$$

Since $e_0 > f$, we have

$$(180)\qquad \sum_{c \leqq z} \frac{1}{c} \sum_{f \leqq z} \sum_{\substack{r \leqq z \\ h|r}} \frac{1}{r e_0} \leqq \sum_{c \leqq z} \frac{1}{c} \sum_{f \leqq z} \frac{1}{f} \sum_{r \leqq z} \frac{1}{r} = O(\log^3 z).$$

In order to estimate the second sum in (179) we put

$$(181)\qquad c = hc', \qquad f = hf', \qquad r = hr'.$$

Then

$$(182)\qquad \sum_{c \leqq z} \sum_{f \leqq z} \sum_{\substack{r \leqq z \\ h|r}} \sum_{u \leqq z} \frac{h}{crfu} \leqq \sum_{h \leqq z} \sum_{c' \leqq z} \sum_{f' \leqq z} \sum_{r' \leqq z} \sum_{u \leqq z} \frac{h}{hc'hr'hf'u}$$

$$= \sum_{h \leqq z} \frac{1}{h^2} \left(\sum_{v \leqq z} \frac{1}{v} \right)^4 = O(\log^4 z).$$

From (175), (180), and (182) the assertion follows.

Notation. For the rest of this article, for $0 \leqq t \leqq T$, let

$$(183)\quad A(s) = A(s; T, \chi) = 1 - \sum_{1 \leqq n \leqq kT} \chi(n) n^{-s} \sum_{1 \leqq n \leqq kT} \chi(n) \mu(n) n^{-s}.$$

Lemma 22. *For $T \geqq 3$, $\frac{1}{2} \leqq \sigma \leqq 1$ we have*

$$(184)\qquad \int_0^T |A(s)|^2 \, dt = O(T^{4(1-\sigma)} \log^4 T).$$

Proof. It is well known that for integral $v > 1$ we have

(185)
$$\sum_{u|v} \chi\left(\frac{v}{u}\right)\chi(u)\mu(u) = \chi(v)\sum_{u|v}\mu(u) = 0,$$

and hence by (183) with suitable $e(n)$,

(186)
$$A(s) = \sum_{kT<n\leqq(kT)^2} e(n)n^{-s},$$

where

$$e(n) = e(n; T, \chi), \qquad |e(n)| \leqq d(n).$$

Therefore

(187)

$$\int_0^T |A(s)|^2\, dt = \int_0^T \left(\sum_{kT<m\leqq(kT)^2} e(m)m^{-\sigma-it} \sum_{kT<n\leqq(kT)^2} \overline{e(n)}\, n^{-\sigma+it} \right) dt$$

$$= T \sum_{kT<m\leqq(kT)^2} |e(m)|^2 m^{-2\sigma} + \sum_{\substack{kT<m\leqq(kT)^2 \\ kT<n\leqq(kT)^2 \\ m\neq n}} e(m)\overline{e(n)}\,(mn)^{-\sigma} \int_0^T \left(\frac{n}{m}\right)^{it} dt$$

$$\leqq T \sum_{kT<m\leqq(kT)^2} \frac{d^2(m)}{m} m^{1-2\sigma} + \sum_{\substack{kT<m\leqq(kT)^2 \\ kT<n\leqq(kT)^2 \\ m\neq n}} \frac{d(m)d(n)}{mn}(mn)^{1-\sigma} \frac{2}{\left|\log\dfrac{m}{n}\right|}$$

$$\leqq T(kT)^{1-2\sigma} \sum_{kT<m\leqq(kT)^2} \frac{d^2(m)}{m} + 4(kT)^{4(1-\sigma)} \sum_{kT<n<m\leqq(kT)^2} \frac{d(m)d(n)}{mn\log\dfrac{m}{n}}$$

$$\leqq k^{1-2\sigma}T^{2(1-\sigma)} \sum_{1\leqq m\leqq(kT)^2} \frac{d^2(m)}{m} + 4k^{4(1-\sigma)}T^{4(1-\sigma)} \sum_{1\leqq n<m\leqq(kT)^2} \frac{d(m)d(n)}{mn\log\dfrac{m}{n}}$$

$$= O\left(T^{4(1-\sigma)}\log^4 T\right)$$

by Lemmas 20 and 21.

§4. Asymptotic expansions of $\dfrac{1}{L(s,x)}$ and $\dfrac{-L'(s,x)}{L(s,x)}$

In this section always assume that

$$(3 \leqq b_2 \leqq) b_3 \leqq b_4 \leqq (\log T)^{a_8+2} \leqq t \leqq T, \tag{188}$$

$$\sigma = 1 - a_8 \frac{\log\log T}{\log T}. \tag{189}$$

Let $b_4 \geqq b_3$ be chosen so large that for $b_4 \leqq \tau_1 \leqq \tau_2$

$$\frac{\log\log \tau_1}{\log \tau_1} \geqq \frac{\log\log \tau_2}{\log \tau_2} \tag{190}$$

and that (188) and (189) imply the following two inequalities:

$$\frac{4}{5} \leqq \sigma < 1, \tag{191}$$

$$\left|\Sigma \chi(n) n^{-s}\right| \geqq \frac{1}{\log T}, \tag{192}$$

where, as in the rest of this paragraph, Σ is the abbreviation of $\sum\limits_{1 \leqq n \leqq kT}$. That (190) and (191) can be attained by choosing a sufficiently large b_4 is immediately clear. From (188), (189), (190), and (191) it follows that Lemmas 4, 5, 17, and 18 are applicable for our s; this will be used repeatedly in this section.

By Lemmas 4 and 18,

(193)

$$\left|\Sigma \chi(n) n^{-s}\right| \geqq |L(s,\chi)| + O\left(\frac{T^{1-\sigma}}{t}\right) \geqq b_5 \frac{\log\log t}{\log t} + O\left(\frac{e^{\log T a_8 \frac{\log\log T}{\log T}}}{(\log T)^{a_8+2}}\right)$$

$$\geqq b_5 \frac{\log\log T}{\log T} + O\left(\frac{1}{\log^2 T}\right) \geqq \frac{1}{\log T}$$

with suitably large b_4. Herewith (192) is proved.

Lemma 23.

$$\frac{1}{L(s,\chi)} = \sum_{1 \leqq n \leqq (kT)^3} c(n) n^{-s} + O\left(\log^2 T \frac{T^{1-\sigma}}{t}\right) + O(\log T) |A(s)|^2, \tag{194}$$

where

$$c(n) = c(n; T, \chi), \qquad |c(n)| \leqq 3 d_3(n). \tag{195}$$

Proof.

$$\frac{1}{L(s,\chi)} = \frac{1}{\Sigma \chi(n) n^{-s}} + \frac{\Sigma \chi(n) n^{-s} - L(s,\chi)}{L(s,\chi) \Sigma \chi(n) n^{-s}}, \tag{196}$$

therefore, by Lemmas 4 and 18 and (192),

$$\begin{aligned} \frac{1}{L(s,\chi)} &= \frac{1}{\Sigma \chi(n) n^{-s}} + O\left(\log^2 T \frac{T^{1-\sigma}}{t}\right) \\ &= \frac{A^2(s)}{\Sigma \chi(n) n^{-s}} + 2 \sum \chi(n) \mu(n) n^{-s} \\ &\quad - \sum \chi(n) n^{-s} \left(\sum \chi(n) \mu(n) n^{-s} \right)^2 \\ &\quad + O\left(\log^2 T \frac{T^{1-\sigma}}{t}\right), \end{aligned} \tag{197}$$

whence by (192) the assertion follows, if one puts

$$c(n) = - \sum_{\substack{uvw = n \\ 1 \leqq u \leqq kT \\ 1 \leqq v \leqq kT \\ 1 \leqq w \leqq kT}} \chi(u) \chi(v) \mu(v) \chi(w) \mu(w) + \begin{cases} 2\chi(n)\mu(n) & \text{for } 1 \leqq n \leqq kT, \\ 0 & \text{for } kT < n \leqq (kT)^3. \end{cases} \tag{198}$$

Lemma 24.

$$-\frac{L'}{L}(s,\chi) = \sum_{1 \leqq n \leqq (kT)^4} g(n) n^{-s} + O\left(\log^2 T \frac{T^{1-\sigma}}{t}\right) + O(\log T) |A(s)|^2, \tag{199}$$

where

$$g(n) = g(n; T, \chi), \qquad |g(n)| \leqq 3(\log(kT))\, d_4(n). \tag{200}$$

Proof.

$$\frac{-L'}{L}(s,\chi) = \frac{\Sigma\chi(n)\log n\, n^{-s}}{\Sigma\chi(n)n^{-s}} + \frac{L'}{L}(s,\chi)\frac{L(s,\chi) - \Sigma\chi(n)n^{-s}}{\Sigma\chi(n)n^{-s}} - \frac{L'(s,\chi) + \Sigma\chi(n)\log n\, n^{-s}}{\Sigma\chi(n)n^{-s}} = \frac{\Sigma\chi(n)\log n\, n^{-s}}{\Sigma\chi(n)n^{-s}} + O\left(\log^2 T \frac{T^{1-\sigma}}{t}\right) \tag{201}$$

by (192) and Lemmas 4, 5, and 17. Since by the calculation carried out in (193)

$$\log^2 T \frac{T^{1-\sigma}}{t} \leqq 1, \tag{202}$$

it follows from (201) and Lemma 17 that

$$\frac{\Sigma\chi(n)\log n\, n^{-s}}{\Sigma\chi(n)n^{-s}} = O(\log T). \tag{203}$$

Moreover, it follows from (201) that

$$-\frac{L'}{L}(s,\chi) = A^2(s)\frac{\Sigma\chi(n)\log n\, n^{-s}}{\Sigma\chi(n)n^{-s}} + 2\sum\chi(n)\log n\, n^{-s}\sum\chi(n)\mu(n)n^{-s} - \sum\chi(n)\log n\, n^{-s}\sum\chi(n)n^{-s}\left(\sum\chi(n)\mu(n)n^{-s}\right)^2 \tag{204}$$

$$+O\left(\log^2 T \frac{T^{1-\sigma}}{t}\right).$$

The assertion follows from (204) and (203), if one puts

(205)

$$g(n) = 2 \sum_{\substack{uv=n \\ u \leqq kT \\ v \leqq kT}} \chi(u) \log u \chi(v) \mu(v) - \sum_{\substack{uvwq=n \\ u \leqq kT \\ v \leqq kT \\ w \leqq kT \\ q \leqq kT}} \chi(u) \log u \chi(v) \chi(w) \mu(w) \chi(q) \mu(q).$$

§5. Proof of the main theorems

We put

(206) $$a_{10} = 249.8.$$

Then

(207) $$a_8(4 - a_{10}) + 5 = \frac{1}{49.15}(-245.8) + 5 = \frac{-0.05}{49.15} < 0.$$

Since it is well known that $L(1 + it) \neq 0$, there exists a $b_6 < 1$ such that

(208) $$L(s, \chi) \neq 0 \quad \text{for } 0 \leqq t \leqq b_4, \qquad \sigma \geqq b_6.$$

Let $x > 0$ be the variable appearing in Theorems 3 and 4. We choose $T > 0$ such that

(209) $$T^{a_{10}} = x$$

and put

(210) $$\sigma_0 = 1 - a_8 \frac{\log \log T}{\log T}.$$

Since T tends to ∞ with x, we can immediately assume that x is so large that, first,

(211) $$b_4 \leqq (\log T)^{a_8 + 2} \leqq T,$$

second,

$$\sigma_0 \geqq b_6, \tag{212}$$

third,

$$T \leqq \frac{x}{(kT)^4}. \tag{213}$$

Then, because of (211), §4 is applicable with

$$\sigma = \sigma_0, \qquad (\log T)^{a_8+2} \leqq t \leqq T. \tag{214}$$

Therefore, by (190) we have, in particular for $b_4 \leqq t \leqq T$,

$$a_5 \frac{\log\log t}{\log t} > a_8 \frac{\log\log t}{\log t} \geqq a_8 \frac{\log\log T}{\log T} = 1 - \sigma_0. \tag{215}$$

From (215), (190), and Lemma 16 on the one hand, and (208) and (212) on the other hand, it follows that

$$L(s, \chi) \neq 0 \quad \text{for } \sigma \geqq \sigma_0, \qquad 0 \leqq t \leqq T. \tag{216}$$

Under these assumptions the following lemmas are valid for $y \geqq x$ while $x \to \infty$:

Lemma 25. $\displaystyle\int_{\sigma_0}^{\sigma_0+iT} y^{s-1} \frac{-L'}{L}(s, \chi)\, ds = o(1).$

Lemma 26. $\displaystyle\int_{\sigma_0}^{\sigma_0+iT} y^{s-1} \frac{1}{L(s, \chi)}\, ds = o(1).$

Proof.

$$T^{1-\sigma_0} = T^{a_8 \frac{\log\log T}{\log T}} = (\log T)^{a_8}, \tag{217}$$

$$y^{\sigma_0-1} \leqq x^{\sigma_0-1} = T^{a_{10}(\sigma_0-1)} = (\log T)^{-a_8 a_{10}}, \tag{218}$$

hence by (207)

$$y^{\sigma_0-1}(T^{1-\sigma_0})^4 \log^5 T \leqq (\log T)^{-a_8 a_{10}+4a_8+5} = o(1). \tag{219}$$

We put, as an abbreviation,

$$v = (\log T)^{a_8+2}. \tag{220}$$

First Step $\left(\int_{\sigma_0}^{\sigma_0+iv}\right)$. For $\sigma = \sigma_0$, $0 \leqq t \leqq v$,

$$-\frac{L'}{L}(s, \chi) = O(\log T), \tag{221}$$

$$\frac{1}{L(s, \chi)} = O(\log T). \tag{222}$$

For, if $t \geqq b_4$, then (221) [respectively, (222)] follows from Lemma 17 (respectively, Lemma 18) and (215). If $0 \leqq t \leqq b_4$, then we even have

$$-\frac{L'}{L}(s, \chi) = O(1) \quad \text{for } \chi \neq \chi_0 \ (\chi_0 \text{ the principal character}), \tag{223}$$

$$\frac{1}{L(s, \chi)} = O(1) \tag{224}$$

because these functions are regular for $\sigma = 1$. Moreover,

$$-\frac{L'}{L}(s, \chi_0) = O\left(\frac{1}{1-\sigma_0}\right) = O(\log T) \tag{225}$$

since this function has on the line $\sigma = 1$ a pole of the first order at $s = 1$ as its only singularity. This proves (221) and (222). Therefore it follows that

(226)

$$\left|\int_{\sigma_0}^{\sigma_0+iv} y^{s-1} \frac{-L'(s, \chi)}{L(s, \chi)}\, ds\right| \leqq v y^{\sigma_0-1} O(\log T)$$

$$= O\left((\log T)^{a_8+2-a_8a_{10}+1}\right) \quad [\text{by (218) and (220)}]$$

$$= o(1) \quad \text{by (207)};$$

and, similarly, that

$$\left|\int_{\sigma_0}^{\sigma_0+iv} y^{s-1} \frac{1}{L(s, \chi)}\, ds\right| = o(1). \tag{227}$$

Second Step (Treatment of the first error term in Lemmas 23 and 24). We have

$$(228)\quad \left|\int_{\sigma_0+iv}^{\sigma_0+iT} y^{s-1}\log^2 T\frac{T^{1-\sigma}}{t}\,ds\right| \leqq y^{\sigma_0-1}\log^2 T\cdot T^{1-\sigma_0}\int_1^T \frac{dt}{t}$$

$$= y^{\sigma_0-1}T^{1-\sigma_0}\log^3 T = o(1) \quad \text{by (219)}.$$

Third Step (Treatment of the second error term in Lemmas 23 and 24). We have

(229)

$$\left|\int_{\sigma_0+iv}^{\sigma_0+iT} y^{s-1}\log T|A(s)|^2\,ds\right| \leqq y^{\sigma_0-1}\log T\int_0^T |A(\sigma_0+iT)|^2\,dt$$

$$= y^{\sigma_0-1}\log T\cdot O\left(T^{4(1-\sigma_0)}\log^4 T\right) \quad \text{(Lemma 22)}$$

$$= o(1) \quad \text{by (219)}.$$

Fourth Step (Proof of Lemma 25). By Lemma 24, (226), (228), and (229) it is enough to prove

$$(230)\quad \int_{\sigma_0+iv}^{\sigma_0+iT} y^{s-1}\sum_{1\leqq n\leqq (kT)^4} g(n)n^{-s}\,ds = o(1).$$

We have

$$(231)\quad \int_{\sigma_0+iv}^{\sigma_0+iT} y^{s-1}\sum_{1\leqq n\leqq (kT)^4} g(n)n^{-s}\,ds$$

$$= i\sum_{1\leqq n\leqq (kT)^4}\frac{g(n)}{n}\left(\frac{y}{n}\right)^{\sigma_0-1}\int_v^T\left(\frac{y}{n}\right)^{it}dt$$

$$= O\left(\sum_{1\leqq n\leqq (kT)^4}\frac{d_4(n)}{n}\log(kT)\left(\frac{y}{n}\right)^{\sigma_0-1}\frac{1}{\log\frac{y}{n}}\right) \quad \text{(by (200))}$$

$$= O\left(\log^4 T \log T\left(\frac{y}{(kT)^4}\right)^{\sigma_0 - 1}\frac{1}{\log T}\right) \quad \text{(by Lemma 19 and (213))}$$

$$= O\left(\left(\frac{y}{T^4}\right)^{\sigma_0 - 1}\log^4 T\right) = o(1) \quad \text{by (219).}$$

This proves Lemma 25.

Fifth Step (Proof of Lemma 26). By Lemma 23, (227), (228), and (229) it is enough to prove

$$(232) \qquad \int_{\sigma_0 + iv}^{\sigma_0 + iT} y^{s-1} \sum_{1 \leqq n \leqq (kT)^3} c(n) n^{-s}\, ds = o(1).$$

We have

$$(233) \qquad \int_{\sigma_0 + iv}^{\sigma_0 + iT} y^{s-1} \sum_{1 \leqq n \leqq (kT)^3} c(n) n^{-s}\, ds$$

$$= i \sum_{i \leqq n \leqq (kT)^3} \frac{c(n)}{n}\left(\frac{y}{n}\right)^{\sigma_0 - 1}\int_v^T \left(\frac{y}{n}\right)^{it} dt$$

$$= O\left(\sum_{1 \leqq n \leqq (kT)^3} \frac{d_3(n)}{n}\left(\frac{y}{n}\right)^{\sigma_0 - 1}\frac{1}{\log\frac{y}{n}}\right) \quad \text{(by (195))}$$

$$= O\left(\log^3 T\left(\frac{y}{(kT)^3}\right)^{\sigma_0 - 1}\frac{1}{\log T}\right) \quad \text{(by Lemma 19 and (213))}$$

$$= O\left(\log^2 T y^{\sigma_0 - 1} T^{3(1-\sigma_0)}\right) = o(1) \quad \text{by (219).}$$

With this, Lemma 26 is proved as well.

Lemma 27. *Suppose $|h(n)| \leq \log n$ for $n \geqq 3$, i.e., suppose that*

$$(234) \qquad H(s) = \sum_{n=1}^{\infty} h(n) n^{-s}$$

is absolutely convergent for $\sigma > 1$. Moreover, suppose that for $1 < \sigma \leqq 2$

(235) $$\sum_{n=1}^{\infty} |h(n)| n^{-\sigma} \leqq a_{11} \frac{1}{\sigma - 1}.$$

Then for $z \geqq T \geqq 3$, $1 < \eta < 2$:

(236) $$\left| \frac{1}{2\pi i} \int_{\eta - iT}^{\eta + iT} \frac{z^s H(s)}{s} \, ds - \sum_{1 \leqq n \leqq z} h(n) \right| \leqq a_{12} \frac{z^\eta \log^2 z}{(\eta - 1) T}.$$

Proof. See Landau, Über einige Summen, die von Nullstellen der Riemann'schen Zetafunktion abhängen, *Acta Mathematica* **35** (1912), 271–294, Hilfssatz 3.[17]

Now we put

(237) $$H_1(s) = \frac{1}{\varphi(k)} \sum_{\chi} \frac{1}{\chi(l)} \frac{-L'}{L}(s, \chi),$$

(238) $$H_2(s) = \frac{1}{\varphi(k)} \sum_{\chi} \frac{1}{\chi(l)} \frac{1}{L(s, \chi)},$$

in which we sum over all characters χ mod k. For $\sigma > 1$, the functions $H_1(s)$ and $H_2(s)$ admit the following expansions:

(239) $$H_1(s) = \sum_{\substack{n=1 \\ n \equiv l \, (\mathrm{mod}\, k)}}^{\infty} \Lambda(n) n^{-s},$$

(240) $$H_2(s) = \sum_{\substack{n=1 \\ n \equiv l \, (\mathrm{mod}\, k)}}^{\infty} \mu(n) n^{-s},$$

which follow immediately from (237) and (238) on the one hand and from (17) and (18) on the other hand. Since by (239) and (240) the functions $H_1(s)$ and $H_2(s)$ satisfy the hypotheses of Lemma 27, we have for

(241) $$1 < \eta = 1 + \frac{1}{\log x} < 2$$

that

[17] Landau proves the theorem only in the weaker form in which a_{12} is replaced by a constant independent of z, η, and T. However, his proof also yields exactly the above lemma.

$$(242)\quad \left|\sum_{\substack{x<n\leqq x+x^{a_4}\\ n\equiv l\ (\mathrm{mod}\ k)}} \Lambda(n) - \frac{1}{2\pi i}\int_{\eta-iT}^{\eta+iT}\{(x+x^{a_4})^s - x^s\}H_1(s)\frac{ds}{s}\right|$$

$$\leqq \frac{a_{12}}{(\eta-1)T}\left((x+x^{a_4})^{1+\frac{1}{\log x}}\log^2(x+x^{a_4}) + x^{1+\frac{1}{\log x}}\log^2 x\right)$$

$$\leqq \frac{a_{13}}{T}\log x\, x\log^2 x = o(x^{a_4})$$

because

$$(243)\quad \frac{x}{T} = x^{1-\frac{1}{a_{10}}},\qquad 1-\frac{1}{a_{10}} = 1-\frac{1}{249.8} < 1-\frac{1}{250} = a_4.$$

Similarly, one obtains

$$(244)\quad \left|\sum_{\substack{x<n\leqq x+x^{a_4}\\ n\equiv l\ (\mathrm{mod}\ k)}} \mu(n) - \frac{1}{2\pi i}\int_{\eta-iT}^{\eta+iT}\{(x+x^{a_4})^s - x^s\}H_2(s)\frac{ds}{s}\right| = o(x^{a_4}).$$

In the closed rectangle

$$(245)\qquad \sigma_0 \leqq \sigma \leqq \eta,\qquad |t| \leqq T$$

the functions

$$(246)\qquad \{(x+x^{a_4})^s - x^s\}H_\nu(s)\frac{1}{s}\qquad (\nu = 1,2)$$

are regular with the exception of a pole at $s = 1$ of at most order 1 with residue $(2-\nu)\dfrac{x^{a_4}}{\varphi(k)}$, since by (216), $L(s,\chi) \neq 0$ throughout (245). In order to treat the integrals appearing in (242) and (244), we apply Cauchy's integral theorem to the rectangle (245). We want to show that

$$(247)\quad \frac{1}{2\pi i}\int_{\eta-iT}^{\eta+iT} - (2-\nu)\frac{x^{a_4}}{\varphi(k)} = \frac{1}{2\pi i}\left(\int_{\eta-iT}^{\sigma_0-iT} + \int_{\sigma_0-iT}^{\sigma_0+iT} + \int_{\sigma_0+iT}^{\eta+iT}\right) = o(x^{a_4}),$$

where the integrand is the function (246). Since this is a real for real s, it is enough to show that

$$\int_{\sigma_0}^{\sigma_0+iT} \{(x+x^{a_4})^s - x^s\} H_\nu(s) \frac{ds}{s} = o(x^{a_4}) \tag{248}$$

and

$$\int_{\sigma_0+iT}^{\eta+iT} \{(x+x^{a_4})^s - x^s\} H_\nu(s) \frac{ds}{s} = o(x^{a_4}). \tag{249}$$

For $t = T$, $\sigma_0 \leqq \sigma \leqq \eta$, we have by Lemma 17 (respectively, Lemma 18) that

$$H_\nu(s) = O(\log T), \tag{250}$$

from which it follows immediately that

$$\begin{aligned} \int_{\sigma_0+iT}^{\eta+iT} \{(x+x^{a_4})^s - x^s\} H_\nu(s) \frac{ds}{s} &= (\eta - \sigma_0) O(x) O(\log T) \frac{1}{T} \\ &= O\left(\frac{x \log T}{T}\right) = o(x^{a_4}) \quad \text{(by (243))} \end{aligned} \tag{251}$$

Moreover,

$$\begin{aligned} \int_{\sigma_0}^{\sigma_0+iT} \{(x+x^{a_4})^s - x^s\} H_\nu(s) \frac{ds}{s} &= \int_x^{x+x^{a_4}} dy \int_{\sigma_0}^{\sigma_0+iT} y^{s-1} H_\nu(s)\, ds \\ &= \int_x^{x+x^{a_4}} o(1)\, dy = o(x^{a_4}) \quad \text{by Lemmas 25 and 26.} \end{aligned} \tag{252}$$

With this, (248) and (249) and hence also (247) are proved. From (247) with $\nu = 2$ and (244), Theorem 2 follows. From (247) with $\nu = 1$ and (242) follows

$$\sum_{\substack{x < n \leqq x + x^{a_4} \\ n \equiv l \pmod{k}}} \Lambda(n) = \frac{x^{a_4}}{\varphi(k)} + o(x^{a_4}). \tag{253}$$

From this, however, one immediately obtains Theorem 3, because

$$(254)\qquad \pi(x + x^{a_4}; k, l) - \pi(x; k, l)$$

$$\leqq \frac{1}{\log x} \sum_{\substack{x < n \leqq x + x^{a_4} \\ n \equiv l \ (\mathrm{mod}\ k)}} \Lambda(n) = \frac{1}{\varphi(k)} \frac{x^{a_4}}{\log x}(1 + o(1)),$$

(255)

$$\pi(x + x^{a_4}; k, l) - \pi(x; k, l) \geqq \sum_{\substack{x < p^m \leqq x + x^{a_4} \\ p^m \equiv l \ (\mathrm{mod}\ k)}} 1 + O(\sqrt{x})$$

$$\geqq \frac{1}{\log(x + x^{a_4})} \sum_{\substack{x < n \leqq x + x^{a_4} \\ n \equiv l \ (\mathrm{mod}\ k)}} \Lambda(n) + O(\sqrt{x})$$

$$= \frac{1}{\varphi(k)} \frac{x^{a_4}}{\log x}(1 + o(1)).$$

(Received 28 January 1932)

3.

Bemerkungen zur vorstehenden Arbeit von Herrn Bochner

(gemeinsam mit E. Landau)

Mathematische Zeitschrift 37 (1933), 10–16
[Commentary on p. 567]

Einleitung.

Jedes $P(\ldots)$ ist > 0 und hängt nur von dem ab, was in der Klammer steht.

Es konvergiere

$$f(s) = \sum_{n=1}^{\infty} \frac{a_n}{n^s}$$

für $s > 1$; *die* a_n *seien* $\geqq 0$. *Es sei*

$$f(s) - \frac{1}{s-1}$$

in $s = 1$ *regulär. Dann ist*

$$\sum_{n=1}^{m} a_n = O(m).$$

Wie beweist man das? Nicht so: Für ganzes $m > 2$ und $1 < s < 2$ ist

$$\frac{1}{m^s} \sum_{n=1}^{m} a_n \leqq \sum_{n=1}^{m} \frac{a_n}{n^s} \leqq f(s) < \frac{P_1(f)}{s-1},$$

$$\sum_{n=1}^{m} a_n < P_1(f) \frac{m^s}{s-1};$$

denn jetzt ist die günstigste s-Wahl

$$s = 1 + \frac{1}{\log m},$$

was nur

$$\sum_{n=1}^{m} a_n = O(m \log m)$$

gibt.

Vielmehr kamen wir auf obigen Satz folgendermaßen. Aus der vorstehenden Arbeit von Herrn Bochner lernten wir die Wienersche Beweismethode des sogenannten Ikeharaschen Satzes. Da Landau fürs kommende Semester eine Vorlesung über Dirichletsche Reihen angekündigt hat, interessierte uns nur der Spezialfall der Dirichletschen Reihen. Bei dieser Spezialisierung der Bochnerschen Beweisanordnung bemerkten wir zu unserem Erstaunen: Ohne viele Änderungen kommt ein Wortlaut heraus, der die Hauptvoraussetzung statt auf der vollen Geraden $\sigma = 1$ nur auf einer beliebig kurzen, $s = 1$ enthaltenden Teilstrecke macht — bei entsprechender Abschwächung der Behauptung, aber so, daß der Ikeharasche Wortlaut für Dirichletsche Reihen in unserem enthalten ist.

Es lautet der neue (den Satz zu Anfang dieser Einleitung reichlich enthaltende)

Hauptsatz: *Es sei*

$$\lambda_1 < \lambda_2 < \ldots < \lambda_n < \ldots, \qquad \lambda_n \to \infty.$$

$$f(s) = \sum_{n=1}^{\infty} a_n e^{-\lambda_n s}$$

konvergiere für $\sigma > 1$; *die* a_n *seien* $\geqq 0$. *Es sei* $\lambda > 0$; *für* $|t| \leqq 2\lambda$ *sei bei* $\varepsilon \to +0$ *gleichmäßig*

$$h_\varepsilon(t) = f(1 + \varepsilon + ti) - \frac{1}{\varepsilon + ti} \to h(t).$$

Dann ist

$$\overline{\lim_{y=\infty}}\, e^{-y} \sum_{\lambda_n \leqq y} a_n \leqq P_2(\lambda),$$

$$\underline{\lim_{y=\infty}}\, e^{-y} \sum_{\lambda_n \leqq y} a_n \geqq P_3(\lambda),$$

wo

$$\lim_{\lambda=\infty} P_2(\lambda) = 1,$$

$$\lim_{\lambda=\infty} P_3(\lambda) = 1.$$

(Wir wiederholen: $P_2(\lambda)$ und $P_3(\lambda)$ sind von f unabhängig.) Insbesondere ist

$$\sum_{\lambda_n \leqq y} a_n = O(e^y),$$

$$\underline{\lim_{y=\infty}}\, e^{-y} \sum_{\lambda_n \leqq y} a_n > 0.$$

Wir beweisen den Hauptsatz nach dem modifizierten Bochnerschen Paradigma ab ovo, so daß der Leser von der Literatur (Landau —

Hardy + Littlewood — Ikehara — Wiener — Bochner) nichts zu kennen braucht.

Die Theorie der Funktionen komplexer Variabler kommt übrigens nirgends zur Anwendung.

§ 1.

Hilfssätze.

Definition 1:

$$G_\lambda(t, y) = \frac{1}{\pi}\left(1 - \frac{|t|}{2\lambda}\right)e^{tyi},$$

also

$$G_\lambda(t, y)\, e^{-txi} = G_\lambda(t, y - x).$$

Definition 2:

$$K_\lambda(y) = \tfrac{1}{2}\int_{-2\lambda}^{2\lambda} G_\lambda(t, y)\, dt,$$

also für $y \gtrless 0$

$$K_\lambda(y) = \frac{1}{\pi}\int_0^{2\lambda}\left(1 - \frac{t}{2\lambda}\right)\cos t y\, dt$$

$$= \frac{1}{\pi}\left\{\left(1 - \frac{t}{2\lambda}\right)\frac{\sin t y}{y} - \frac{\cos t y}{2\lambda y^2}\right\}_0^{2\lambda} = \frac{\sin^2 \lambda y}{\pi \lambda y^2},$$

$$K_\lambda(0) = K_\lambda(+0) = \frac{\lambda}{\pi},$$

$$\int_{-\infty}^{\infty} K_\lambda(y)\, dy = \frac{1}{\pi}\int_{-\infty}^{\infty}\frac{\sin^2 v}{v^2}\, dv = 1.$$

Definition 3:

$$b_\lambda(\alpha) = \frac{1}{\pi}\int_{-\infty}^{\lambda\alpha} e^{v/\lambda}\,\frac{\sin^2 v}{v^2}\, dv = \int_{-\infty}^{\alpha} e^x K_\lambda(x)\, dx \quad \textit{für} \quad \alpha \gtreqless 0.$$

Satz 1: 1)

$$\sum_{n=1}^{\infty} a_n e^{-\lambda_n} K_\lambda(y - \lambda_n) = \tfrac{1}{2}\int_{-2\lambda}^{2\lambda} G_\lambda(t, y)\, h(t)\, dt + \int_{-\infty}^{y} K_\lambda(z)\, dz \quad \textit{für} \quad y \gtreqless 0.$$

2)

$$\sum_{n=1}^{\infty} a_n e^{-\lambda_n} K_\lambda(y - \lambda_n) \to 1 \quad \textit{bei} \quad y \to \infty.$$

3)

$$\sum_{n=1}^{\infty} a_n e^{-\lambda_n} K_\lambda(y - \lambda_n) < P_4(\lambda, f) \quad \textit{für} \quad y \gtreqless 0.$$

4)

$$\sum_{z-\frac{1}{\lambda}\leq\lambda_n\leq z} a_n e^{-\lambda_n} < \frac{P_5}{\lambda} \quad \textit{für} \quad z > P_6(\lambda, f).$$

5)

$$\sum_{|\lambda_n - y| > \delta} \frac{a_n e^{-\lambda_n}}{(\lambda_n - y)^2} < \frac{P_7}{\delta}\left(\frac{1}{\lambda\,\delta} + 1\right) \quad \textit{für} \quad \delta > 0, \quad y > P_8(\lambda, \delta, f).$$

Beweis: 1) Für $\varepsilon > 0$ ist

$$\frac{1}{2}\int_{-2\lambda}^{2\lambda} G_\lambda(t, y)\, h_\varepsilon(t)\, dt = \frac{1}{2}\int_{-2\lambda}^{2\lambda} G_\lambda(t, y) \sum_{n=1}^{\infty} a_n e^{-\lambda_n(1+\varepsilon+ti)}\, dt$$

$$- \frac{1}{2}\int_{-2\lambda}^{2\lambda} G_\lambda(t, y)\, dt \int_0^{\infty} e^{-(\varepsilon+ti)x}\, dx$$

$$= \frac{1}{2}\sum_{n=1}^{\infty} a_n e^{-\lambda_n(1+\varepsilon)} \int_{-2\lambda}^{2\lambda} G_\lambda(t, y-\lambda_n)\, dt - \frac{1}{2}\int_0^{\infty} e^{-\varepsilon x}\, dx \int_{-2\lambda}^{2\lambda} G_\lambda(t, y-x)\, dt$$

$$= \sum_{n=1}^{\infty} a_n e^{-\lambda_n(1+\varepsilon)} K_\lambda(y-\lambda_n) - \int_0^{\infty} e^{-\varepsilon x} K_\lambda(y-x)\, dx.$$

Bei $\varepsilon \to 0$ ist

$$\frac{1}{2}\int_{-2\lambda}^{2\lambda} G_\lambda(t, y)\, h_\varepsilon(t)\, dt \to \frac{1}{2}\int_{-2\lambda}^{2\lambda} G_\lambda(t, y)\, h(t)\, dt,$$

$$\int_0^{\infty} e^{-\varepsilon x} K_\lambda(y-x)\, dx \to \int_0^{\infty} K_\lambda(y-x)\, dx = \int_{-\infty}^{y} K_\lambda(z)\, dz.$$

Also existiert

$$\lim_{\varepsilon=0} \sum_{n=1}^{\infty} a_n e^{-\lambda_n(1+\varepsilon)} K_\lambda(y-\lambda_n).$$

Also konvergiert

$$\sum_{n=1}^{\infty} a_n e^{-\lambda_n} K_\lambda(y-\lambda_n)$$

und hat den behaupteten Wert.

2) In 1) strebt bei $y \to \infty$ das erste Integral als Fourierkonstante einer stetigen Funktion gegen 0; das zweite strebt gegen 1.

3) Nach 1) ist

$$\sum_{n=1}^{\infty} a_n e^{-\lambda_n} K_\lambda(y-\lambda_n) < \frac{1}{2\pi}\int_{-2\lambda}^{2\lambda} |h(t)|\, dt + 1 = P_4(\lambda, f).$$

4) $$K_\lambda(u) \geqq P_9 \lambda \quad \text{für} \quad 0 \leqq u \leqq \frac{1}{\lambda},$$

$$\sum_{z-\frac{1}{\lambda} \leqq \lambda_n \leqq z} a_n e^{-\lambda_n} \leqq \frac{1}{P_9 \lambda} \sum_{z-\frac{1}{\lambda} \leqq \lambda_n \leqq z} a_n e^{-\lambda_n} K_\lambda(z-\lambda_n)$$

$$\leqq \frac{1}{P_9 \lambda} \sum_{n=1}^{\infty} a_n e^{-\lambda_n} K_\lambda(z-\lambda_n),$$

also nach 2)

$$< \frac{P_5}{\lambda} \quad \text{für} \quad z > P_6(\lambda, f).$$

5) Für $y > P_6(\lambda, f)$ ist

$$\sum_{\substack{|\lambda_n - y| > \delta \\ \lambda_n \leqq P_6(\lambda, f)}} \frac{a_n e^{-\lambda_n}}{(\lambda_n - y)^2} \leqq \frac{1}{(y - P_6(\lambda, f))^2} \sum_{\lambda_n \leqq P_6(\lambda, f)} a_n e^{-\lambda_n},$$

also

$$< \frac{P_{10}}{\delta} \quad \text{für } y > P_{11}(\lambda, \delta, f).$$

Für $y \gtreqless 0$ ist nach 4)

$$\sum_{\substack{|\lambda_n - y| > \delta \\ \lambda_n > P_6(\lambda, f)}} \frac{a_n e^{-\lambda_n}}{(\lambda_n - y)^2} \leqq \sum_{\varkappa=0}^{\infty} \frac{1}{\left(\delta + \frac{\varkappa}{\lambda}\right)^2} \sum_{\substack{\delta + \frac{\varkappa}{\lambda} < |\lambda_n - y| \leqq \delta + \frac{\varkappa+1}{\lambda} \\ \lambda_n > P_6(\lambda, f)}} a_n e^{-\lambda_n}$$

$$< \sum_{\varkappa=0}^{\infty} \frac{1}{\left(\delta + \frac{\varkappa}{\lambda}\right)^2} \frac{2P_5}{\lambda} < \frac{2P_5}{\lambda}\left(\frac{1}{\delta^2} + \int_0^\infty \frac{du}{\left(\delta + \frac{u}{\lambda}\right)^2}\right) = \frac{2P_5}{\delta}\left(\frac{1}{\lambda\delta} + 1\right).$$

Satz 2:

$$e^{-y} \sum_{n=1}^{\infty} a_n b_\lambda(y - \lambda_n) \to 1 \quad \textit{für} \quad y \to \infty.$$

Beweis: Nach Satz 1, 3) konvergiert

$$\int_0^\infty e^{-z} \sum_{n=1}^{\infty} a_n e^{-\lambda_n} K_\lambda(y - z - \lambda_n)\, dz$$

für $y \gtreqless 0$ gleichmäßig. Nach Satz 1, 2) ist für $\omega > 0$

$$\lim_{y=\infty} \int_0^\omega e^{-z} \sum_{n=1}^{\infty} a_n e^{-\lambda_n} K_\lambda(y - z - \lambda_n)\, dz = \int_0^\omega e^{-z}\, dz.$$

Also ist

$$1 = \lim_{y=\infty} \int_0^\infty e^{-z} \sum_{n=1}^{\infty} a_n e^{-\lambda_n} K_\lambda(y - z - \lambda_n)\, dz$$

$$= \lim_{y=\infty} \sum_{n=1}^{\infty} a_n e^{-\lambda_n} \int_0^\infty e^{-z} K_\lambda(y - z - \lambda_n)\, dz;$$

hierin ist

$$\int_0^\infty e^{-z} K_\lambda(y - z - \lambda_n)\, dz = e^{\lambda_n - y} \int_{-\infty}^{y - \lambda_n} e^x K_\lambda(x)\, dx = e^{\lambda_n - y} b_\lambda(y - \lambda_n).$$

Satz 3: 1)

$$b_\lambda(\alpha) > \frac{2}{\pi}\int_0^{\lambda\delta} \frac{\sin^2 v}{v^2}\,dv \quad \textit{für} \quad \alpha \geqq \delta > 0.$$

2) a)

$$b_\lambda(\alpha) < \frac{e^\alpha}{\lambda\alpha^2} \quad \textit{für} \quad \alpha < 0.$$

b)

$$b_\lambda(\alpha) < e^\delta \quad \textit{für} \quad \delta > 0, \quad \alpha \leqq \delta.$$

c)

$$b_\lambda(\alpha) < 1 + P_{12}\frac{\log(\lambda+2)}{\lambda}\frac{e^\alpha}{\alpha^2} \quad \textit{für} \quad \alpha > 0.$$

Beweis: 1)

$$b_\lambda(\alpha) > \frac{1}{\pi}\int_0^{\lambda\alpha} (e^{v/\lambda} + e^{-v/\lambda})\frac{\sin^2 v}{v^2}\,dv \geqq \frac{2}{\pi}\int_0^{\lambda\delta} \frac{\sin^2 v}{v^2}\,dv.$$

2) a)

$$b_\lambda(\alpha) < \frac{1}{\lambda^2\alpha^2}\int_{-\infty}^{\lambda\alpha} e^{v/\lambda}\,dv = \frac{e^\alpha}{\lambda\alpha^2}.$$

b)

$$b_\lambda(\alpha) < \frac{1}{\pi}e^\alpha\int_{-\infty}^{\lambda\alpha} \frac{\sin^2 v}{v^2}\,dv < \frac{1}{\pi}e^\delta\int_{-\infty}^{\infty} \frac{\sin^2 v}{v^2}\,dv = e^\delta.$$

c)

$$b_\lambda(\alpha) - 1 < \frac{1}{\pi}\int_{-\infty}^{\lambda\alpha} (e^{v/\lambda} - 1)\frac{\sin^2 v}{v^2}\,dv < \frac{1}{\pi}\int_0^{\lambda\alpha} (e^{v/\lambda} - 1)\frac{\sin^2 v}{v^2}\,dv.$$

Es ist

$$\int_0^{4\lambda} (e^{v/\lambda} - 1)\frac{\sin^2 v}{v^2}\,dv < \frac{P_{13}}{\lambda}\int_0^{4\lambda} \frac{\sin^2 v}{v}\,dv < \frac{P_{14}}{\lambda}\log(\lambda+2)$$

$$< P_{15}\frac{\log(\lambda+2)}{\lambda}\frac{e^\alpha}{\alpha^2}.$$

Also sei o. B. d. A. $\alpha > 4$. Dann ist

$$\int_{4\lambda}^{\lambda\alpha} (e^{v/\lambda} - 1)\frac{\sin^2 v}{v^2}\,dv < \int_{4\lambda}^{\lambda\alpha} \frac{e^{v/\lambda}}{v^2}\,dv = \frac{1}{\lambda}\int_4^{\alpha} \frac{e^w}{w^2}\,dw \leqq \frac{1}{\lambda}\int_4^{\alpha} \left(2\frac{e^w}{w^2} - 4\frac{e^w}{w^3}\right)dw$$

$$= \frac{1}{\lambda}\left(\frac{2e^\alpha}{\alpha^2} - \frac{2e^4}{4^2}\right) < P_{16}\frac{\log(\lambda-2)}{\lambda}\frac{e^\alpha}{\alpha^2}.$$

§ 2.

Beweis des Hauptsatzes.

1) Nach Satz 2 und Satz 3, 1) ist für $\delta > 0$

$$1 = \overline{\lim_{y=\infty}}\, e^{-y} \sum_{n=1}^{\infty} a_n b_\lambda (y - \lambda_n) \geqq \overline{\lim_{y=\infty}}\, e^{-y} \sum_{\lambda_n \leqq y-\delta} a_n b_\lambda (y - \lambda_n)$$

$$\geqq \frac{2}{\pi} \int_0^{\lambda\delta} \frac{\sin^2 v}{v^2}\, dv\, \overline{\lim_{y=\infty}}\, e^{-y} \sum_{\lambda_n \leqq y-\delta} a_n,$$

$$\overline{\lim_{y=\infty}}\, e^{-y} \sum_{\lambda_n \leqq y} a_n = \overline{\lim_{y=\infty}}\, e^{-(y-\delta)} \sum_{\lambda_n \leqq y-\delta} a_n \leqq \frac{e^{\delta}}{\frac{2}{\pi} \int_0^{\lambda\delta} \frac{\sin^2 v}{v^2}\, dv}.$$

Setzt man $\delta = \lambda^{-1/2}$, so ist die rechte Seite $P_2(\lambda)$ mit

$$\lim_{\lambda=\infty} P_2(\lambda) = 1.$$

2) Nach Satz 2, Satz 3, 2) und Satz 1, 5) ist für $\delta > 0$

$$1 = \underline{\lim_{y=\infty}}\, e^{-y} \sum_{n=1}^{\infty} a_n b_\lambda (y - \lambda_n)$$

$$\leqq \underline{\lim_{y=\infty}}\, e^{-y} \Bigg(\sum_{\lambda_n < y-\delta} a_n \left(1 + P_{12} \frac{\log(\lambda+2)}{\lambda} \frac{e^{y-\lambda_n}}{(y-\lambda_n)^2}\right)$$

$$+ \sum_{|\lambda_n - y| \leqq \delta} a_n e^{\delta} + \sum_{\lambda_n > y+\delta} a_n \frac{1}{\lambda} \frac{e^{y-\lambda_n}}{(y-\lambda_n)^2}\Bigg)$$

$$\leqq \underline{\lim_{y=\infty}}\, e^{-(y+\delta)} e^{2\delta} \sum_{\lambda_n \leqq y+\delta} a_n + P_{17} \frac{\log(\lambda+2)}{\lambda\delta} \left(\frac{1}{\lambda\delta} + 1\right),$$

$$\underline{\lim_{y=\infty}}\, e^{-y} \sum_{\lambda_n \leqq y} a_n = \underline{\lim_{y=\infty}}\, e^{-(y+\delta)} \sum_{\lambda_n \leqq y+\delta} a_n$$

$$\geqq e^{-2\delta} \left(1 - P_{17} \frac{\log(\lambda+2)}{\lambda\delta} \left(\frac{1}{\lambda\delta} + 1\right)\right).$$

Für $\delta = P_{18}(\lambda)$ ist die Klammer rechts > 0, also die rechte Seite > 0. Für $\delta = \lambda^{-1/2}$ strebt die rechte Seite gegen 1 bei $\lambda \to \infty$. Also ist die rechte Seite für $\delta = P_{19}(\lambda)$ jedenfalls $P_3(\lambda)$ mit

$$\lim_{\lambda=\infty} P_3(\lambda) = 1.$$

Göttingen, den 17. Oktober 1932.

(Eingegangen am 18. Oktober 1932.)

3t.

Remarks on the Above Article of Bochner

(With E. Landau)

Translation of
Mathematische Zeitschrift 37 (1933), 10–16
[Commentary on p. 567]

Introduction

Every $P(\ldots)$ is >0 and depends only on that which is written in the parentheses.

Suppose that

$$f(s)=\sum_{n=1}^{\infty}\frac{a_n}{n^s}$$

converges for $s>1$, *and that* $a_n \geqq 0$. *Assume that*

$$f(s)-\frac{1}{s-1}$$

is regular at $s=1$. *Then*

$$\sum_{n=1}^{m} a_n = O(m).$$

How does one prove this? Not like this: For integral $m>2$ and $1<s<2$

$$\frac{1}{m^s}\sum_{n=1}^{m} a_n \leqq \sum_{n=1}^{m}\frac{a_n}{n^s} \leqq f(s) < \frac{P_1(f)}{s-1},$$

$$\sum_{n=1}^{m} a_n < P_1(f)\frac{m^s}{s-1};$$

because now the most advantageous choice of s is

$$s = 1 + \frac{1}{\log m},$$

which yields only

$$\sum_{n=1}^{m} a_n = O(m \log m).$$

On the contrary, we hit upon the above theorem as follows. From the preceding article of Bochner we learned Wiener's method of proof of the so-called Theorem of Ikehara. Since Landau had announced a course on Dirichlet series for the coming semester, only the special case of Dirichlet series was of interest to us. In this specialization of Bochner's arrangement of the proof we noticed to our amazement: without much change this results in a wording which, by a suitable weakening of the assertion—but such that Ikehara's wording for Dirichlet series is contained in ours—requires that the main hypothesis need not hold for the entire line $\sigma = 1$ but only for an arbitrarily short subsection containing $s = 1$.

The new theorem (which amply contains the theorem at the beginning of this introduction) reads as follows.

Main Theorem. *Let*

$$\lambda_1 < \lambda_2 < \cdots < \lambda_n < \cdots, \qquad \lambda_n \to \infty.$$

Assume that

$$f(s) = \sum_{n=1}^{\infty} a_n e^{-\lambda_n s}$$

converges for $\sigma > 1$ *and that* $a_n \geqq 0$. *Let* $\lambda > 0$. *Suppose that for* $|t| \leqq 2\lambda$

$$h_\varepsilon(t) = f(1 + \varepsilon + ti) - \frac{1}{\varepsilon + ti} \to h(t)$$

uniformly with $\varepsilon \to +0$. *Then*

$$\overline{\lim_{y=\infty}}\, e^{-y} \sum_{\lambda_n \leqq y} a_n \leqq P_2(\lambda),$$

$$\underline{\lim_{y=\infty}}\, e^{-y} \sum_{\lambda_n \leqq y} a_n \geqq P_3(\lambda),$$

where

$$\lim_{\lambda=\infty} P_2(\lambda) = 1,$$

$$\lim_{\lambda=\infty} P_3(\lambda) = 1.$$

(We repeat: $P_2(\lambda)$ and $P_3(\lambda)$ are independent of f.)
In particular,

$$\sum_{\lambda_n \leqq y} a_n = O(e^y),$$

$$\underline{\lim}_{y=\infty} e^{-y} \sum_{\lambda_n \leqq y} a_n > 0.$$

Following Bochner's modified *paradigma*, we prove the main theorem *ab ovo*, in order that the reader does not require any knowledge about the literature (Landau–Hardy + Littlewood–Ikehara–Wiener–Bochner).
Incidentally, no use is made of the theory of functions of a complex variable.

§1. Lemmas

Definition 1.

$$G_\lambda(t, y) = \frac{1}{\pi}\left(1 - \frac{|t|}{2\lambda}\right)e^{tyi},$$

hence

$$G_\lambda(t, y)e^{-txi} = G_\lambda(t, y - x).$$

Definition 2.

$$K_\lambda(y) = \frac{1}{2}\int_{-2\lambda}^{2\lambda} G_\lambda(t, y)\, dt,$$

hence for $y \gtrless 0$

$$K_\lambda(y) = \frac{1}{\pi}\int_0^{2\lambda}\left(1 - \frac{t}{2\lambda}\right)\cos ty\, dt$$

$$= \frac{1}{\pi}\left\{\left(1-\frac{t}{2\lambda}\right)\frac{\sin ty}{y} - \frac{\cos ty}{2\lambda y^2}\right\}_0^{2\lambda} = \frac{\sin^2\lambda y}{\pi\lambda y^2},$$

$$K_\lambda(0) = K_\lambda(+0) = \frac{\lambda}{\pi},$$

$$\int_{-\infty}^{\infty} K_\lambda(y)\,dy = \frac{1}{\pi}\int_{-\infty}^{\infty}\frac{\sin^2 v}{v^2}\,dv = 1.$$

Definition 3.

$$b_\lambda(\alpha) = \frac{1}{\pi}\int_{-\infty}^{\lambda\alpha} e^{v/\lambda}\frac{\sin^2 v}{v^2}\,dv = \int_{-\infty}^{\alpha} e^x K_\lambda(x)\,dx \quad \textit{for } \alpha \gtreqless 0.$$

Theorem 1. 1.

$$\sum_{n=1}^{\infty} a_n e^{-\lambda_n} K_\lambda(y-\lambda_n) = \frac{1}{2}\int_{-2\lambda}^{2\lambda} G_\lambda(t,y)h(t)\,dt + \int_{-\infty}^{y} K_\lambda(z)\,dz \quad \textit{for } y \gtreqless 0.$$

2.

$$\sum_{n=1}^{\infty} a_n e^{-\lambda_n} K_\lambda(y-\lambda_n) \to 1 \quad \textit{as } y \to \infty.$$

3.

$$\sum_{n=1}^{\infty} a_n e^{-\lambda_n} K_\lambda(y-\lambda_n) < P_4(\lambda, f) \quad \textit{for } y \gtreqless 0.$$

4.

$$\sum_{z-\frac{1}{\lambda}\leqq\lambda_n\leqq z} a_n e^{-\lambda_n} < \frac{P_5}{\lambda} \quad \textit{for } z > P_6(\lambda, f).$$

5.

$$\sum_{|\lambda_n - y|>\delta} \frac{a_n e^{-\lambda_n}}{(\lambda_n - y)^2} < \frac{P_7}{\delta}\left(\frac{1}{\lambda\delta}+1\right) \quad \textit{for } \delta > 0, \quad y > P_8(\lambda,\delta,f).$$

Proof. 1. For $\varepsilon > 0$,

$$\frac{1}{2}\int_{-2\lambda}^{2\lambda} G_\lambda(t, y)h_\varepsilon(t)\,dt$$

$$= \frac{1}{2}\int_{-2\lambda}^{2\lambda} G_\lambda(t, y)\sum_{n=1}^{\infty} a_n e^{-\lambda_n(1+\varepsilon+ti)}\,dt - \frac{1}{2}\int_{-2\lambda}^{2\lambda} G_\lambda(t, y)\,dt\int_0^{\infty} e^{-(\varepsilon+ti)x}\,dx$$

$$= \frac{1}{2}\sum_{n=1}^{\infty} a_n e^{-\lambda_n(1+\varepsilon)}\int_{-2\lambda}^{2\lambda} G_\lambda(t, y-\lambda_n)\,dt - \frac{1}{2}\int_0^{\infty} e^{-\varepsilon x}\,dx\int_{-2\lambda}^{2\lambda} G_\lambda(t, y-x)\,dt$$

$$= \sum_{n=1}^{\infty} a_n e^{-\lambda_n(1+\varepsilon)}K_\lambda(y-\lambda_n) - \int_0^{\infty} e^{-\varepsilon x}K_\lambda(y-x)\,dx.$$

Letting $\varepsilon \to 0$, we have

$$\frac{1}{2}\int_{-2\lambda}^{2\lambda} G_\lambda(t, y)h_\varepsilon(t)\,dt \to \frac{1}{2}\int_{-2\lambda}^{2\lambda} G_\lambda(t, y)h(t)\,dt,$$

$$\int_0^{\infty} e^{-\varepsilon x}K_\lambda(y-x)\,dx \to \int_0^{\infty} K_\lambda(y-x)\,dx = \int_{-\infty}^{y} K_\lambda(z)\,dz.$$

Therefore the limit

$$\lim_{\varepsilon=0}\sum_{n=1}^{\infty} a_n e^{-\lambda_n(1+\varepsilon)}K_\lambda(y-\lambda_n)$$

exists. Hence

$$\sum_{n=1}^{\infty} a_n e^{-\lambda_n}K_\lambda(y-\lambda_n)$$

converges and has the asserted value.

2. In (1), the first integral, being a Fourier coefficient of a continuous function, tends to 0 as $y \to \infty$; the second integral tends to 1.

3. By (1),

$$\sum_{n=1}^{\infty} a_n e^{-\lambda_n}K_\lambda(y-\lambda_n) < \frac{1}{2\pi}\int_{-2\lambda}^{2\lambda} |h(t)|\,dt + 1 = P_4(\lambda, f).$$

4.

$$K_\lambda(u) \geqq P_9\lambda \quad \text{for } 0 \leqq u \leqq \frac{1}{\lambda},$$

$$\sum_{z-\frac{1}{\lambda}\leqq\lambda_n\leqq z} a_n e^{-\lambda_n} \leqq \frac{1}{P_9\lambda} \sum_{z-\frac{1}{\lambda}\leqq\lambda_n\leqq z} a_n e^{-\lambda_n} K_\lambda(z-\lambda_n)$$

$$\leqq \frac{1}{P_9\lambda} \sum_{n=1}^{\infty} a_n e^{-\lambda_n} K_\lambda(z-\lambda_n),$$

hence by (2)

$$< \frac{P_5}{\lambda} \quad \text{for } z > P_6(\lambda, f).$$

5. If $y > P_6(\lambda, f)$ then

$$\sum_{\substack{|\lambda_n - y|>\delta \\ \lambda_n \leqq P_6(\lambda, f)}} \frac{a_n e^{-\lambda_n}}{(\lambda_n - y)^2} \leqq \frac{1}{(y - P_6(\lambda, f))^2} \sum_{\lambda_n \leqq P_6(\lambda, f)} a_n e^{-\lambda_n},$$

hence

$$< \frac{P_{10}}{\delta} \quad \text{for } y > P_{11}(\lambda, \delta, f).$$

If $y \gtreqless 0$, then by (4)

$$\sum_{\substack{|\lambda_n - y|>\delta \\ \lambda_n > P_6(\lambda, f)}} \frac{a_n e^{-\lambda_n}}{(\lambda_n - y)^2} \leqq \sum_{\kappa=0}^{\infty} \frac{1}{\left(\delta + \frac{\kappa}{\lambda}\right)^2} \sum_{\substack{\delta + \frac{\kappa}{\lambda} < |\lambda_n - y| \leqq \delta + \frac{\kappa+1}{\lambda} \\ \lambda_n > P_6(\lambda, f)}} a_n e^{-\lambda_n}$$

$$< \sum_{\kappa=0}^{\infty} \frac{1}{\left(\delta + \frac{\kappa}{\lambda}\right)^2} \frac{2P_5}{\lambda} < \frac{2P_5}{\lambda}\left(\frac{1}{\delta^2} + \int_0^\infty \frac{du}{\left(\delta + \frac{u}{\lambda}\right)^2}\right)$$

$$= \frac{2P_5}{\delta}\left(\frac{1}{\lambda\delta} + 1\right).$$

***Theorem* 2.**

$$e^{-y}\sum_{n=1}^{\infty} a_n b_\lambda(y-\lambda_n) \to 1 \quad \textit{for } y \to \infty.$$

Proof. By Theorem 1, (3)

$$\int_0^{\infty} e^{-z} \sum_{n=1}^{\infty} a_n e^{-\lambda_n} K_\lambda(y-z-\lambda_n)\, dz$$

converges uniformly for $y \gtreqless 0$. By Theorem 1, (2) we have for $\omega > 0$,

$$\lim_{y=\infty} \int_0^{\omega} e^{-z} \sum_{n=1}^{\infty} a_n e^{-\lambda_n} K_\lambda(y-z-\lambda_n)\, dz = \int_0^{\omega} e^{-z}\, dz.$$

Therefore

$$1 = \lim_{y=\infty} \int_0^{\infty} e^{-z} \sum_{n=1}^{\infty} a_n e^{-\lambda_n} K_\lambda(y-z-\lambda_n)\, dz$$

$$= \lim_{y=\infty} \sum_{n=1}^{\infty} a_n e^{-\lambda_n} \int_0^{\infty} e^{-z} K_\lambda(y-z-\lambda_n)\, dz,$$

where

$$\int_0^{\infty} e^{-z} K_\lambda(y-z-\lambda_n)\, dz = e^{\lambda_n - y} \int_{-\infty}^{y-\lambda_n} e^{x} K_\lambda(x)\, dx = e^{\lambda_n - y} b_\lambda(y-\lambda_n).$$

***Theorem* 3.** 1.

$$b_\lambda(\alpha) > \frac{2}{\pi} \int_0^{\lambda\delta} \frac{\sin^2 v}{v^2}\, dv \quad \text{for } \alpha \geqq \delta > 0.$$

2a.

$$b_\lambda(\alpha) < \frac{e^{\alpha}}{\lambda\alpha^2} \quad \text{for } \alpha < 0.$$

b.

$$b_\lambda(\alpha) < e^{\delta} \quad \text{for } \delta > 0,\ \alpha \leqq \delta.$$

c.

$$b_\lambda(\alpha) < 1 + P_{12}\frac{\log(\lambda+2)}{\lambda}\frac{e^\alpha}{\alpha^2} \quad \text{for } \alpha > 0.$$

Proof. 1.

$$b_\lambda(\alpha) > \frac{1}{\pi}\int_0^{\lambda\alpha}(e^{v/\lambda}+e^{-v/\lambda})\frac{\sin^2 v}{v^2}\,dv \geqq \frac{2}{\pi}\int_0^{\lambda\delta}\frac{\sin^2 v}{v^2}\,dv.$$

2a.

$$b_\lambda(\alpha) < \frac{1}{\lambda^2\alpha^2}\int_{-\infty}^{\lambda\alpha} e^{v/\lambda}\,dv = \frac{e^\alpha}{\lambda\alpha^2}.$$

b.

$$b_\lambda(\alpha) < \frac{1}{\pi}e^\alpha\int_{-\infty}^{\lambda\alpha}\frac{\sin^2 v}{v^2}\,dv < \frac{1}{\pi}e^\delta\int_{-\infty}^{\infty}\frac{\sin^2 v}{v^2}\,dv = e^\delta.$$

c.

$$b_\lambda(\alpha) - 1 < \frac{1}{\pi}\int_{-\infty}^{\lambda\alpha}(e^{v/\lambda}-1)\frac{\sin^2 v}{v^2}\,dv < \frac{1}{\pi}\int_0^{\lambda\alpha}(e^{v/\lambda}-1)\frac{\sin^2 v}{v^2}\,dv.$$

We have

$$\int_0^{4\lambda}(e^{v/\lambda}-1)\frac{\sin^2 v}{v^2}\,dv < \frac{P_{13}}{\lambda}\int_0^{4\lambda}\frac{\sin^2 v}{v}\,dv < \frac{P_{14}}{\lambda}\log(\lambda+2)$$

$$< P_{15}\frac{\log(\lambda+2)}{\lambda}\frac{e^\alpha}{\alpha^2}.$$

Therefore, without loss of generality, let $\alpha > 4$. Then

$$\int_{4\lambda}^{\lambda\alpha}(e^{v/\lambda}-1)\frac{\sin^2 v}{v^2}\,dv < \int_{4\lambda}^{\lambda\alpha}\frac{e^{v/\lambda}}{v^2}\,dv = \frac{1}{\lambda}\int_4^\alpha\frac{e^w}{w^2}\,dw \leqq \frac{1}{\lambda}\int_4^\alpha\left(2\frac{e^w}{w^2}-4\frac{e^w}{w^3}\right)dw$$

$$= \frac{1}{\lambda}\left(\frac{2e^\alpha}{\alpha^2}-\frac{2e^4}{4^2}\right) < P_{16}\frac{\log(\lambda+2)}{\lambda}\frac{e^\alpha}{\alpha^2}.$$

§2. Proof of the Main Theorem

1. By Theorem 2 and Theorem 3, (1), we have for $\delta > 0$

$$1 = \overline{\lim_{y=\infty}}\, e^{-y} \sum_{n=1}^{\infty} a_n b_\lambda(y - \lambda_n) \geqq \overline{\lim_{y=\infty}}\, e^{-y} \sum_{\lambda_n \leqq y-\delta} a_n b_\lambda(y-\lambda_n)$$

$$\geqq \frac{2}{\pi} \int_0^{\lambda\delta} \frac{\sin^2 v}{v^2}\, dv\; \overline{\lim_{y=\infty}}\, e^{-y} \sum_{\lambda_n \leqq y-\delta} a_n,$$

$$\overline{\lim_{y=\infty}}\, e^{-y} \sum_{\lambda_n \leqq y} a_n = \overline{\lim_{y=\infty}}\, e^{-(y-\delta)} \sum_{\lambda_n \leqq y-\delta} a_n \leqq \frac{e^{\delta}}{\dfrac{2}{\pi}\displaystyle\int_0^{\lambda\delta} \frac{\sin^2 v}{v^2}\, dv}.$$

If one puts $\delta = \lambda^{-1/2}$, then the right-hand side is $P_2(\lambda)$ with

$$\lim_{\lambda=\infty} P_2(\lambda) = 1.$$

2. By Theorem 2, Theorem 3, (2), and Theorem 1, (5), we have for $\delta > 0$

$$1 = \underline{\lim}_{y=\infty}\, e^{-y} \sum_{n=1}^{\infty} a_n b_\lambda(y-\lambda_n)$$

$$\leqq \underline{\lim}_{y=\infty}\, e^{-y} \left(\sum_{\lambda_n < y-\delta} a_n \left| 1 + P_{12} \frac{\log(\lambda+2)}{\lambda} \frac{e^{y-\lambda_n}}{(y-\lambda_n)^2} \right| \right.$$

$$\left. + \sum_{|\lambda_n - y| \leqq \delta} a_n e^{\delta} + \sum_{\lambda_n > y+\delta} a_n \frac{1}{\lambda} \frac{e^{y-\lambda_n}}{(y-\lambda_n)^2} \right)$$

$$\leqq \underline{\lim}_{y=\infty}\, e^{-(y+\delta)} e^{2\delta} \sum_{\lambda_n \leqq y+\delta} a_n + P_{17} \frac{\log(\lambda+2)}{\lambda\delta} \left(\frac{1}{\lambda\delta} + 1 \right),$$

$$\underline{\lim}_{y=\infty}\, e^{-y} \sum_{\lambda_n \leqq y} a_n = \underline{\lim}_{y=\infty}\, e^{-(y+\delta)} \sum_{\lambda_n \leqq y+\delta} a_n$$

$$\geqq e^{-2\delta} \left(1 - P_{17} \frac{\log(\lambda+2)}{\lambda\delta} \left(\frac{1}{\lambda\delta} + 1 \right) \right).$$

For $\delta = P_{18}(\lambda)$ the bracket on the right is > 0, and hence the right-hand side > 0. For $\delta = \lambda^{-1/2}$, the right-hand side tends to 1 as $\lambda \to \infty$. Therefore, for $\delta = P_{19}(\lambda)$ the right-hand side is in any case $P_3(\lambda)$ with

$$\lim_{\lambda=\infty} P_3(\lambda) = 1.$$

Göttingen, 17 October 1932.
(Received 18 October 1932)

4.

Ein Satz über Potenzreihen

(gemeinsam mit E. Landau)

Mathematische Zeitschrift 37 (1933), 17
[Commentary on p. 567]

Wir beweisen möglichst kurz einen Satz, der in dem Spezialfall $\lambda_n = n - 1$ des Hauptsatzes unserer vorstehenden Arbeit enthalten ist, uns aber bisher unbekannt war.

Satz:

$$f(x) = \sum_{n=0}^{\infty} c_n x^n$$

konvergiere für $|x| < 1$; *die* c_n *seien* $\geqq 0$;

$$F(x) = f(x) - \frac{1}{1-x}$$

sei regulär in $x = 1$. *Dann ist*

$$c_n = O(1)$$

(also nach einem Fatouschen Satz in Acta Mathematica **30**, S. 391, sogar

$$\sum_{\nu=0}^{n} c_\nu = n + O(1)).$$

Beweis: Man wähle $\lambda > 0$ und P so, daß

$$|F(re^{\varphi i})| < P \quad \text{für} \quad 0 \leqq r < 1, \quad -\lambda \leqq \varphi \leqq \lambda.$$

$$\int_{-\lambda}^{\lambda} \left(1 - \frac{|\varphi|}{\lambda}\right) e^{\alpha \varphi i}\, d\varphi = \frac{4 \sin^2 \frac{\alpha \lambda}{2}}{\alpha^2 \lambda} \leqq \frac{4}{\alpha^2 \lambda} \quad \text{für } \alpha \gtrless 0.$$

$$2\lambda P \geqq \int_{-\lambda}^{\lambda} \left(1 - \frac{|\varphi|}{\lambda}\right) e^{-n\varphi i} F(re^{\varphi i})\, d\varphi = \sum_{\nu=0}^{\infty} (c_\nu - 1) r^\nu \int_{-\lambda}^{\lambda} \left(1 - \frac{|\varphi|}{\lambda}\right) e^{(\nu - n)\varphi i} d\varphi$$

Article was originally printed on one page.

$$\geqq (c_n - 1)\, r^n \lambda - \frac{4}{\lambda} \sum_{\substack{\nu=0 \\ \nu \neq n}}^{\infty} \frac{1}{(\nu - n)^2};$$

$r \to 1$ ergibt die Behauptung.

Göttingen, den 26. Oktober 1932.

(Eingegangen am 27. Oktober 1932.)

4t.

A Theorem on Power Series

(With E. Landau)

Translation of
Mathematische Zeitschrift 37 (1933), 17
[Commentary on p. 567]

We shall prove as briefly as possible a theorem which is contained as the special case $\lambda_n = n - 1$ in the main theorem of our article above, but which was hitherto unknown to us.

Theorem. *Suppose that*

$$f(x) = \sum_{n=0}^{\infty} c_n x^n$$

converges for $|x| < 1$ *and that* $c_n \geqq 0$; *moreover, suppose that*

$$F(x) = f(x) - \frac{1}{1-x}$$

is regular at $x = 1$. *Then*

$$c_n = O(1)$$

(and hence by a theorem of Fatou in *Acta Mathematica* **30**, p. 391, even

$$\sum_{\nu=0}^{n} c_\nu = n + O(1)).$$

Proof. Choose $\lambda > 0$ and P such that

$$|F(re^{\varphi i})| < P \quad \text{for } 0 \leqq r < 1, \qquad -\lambda \leqq \varphi \leqq \lambda.$$

$$\int_{-\lambda}^{\lambda}\left(1-\frac{|\varphi|}{\lambda}\right)e^{\alpha\varphi i}\,d\varphi=\frac{4\sin^2\dfrac{\alpha\lambda}{2}}{\alpha^2\lambda}\leqq\frac{4}{\alpha^2\lambda}\qquad\text{for } \alpha\gtrless 0.$$

$$\begin{aligned}2\lambda P&\geqq\int_{-\lambda}^{\lambda}\left(1-\frac{|\varphi|}{\lambda}\right)e^{-n\varphi i}F(re^{\varphi i})\,d\varphi\\&=\sum_{\nu=0}^{\infty}(c_\nu-1)r^\nu\int_{-\lambda}^{\lambda}\left(1-\frac{|\varphi|}{\lambda}\right)e^{(\nu-n)\varphi i}\,d\varphi\\&\geqq(c_n-1)r^n\lambda-\frac{4}{\lambda}\sum_{\substack{\nu=0\\ \nu\neq n}}^{\infty}\frac{1}{(\nu-n)^2}.\end{aligned}$$

Letting $r \to 1$ yields the claim.

Göttingen, 26 October 1932.
(Received 27 October 1932)

5.

Anwendungen der N. Wienerschen Methode

(gemeinsam mit E. Landau)

Mathematische Zeitschrift 37 (1933), 18–21
[Commentary on p. 567]

In unserer vorvorstehenden Arbeit bewiesen wir u. a.:

Die Dirichletsche Reihe

$$f(s) = \sum_{n=1}^{\infty} a_n e^{-\lambda_n s}$$

konvergiere für $\sigma > 1$; *die* a_n *seien* $\geqq 0$. *Es sei* $f(s) - \frac{1}{s-1}$ *für* $s = 1$ *regulär. Dann ist*

$$A(y) = \sum_{\lambda_n \leqq y} a_n = O(e^y). \tag{1}$$

[Übrigens lehrt ein Blick auf unseren Beweis, daß für diesen Spezialfall unseres Hauptsatzes die schwierigsten Hilfsbetrachtungen, insbesondere Satz 1, 4) und 5) und Satz 3, 2) unnötig sind.]

Aus (1) folgt

$$\sum_{n=1}^{m} a_n e^{-\lambda_n} = \int_{\lambda_1}^{\lambda_m} A(u) e^{-u} du + A(\lambda_m) e^{-\lambda_m} = O(\lambda_m),$$

$$\sum_{\lambda_n \leqq y} a_n e^{-\lambda_n} = O(y).$$

Wir werden aber aus (1) und den Voraussetzungen sogar

$$\sum_{\lambda_n \leqq y} a_n e^{-\lambda_n} = y + O(1) \tag{2}$$

folgern.

Es ist erstaunlich, daß ein Satz sich selbst verschärfen kann. Unser Satz tut es gewissermaßen durch Kombination mit Teil 1) des folgenden Satzes von Herrn M. Riesz:

Es sei

$$\lambda_1 < \lambda_2 < \ldots, \quad \lambda_n \to \infty,$$

$$B(y) = \sum_{\lambda_n \leqq y} b_n = O(e^y),$$

also

$$g(s) = \sum_{n=1}^{\infty} b_n e^{-\lambda_n s}$$

für $\sigma > 1$ *konvergent.* $g(s)$ *sei in* $s = 1$ *regulär.*

1) *Dann ist*

$$\sum_{\lambda_n \leqq y} b_n e^{-\lambda_n} = O(1).$$

2) *Wird sogar*

$$B(y) = o(e^y)$$

vorausgesetzt, so konvergiert

$$\sum_{n=1}^{\infty} b_n e^{-\lambda_n}.$$

Herr Riesz hat Teil 2) auf S. 350—354 des Bandes **40** der Acta Mathematica bewiesen. Teil 1) ergibt sich entsprechend einfacher aus seiner Methode; um es dem Leser leicht zu machen, führen wir dies im Anhang aus.

Teil 2) dient dazu, unter der ferneren Annahme, daß $f(s) - \frac{1}{s-1}$ für $\sigma = 1$ regulär ist, die Existenz von

$$\lim_{y=\infty} \left(\sum_{\lambda_n \leqq y} a_n e^{-\lambda_n} - y \right) \tag{3}$$

zu beweisen. Hierbei kommt statt (1) der Ikeharasche Spezialfall des Hauptsatzes unserer vorvorstehenden Arbeit zur Anwendung.

§ 1.

Beweis von (2).

O. B. d. A. kommen unter den λ_n alle ganzen positiven Zahlen vor (sonst füge man die fehlenden mit verschwindenden Koeffizienten hinzu). Dann ist

$$g(s) = f(s) - \frac{1}{e^{s-1} - 1} = f(s) - \sum_{\nu=1}^{\infty} e^{\nu} e^{-\nu s} = \sum_{n=1}^{\infty} b_n e^{-\lambda_n s}$$

für $\sigma > 1$ konvergent und in $s = 1$ regulär. Nach (1) ist

$$\sum_{\lambda_n \leqq y} b_n = \sum_{\lambda_n \leqq y} a_n - \sum_{1 \leqq \nu \leqq y} e^{\nu} = O(e^y),$$

nach Teil 1) des Rieszschen Satzes also

$$\sum_{\lambda_n \leqq y} a_n e^{-\lambda_n} = \sum_{\lambda_n \leqq y} b_n e^{-\lambda_n} + \sum_{1 \leqq \nu \leqq y} 1 = y + O(1).$$

§ 2.

Beweis von (3).

O. B. d. A. kommen unter den λ_n die Logarithmen aller ganzen positiven Zahlen vor. Dann ist

$$g(s) = f(s) - \zeta(s) = f(s) - \sum_{\nu=1}^{\infty} e^{-\log \nu \cdot s} = \sum_{n=1}^{\infty} b_n e^{-\lambda_n s}$$

für $\sigma > 1$ konvergent und für $\sigma = 1$ regulär. Nach Ikehara ist also

$$\sum_{\lambda_n \leqq y} b_n = \sum_{\lambda_n \leqq y} a_n - \sum_{1 \leqq \nu \leqq e^y} 1 = o(e^y).$$

Nach Teil 2) des Rieszschen Satzes konvergiert also

$$\sum_{n=1}^{\infty} b_n e^{-\lambda_n} = c.$$

Also ist

$$\sum_{\lambda_n \leqq y} a_n e^{-\lambda_n} = \sum_{\lambda_n \leqq y} b_n e^{-\lambda_n} + \sum_{1 \leqq \nu \leqq e^y} \frac{1}{\nu} = y + c + C + o(1).$$

Anhang.

Beweis von Teil 1) des Rieszschen Satzes.

Alle c-Konstanten sind > 0 und nur von g abhängig.

$$|B(y)| \leqq c_1 e^y \qquad \text{für} \qquad y \gtreqless 0.$$

$g(s)$ ist in einem Rechteck

$$(R) \qquad 1 - c_2 \leqq \sigma \leqq 2, \quad |t| \leqq c_3$$

regulär. Für ganzes $m \geqq 1$ setze man auf R

$$g_m(s) = \left(g(s) - \sum_{n=1}^{m} b_n e^{-\lambda_n s}\right) e^{\lambda_m (s-1)} (s - 1 - c_3 i)(s - 1 + c_3 i).$$

Auf dem Rand von R ist

$$|g_m(s)| \leqq \left| g(s) - \sum_{n=1}^{m} b_n e^{-\lambda_n s} \right| e^{\lambda_m (\sigma - 1)} c_4 |\sigma - 1|.$$

Auf dem zu $\sigma > 1$ gehörigen Randteil ist

$$\left| g(s) - \sum_{n=1}^{m} b_n e^{-\lambda_n s} \right| = \left| -B(\lambda_m) e^{-\lambda_m s} + s \int_{\lambda_m}^{\infty} B(u) e^{-us} du \right|$$

$$\leqq c_1 \left(e^{-\lambda_m (\sigma-1)} + |s| \int_{\lambda_m}^{\infty} e^{-u(\sigma-1)} du\right) < \frac{c_5}{\sigma - 1} e^{-\lambda_m (\sigma - 1)}.$$

Auf dem zu $\sigma < 1$ gehörigen Randteil ist

$$\left|\sum_{n=1}^{m} b_n e^{-\lambda_n s}\right| = \left|B(\lambda_m)\, e^{-\lambda_m s} + s\int_{-\infty}^{\lambda_m} B(u)\, e^{-us}\, du\right|$$

$$\leqq c_1\left(e^{-\lambda_m(\sigma-1)} + |s|\int_{-\infty}^{\lambda_m} e^{-u(\sigma-1)}\, du\right) < \frac{c_6}{1-\sigma}\, e^{-\lambda_m(\sigma-1)},$$

$$\left|g(s) - \sum_{n=1}^{m} b_n e^{-\lambda_n s}\right| < c_7 + \frac{c_6}{1-\sigma}\, e^{-\lambda_m(\sigma-1)} < \frac{c_8}{1-\sigma}\, e^{-\lambda_m(\sigma-1)}.$$

Auf dem ganzen Rand von R ist also

$$|g_m(s)| < c_9.$$

Daher ist

$$\left|g(1) - \sum_{n=1}^{m} b_n e^{-\lambda_n}\right| c_3^2 = |g_m(1)| < c_9,$$

$$\left|\sum_{n=1}^{m} b_n e^{-\lambda_n}\right| < c_{10}.$$

Göttingen, den 26. Oktober 1932.

(Eingegangen am 27. Oktober 1932.)

5t.

Applications of the Method of N. Wiener

(With E. Landau)

Translation of
Mathematische Zeitschrift 37 (1933), 18–21
[Commentary on p. 567]

In the article prior to the one above we proved, among other things:
Suppose that the Dirichlet series

$$f(s) = \sum_{n=1}^{\infty} a_n e^{-\lambda_n s}$$

converges for $\sigma > 1$ *and that* $a_n \geqq 0$. *Assume that* $f(s) - \dfrac{1}{s-1}$ *is regular at* $s = 1$. *Then*

$$A(y) = \sum_{\lambda_n \leqq y} a_n = O(e^y). \tag{1}$$

[Incidentally, a glance at our proof shows that, for this special case of our main theorem, the most difficult of the auxiliary considerations, in particular, Theorem 1, (4) and (5) and Theorem 3, (2), are unnecessary.]

From (1) follows

$$\sum_{n=1}^{m} a_n e^{-\lambda_n} = \int_{\lambda_1}^{\lambda_m} A(u) e^{-u}\, du + A(\lambda_m) e^{-\lambda_m} = O(\lambda_m),$$

$$\sum_{\lambda_n \leqq y} a_n e^{-\lambda_n} = O(y).$$

We will, however, deduce from (1) and the hypotheses that even

$$\sum_{\lambda_n \leqq y} a_n e^{-\lambda_n} = y + O(1) \tag{2}$$

holds.

It is astonishing that a theorem can sharpen itself. Our theorem does this in a way by combination with part 1 of the following theorem of M. Riesz:

Assume that

$$\lambda_1 < \lambda_2 < \cdots, \qquad \lambda_n \to \infty,$$

$$B(y) = \sum_{\lambda_n \leqq y} b_n = O(e^y),$$

hence that

$$g(s) = \sum_{n=1}^{\infty} b_n e^{-\lambda_n s}$$

is convergent for $\sigma > 1$. Suppose that $g(s)$ is regular at $s = 1$.

1. *Then*

$$\sum_{\lambda_n \leqq y} b_n e^{-\lambda_n} = O(1).$$

2. *If one even assumes that*

$$B(y) = o(e^y),$$

then

$$\sum_{n=1}^{\infty} b_n e^{-\lambda_n}$$

converges.

Riesz proved part 2 on pp. 350–354 of volume 40 of *Acta Mathematica*. Part 1 follows in a corresponding manner more simply from his method; in order to make it easier for the reader, we carry this out in the appendix.

Part 2 serves to prove, under the additional hypothesis that $f(s) - \dfrac{1}{s-1}$ is regular at $\sigma = 1$, the existence of

$$\lim_{y=\infty} \Big(\sum_{\lambda_n \leqq y} a_n e^{-\lambda_n} - y \Big). \tag{3}$$

Hereby Ikehara's special case of the main theorem of our article is used in place of (1).

§1. Proof of (2)

Without loss of generality, all positive integers appear among the λ_n (otherwise, add the missing ones with vanishing coefficients). Then

$$g(s) = f(s) - \frac{1}{e^{s-1} - 1} = f(s) - \sum_{\nu=1}^{\infty} e^{\nu} e^{-\nu s} = \sum_{n=1}^{\infty} b_n e^{-\lambda_n s}$$

is convergent for $\sigma > 1$ and regular at $s = 1$. By (1),

$$\sum_{\lambda_n \leqq y} b_n = \sum_{\lambda_n \leqq y} a_n - \sum_{1 \leqq \nu \leqq y} e^{\nu} = O(e^{y}),$$

hence by part 1 of Riesz's theorem,

$$\sum_{\lambda_n \leqq y} a_n e^{-\lambda_n} = \sum_{\lambda_n \leqq y} b_n e^{-\lambda_n} + \sum_{1 \leqq \nu \leqq y} 1 = y + O(1).$$

§2. Proof of (3)

Without loss of generality, the logarithms of all positive integers appear among the λ_n. Then

$$g(s) = f(s) - \zeta(s) = f(s) - \sum_{\nu=1}^{\infty} e^{-\log \nu \cdot s} = \sum_{n=1}^{\infty} b_n e^{-\lambda_n s}$$

is convergent for $\sigma > 1$ and regular for $\sigma = 1$. Hence, by Ikehara,

$$\sum_{\lambda_n \leqq y} b_n = \sum_{\lambda_n \leqq y} a_n - \sum_{1 \leqq \nu \leqq e^{y}} 1 = o(e^{y}).$$

Therefore, by part 2 of the theorem of Riesz,

$$\sum_{n=1}^{\infty} b_n e^{-\lambda_n} = c$$

converges. Thus

$$\sum_{\lambda_n \leqq y} a_n e^{-\lambda_n} = \sum_{\lambda_n \leqq y} b_n e^{-\lambda_n} + \sum_{1 \leqq \nu \leqq e^{y}} \frac{1}{\nu} = y + c + C + o(1).$$

Appendix: Proof of part 1 of Riesz's Theorem

All c-constants are >0 and depend only on g.

$$|B(y)| \leqq c_1 e^{y} \quad \text{for } y \gtreqless 0.$$

$g(s)$ is regular in a rectangle

$$(R) \qquad 1 - c_2 \leqq \sigma \leqq 2, \qquad |t| \leqq c_3.$$

For integral $m \geqq 1$ define g_m on R by

$$g_m(s) = \left(g(s) - \sum_{n=1}^{m} b_n e^{-\lambda_n s} \right) e^{\lambda_m (s-1)} (s - 1 - c_3 i)(s - 1 + c_3 i).$$

On the boundary of R we have

$$|g_m(s)| \leqq \left| g(s) - \sum_{n=1}^{m} b_n e^{-\lambda_n s} \right| e^{\lambda_m(\sigma-1)} c_4 |\sigma - 1|.$$

On the part of the boundary belonging to $\sigma > 1$ we have

$$\left| g(s) - \sum_{n=1}^{m} b_n e^{-\lambda_n s} \right| = \left| -B(\lambda_m) e^{-\lambda_m s} + s \int_{\lambda_m}^{\infty} B(u) e^{-us}\, du \right|$$

$$\leqq c_1 \left(e^{-\lambda_m(\sigma-1)} + |s| \int_{\lambda_m}^{\infty} e^{-u(\sigma-1)}\, du \right)$$

$$< \frac{c_5}{\sigma - 1} e^{-\lambda_m(\sigma-1)}.$$

On the part of the boundary belonging to $\sigma < 1$ we have

$$\left| \sum_{n=1}^{m} b_n e^{-\lambda_n s} \right| = \left| B(\lambda_m) e^{-\lambda_m s} + s \int_{-\infty}^{\lambda_m} B(u) e^{-us}\, du \right|$$

$$\leqq c_1 \left(e^{-\lambda_m(\sigma-1)} + |s| \int_{-\infty}^{\lambda_m} e^{-u(\sigma-1)}\, du \right) < \frac{c_6}{1 - \sigma} e^{-\lambda_m(\sigma-1)},$$

$$\left| g(s) - \sum_{n=1}^{m} b_n e^{-\lambda_n s} \right| < c_7 + \frac{c_6}{1-\sigma} e^{-\lambda_m(\sigma-1)} < \frac{c_8}{1-\sigma} e^{-\lambda_m(\sigma-1)}.$$

On the entire boundary we therefore have

$$|g_m(s)| < c_9.$$

Thus

$$\left| g(1) - \sum_{n=1}^{m} b_n e^{-\lambda_n} \right| c_3^2 = |g_m(1)| < c_9,$$

$$\left| \sum_{n=1}^{m} b_n e^{-\lambda_n} \right| < c_{10}.$$

Göttingen, 26 October 1932.
(Received 27 October 1932)

6.

Zu dem Integralsatz von Cauchy

Mathematische Zeitschrift 37 (1933), 37–38
[Commentary on p. 568]

In einer kürzlich erschienenen gleichnamigen Arbeit[1]) macht Herr Kamke darauf aufmerksam, daß bisher für die folgende Fassung des Cauchyschen Satzes noch kein Beweis bekannt ist.

Satz: *Ist $\mathfrak{C}$ eine geschlossene rektifizierbare Jordankurve und ist die Funktion $f(z)$ innerhalb $\mathfrak{C}$ regulär und auf dem aus dem Innern von $\mathfrak{C}$ und $\mathfrak{C}$ selber bestehenden abgeschlossenen Bereich stetig, so gilt*

$$\int_{\mathfrak{C}} f(z)\,dz = 0,$$

wenn die Kurve $\mathfrak{C}$ in positivem Sinn durchlaufen wird.

Für diesen Satz gebe ich im folgenden einen Beweis, der sich stützt

1. auf den Riemannschen Abbildungssatz mit der Carathéodoryschen Verschärfung über die Ränderzuordnung[2]),

2. auf den Fejérschen Satz, daß eine für $|x| \leqq 1$ stetige und für $|x| < 1$ reguläre Funktion für $|x| = 1$ gleichmäßig durch Polynome approximiert werden kann[3]),

3. auf die einfachsten Sätze über Stieltjes-Integrale[4]).

Beweis: Es sei G das Innere von $\mathfrak{C}$ und $\overline{G} = G + \mathfrak{C}$. Dann gibt es nach Riemann, Carathéodory eine Funktion $g(x)$, die den abgeschlossenen Einheitskreis schlicht und stetig auf $\overline{G}$ und den offenen Einheitskreis

---

[1]) Mathem. Zeitschr. Bd. 35 (1932), S. 539—543.

[2]) Siehe etwa Hurwitz-Courant, Vorlesungen über allgemeine Funktionentheorie, 3. Aufl. (1929), S. 400—405.

[3]) Siehe Fejér, Untersuchungen über Fouriersche Reihen, Mathem. Ann., Bd. 58 (1904), S. 51—69. Der dort bewiesene Satz besagt: Die Fouriersche Reihe einer stetigen Funktion der Periode 2π ist gleichmäßig (C, 1) summabel. Wegen der formalen Gleichheit der Fourierreihe mit der Potenzreihe folgt hieraus, daß in unserem Falle die Potenzreihe auf dem Rand des Einheitskreises gleichmäßig (C, 1) summabel ist, was den im Text genannten Satz enthält.

[4]) Die gebrauchten Sätze stehen z. B. bei Perron, Die Lehre von den Kettenbrüchen, 2. Aufl. (1929), S. 362—367.

konform auf G abbildet. Also ist $f(g(x))$ eine Funktion, die für $|x| \leqq 1$ stetig und für $|x| < 1$ regulär ist. Nach der Definition des komplexen Integrals ist

$$\int_{\mathfrak{C}} f(z)\, dz = \int_{|x|=1} f(g(x))\, d\, g(x),$$

wo das rechtsstehende Stieltjes-Integral über den Einheitskreis zu nehmen ist. Die Frage, in welcher Richtung der Einheitskreis durchlaufen wird, ist für das Folgende belanglos, doch sei der Umlaufssinn von nun an fest. Nach dem Gesagten ist $f(g(x))$ für $|x| = 1$ gleichmäßig durch Polynome in x approximierbar, es genügt also zu zeigen, für jedes Polynom $P(x)$ ist

$$\int_{|x|=1} P(x)\, d\, g(x) = 0.$$

Da $P(x)$ und $g(x)$ stetige Funktionen von beschränkter Variation sind, ist

$$\int_{|x|=1} P(x)\, d\, g(x) = -\int_{|x|=1} g(x)\, d\, P(x) = -\int_{|x|=1} g(x)\, P'(x)\, d\, x = 0,$$

da $g(x)$ nach Fejér für $|x| = 1$ gleichmäßig durch Polynome $Q(x)$ approximierbar ist, und der Cauchysche Satz für das Polynom $Q(x)\, P'(x)$ trivial ist.

Zusatz bei der Korrektur: Ich bemerke, daß der im Text bewiesene Satz bereits bei Herrn Pollard, On the conditions for Cauchys theorem, Proc. of the Lond. Math. Soc. **21** (1923), S. 456—482, steht. Der dort angegebene Beweis ist elementar, aber komplizierter als der meinige.

(Eingegangen am 12. Juli 1932.)

[Göttingen]

6t.

On the Integral Theorem of Cauchy

Translation of
Mathematische Zeitschrift 37 (1933), 37–38
[Commentary on p. 568]

In an article of the same name,[1] Kamke points out that up to now no proof has been known for the following version of Cauchy's theorem.

Theorem. *If $\mathfrak{C}$ is a closed rectifiable Jordan curve and if the function $f(z)$ is continuous on the closed domain consisting of the interior of $\mathfrak{C}$ and $\mathfrak{C}$ itself, then*

$$\int_{\mathfrak{C}} f(z)\, dz = 0,$$

if the curve $\mathfrak{C}$ is traversed in a positive sense.

For this theorem I shall give below a proof which depends

1. on Riemann's mapping theorem, together with Carathéodory's sharpening concerning the mapping of the boundary,[2]
2. on Fejér's theorem that a function which is continuous on $|x| \leqq 1$ and regular in $|x| < 1$ can be approximated uniformly in $|x| = 1$ by polynomials,[3]
3. on the simplest theorems about Stieltjes integrals.[4]

[1] *Mathem. Zeitschr.* **35** (1932), 539–543.

[2] See, e.g., Hurwitz-Courant, *Vorlesungen über allgemeine Funktionentheorie*, 3rd ed., 1929, pp. 400–405.

[3] See Fejér, Untersuchungen über Fouriersche Reihen, *Mathem. Ann.* **58** (1904), 51–69. The theorem proved there states: The Fourier series of a continuous function of period 2π is uniformly $(C, 1)$-summable. Because of the formal equality between Fourier series and power series it follows from this that in our case the power series is uniformly $(C, 1)$-summable on the boundary. This statement includes the theorem mentioned in the text.

[4] The theorems employed can be found in, e.g., Perron, *Die Lehre von den Kettenbrüchen*, 2nd ed., 1929, pp. 362–367.

Proof. Let G be the interior of $\mathfrak{C}$ and $\overline{G} = G \cup \mathfrak{C}$. Then by Riemann and Carathéodory there exists a function $g(x)$ which maps the closed unit disc univalently and continuously onto $\overline{G}$ and the open unit disc conformally onto G. Then $f(g(x))$ is a function which is continuous on $|x| \leqq 1$ and regular on $|x| < 1$. By definition of the complex integral we have

$$\int_{\mathfrak{C}} f(z)\, dz = \int_{|x|=1} f(g(x))\, dg(x),$$

where the Stieltjes integral on the right-hand side is to be taken over the unit circle. The direction in which the unit circle is traversed is unimportant for what follows, but from now on let this be fixed. By what has been said, $f(g(x))$ can be approximated uniformly on $|x| = 1$ by polynomials in x; it is therefore sufficient to show that for every polynomial $P(x)$ we have

$$\int_{|x|=1} P(x)\, dg(x) = 0$$

Since $P(x)$ and $g(x)$ are continuous functions of bounded variation, we have

$$\int_{|x|=1} P(x)\, dg(x) = -\int_{|x|=1} g(x)\, dP(x) = -\int_{|x|=1} g(x)P'(x)\, dx = 0,$$

because, by Fejér, $g(x)$ can be uniformly approximated on $|x| = 1$ by polynomials $Q(x)$ and because Cauchy's theorem is trivial for the polynomial $Q(x)P'(x)$.

Added in proof. I notice that the theorem proved in the text is also contained in Pollard, On the conditions for Cauchy's theorem, *Proc. London Math. Soc.* **21** (1923), 456–482. The proof given there is elementary, but more complicated than my own.

(Received 12 July 1932)
[Göttingen]

7.

On the Class-Number in Imaginary Quadratic Fields

The Quarterly Journal of Mathematics, Oxford Series, 5 (1934), 150–160.
[Commentary on p. 571]

THERE is an old conjecture of Gauss* that the number of classes of definite binary quadratic forms

$$Q(x,y) = ax^2+bxy+cy^2 \qquad (a > 0, \quad (a,b,c) = 1)$$

of discriminant

$$d = b^2-4ac$$

tends to infinity as $d \to -\infty$ through even values. The latter restriction is due to the fact that Gauss considered only forms with an even b.

In this paper the following slightly more general theorem will be proved.

THEOREM I. *If $h(d)$ denotes the number of classes of non-equivalent forms of discriminant d, then*

$$h(d) \to \infty \quad as \quad d \to -\infty. \tag{1}$$

Theorem I is equivalent to both of the following theorems:

THEOREM II. *If d runs through all negative fundamental† discriminants, then* (1) *is true.*

THEOREM III. *The number of ideal classes in the imaginary quadratic field $\mathrm{P}(\sqrt{d})$ of discriminant d tends to infinity as $d \to -\infty$.*

To establish the equivalence of Theorems I, II, and III we need two lemmas:

LEMMA I.‡ *If d_1 and d_2 are discriminants, and if there is an integer g such that*

$$d_1 = g^2 d_2,$$

then

$$h(d_1) = |g|h(d_2) \prod_{p|g} \left\{1-\left(\frac{d_2}{p}\right)\frac{1}{p}\right\} \geqslant h(d_2).$$

LEMMA II.§ *To every ideal class in the field $\mathrm{P}(\sqrt{d})$ of discriminant $d < -4$ corresponds one class of forms of discriminant d representing*

* *Disquisitiones Arithmeticae* (1801), Art. 303.

† A discriminant d is called fundamental if for every discriminant $d' \neq d$ the number $\sqrt{(d/d')}$ is not a rational integer.

‡ Landau, *Vorlesungen über Zahlentheorie*, Bd. 1, Satz 214 and Satz 209.

§ Ibid., Bd. 3, Teil XI, Kap. 3.

twice each the norms of all integer ideals in that class. To the principal class corresponds the form

$$x^2+xy+\tfrac{1}{4}(1-d)y^2 \quad \textit{if} \quad d \equiv 1 \pmod 4,$$
$$x^2-\tfrac{1}{4}dy^2 \quad \textit{if} \quad d \equiv 0 \pmod 4.$$

This correspondence can be made one-to-one.

Now we can prove the equivalence of the three theorems. As every fundamental discriminant (of a form) is also the discriminant of a field and vice versa, it is obvious by Lemma II that Theorems II and III are equivalent. Theorem II is contained in Theorem I, and Theorem I is a consequence of Theorem II and Lemma I.

Therefore it is sufficient to prove Theorem II or Theorem III only. We shall prove them both together, using sometimes the language of forms, sometimes the language of algebraic numbers, as is most convenient.

Many important steps have already been made towards proving the Gauss hypothesis. Two of them are relevant here.

THEOREM IV (Hecke).* *If*

$$L(s,\chi) = \sum_{n=1}^{\infty} \chi(n) n^{-s} \neq 0$$

for $\sigma > \frac{1}{2}$ *and for all real characters* χ *different from the principal character, then* (1) *holds.*

This theorem seems to be a very natural one; on the other hand the following surprising result has been obtained by Deuring.

THEOREM V (Deuring).† *If Riemann's function* $\zeta(s)$ *has at least one zero in the half-plane* $\sigma > \frac{1}{2}$, *then*

$$\varlimsup_{d\to-\infty} h(d) \geqslant 2.$$

Quite recently Mordell‡ has proved

THEOREM VI. (1) *holds under the assumption of Theorem V.*

In this paper we shall prove a generalization of Theorem VI.

THEOREM VII. *If there is, to modulus m, at least one real character* χ, *principal or not, so that*

$$L(\rho,\chi) = 0$$

for at least one ρ *in the half-plane* $\sigma > \frac{1}{2}$, *then* (1) *is true.*

* A proof is published in Landau's paper: 'Über die Klassenzahl imaginär-quadratischer Zahlkörper': *Göttinger Nachrichten* (1918), 285–95.

† Deuring, 'Imaginär-quadratische Zahlkörper mit der Klassenzahl (1)': *Math. Zeits.* 37 (1933), 405–15.

‡ *Proc. London Math. Soc.*, in press.

As the assumptions of either Theorem IV or Theorem VII are true, it is sufficient to prove Theorem VII. We shall assume throughout the whole paper that (1) does not hold and that the assumption of Theorem VII is satisfied which will give us a contradiction.

Notations

All *italic* letters except o, O, s, L, Q denote rational integers. $\chi(n)$ denotes a real character mod m $(m > 0)$ such that

$$L_0(s) = L(s, \chi) = \sum_{n=1}^{\infty} \chi(n) n^{-s}$$

vanishes for at least one ρ in the half-plane $\sigma > \frac{1}{2}$. We put

$$\sigma_m = \tfrac{1}{2}(\tfrac{1}{2} + \Re\rho).$$

Obviously

$$\tfrac{1}{2} < \sigma_m < \Re\rho < 1.$$

We suppose that m, χ, ρ, and σ_m are fixed throughout the paper.

We assume the existence of a positive H fixed throughout the paper such that there is an infinite number of discriminants d satisfying the relation

$$h(d) = H.$$

Being discriminants of quadratic fields, the numbers d satisfy the conditions

$$d \equiv 1, 5, 8, 9, 12, 13 \pmod{16};$$

$$\mu(-d) \neq 0 \quad \text{if} \quad d \equiv 1 \pmod 4,$$

$$\mu(-\tfrac{1}{4}d) \neq 0 \quad \text{if} \quad d \equiv 0 \pmod 4;$$

$$d \leqslant -3.$$

It is convenient to assume at once

$$d < -4.$$

All discriminants mentioned in the following text of this paper are supposed to satisfy all these conditions.

$$Q(x, y) = ax^2 + bxy + cy^2 = \frac{1}{4a}\{(2ax + by)^2 + |d| y^2\}$$

is called a 'reduced quadratic form' (shortly a 'form') of discriminant d if the following conditions are satisfied:

$$d = b^2 - 4ac,$$

$$-a < b \leqslant a \leqslant c,$$

$$b + 1 + |b + 1| + c - a > 0.$$

Only forms of this type will occur in this paper.

By a well-known theorem* no two of these H forms are equivalent. We introduce the following Dirichlet series:

$$\zeta(s) = \sum_{n=1}^{\infty} n^{-s},$$

$$\zeta(s,\omega) = \sum_{n=0}^{\infty} (n+\omega)^{-s} \quad (0 < \omega \leqslant 1),$$

$$\zeta_Q(s) = \tfrac{1}{2} \sum_{x,y \neq 0,0} Q(x,y)^{-s},$$

$$\zeta_d(s) = \sum_{Q} \zeta_Q(s),$$

where Q runs through the H forms of discriminant d;

$$L_0(s) = \sum_{n=1}^{\infty} \chi(n) n^{-s},$$

$$L_1(s) = \sum_{n=1}^{\infty} \left(\frac{d}{n}\right) n^{-s},$$

$$L_2(s) = \sum_{n=1}^{\infty} \chi(n) \left(\frac{d}{n}\right) n^{-s}.$$

Every series belonging to one of these seven types is absolutely convergent and regular for $\sigma > 1$.

The constants implied in the signs O and o depend only on H, m, ρ, σ_m, but they are independent of s, d, Q.

Some old lemmas

LEMMA III.† *If d is divisible by exactly t different primes, then*

$$2^{t-1} \mid h(d).$$

LEMMA IV.‡ *The functions*

$$\zeta(s),\ \zeta(s,\omega),\ \zeta_Q(s),\ \zeta_d(s),\ L_0(s),\ L_1(s),\ L_2(s)$$

are regular for $\sigma > 0$, $s \neq 1$.

LEMMA V.§ $\zeta_d(s) = \zeta(s) L_1(s)$ *for* $\sigma > 0$, $s \neq 1$.

LEMMA VI.|| $3a^2 \leqslant |d|$.

LEMMA VII.** $Q(x,y) \geqslant a$ *for x, y not both zero.*

* Landau, *Vorlesungen über Zahlentheorie*, Bd. 1, Satz 198.

† Gauss, *Disquisitiones Arithmeticae*, Art. 257.

‡ Landau, *Vorlesungen über Zahlentheorie*, Bd. 2, Satz 369; and *Einführung in die elementare und analytische Theorie der algebraischen Zahlen und Ideale*, Satz 154 (1).

§ Dirichlet-Dedekind, *Vorlesungen über Zahlentheorie*, 3. Aufl. (1879), § 90.

|| $3a^2 \leqslant 4ac - a^2 \leqslant 4ac - b^2 = |d|$.

** $Q(x,y) \geqslant ax^2 - |bxy| + cy^2 \geqslant a(x^2 - |xy| + y^2) \geqslant a$ for $x^2 + y^2 > 0$.

Further analytical lemmas

LEMMA VIII.

$$\gamma(s) = \gamma(s, m, l_1, y, Q) = \sum_{\substack{x=-\infty \\ x \equiv l_1 (m)}}^{\infty} Q(x, y)^{-s}$$

$$= 2^{2s-1} a^{s-1} m^{-1} |d|^{\frac{1}{2}-s} y^{1-2s} \int_{-\infty}^{\infty} (\eta^2+1)^{-s} \, d\eta + O(|s| a^{\sigma} |d|^{-\sigma} y^{-2\sigma})$$

for $y \geqslant 1$, $\frac{1}{2} < \sigma < 2$.

Proof. We put

$$x = mz + l_1,$$

$$\alpha = (2am)^{-1}(2al_1 + by),$$

$$\beta = (2am)^{-1} |d|^{\frac{1}{2}} y.$$

Then

$$2ax + by = 2amz + 2al_1 + by$$

$$= 2am(z + \alpha).$$

Hence

$$Q(x, y) = \frac{1}{4a}\{(2ax+by)^2 + |d| y^2\}$$

$$= am^2\{(z+\alpha)^2 + \beta^2\},$$

$$a^s m^{2s} \gamma(s) = \sum_{z=-\infty}^{\infty} \{(z+\alpha)^2 + \beta^2\}^{-s}.$$

Hence we get, by the Euler-Maclaurin formula,

$$a^s m^{2s} \gamma(s) = \int_{-\infty}^{\infty} \{(\xi+\alpha)^2 + \beta^2\}^{-s} \, d\xi + \int_{-\infty}^{\infty} (\xi - [\xi] - \tfrac{1}{2}) \frac{d}{d\xi} \{(\xi-\alpha)^2 + \beta^2\}^{-s} \, d\xi$$

$$= \beta^{1-2s} \int_{-\infty}^{\infty} (\eta^2+1)^{-s} \, d\eta + \beta^{-2s} \int_{-\infty}^{\infty} (\beta\eta - \alpha - [\beta\eta - \alpha] - \tfrac{1}{2}) \frac{d}{d\eta} (\eta^2+1)^{-s} \, d\eta.$$

Using the inequality

$$\left| \int_{-\infty}^{\infty} (\beta\eta - \alpha - [\beta\eta - \alpha] - \tfrac{1}{2}) \frac{d}{d\eta} (\eta^2+1)^{-s} \, d\eta \right.$$

$$\leqslant \frac{1}{2} \int_{-\infty}^{\infty} \left| \frac{d}{d\eta} (\eta^2+1)^{-s} \right| d\eta$$

$$= \frac{|s|}{\sigma} \int_{0}^{\infty} \frac{d}{d\eta} (\eta^2+1)^{-\sigma} \, d\eta$$

$$= \frac{|s|}{\sigma} = O|s|$$

we get

$$\gamma(s) = a^{-s}m^{-2s}\beta^{1-2s}\int_{-\infty}^{\infty}(\eta^2+1)^{-s}\,d\eta + O(|s|a^{-\sigma}m^{-2\sigma}\beta^{-2\sigma})$$

$$= 2^{2s-1}a^{s-1}m^{-1}|d|^{\frac{1}{2}-s}y^{1-2s}\int_{-\infty}^{\infty}(\eta^2+1)^{-s}\,d\eta + O(|s|a^{\sigma}|d|^{-\sigma}y^{-2\sigma}).$$

LEMMA IX. *If* $1 \leqslant l_2 \leqslant m$, *and if, for* $\sigma > 1$,

$$\phi(s) = \phi(s,m,l_1,l_2,Q) = \sum_{\substack{y=1\\ y\equiv l_2(m)}}^{\infty}\ \sum_{\substack{x=-\infty\\ x\equiv l_1(m)}}^{\infty} Q(x,y)^{-s} = \sum_{\substack{y=1\\ y\equiv l_2(m)}}^{\infty}\gamma(s,l_1,m,y,Q),$$

then $\phi(s)$ *is regular for* $\sigma_m < \sigma < 2$, $s \neq 1$, *and*

$$\phi(s) = o|s| + o\left|\frac{1}{s-1}\right|$$

as $d \to -\infty$.

Proof. We put, for $\sigma > \frac{1}{2}$,

$$\psi(s,y) = \psi(s,m,l_1,y,Q)$$

$$= \gamma(s,m,l_1,y,Q) - 2^{2s-1}a^{s-1}m^{-1}|d|^{\frac{1}{2}-s}y^{1-2s}\int_{-\infty}^{\infty}(\eta^2+1)^{-s}\,d\eta.$$

Then, by Lemma VIII,

$$\sum_{\substack{y=1\\ y\equiv l_2(m)}}^{\infty}\psi(s,y) = \Psi(s) = \Psi(s,m,l_1,l_2,Q)$$

is uniformly convergent in every bounded region belonging to the strip $\sigma_m < \sigma < 2$ and therefore $\psi(s)$ is regular for $\sigma_m < \sigma < 2$.

Further, we get by Lemma VIII

$$\Psi(s) = O\{|s|a^{\sigma}|d|^{-\sigma}\zeta(2\sigma)\} = O(|s|a^{\sigma}|d|^{-\sigma}).$$

The identity

$$\phi(s) - \Psi(s) = 2^{2s-1}a^{s-1}m^{-1}|d|^{\frac{1}{2}-s}\int_{-\infty}^{\infty}(\eta^2+1)^{-s}\,d\eta \sum_{\substack{y=1\\ y\equiv l_2(m)}}^{\infty} y^{1-2s}$$

$$= 2^{2s-1}a^{s-1}m^{-2s}|d|^{\frac{1}{2}-s}\int_{-\infty}^{\infty}(\eta^2+1)^{-s}\,d\eta\ \zeta\left(s,\frac{l_2}{m}\right)$$

holds for the whole strip $\sigma_m < \sigma < 2$ except $s = 1$, since all functions are regular there. Using the inequalities

$$\int_{-\infty}^{\infty}(\eta^2+1)^{-s}\,d\eta = O(1),$$

$$\zeta\left(s,\frac{l_2}{m}\right) = O|s| + O\left|\frac{1}{s-1}\right|,$$

both valid for $\sigma > \sigma_m$, we obtain, for $\sigma_m < \sigma < 2$, $s \neq 1$,

$$\begin{aligned}\phi(s) &= O(|s|a^{\sigma}|d|^{-\sigma})+O\left\{a^{\sigma-1}|d|^{\frac{1}{2}-\sigma}\left(|s|+\frac{1}{|s-1|}\right)\right\}\\ &= O\left\{\left(|s|+\frac{1}{|s-1|}\right)\left|\frac{d}{a^2}\right|^{\frac{1}{2}-\sigma}(a^{1-\sigma}|d|^{-\frac{1}{2}}+a^{-\sigma})\right\}\\ &= O\left\{\left(|s|+\frac{1}{|s-1|}\right)\left|\frac{d}{a^2}\right|^{\frac{1}{2}-\sigma}a^{-\sigma}\left(\left|\frac{d}{a^2}\right|^{-\frac{1}{2}}+1\right)\right\}\\ &= o\left(|s|+\frac{1}{|s-1|}\right),\end{aligned}$$

by Lemma VI.

Lemma X. *For* $\sigma_m < \sigma < 2$, $s \neq 1$,

$$L_0(s)L_2(s) = \zeta(2s)\prod_{p|m}(1-p^{-2s})\sum_a \chi(a)a^{-s}+o\left(|s|+\frac{1}{|s-1|}\right)$$

as $d \to -\infty$, *where* a *runs through the minima of the* H *forms of discriminant* d.

Proof. Let $e_{n,Q}$ denote the number of representations of n by the form Q. Then we know by Lemma V, for $\sigma > 1$,

$$\begin{aligned}\sum_Q \tfrac{1}{2}\sum_{n=1}^{\infty} e_{n,Q}\, n^{-s} = \sum_Q \zeta_Q(s) = \zeta_d(s) = \zeta(s)L_1(s) &= \sum_{n=1}^{\infty} n^{-s}\sum_{n=1}^{\infty}\left(\frac{d}{n}\right)n^{-s}\\ &= \sum_{n=1}^{\infty} n^{-s}\sum_{g|n}\left(\frac{d}{g}\right).\end{aligned}$$

Hence, by the uniqueness of coefficients of Dirichlet's series,

$$\sum_{g|n}\left(\frac{d}{g}\right) = \tfrac{1}{2}\sum_Q e_{n,Q},$$

$$\begin{aligned}L_0(s)L_2(s) &= \sum_{n=1}^{\infty}\chi(n)n^{-s}\sum_{n=1}^{\infty}\chi(n)\left(\frac{d}{n}\right)n^{-s}\\ &= \sum_{n=1}^{\infty}\chi(n)n^{-s}\sum_{g|n}\left(\frac{d}{g}\right)\\ &= \sum_{n=1}^{\infty}\chi(n)n^{-s}\tfrac{1}{2}\sum_Q e_{n,Q}\\ &= \sum_Q \tfrac{1}{2}\sum_{x,y\neq 0,0}\chi\{Q(x,y)\}Q(x,y)^{-s}\\ &= \sum_Q\sum_{x=1}^{\infty}\chi\{Q(x,0)\}Q(x,0)^{-s}+\sum_Q\sum_{y=1}^{\infty}\sum_{x=-\infty}^{\infty}\chi\{Q(x,y)\}Q(x,y)^{-s}\end{aligned}$$

$$= \sum_{Q} \sum_{x=1} \chi(ax^2)(ax^2)^{-s} + \sum_{Q} \sum_{l_1=1}^{m} \sum_{l_2=1}^{m} \sum_{\substack{y=1 \\ y\equiv l_2(m)}}^{\infty} \sum_{\substack{x=-\infty \\ x\equiv l_1(m)}}^{\infty} \chi\{Q(x,y)\}Q(x,y)^{-s}$$

$$= \sum_{Q} \chi(a)a^{-s} \sum_{\substack{x=1 \\ (x,m)=1}}^{\infty} x^{-2s} + \sum_{Q} \sum_{l_1=1}^{m} \sum_{l_2=1}^{m} \chi\{Q(l_1,l_2)\}\phi(s,m,l_1,l_2,Q),$$

and this identity holds for $\sigma_m < \sigma < 2$ by Lemma IX. Using the second part of Lemma IX and the relation

$$\sum_{\substack{x=1 \\ (x,m)=1}}^{\infty} x^{-2s} = \zeta(2s) \prod_{p|m} (1-p^{-2s}) \quad \text{for} \quad \sigma > \tfrac{1}{2}$$

we get the desired result.

Further arithmetical lemmas

Lemma XI. *If $a \mid d^k$ for some $k > 0$, then $\mu(a) \neq 0$, and therefore $a \mid d$.*

Proof. Otherwise there would be a prime p such that

$$\begin{aligned} &p^2 \mid a \mid d^k, \\ &p \mid d, \\ &p \mid (d+4ac), \\ &p \mid b^2, \\ &p^2 \mid b^2, \\ &p^2 \mid (b^2-4ac) \mid d, \\ &p = 2; \\ &d \equiv 0 \pmod 4, \\ &d \equiv b^2 \pmod{16}, \\ &d \equiv 0 \text{ or } 4 \pmod{16}, \end{aligned}$$

which is a contradiction.

Lemma XII. *If $A > 0$, $A \mid d$, $\mu(A) \neq 0$, then there is at most one form of discriminant d with minimum A.*

Proof. Let (a,b,c) and (a',b',c') be the coefficients of two different forms of this type. Then

$$a = a' = A, \qquad b \neq b', \qquad c \neq c', \qquad b^2-4ac = b'^2-4a'c' = d,$$
$$-A < b \leqslant A, \qquad -A < b' \leqslant A.$$

The congruence $\qquad x^2 \equiv d \equiv 0 \pmod A$

has only the solutions $x = 0$ and $x = A$ for $-A < x \leqslant A$. Hence, without restriction of generality,

$$b = 0, \qquad b' = A, \qquad c = \frac{-d}{4A}, \qquad c' = \frac{A^2-d}{4A}$$

which is a contradiction since

$$c-c' = -\tfrac{1}{4}A$$

is not an integer.

LEMMA XIII. *If under the assumptions of Lemma XII*

$$A \leqslant \tfrac{1}{2}|d|^{\frac{1}{2}},$$

then there is at least one form of discriminant d with minimum A.

Proof. We put

$$\text{(i)} \qquad Q(x,y) = Ax^2+Axy+\frac{A^2-d}{4A}y^2$$

if $d \equiv 1 \pmod 4$ and if $d \equiv 12 \pmod{16}$, $A \equiv 0 \pmod 2$;

$$\text{(ii)} \qquad Q(x,y) = Ax^2-\frac{d}{4A}y^2$$

if $d \equiv 12 \pmod{16}$, $A \equiv 1 \pmod 2$, and if $d \equiv 8 \pmod{16}$. In both cases the coefficients are integers since

$$A \mid d, \qquad 4 \nmid A;$$

the discriminant is d and the form is reduced since

$$0 < A = \frac{4A^2}{4A} \leqslant \frac{-d}{4A} < \frac{A^2-d}{4A}.$$

LEMMA XIV. *If $a \nmid d^k$ for any $k > 0$, then $a \geqslant \frac{1}{4}|d|^{1/H}$.*

Proof. Let $\mathfrak{C}$ denote the class of ideals corresponding to the form

$$ax^2+bxy+cy^2$$

by Lemma II. Then we know the existence of an integer ideal $\mathfrak{a}$ in $\mathfrak{C}$ such that

$$N\mathfrak{a} = a.$$

Let p be a prime dividing a but not dividing d. Then

$$p \nmid \mathfrak{a}.$$

Otherwise the ideal $p^{-1}\mathfrak{a}$ would be an integer ideal in $\mathfrak{C}$ and

$$p^{-2}a = N(p^{-1}\mathfrak{a})$$

would be representable by our form which, by Lemma VII, is impossible. So we know, $\mathfrak{a}'$ denoting the conjugate ideal to $\mathfrak{a}$, that

$$p \nmid \mathfrak{a}, \qquad p \nmid \mathfrak{a}', \qquad p \mid \mathfrak{a}\mathfrak{a}'.$$

Therefore p splits up in two prime ideals $\mathfrak{p}$ and $\mathfrak{p}'$ such that

$$\mathfrak{p} \mid \mathfrak{a}, \qquad \mathfrak{p}' \mid \mathfrak{a}', \qquad \mathfrak{p}\mathfrak{p}' \nmid \mathfrak{a}, \qquad \mathfrak{p}\mathfrak{p}' \nmid \mathfrak{a}'.$$

As $p \nmid d$ the ideals $\mathfrak{p}$ and $\mathfrak{p}'$ are different. Therefore

$$\mathfrak{p}' \nmid \mathfrak{a},$$
$$\mathfrak{a} \neq \mathfrak{a}',$$
$$\mathfrak{a}^H \neq \mathfrak{a}'^H.$$

As H is the class-number of our field $P(\sqrt{d})$, these two different ideals are principal ideals and their norm a^H is at least four times representable by the principal form, i.e. there is at least one representation by the principal form with $y > 0$:

$$a^H = x^2+xy+\tfrac{1}{4}(1-d)y^2$$
$$= \tfrac{1}{4}\{(2x+y)^2+|d|y^2\} \geqslant \tfrac{1}{4}|d| \quad \text{if} \quad d \equiv 1 \pmod 4,$$
$$a^H = x^2-\tfrac{1}{4}dy^2 \geqslant \tfrac{1}{4}|d| \quad \text{if} \quad d \equiv 0 \pmod 4).$$
$$a \geqslant \tfrac{1}{4}|d|^{1/H}.$$

LEMMA XV. *If a runs through the minima of the H quadratic forms belonging to d, then, if $\sigma \geqslant \frac{1}{2}$,*

$$\Big|\sum_a \chi(a)a^{-s}\Big| \geqslant \tfrac{1}{4}H^2+o(1),$$

for $d \to -\infty$.

Proof. Let t denote the number of different prime divisors of d. Then, by Lemma III,

$$2^t \leqslant 2H.$$

Using Lemmas XI, XII, XIII, and XIV, we get

$$\Big|\sum_a \chi(a)a^{-s}-\sum_{n|d} \chi(n)\mu^2(n)n^{-s}\Big|$$
$$= \Big|\sum_{a \nmid d} \chi(a)a^{-s}-\sum_{\substack{n|d\\ n\neq a}} \chi(n)\mu^2(n)n^{-s}\Big|$$
$$\leqslant (\tfrac{1}{4}|d|^{1/H})^{-\sigma}\sum_a 1+(\tfrac{1}{2}|d|^{\frac{1}{2}})^{-\sigma}\sum_{n|d}\mu^2(n)$$
$$= o(1)H+o(1)2^t$$
$$= o(1).$$

Hence

$$\Big|\sum_a \chi(a)a^{-s}\Big| \geqslant o(1)+\Big|\sum_{n|d}\chi(n)\mu^2(n)n^{-s}\Big|$$
$$= o(1)+\Big|\prod_{p|d}(1+\chi(p)p^{-s})\Big|$$
$$\geqslant o(1)+\prod_{p|d}(1-2^{-\frac{1}{2}})$$
$$= o(1)+\left(\frac{1-2^{-1}}{1+2^{-\frac{1}{2}}}\right)^t$$
$$\geqslant o(1)+4^{-t}$$
$$\geqslant \tfrac{1}{4}H^{-2}+o(1).$$

Proof of Theorem VII

We put $s = \rho$ in Lemma X and we let d tend to $-\infty$. Then we get

$$0 = \lim_{d \to -\infty} \zeta(2\rho) \prod_{p \mid m} (1-p^{-2\rho}) \sum_{a} \chi(a) a^{-\rho},$$

and, since

$$\zeta(2\rho) \prod_{p \mid m} (1-p^{-2\rho})$$

does not vanish and is independent of d,

$$\lim_{d \to -\infty} \sum_{a} \chi(a) a^{-\rho} = 0,$$

which is a contradiction to Lemma XV.

[Received 24 March 1934] [Bristol]

8.

On the Imaginary Quadratic Corpora of Class-Number One

(With E. H. Linfoot)

The Quarterly Journal of Mathematics, Oxford Series, 5 (1934), 293–301.
[Commentary on p. 571]

LET $h(d)$ denote the number of classes of positive non-equivalent forms

$$ax^2+bxy+cy^2$$

with integral coefficients and discriminant

$$d = b^2-4ac.$$

We shall prove the following

MAIN THEOREM. *There are at most ten negative fundamental discriminants* d for which*

$$h(d) = 1. \tag{1}$$

Nine are known, namely,

$$-3,\ -4,\ -7,\ -8,\ -11,\ -19,\ -43,\ -67,\ -163$$

and Dickson showed† by calculation that there are no others down to -15.10^5, while Lehmer‡ improved this limit last year to -5.10^8.

Up to the present, three methods have been applied with some success to the investigation of the class-number problem and three results stand out as the principal rewards of their application:

1. HECKE'S THEOREM.§ *If $d < 0$ and*

$$L_d(s) = \sum_{n=1}^{\infty} \left(\frac{d}{n}\right) n^{-s} \neq 0$$

for $1-\dfrac{1}{\log|d|} < s < 1$, *then*

$$h(d) > c\frac{|d|^{\frac{1}{2}}}{\log|d|},$$

where c is a positive constant;

2. LANDAU'S THEOREM.|| *To every $w > 1$ corresponds a constant $c > 0$ such that, if*

$$0 > d_1 > d_2, \qquad h(d_1) < c\frac{|d_1|^{\frac{1}{2}}}{\log|d_1|}, \qquad h(d_2) < c\frac{|d_2|^{\frac{1}{2}}}{\log|d_2|},$$

then

$$|d_2| \geqslant |d_1|^w;$$

* A discriminant, i.e. an integer $\equiv 0$ or 1 (mod 4) and not a square, is fundamental, if it contains no squared factor $q^2 > 1$ such that d/q^2 is also a discriminant.

† L. E. Dickson, *Bull. American Math. Soc.* (2) 17 (1911), 534–7.

‡ D. H. Lehmer, *Bull. American Math. Soc.* (2) 39 (1933), 369.

§ A proof has been published by Landau, *Göttinger Nachrichten* (1913), 285–95.

|| E. Landau, loc. cit.

3. DEURING'S THEOREM.* *If $\zeta(s)$ has a zero in $\sigma > \frac{1}{2}$, then $h(d) > 1$ for all sufficiently large negative d.*

The last result has been recently generalized to

THEOREM 3′.† *If for at least one $d < 0$, $L_d(s)$ has a zero in $\sigma > \frac{1}{2}$, then $h(d) \to \infty$ as $d \to -\infty$.*

A proof of this theorem was given by Heilbronn. Theorems 1 and 3′ together show that $h(d) \to \infty$ as $d \to -\infty$, but it seems very difficult to modify the arguments so as to obtain an actual lower bound for $h(d)$ when d is large. However, by introducing also the idea contained in Landau's theorem we are able at least to find an upper bound for the size of the discriminants $d < 0$ for which $h(d) = 1$, with one possible exception. We show that *there is at most one fundamental discriminant $d < -10^4$ for which $h(d) = 1$.*

While the arguments of Hecke, Landau, and Deuring were decidedly different in character, the results in the present paper are obtained by the argument of Deuring alone. Instead of Theorems 1, 2, and 3′ above, we use respectively Lemmas 2, 5, and 4, from which modified forms of these theorems could be deduced.‡ Thus all three results enter in a form which can be proved by the arguments of Deuring alone.

It is easily shown that, if $d < 0$ is a fundamental discriminant other than -4 or -8 for which $h(d) = 1$, then d is a prime.§ Thus it is sufficient to prove that the assumption

$$h(-p) = h(-P) = 1$$

for primes p, P such that $10^4 < p < P$, $p \equiv P \equiv 3 \pmod 4$ leads to a contradiction.

NOTATION. We assume p, P are primes such that

$$10^4 < p < P,$$

* M. Deuring, *Math. Zeitschrift*, 37 (1933), 405–15.

† H. Heilbronn, *Quart. J. of Math. (Oxford)*, 5 (1934), 150–160.

‡ Lemma 5 actually gives a sharper result than Theorem 2 for the particular case $h(d) = 1$.

§ E. Hecke, *Algebraische Zahlen*, p. 179, Satz 132, or directly as follows: Every prime divisor of d is representable by a form of discriminant d and therefore, when $h(d) = 1$, by the principal form $x^2 + \frac{1}{4}|d|y^2$ or $x^2 + xy + \frac{1}{4}(1 + |d|)y^2$. In either case $4p = X^2 + |d|Y^2$, where $Y \neq 0$ since $4p$ is not a square, and hence $4p \geqslant |d|$. If $|d| > 16$, $\frac{1}{4}|d| > \sqrt{|d|}$, this gives $p > \sqrt{|d|}$ for every $p|d$, whence $|d|$ is a prime. If $d = -15$, there are two non-equivalent forms, (1, 1, 4) and (2, 1, 2); while -3, -7, -11 are primes.

$$p \equiv P \equiv 3 \pmod 4,$$
$$h(-p) = h(-P) = 1,$$

and have to obtain a contradiction.

For $\sigma = R(s) > 0$, the series

$$L_p(s) = \sum_{n=1}^{\infty} \left(\frac{-p}{n}\right) n^{-s},$$

$$L_{pP}(s) = \sum_{n=1}^{\infty} \left(\frac{pP}{n}\right) n^{-s}$$

represent regular analytic functions,* $\left(\frac{a}{b}\right)$ being Kronecker's symbol.

We write

$$q(x,y) = x^2+xy+\tfrac{1}{4}(p+1)y^2,$$
$$Q(x,y) = x^2+xy+\tfrac{1}{4}(P+1)y^2.$$

Then† for $\sigma > 1$

$$\zeta(s)\, L_p(s) = \tfrac{1}{2} \sum_{x,y \neq 0,0} q(x,y)^{-s}, \tag{2}$$

$$L_p(s)\, L_{pP}(s) = \frac{1}{2} \sum_{x,y \neq 0,0} \left(\frac{-p}{Q(x,y)}\right) Q(x,y)^{-s}. \tag{3}$$

LEMMA 1.
$$\sum_{l=0}^{p-1} \left(\frac{-p}{Q(l,y)}\right) = \begin{cases} -1 & \text{if } p \nmid y, \\ p-1 & \text{if } p \mid y. \end{cases}$$

Proof.

$$p + \sum_{l=0}^{p-1} \left(\frac{-p}{Q(l,y)}\right) = p + \sum_{l=0}^{p-1} \left(\frac{(2l+y)^2+Py^2}{p}\right)$$
$$= \sum_{l=0}^{p-1} \left\{1 + \left(\frac{l^2+Py^2}{p}\right)\right\}$$
$$= \text{number of solutions of } m^2 \equiv l^2+Py^2 \pmod p$$
$$= \text{number of solutions of } uv \equiv Py^2 \pmod p$$
$$= \begin{cases} p-1, & \text{if } p \nmid y, \\ 2p-1, & \text{if } p \mid y. \end{cases}$$

* The notation differs from that of Theorem 1, in which the first series would be written as $L_{-p}(s)$.

† See Heilbronn, loc. cit.

LEMMA 2. *For $s > \frac{1}{2}$*

$$\left|\zeta(s)\,L_p(s)-\zeta(2s)-2^{2s-1}p^{\frac{1}{2}-s}\zeta(2s-1)\int_{-\infty}^{\infty}(u^2+1)^{-s}\,du\right| \leqslant 4^s p^{-s}\zeta(2s).$$

Proof. From (2), we have, for $\sigma > 1$,

$$\begin{aligned}\zeta(s)\,L_p(s) &= \zeta(2s)+\sum_{y=1}^{\infty}\sum_{x=-\infty}^{\infty} q(x,y)^{-s}\\ &= \zeta(2s)+\sum_{y=1}^{\infty}\int_{-\infty}^{\infty} q(x,y)^{-s}\,dx +\\ &\quad+\sum_{y=1}^{\infty}\int_{-\infty}^{\infty}(x-[x]-\tfrac{1}{2})\frac{d}{dx}\{q(x,y)^{-s}\}\,dx, \qquad (4)\end{aligned}$$

by the Euler-Maclaurin formula. Here

$$\begin{aligned}\sum_{y=1}^{\infty}\int_{-\infty}^{\infty} q(x,y)^{-s}\,dx &= \sum_{y=1}^{\infty}\int_{-\infty}^{\infty}\{(x+\tfrac{1}{2}y)^2+\tfrac{1}{4}py^2\}^{-s}\,dx\\ &= \sum_{y=1}^{\infty}(\tfrac{1}{4}py^2)^{\frac{1}{2}-s}\int_{-\infty}^{\infty}(u^2+1)^{-s}\,du\\ &= 2^{2s-1}p^{\frac{1}{2}-s}\zeta(2s-1)\int_{-\infty}^{\infty}(u^2+1)^{-s}\,du,\end{aligned}$$

and this function of s is regular for $\sigma > \frac{1}{2}$, $s \neq 1$. The last term of (4) is, for $\sigma > \frac{1}{2}$, less in absolute value than

$$\begin{aligned}&\sum_{y=1}^{\infty}\left|\int_{-\infty}^{\infty}(x-[x]-\tfrac{1}{2})\frac{d}{dx}\{q(x,y)^{-s}\}\,dx\right|\\ &\leqslant \frac{1}{2}\sum_{y=1}^{\infty}\int_{-\infty}^{\infty}\left|\frac{d}{dx}\left\{\left(x+\frac{y}{2}\right)^2+\tfrac{1}{4}py^2\right\}^{-s}\right|\,dx\\ &= \frac{|s|}{2\sigma}\sum_{y=1}^{\infty}\int_{-\infty}^{\infty}\left|\frac{d}{dx}\left\{\left(x+\frac{y}{2}\right)^2+\tfrac{1}{4}py^2\right\}^{-\sigma}\right|\,dx\\ &= \frac{|s|}{2\sigma}\sum_{y=1}^{\infty}2(\tfrac{1}{4}py^2)^{-\sigma}\\ &= \frac{|s|}{\sigma}4^{\sigma}p^{-\sigma}\zeta(2\sigma).\end{aligned}$$

The lemma now follows by analytic continuation.

LEMMA 3. $$L_p(1) = \pi p^{-\frac{1}{2}}.$$

Proof. For $s > 1$

$$|(s-1)\zeta(s)\,L_p(s)-2^{2s-1}p^{\frac{1}{2}-s}(s-1)\zeta(2s-1)\int_{-\infty}^{\infty}(u^2+1)^{-s}\,du|$$
$$\leqslant (s-1)(1+4^s p^{-s})\zeta(2s),$$

by Lemma 2. The result follows on taking limits as $s \to 1$.

LEMMA 4. *For* $s > \frac{1}{2}$

$$L_p(s)L_{pP}(s)-\zeta(2s)(1-p^{-2s})+$$
$$+2^{2s-1}P^{\frac{1}{2}-s}p^{-1}(1-p^{2(1-s)})\zeta(2s-1)\int_{-\infty}^{\infty}(u^2+1)^{-s}\,du\Big| \leqslant 4^s pP^{-s}\zeta(2s).$$

Proof. By (3) we have, for $\sigma > 1$,

$$L_p(s)L_{pP}(s) = \frac{1}{2}\sum_{x,y\neq 0,0}\left(\frac{-p}{Q(x,y)}\right)Q(x,y)^{-s}$$
$$= \sum_{x=1}^{\infty}\left(\frac{-p}{x^2}\right)x^{-2s}+\sum_{y=1}^{\infty}\sum_{l=0}^{p-1}\left(\frac{-p}{Q(l,y)}\right)\sum_{\substack{x=-\infty\\ x\equiv l(p)}}^{\infty}Q(x,y)^{-s}$$
$$= \zeta(2s)(1-p^{-2s})+\sum_{y=1}^{\infty}\sum_{l=0}^{p-1}\left(\frac{-p}{Q(l,y)}\right)\sum_{z=-\infty}^{\infty}p^{-2s}\left\{\left(z+\frac{l}{p}+\frac{y}{2p}\right)^2+\frac{P}{4p^2}y^2\right\}^{-s}.$$
$$= \zeta(2s)(1-p^{-2s})+$$
$$+\sum_{y=1}^{\infty}\sum_{l=0}^{p-1}\left(\frac{-p}{Q(l,y)}\right)p^{-2s}\left[\int_{-\infty}^{\infty}\left\{\left(z+\frac{l}{p}+\frac{y}{2p}\right)^2+\frac{P}{4p^2}y^2\right\}^{-s}dz\,+\right.$$
$$\left.+\int_{-\infty}^{\infty}(z-[z]-\tfrac{1}{2})\,\frac{d}{dz}\left\{\left(z+\frac{l}{p}+\frac{y}{2p}\right)^2+\frac{P}{4p^2}y^2\right\}^{-s}dz\right].$$

Now

$$\sum_{y=1}^{\infty}\sum_{l=0}^{p-1}\left(\frac{-p}{Q(l,y)}\right)p^{-2s}\int_{-\infty}^{\infty}\left\{\left(z+\frac{l}{p}+\frac{y}{2p}\right)^2+\frac{P}{4p^2}y^2\right\}^{-s}dz$$
$$= \sum_{y=1}^{\infty}\sum_{l=0}^{p-1}\left(\frac{-p}{Q(l,y)}\right)p^{-2s}\left\{\frac{Py^2}{4p^2}\right\}^{\frac{1}{2}-s}\int_{-\infty}^{\infty}(u^2+1)^{-s}\,du$$
$$= 2^{2s-1}p^{-1}P^{\frac{1}{2}-s}\sum_{y=1}^{\infty}y^{1-2s}\sum_{l=0}^{p-1}\left(\frac{-p}{Q(l,y)}\right)\int_{-\infty}^{\infty}(u^2+1)^{-s}\,du$$
$$= 2^{2s-1}p^{-1}P^{\frac{1}{2}-s}\sum_{y=1}^{\infty}\{-y^{1-2s}+p(py)^{1-2s}\}\int_{-\infty}^{\infty}(u^2+1)^{-s}\,du,$$

$$\sum_y \mathfrak{A}^2(y) \leqq \sum_y \mathfrak{B}^2(y)$$

etc., but rather, following Romanoff, that

$$\sum_y \mathfrak{A}^2(y) \leqq \sum_y \mathfrak{A}(y)\mathfrak{B}(y)$$

etc.

The main idea we therefore owe to Romanoff. Everything else is a laborious deepening of each individual device of the old method. Romanoff had to omit this because he knew neither Landau's article nor a later general investigation on sets of numbers by Khintchine.[6]

From now on the following notations are in effect.

An empty sum always means 0, an empty product 1.

C is Euler's constant.

q is > 0, square-free and prime to 30,

$$\psi(q) = \prod_{p|q}(p-2),$$

$$\alpha = \sum_q \frac{1}{\varphi(q)\psi(q)} = \prod_{p \geqq 7}\left(1 + \frac{1}{(p-1)(p-2)}\right),$$

$$\beta = \prod_{p \geqq 7} \frac{1 - \dfrac{2}{p}}{\left(1 - \dfrac{1}{p}\right)^2},$$

$$\varrho = \alpha\beta,$$

hence

[6] Zur additiven Zahlentheorie [*Matematitscheskij Sbornik* **39** (1932), 27–34], German with Russian summary. However, our reader does not have to know this article, for we need from it only the not very deep theorem:

Let $\mathfrak{N}$ be a set of natural numbers, $N(z)$ the number of n in $\mathfrak{N}$ with $n \leqq z$. Let $h > 0$ and

$$N(z) > \frac{z}{h} \quad \textit{for all } z > 0.$$

Then every natural number can be represented as a sum of at most h terms from $\mathfrak{N}$.

And for this fact one can extract without difficulty a proof from the recently published article of Besicovitch: On the density of the sum of two sequences of integers [*The Journal of the London Mathematical Society* **10** (1935), 246–248].

$$\varrho = \prod_{p \geqq 7} \frac{p^2 - 3p + 3}{(p-1)(p-2)} \frac{(p-2)p}{(p-1)^2}$$

$$= \prod_{p \geqq 7} \frac{p^3 - 3p^2 + 3p}{(p-1)^3} = \prod_{p \geqq 7} \left(1 + \frac{1}{(p-1)^3}\right).$$

First Part: The Main Theorem with a parameter

The objective of this first part is the

Theorem 13. *Let*

(1) $$\Theta > \log \tfrac{7}{5}$$

and

(2) $$\Theta^2 e^{\Theta} \leqq \frac{4}{e^2}$$

hence

$$\Theta < 1$$

and

$$\sum_{m=1}^{\infty} \frac{(\Theta^2 e^{\Theta})^m m^{2m}}{(2m+1)!}$$

is convergent, because

$$\frac{(\Theta^2 e^{\Theta})^m m^{2m}}{(2m+1)!} < \frac{\left(\frac{4}{e^2}\right)^m m^{2m}}{2m(2m)!} < \frac{\left(\frac{4}{e^2}\right)^m m^{2m}}{2m} \frac{e^{2m}}{(2m)^{2m}\sqrt{2\pi}\sqrt{2m}}$$

$$= \frac{1}{4\sqrt{\pi}\, m^{\frac{3}{2}}}.$$

Put

$$\sigma = \sigma(\Theta) = 1 + \Theta \sum_{m=1}^{\infty} \frac{(\Theta^2 e^{\Theta})^m m^{2m}}{(2m+1)!}$$

and

$$h = h(\Theta) = \left[\frac{16}{3}\varrho\sigma e^{-2C}\frac{1}{(1 - e^{-\frac{1}{2}\Theta})^2}\right] + 1.$$

Assume that the Goldbach conjecture is true up to $30h + 1$, *i.e., that every even* x *with* $4 \leqq x \leqq 30h + 1$ *is the sum of two prime numbers, and hence that every* y *with* $2 \leqq y \leqq 30h + 1$ *is the sum of at most* 3 *prime numbers.*

Then Schnirelmann's constant satisfies

$$S \leqq 2h + 3.$$

In the second part, which contains only numerical calculations, we shall produce a Θ with (1) and (2) for which

$$\frac{16}{3}\varrho\sigma e^{-2C}\frac{1}{(1 - e^{-\frac{1}{2}\Theta})^2} < 34.$$

Thus

$$h \leqq 34$$

and hence, since Goldbach's conjecture is known to be true up to 1021,

$$S \leqq 71.$$

Definitions. *For* $x > 0$ *let*

$$A(x) = \sum_{p+p'=x} 1;$$

for $\xi \geqq 0,\ j > 0$, *let*

$$B(\xi, j) = \sum_{\substack{1 \leqq y \leqq \xi \\ y \equiv 0 \ (\mathrm{mod}\ j)}} A(y) \left(= \sum_{\substack{p+p' \leqq \xi \\ p+p' \equiv 0 \ (\mathrm{mod}\ j)}} 1 \right),$$

$$C(\xi, j) = \sum_{\substack{1 \leqq y \leqq \xi \\ y \equiv 0 \ (\mathrm{mod}\ j)}} A(y)y$$

and for every l

$$\pi(\xi, j, l) = \sum_{\substack{p \leqq \xi \\ p \equiv l \ (\mathrm{mod}\ j)}} 1.$$

Theorem 1. 1. *For fixed j, l with $(j, l) = 1$, one has*

$$\pi(\xi, j, l) \sim \frac{1}{\varphi(j)} \frac{\xi}{\log \xi}.$$

2. *For $\xi \geqq 2$, $\xi \geqq j$ one has*

$$\pi(\xi, j, l) < \frac{c_5}{j^{0.1}} \frac{\xi}{\log \xi}.$$

Proof. Both assertions are well known to be true.

1. The first is classical.
2. The second may be derived, e.g., by the Brun-Schnirelmann method, following Romanoff,[7] as follows.

For $u > 0$ let

$$f(u) = \prod_{p|u}\left(1 + \frac{1}{p}\right) = \sum_{r|u} \frac{1}{r},$$

where r runs over all positive, square-free numbers. Let $g(u, \xi)$ denote the number of solutions of

$$u = p - p', \quad p \leqq \xi.$$

By Theorem $L23$ one has for $x \geqq 2$, $1 \leqq u \leqq x$,

$$g(u, x) < c_6 \frac{x}{\log^2 x} f(u).$$

For $\xi \geqq 2$, $1 \leqq u \leqq \xi$, one therefore has

$$g(u, \xi) = g(u, [\xi]) < c_7 \frac{\xi}{\log^2 \xi} f(u).$$

Now for $\eta > 0$ one has

$$\sum_{1 \leqq v \leqq \eta} f(v) = \sum_{1 \leqq v \leqq \eta} \sum_{r|v} \frac{1}{r} = \sum_{r \leqq \eta} \frac{1}{r}\left[\frac{\eta}{r}\right] < \eta \sum_r \frac{1}{r^2} < c_8 \eta \tag{3}$$

[7]Über einige Sätze der additiven Zahlentheorie [*Mathematische Annalen* **109** (1934), 668–678], pp. 675–676.

and for $v > 0$, $w > 0$ evidently

$$(4) \quad f(vw) = \prod_{p \mid vw} \left(1 + \frac{1}{p}\right) \leqq \prod_{p \mid v} \left(1 + \frac{1}{p}\right) \prod_{p \mid w} \left(1 + \frac{1}{p}\right) = f(v) f(w);$$

moreover, it is well known that

$$(5) \qquad f(j) \leqq \prod_{p \mid j} \frac{1}{1 - \frac{1}{p}} = \frac{j}{\varphi(j)} < c_9 j^{0.8}.$$

21. In the case

$$\pi(\xi, j, l) \leqq 1$$

one has trivially

$$\pi(\xi, j, l) < c_{10} \frac{\xi^{0.9}}{\log \xi} \leqq \frac{c_{10}}{j^{0.1}} \frac{\xi}{\log \xi}.$$

22. In the case

$$\pi(\xi, j, l) \geqq 2$$

one has

$$(6) \quad \tfrac{1}{4}\pi^2(\xi, j, l) \leqq \tfrac{1}{2}\left(\pi^2(\xi, j, l) - \pi(\xi, j, l)\right)$$

$$= \text{number of solutions of } p \equiv p' \equiv l \pmod{j}, \quad p' < p \leqq \xi$$

$$\leqq \text{number of solutions of } p - p' \equiv 0 \pmod{j}, \quad p' < p \leqq \xi$$

$$= \sum_{\substack{1 \leqq u \leqq \xi \\ u \equiv 0 \ (\mathrm{mod}\ j)}} g(u, \xi) < c_7 \frac{\xi}{\log^2 \xi} \sum_{\substack{1 \leqq u \leqq \xi \\ u \equiv 0 \ (\mathrm{mod}\ j)}} f(u).$$

Herein one has by (4), (3), and (5)

$$(7) \qquad \sum_{\substack{1 \leqq u \leqq \xi \\ u \equiv 0 \ (\mathrm{mod}\ j)}} f(u) = \sum_{1 \leqq v \leqq \frac{\xi}{j}} f(vj) \leqq f(j) \sum_{1 \leqq v \leqq \frac{\xi}{j}} f(v)$$

$$< c_9 j^{0.8} c_8 \frac{\xi}{j} = c_{11} \frac{\xi}{j^{0.2}}.$$

From (6) and (7) follows

$$\pi^2(\xi, j, l) < 4c_7\frac{\xi}{\log^2\xi}c_{11}\frac{\xi}{j^{0.2}} = \frac{c_{12}}{j^{0.2}}\frac{\xi^2}{\log^2\xi},$$

$$\pi(\xi, j, l) < \frac{c_{12}}{j^{0.1}}\frac{\xi}{\log \xi}.$$

***Theorem* 2.** 1. *For j fixed one has*

$$B(\xi, j) \sim \frac{1}{2\varphi(j)}\frac{\xi^2}{\log^2\xi}.$$

2. *For* $\xi \geqq 2$ *one has*

$$B(\xi, j) < \frac{c_{13}}{j^{0.1}}\frac{\xi^2}{\log^2\xi}.$$

Proof. All congruences are meant to be mod j.

1. For $(a, j) > 1$ one has

$$\sum_{\substack{p+p'\leqq\xi \\ p\equiv -p'\equiv a}} 1 \leqq 1.$$

Therefore

$$B(\xi, j) = \sum_{a=1}^{j} \sum_{\substack{p+p'\leqq\xi \\ p\equiv -p'\equiv a}} 1 \tag{8}$$

$$= \sum_{\substack{a=1 \\ (a,j)=1}}^{j} \sum_{\substack{p+p'\leqq\xi \\ p\equiv -p'\equiv a}} 1 + O(1).$$

Now for $(a, j) = 1$ one has

$$\sum_{\substack{p+p'\leqq\xi \\ p\equiv -p'\equiv a}} 1 = \sum_{\substack{p\leqq\frac{1}{2}\xi \\ p\equiv a}} \pi(\xi - p, j, -a) \tag{9}$$

$$+ \sum_{\substack{p'\leqq\frac{1}{2}\xi \\ p'\equiv -a}} \pi(\xi - p', j, a) - \pi\left(\tfrac{1}{2}\xi, j, -a\right)\pi\left(\tfrac{1}{2}\xi, j, a\right).$$

In the first sum on the right one has uniformly

$$\log(\xi - p) \sim \log \xi;$$

thus, by Theorem 1, part 1, one has

$$\sum_{\substack{p \leqq \frac{1}{2}\xi \\ p \equiv a}} \pi(\xi - p, j, -a) \sim \frac{1}{\varphi(j)} \frac{1}{\log \xi} \sum_{\substack{p \leqq \frac{1}{2}\xi \\ p \equiv a}} (\xi - p)$$

$$\sim \frac{1}{\varphi(j)} \frac{1}{\log \xi} \frac{1}{\varphi(j)} \int_2^{\frac{1}{2}\xi} \frac{\xi - \eta}{\log \eta} \, d\eta$$

$$= \frac{1}{\varphi^2(j)} \frac{1}{\log \xi} \left(\xi \int_2^{\frac{1}{2}\xi} \frac{d\eta}{\log \eta} - \int_2^{\frac{1}{2}\xi} \frac{\eta}{\log \eta} \, d\eta \right)$$

$$\sim \frac{1}{\varphi^2(j)} \frac{1}{\log \xi} \left(\xi \frac{\frac{1}{2}\xi}{\log \xi} - \frac{\left(\frac{1}{2}\xi\right)^2}{2 \log \xi} \right) = \frac{3}{8\varphi^2(j)} \frac{\xi^2}{\log^2 \xi}.$$

By symmetry one therefore has

$$\sum_{\substack{p' \leqq \frac{1}{2}\xi \\ p' \equiv -a}} \pi(\xi - p', j, a) \sim \frac{3}{8\varphi^2(j)} \frac{\xi^2}{\log^2 \xi}.$$

Moreover,

$$\pi\left(\tfrac{1}{2}\xi, j, -a\right) \pi\left(\tfrac{1}{2}\xi, j, a\right) \sim \left(\frac{1}{\varphi(j)} \frac{\frac{1}{2}\xi}{\log \xi} \right)^2 = \frac{1}{4\varphi^2(j)} \frac{\xi^2}{\log^2 \xi}.$$

Since

$$\frac{3}{8} + \frac{3}{8} - \frac{1}{4} = \frac{1}{2},$$

one has therefore by (9)

$$\sum_{\substack{p + p' \leqq \xi \\ p \equiv -p' \equiv a}} 1 \sim \frac{1}{2\varphi^2(j)} \frac{\xi^2}{\log^2 \xi},$$

and (8) gives the assertion.

2. If $\xi < j$, then

$$B(\xi, j) = 0.$$

If $\xi \geqq j$, then by Theorem 1, part 2,

$$B(\xi, j) \leqq \sum_{p \leqq \xi} \pi(\xi, j, -p) \leqq \sum_{p \leqq \xi} \frac{c_5}{j^{0.1}} \frac{\xi}{\log \xi}$$

$$< \frac{c_{13}}{j^{0.1}} \frac{\xi^2}{\log^2 \xi}.$$

Theorem 3. 1. *For j fixed one has*

$$C(\xi, j) \sim \frac{1}{3\varphi(j)} \frac{\xi^3}{\log^2 \xi}.$$

2. *For $\xi \geqq 2$ one has*

$$C(\xi, j) < \frac{c_{13}}{j^{0.1}} \frac{\xi^3}{\log^2 \xi}.$$

Proof. 1. W.l.o.g. let ξ be integral. From Theorem 2, part 1, it follows that

$$C(\xi, j) = \sum_{y=1}^{\xi} (B(y, j) - B(y-1, j)) y$$

$$= -\sum_{y=1}^{\xi-1} B(y, j) + \xi B(\xi, j)$$

$$\sim -\frac{1}{2\varphi(j)} \int_2^{\xi} \frac{\eta^2 \, d\eta}{\log^2 \eta} + \frac{1}{2\varphi(j)} \frac{\xi^3}{\log^2 \xi}$$

$$\sim \left(-\tfrac{1}{2} \cdot \tfrac{1}{3} + \tfrac{1}{2}\right) \frac{1}{\varphi(j)} \frac{\xi^3}{\log^2 \xi} = \frac{1}{3\varphi(j)} \frac{\xi^3}{\log^2 \xi}.$$

2. From Theorem 2, part 2, it follows that

$$C(\xi, j) \leqq \xi B(\xi, j) < \frac{c_{13}}{j^{0.1}} \frac{\xi^3}{\log^2 \xi}.$$

Theorem 4.

$$\sum_{\substack{\frac{\xi}{\log \xi} < y \leqq \xi \\ y \equiv 0 \pmod{30}}} A(y) \frac{y}{\log^2 y} \sum_{q|y} \frac{1}{\psi(q)} \sim \frac{1}{24} \alpha \frac{\xi^3}{\log^4 \xi}.$$

Proof. Since

$$\log y \sim \log \xi,$$

uniformly, it is enough to prove

$$\frac{\log^2 \xi}{\xi^3} \sum_{\substack{\frac{\xi}{\log \xi} < y \leqq \xi \\ y \equiv 0 \pmod{30}}} A(y) y \sum_{q|y} \frac{1}{\psi(q)} \to \frac{1}{24} \alpha. \tag{10}$$

For $\xi \geqq e$ put

$$\Delta(\xi, q) = \frac{\log^2 \xi}{\xi^3} \frac{1}{\psi(q)} \left(C(\xi, 30q) - C\left(\frac{\xi}{\log \xi}, 30q \right) \right).$$

Then the left-hand side of (10) has the value

$$\frac{\log^2 \xi}{\xi^3} \sum_q \frac{1}{\psi(q)} \sum_{\substack{\frac{\xi}{\log \xi} < y \leqq \xi \\ y \equiv 0 \pmod{30q}}} A(y) y = \sum_q \Delta(\xi, q). \tag{11}$$

By Theorem 3, part 2, one has

$$0 \leqq \Delta(\xi, q) \leqq \frac{\log^2 \xi}{\xi^3} \frac{1}{\psi(q)} C(\xi, 30q) \tag{12}$$

$$< \frac{\log^2 \xi}{\xi^3} \frac{1}{\psi(q)} \frac{c_{13}}{q^{0.1}} \frac{\xi^3}{\log^2 \xi} = \frac{c_{13}}{q^{0.1} \psi(q)}.$$

The series

$$\sum_q \frac{1}{q^{0.1} \psi(q)} = \prod_{p \geqq 7} \left(1 + \frac{1}{p^{0.1}(p-2)} \right)$$

converges; thus, because of (12), the series on the right-hand side of (11) converges uniformly.

By Theorem 3, part 1, one has for q fixed

$$\lim_{\xi=\infty} \Delta(\xi, q) = \frac{1}{\psi(q) \cdot 3\varphi(30q)} = \frac{1}{24\varphi(q)\psi(q)}.$$

Therefore, the left-hand side of (11) tends to

$$\frac{1}{24}\sum_{q} \frac{1}{\varphi(q)\psi(q)} = \frac{1}{24}\alpha$$

as $\xi \to \infty$.

Theorem 5. *If*

$$\overline{\lim_{\substack{x=\infty \\ x\equiv 0 \ (\mathrm{mod}\ 30)}}} \frac{A(x)}{\dfrac{x}{\log^2 x}\displaystyle\sum_{q|x}\frac{1}{\psi(q)}} \leqq A, \tag{13}$$

then

$$\overline{\lim_{\xi=\infty}} \frac{\log^4 \xi}{\xi^3} \sum_{\substack{1 \leqq y \leqq \xi \\ y \equiv 0 \ (\mathrm{mod}\ 30)}} A^2(y) \leqq \frac{1}{24}\alpha A. \tag{14}$$

Proof. Let B be any number $> A$. Then in the end one has

$$A(y) < B\frac{y}{\log^2 y}\sum_{q|y}\frac{1}{\psi(q)} \qquad \text{for } 30|y.$$

Since

$$A(y) \leqq \pi(y) = O\left(\frac{y}{\log y}\right)$$

one has therefore

$$\sum_{\substack{1 \leqq y \leqq \xi \\ y \equiv 0 \ (\mathrm{mod}\ 30)}} A^2(y) \leqq O\left(\sum_{2 \leqq y \leqq \frac{\xi}{\log \xi}} \frac{y^2}{\log^2 y} + B \sum_{\substack{\frac{\xi}{\log \xi} < y \leqq \xi \\ y \equiv 0 \ (\mathrm{mod}\ 30)}} A(y)\frac{y}{\log^2 y}\sum_{q|y}\frac{1}{\psi(q)}\right).$$

The first term on the right is

$$O\left(\frac{\left(\frac{\xi}{\log \xi}\right)^3}{\log^2 \xi}\right) = O\left(\frac{\xi^3}{\log^5 \xi}\right).$$

Thus, by Theorem 4 the left-hand side of (14) is at most $B\frac{1}{24}\alpha$, and $B \to A$ gives the assertion.

Theorem 6. *Let $\mathfrak{M}$ be a set of natural numbers m and $M(x)$ the number of $m \leqq x$. Let*

$$h > 0,$$

$$\varliminf_{x=\infty} \frac{M(x)}{x} > \frac{1}{h}. \tag{15}$$

Then all large u can be represented as a sum of h numbers m and at most $h - 1$ ones.

Preliminary Remark. If $\mathfrak{M}$ is the set of m with

$$30m = p + p'$$

and (15) holds, then one has therefore that every large

$$u = \sum_{i=1}^{h} m_i + v, \quad 0 \leqq v \leqq h - 1,$$

hence for every j with $0 \leqq j \leqq 29$, every large

$$30u + j = \sum_{i=1}^{h} (p_i + p_i') + 30v + j,$$

hence every large

$$w - 2 = \sum_{i=1}^{h} (p_i + p_i') + z, \quad 0 \leqq z \leqq 30h - 1,$$

and hence every large

$$w = \sum_{i=1}^{h} (p_i + p_i') + t, \quad 2 \leqq t \leqq 30h + 1.$$

Thus, if (15) holds and Goldbach's conjecture is true up to $30h + 1$, then

$$S \leqq 2h + 3.$$

Proof. There exists a $y \geqq 0$ with

$$\begin{cases} M(y) \leqq \dfrac{y}{h}, \\ M(x) > \dfrac{x}{h} \quad \text{for } x > y. \end{cases}$$

$y + 1$ is therefore an m, and one has

$$M(x) - M(y) > \frac{x - y}{h} \quad \text{for } x > y.$$

Applying Khintchine's Theorem, which was cited in footnote 6, to the set of the $m - y$ with $m > y$ yields:

Every $t > 0$ has the form

$$t = \sum_{i=1}^{s} (m_i - y), \quad 1 \leqq s \leqq h;$$

every $u > hy + h - 1$ therefore has the form

$$u = (u - hy - h + 1) + hy + h - 1 = \sum_{i=1}^{s} (m_i - y) + hy + h - 1$$

$$= \sum_{i=1}^{s} m_i + \sum_{i=s+1}^{h} (y + 1) + s - 1 = \sum_{i=1}^{h} m_i + v, \quad 0 \leqq v \leqq h - 1.$$

***Theorem* 7.** *Let $\mathfrak{M}$ be the set of m with*

$$30m = p + p'.$$

Assume A satisfies (13). *Then* (15) *holds with*

$$h = \left[\tfrac{16}{45} A\alpha\right] + 1.$$

Preliminary Remark. If A satisfies (13) and Goldbach's conjecture is true up to $30[\frac{16}{45}A\alpha] + 31$, then by the preliminary remark to Theorem 6,

$$S \leqq 2\left[\tfrac{16}{45} A\alpha\right] + 5.$$

Proof. Let $\mathfrak{M}'$ be the set of $p + p'$ which are divisible by 30, and $M'(x)$ the number of these numbers $\leqq x$. Then

$$M'(30x) = M(x).$$

One has

$$\left(\sum_{\substack{y=1 \\ y \equiv 0 \pmod{30}}}^{30x} A(y) \right)^2 \leqq M'(30x) \sum_{\substack{y=1 \\ y \equiv 0 \pmod{30}}}^{30x} A^2(y). \tag{16}$$

By Theorem 2, part 1, with $j = 30$, one has

$$\sum_{\substack{y=1 \\ y \equiv 0 \pmod{30}}}^{30x} A(y) = B(30x, 30) \sim \frac{1}{2\varphi(30)} \frac{(30x)^2}{\log^2 x} = \frac{225}{4} \frac{x^2}{\log^2 x}.$$

By (16) and Theorem 5 one therefore has

$$\underline{\lim}_{x=\infty} \frac{M(x)}{x} = \underline{\lim}_{x=\infty} \frac{M'(30x)}{x}$$

$$\geqq \frac{\underline{\lim}_{x=\infty} \dfrac{\log^4 x}{x^4} \left(\sum_{\substack{y=1 \\ y \equiv 0 \pmod{30}}}^{30x} A(y) \right)^2}{\overline{\lim}_{x=\infty} \dfrac{\log^4 x}{x^3} \sum_{\substack{y=1 \\ y \equiv 0 \pmod{30}}}^{30x} A^2(y)}$$

$$\geqq \frac{\dfrac{225^2}{4^2}}{\frac{1}{24}\alpha A \cdot 30^3} = \frac{1}{\frac{16}{45} A\alpha} > \frac{1}{\left[\frac{16}{45} A\alpha\right] + 1}.$$

*

We shall now prove relation (13) with

$$A = 15\sigma e^{-2C} \frac{1}{(1 - e^{-\frac{1}{2}\Theta})^2} \beta$$

for every Θ with (1), (2). By Theorem 7, this then yields immediately Theorem 13 which was stated at the beginning of this first part.

Theorem 8. *Let $m > 0$, all summation letters $j_l \geqq 0$,*

$$T = \sum_{\substack{\sum_1^m j_l = 2m+1 \\ \sum_1^r j_l \leqq 2r \text{ for } 1 \leqq r < m}} \frac{(2m+1)!}{j_1! j_2! \ldots j_m!}.$$

Then

$$T \leqq m^{2m}.$$

Proof. Define j_l for all l with period m.

For $1 \leqq k \leqq m$ let T_k arise from T in that every j_l is replaced by j_{l+k-1}. T_1 is therefore T; T_k differs from T only in the designation of the summation letters (since the old summand was symmetrical in $j_1, \ldots, j_m$ and since $j_k, \ldots, j_{m+k-1}$ is a cyclic permutation of these numbers); T_k thus has the same value as T.

It suffices to show

$$\sum_{k=1}^{m} T_k \leqq \sum_{\sum_1^m j_l = 2m+1} \frac{(2m+1)!}{j_1! j_2! \ldots j_m!}; \tag{17}$$

for the left side $= mT$, the right side

$$= \left(\sum_{l=1}^{m} 1 \right)^{2m+1} = m^{2m+1}.$$

(17) will be proved as follows. Every system $j_1, \ldots, j_m$ which appears in a T_k also appears on the right. It is therefore enough to show that every system $j_1, \ldots, j_m$ which appears in a T_{k_0} does not appear in a T_k with $k \neq k_0$. For reasons of symmetry one can assume that $j_1, \ldots, j_m$ appears in T_1 and then one must show that $j_1, \ldots, j_m$ does not appear in T_k for every k with $2 \leqq k \leqq m$. Indeed,

$$\sum_1^{k-1} j_l \leqq 2(k-1),$$

because

$$\sum_1^m j_l = 2m + 1 > 2m$$

hence

$$\sum_k^m j_l > 2(m - k + 1).$$

However, if $j_1, \ldots, j_m$ were to appear in T_k, then one would have

$$\sum_k^m j_l \leqq 2(m - k + 1).$$

Theorem 9. *With the notations of Theorem L2 one has for* $0 \leqq m \leqq t$, $0 \leqq i \leqq 2m$:

$$\varrho_i^{(m)} = \sum_{\substack{\Sigma_1^m j_l = i \\ \Sigma_1^r j_l \leqq 2r \text{ for } 1 \leqq r < m}} \sigma_{j_1}^{(1)} \sigma_{j_2}^{(2)} \ldots \sigma_{j_m}^{(m)}, \tag{18}$$

where the $j_l \geqq 0$ *and the right side is* 1 *in the case* $m = 0$.

Proof. 1. For $m = 0$ one has

$$i = 0,$$

$$\varrho_i^{(m)} = \varrho_0^{(0)} = 1.$$

2. For $m = 1$ one has

$$0 \leqq i \leqq 2.$$

21. For $i = 0$ the left side of the assertion is 1, and the right side is

$$\sigma_0^{(1)} = 1.$$

22. For $i = 1$ the left side is

$$\varrho_1^{(1)} = \sum_{k_0 \geqq s > k_1} \gamma_s = \sigma_1^{(1)} = \text{right side.}$$

23. For $i = 2$ the left side is

$$\varrho_2^{(1)} = \sum_{k_0 \geqq s_1 > s_2 > k_1} \gamma_{s_1} \gamma_{s_2} = \sigma_2^{(1)} = \text{right side.}$$

3. The induction step from $m - 1$ to m for $m \geqq 2$ runs as follows. By formula (9), Theorem $L2$ one has

$$\varrho_i^{(m)} = \sum_{n=0}^{i} \varrho_n^{(m-1)} \sigma_{i-n}(m) = \sum_{n=0}^{\text{Min}(i, 2(m-1))} \varrho_n^{(m-1)} \sigma_{i-n}(m),$$

because for $i \geqq n > 2(m - 1)$ one has by a remark made in the wording of Theorem $L2$ that

$$\varrho_n^{(m-1)} = 0$$

since $n \leqq i \leqq 2m \leqq 2t$.

By formula (18), applied with n in place of i and $m-1$ in place of m (note $n \leqq 2(m-1)$), one has therefore

$$\varrho_i^{(m)} = \sum_{n=0}^{\operatorname{Min}(i,2(m-1))} \sigma_{i-n}^{(m)} \sum_{\substack{\sum_1^{m-1} j_l = n \\ \sum_1^r j_l \leqq 2r \text{ for } 1 \leqq r < m-1}} \sigma_{j_1}^{(1)} \ldots \sigma_{j_{m-1}}^{(m-1)},$$

Thus, putting

$$i - n = j_m$$

(one has $j_m \geqq 0$),

$$\varrho_i^{(m)} = \sum_{j_1, \ldots, j_m} \sigma_{j_1}^{(1)} \ldots \sigma_{j_m}^{(m)},$$

where the summation extends over all systems of values $j_l \geqq 0$ for which

$$\left\{\begin{array}{ll} \sum_1^m j_l = i, & (19) \\ \sum_1^r j_l \leqq 2r \quad \text{for } 1 \leqq r < m-1, & (20) \\ i - 2(m-1) \leqq j_m. & (21) \end{array}\right.$$

And this means exactly

$$\left\{\begin{array}{l} \sum_1^m j_l = i, \\ \sum_1^r j_l \leqq 2r \quad \text{for } 1 \leqq r < m \end{array}\right.$$

because from (19), (20), (21) follows

$$\sum_1^{m-1} j_l = i - j_m \leqq 2(m-1),$$

and from (19), (20) together with

$$\sum_1^{m-1} j_l \leqq 2(m-1)$$

follows

$$j_m = i - \sum_{1}^{m-1} j_l \geqq i - 2(m-1).$$

Theorem 10. *Let $t, k_0, \ldots, k_t, \varrho^{(t)}$ have the same meaning as in Theorem L2. Suppose that Θ satisfies* (1) *and* (2) *and let $\sigma(\Theta)$ be defined as in Theorem* 13. *Let*

$$0 < \gamma_s < 1 \quad \textit{for } 1 \leqq s \leqq k_0$$

and

$$L_m = \prod_{k_m < s \leqq k_{m-1}} (1 - \gamma_s) \geqq e^{-\Theta} \quad \textit{for } 1 \leqq m \leqq t. \tag{22}$$

Then

$$|\varrho^{(t)}| < \sigma \prod_{i=1}^{t} L_i.$$

Proof. For $1 \leqq m \leqq t$ one has by (22)

$$\sigma_1^{(m)} = \sum_{k_m < s \leqq k_{m-1}} \gamma_s < -\log L_m \leqq \Theta < 1;$$

by formula (5) of Theorem $L1$ one therefore has for $u \geqq 1$

$$\sigma_u^{(m)} \leqq \frac{\sigma_1^{(m)}}{u} \sigma_{u-1}^{(m)} \leqq \sigma_{u-1}^{(m)},$$

$$\Theta > \sigma_1^{(m)} \geqq \sigma_2^{(m)} \geqq \cdots \text{ ad inf.}; \tag{23}$$

by formula (6) of Theorem $L1$ one has for $k \geqq 1,\ j \geqq 0$

$$\sigma_j^{(k)} \leqq \frac{\left(\sigma_1^{(k)}\right)^j}{j!} \leqq \frac{\Theta^j}{j!}. \tag{24}$$

From formula (10) of Theorem 10 and (23) follows

$$\left|\varrho^{(m)} - \varrho^{(m-1)} L_m\right| = \left| \sum_{n=0}^{2m-2} (-1)^n \varrho_n^{(m-1)} \sum_{u=2m-n+1}^{\infty} (-1)^u \sigma_u^{(m)} \right|$$

$$\leqq \sum_{n=0}^{2m-2} \varrho_n^{(m-1)} \sigma_{2m-n+1}^{(m)}.$$

By Theorem 9 (with n in place of i, and $m - 1$ in place of m) one therefore has

$$\left|\varrho^{(m)} - \varrho^{(m-1)}L_m\right| \leqq \sum_{n=0}^{2m-2} \sigma_{2m-n+1}^{(m)} \sum_{\substack{\Sigma_1^{m-1} j_l = n \\ \Sigma_1^r j_l \leqq 2r \text{ for } 1 \leqq r < m-1}} \sigma_{j_1}^{(1)} \cdots \sigma_{j_{m-1}}^{(m-1)}.$$

Introducing

$$2m - n + 1 = j_m,$$

then one therefore has

$$\left|\varrho^{(m)} - \varrho^{(m-1)}L_m\right| \leqq \sum_{\substack{\Sigma_1^m j_l = 2m+1 \\ \Sigma_1^r j_l \leqq 2r \text{ for } 1 \leqq r < m-1}} \sigma_{j_1}^{(1)} \cdots \sigma_{j_m}^{(m)}$$

$$\leqq \sum_{\substack{\Sigma_1^m j_l = 2m+1 \\ \Sigma_1^r j_l \leqq 2r \text{ for } 1 \leqq r < m}} \sigma_{j_1}^{(1)} \cdots \sigma_{j_m}^{(m)};$$

because

$$\sum_1^{m-1} j_l = 2m + 1 - j_m \leqq 2(m - 1).$$

By Theorem 8 and (24) one therefore has

$$\left|\varrho^{(m)} - \varrho^{(m-1)}L_m\right| \leqq \sum_{\substack{\Sigma_1^m j_l = 2m+1 \\ \Sigma_1^r j_l \leqq 2r \text{ for } 1 \leqq r < m}} \frac{\Theta^{j_1} \cdots \Theta^{j_m}}{j_1! \cdots j_m!}$$

$$= \frac{\Theta^{2m+1}}{(2m+1)!} T \leqq \frac{\Theta^{2m+1}}{(2m+1)!} m^{2m},$$

hence by (22)

$$\left| \frac{\varrho^{(m)}}{\prod_1^m L_i} - \frac{\varrho^{(m-1)}}{\prod_1^{m-1} L_i} \right| \leqq e^{m\Theta} \frac{\Theta^{2m+1}}{(2m+1)!} m^{2m}.$$

Thus

$$\left| \frac{\varrho^{(t)}}{\prod_1^t L_i} - 1 \right| = \left| \sum_{m=1}^{t} \left(\frac{\varrho^{(m)}}{\prod_1^m L_i} - \frac{\varrho^{(m-1)}}{\prod_1^{m-1} L_i} \right) \right|.$$

$$< \sum_{m=1}^{\infty} e^{m\Theta} \frac{\Theta^{2m+1}}{(2m+1)!} m^{2m} = \sigma - 1,$$

$$\frac{|\varrho^{(t)}|}{\prod_1^t L_i} < \sigma,$$

$$|\varrho^{(t)}| < \sigma \prod_{i=1}^{t} L_i.$$

*

Now please read §2 of Landau's article up to Theorem $L7$ (inclusive). Note that if one replaces in Theorem $L8$ all the first arguments of F (i.e., all arguments which do not follow a semicolon) by its 30-fold, then the method of proof also yields the respective statement.

A fortiori one has, therefore, if all $p_s \geqq 7$,

$$F(30; p_1, \ldots, p_k) \leqq F(30) + \sum_{i=1}^{2t} (-1)^i \sum_{r_1, \ldots, r_i} v_{r_1} \ldots v_{r_i} F(30 p_{r_1} \ldots p_{r_i}),$$

where the r_i are subject to conditions (15) of Theorem $L8$.

The method of proof of Theorem $L9$ therefore gives

$$F(30; p_1, \ldots, p_k) < \frac{y\tau}{30} + 2 \prod_{m=0}^{t-1} (2k_m)^2, \tag{25}$$

where τ is defined as in Theorem $L9$.

Theorem 11. *If*

$$p_1 < \ldots < p_k$$

are the p with

$$7 \leqq p \leqq \sqrt{y},$$

then (*for every* a)

$$\varlimsup_{\substack{y=\infty \\ y\equiv 0 \,(\mathrm{mod}\, 30)}} \frac{F(30;\, p_1,\ldots,p_k)}{\dfrac{y}{\log^2 y}\displaystyle\sum_{q|y}\frac{1}{\psi(q)}} \leqq \frac{15}{8}\sigma e^{-2C}\frac{1}{(1-e^{-\frac{1}{2}\Theta})^2}\beta.$$

Proof. It is well known that

$$\prod_{p\leqq\xi}\left(1-\frac{1}{p}\right) \sim \frac{e^{-C}}{\log\xi},$$

hence that

$$\prod_{7\leqq p\leqq\xi}\left(1-\frac{1}{p}\right) \sim \frac{30}{\varphi(30)}\frac{e^{-C}}{\log\xi} = \frac{15}{4}\frac{e^{-C}}{\log\xi}. \tag{26}$$

It clearly suffices to find for each

$$\delta > \frac{15}{8}\sigma e^{-2C}\frac{1}{(1-e^{-\frac{1}{2}\Theta})^2}$$

a positive number $\gamma \leqq \frac{1}{2}$ depending only on Θ and δ such that if

$$p_1 < \cdots < p_k$$

are the p with

$$7 \leqq p \leqq y^{\gamma},$$

then

$$\varlimsup_{\substack{y=\infty \\ y\equiv 0 \,(\mathrm{mod}\, 30)}} \frac{F(30;\, p_1,\ldots,p_k)}{\dfrac{y}{\log^2 y}\displaystyle\sum_{q|y}\frac{1}{\psi(q)}} \leqq \delta\beta$$

holds. [For $F(30; p_1,\ldots,p_k)$ does not increase if γ increases and the left-hand side of the statement of the theorem stays $\leqq \delta\beta$, hence $\leqq$ right-hand side of the statement.]

Choose γ with

$$0 < \gamma < \tfrac{1}{2}(1-e^{-\frac{1}{2}\Theta}),$$

$$\frac{15}{32}\sigma e^{-2C}\frac{1}{\gamma^2} < \delta$$

and then a number ε depending on Θ, δ, and γ with

$$\varepsilon > 1,$$

$$\varepsilon e^{-\frac{1}{2}\Theta} < 1, \tag{27}$$

$$\varepsilon \frac{15}{32} \sigma e^{-2C} \frac{1}{\gamma^2} < \delta, \tag{28}$$

$$\zeta = \frac{\varepsilon^2 \gamma}{1 - \varepsilon e^{-\frac{1}{2}\Theta}} + \varepsilon - 1 < \frac{1}{2}. \tag{29}$$

Let

$$y \equiv 0 \pmod{30}$$

and

$$y \geqq 7^{1/\gamma},$$

hence

$$k > 0.$$

We put for $1 \leqq i \leqq k$

$$\gamma_i = \frac{v_i}{p_i}.$$

Then by (1)

$$1 - \gamma_i \geqq 1 - \frac{2}{p_i} \geqq \frac{5}{7} > e^{-\Theta}.$$

Now comes the analogue of the old construction (which is familiar from the proof of Theorem $L10$) of the numbers

$$t > 0, k_0, \ldots, k_t$$

(now depending on y and Θ) with

$$\begin{cases} k = k_0 > k_1 > \cdots > k_{t-1} > k_t = 0, \\ L_i = \prod\limits_{k_i < s \leqq k_{i-1}} (1 - \gamma_s) \text{ just} \geqq e^{-\Theta}, \end{cases}$$

where $k_1, k_2, \ldots$ are successively chosen to be minimal. Because

$$\lim_{p=\infty} \frac{2}{p} = 0,$$

one has hereby

$$L_i < \varepsilon e^{-\Theta} \quad \text{for } 1 \leqq i \leqq t - w_1,$$

where w_1 (similarly later w_2 and w_3) is positive and depends only on Θ, δ, γ, and ε.

Clearly one has

$$\tau = \varrho^{(t)},$$

hence by Theorem 10

$$|\tau| < \sigma \prod_{i=1}^{t} L_i,$$

thus by (25)

$$F(30; p_1, \ldots, p_k) < y \frac{1}{30} \sigma \prod_{i=1}^{t} L_i + 2 \prod_{m=0}^{t-1} (2k_m)^2. \tag{30}$$

Put

$$\lambda = \lambda(y) = \prod_{s=1}^{k} \left(1 - \frac{1}{p_s}\right);$$

then by (26)

$$\lambda \sim \frac{15}{4} \frac{e^{-C}}{\log(y^\gamma)} = \frac{15}{4} \frac{e^{-C}}{\gamma \log y}. \tag{31}$$

Therefore we have

$$\prod_{i=1}^{t} L_i = \prod_{s=1}^{k} \left(1 - \frac{v_s}{p_s}\right)$$

$$= \prod_{s=1}^{k} \left(1 - \frac{1}{p_s}\right)^2 \prod_{s=1}^{k} \frac{1 - \frac{2}{p_s}}{\left(1 - \frac{1}{p_s}\right)^2} \prod_{\substack{s=1 \\ p_s | y}}^{k} \frac{1 - \frac{1}{p_s}}{1 - \frac{2}{p_s}}$$

$$\leqq \lambda^2 \prod_{s=1}^{k} \frac{1 - \dfrac{2}{p_s}}{\left(1 - \dfrac{1}{p_s}\right)^2} \prod_{\substack{p|y \\ p>5}} \left(1 + \frac{1}{p-2}\right)$$

$$\sim \frac{225}{16} \frac{e^{-2C}}{\gamma^2 \log^2 y} \beta \sum_{q|y} \frac{1}{\psi(q)}.$$

In the end (i.e., for all large y) one therefore has by (28)

$$y \frac{1}{30} \sigma \prod_{i=1}^{t} L_i < \varepsilon y \frac{1}{30} \sigma \frac{225}{16} \frac{e^{-2C}}{\gamma^2 \log^2 y} \beta \sum_{q|y} \frac{1}{\psi(q)}$$

$$= \varepsilon \frac{15}{32} \sigma \frac{e^{-2C}}{\gamma^2} \beta \frac{y}{\log^2 y} \sum_{q|y} \frac{1}{\psi(q)}$$

$$< \delta \beta \frac{y}{\log^2 y} \sum_{q|y} \frac{1}{\psi(q)}.$$

Since $\xi < \frac{1}{2}$, it is therefore by (30) enough to prove that

$$\prod_{m=0}^{t-1} 2k_m = O(y^{\xi}).$$

By (31) one has in the end

$$\frac{\frac{15}{4} e^{-C}}{\lambda} < \varepsilon \gamma \log y. \tag{32}$$

Now one has that

$$\left(1 - \frac{1}{p_s}\right)^2 \leqq 1 - \frac{v_s}{p_s} + \frac{1}{p_s^2} < \left(1 - \frac{v_s}{p_s}\right)\left(1 + \frac{2}{p_s^2}\right) \prod_{p} \left(1 + \frac{2}{p^2}\right)$$

converges; thus, if we put for $1 \leqq i \leqq t$

$$M_i = \prod_{k_i < s \leqq k_{i-1}} \left(1 - \frac{1}{p_s}\right),$$

then for $1 \leqq i \leqq t - w_2$ one has in the end

$$M_i \leqq \sqrt{\varepsilon \prod_{k_i < s \leqq k_{i-1}} \left(1 - \frac{v_s}{p_s}\right)} = \sqrt{\varepsilon L_i} < \sqrt{\varepsilon \cdot \varepsilon e^{-\Theta}} = \varepsilon e^{-\frac{1}{2}\Theta}.$$

Since $p_{k_m} \geqq 2k_m$, one has by (26) and (32) in the end for $0 \leqq m \leqq t - w_3$

$$\text{(33)} \qquad \log(2k_m) < \varepsilon \frac{15}{4} e^{-C} \prod_{7 \leqq p \leqq 2k_m} \left(1 - \frac{1}{p}\right)^{-1}$$

$$\leqq \varepsilon \frac{15}{4} e^{-C} \prod_{7 \leqq p \leqq p_{k_m}} \left(1 - \frac{1}{p}\right)^{-1} = \varepsilon \frac{15}{4} e^{-C} \frac{1}{\lambda} \prod_{i=1}^{m} M_i$$

$$< \varepsilon^2 \gamma \log y (\varepsilon e^{-\frac{1}{2}\Theta})^m.$$

Let y be then so large that $t \geqq w_3$.
For $t - w_3 < m \leqq t$ one has by (33) in the end

$$\log(2k_m) < \log(2k_{t-w_3}) < \varepsilon^2 \gamma \log y (\varepsilon e^{-\frac{1}{2}\Theta})^{t-w_3};$$

one therefore has in the end

$$\sum_{m=t-w_3+1}^{t-1} \log(2k_m) < w_3 \varepsilon^2 \gamma \log y (\varepsilon e^{-\frac{1}{2}\Theta})^{t-w_3};$$

hence, because of (27), one has in the end

$$\text{(34)} \qquad \sum_{m=t-w_3+1}^{t-1} \log(2k_m) < (\varepsilon - 1) \log y.$$

From (29), (33), and (34) it follows that in the end

$$\sum_{m=0}^{t-1} \log(2k_m) < \left(\varepsilon^2 \gamma \sum_{m=0}^{\infty} (\varepsilon e^{-\frac{1}{2}\Theta})^m + \varepsilon - 1\right) \log y = \zeta \log y,$$

$$\prod_{m=0}^{t-1} 2k_m < y^{\zeta}.$$

Theorem 12.

$$\overline{\lim_{\substack{y=\infty \\ y \equiv 0 \,(\mathrm{mod}\, 30)}}} \frac{A(y)}{\dfrac{y}{\log^2 y} \displaystyle\sum_{q|y} \frac{1}{\psi(q)}} \leqq 15 \sigma e^{-2C} \frac{1}{(1 - e^{-\frac{1}{2}\Theta})^2} \beta.$$

Proof. Let

$$y > 0, \quad y \equiv 0 \pmod{30}.$$

The number of solutions of

$$y = p + p',\ p \leqq \sqrt{y} \quad \text{or} \quad p' \leqq \sqrt{y}$$

is

$$\leqq 2\sqrt{y}.$$

The number of solutions of

$$y = p + p', \quad p > \sqrt{y}, \quad p' > \sqrt{y}$$

is, since every p_s with $7 \leqq p_s \leqq \sqrt{y}$ divides neither p nor $p - y = -p'$ and since for $p \equiv a \pmod{30}$ certainly $(a, 30) = 1$ because $p > \sqrt{30} > 5$,

$$\leqq \sum_{\substack{a=1 \\ (a,30)=1}}^{30} F(30; p_1, \ldots, p_k),$$

where the $\varphi(30) = 8$ terms F refer to the respective a.

Thus the assertion follows from Theorem 11.

Second Part: Numerical calculations

Main Theorem. $S \leqq 71$.

Proof. We put

$$\Theta = 0.54.$$

Since Goldbach's conjecture is true up to 1021, it suffices by Theorem 13 to prove (1), (2), and

$$\frac{16}{3}\varrho\sigma(\Theta)e^{-2C}\frac{1}{(1 - e^{-\frac{1}{2}\Theta})^2} < 34$$

We carry out the following calculations without the use of logarithmic or other tables.

For our Θ we have

$$\Theta^2 e^{\Theta} < 0.5004,$$

hence (2) holds; (1) is clear.

Accordingly we have

$$\sigma(\Theta) < 1 + \Theta \sum_{m=1}^{\infty} \frac{0.5004^m m^{2m}}{(2m+1)!}$$

If we use for $m \geqq 3$ the estimate

$$\frac{m^{2m}}{(2m+1)!} < \frac{1}{4\sqrt{\pi}} \left(\tfrac{1}{4}e^2\right)^m \frac{1}{m^{\frac{3}{2}}} \leqq \frac{1}{4\sqrt{3\pi}} \frac{\left(\tfrac{1}{4}e^2\right)^m}{m}$$

then we find

$$\sigma(\Theta) < 1.1174.$$

On the other hand we have

$$e^{\sum\limits_{p \geqq 13} \frac{1}{(p-1)^3}} < e^{\sum\limits_{m=6}^{\infty} \frac{1}{(2m)^3}} < e^{\frac{1}{8}\frac{1}{2}\frac{1}{5^2}} = e^{\frac{1}{400}},$$

hence

$$\varrho < \left(1 + \frac{1}{6^3}\right)\left(1 + \frac{1}{10^3}\right) e^{\sum\limits_{p \geqq 13} \frac{1}{(p-1)^3}} < \frac{123}{122}.$$

Moreover,

$$C > 0.577,$$

$$e^{2C} > 3.17,$$

$$\left(1 - e^{-\frac{1}{2}\Theta}\right)^2 > 0.0559,$$

hence

$$34 e^{2C}\left(1 - e^{-\frac{1}{2}\Theta}\right)^2 > 6.009.$$

From this follows

$$\frac{\frac{16}{3}\varrho\sigma(\Theta)}{34 e^{2C}\left(1 - e^{-\frac{1}{2}\Theta}\right)^2} < \frac{\frac{16}{3} \cdot \frac{123}{122} \cdot 1.1174}{6.009} < 1.$$

Cambridge, Berlin, and Prague, 14 January 1936.

[Received 14 January 1936]

All Large Numbers Can be Represented As a Sum of at Most 71 Primes*
(Content of the previous article)

Schnirelmann has proved the following theorem. There exists a number $c_1 > 0$ such that every integer $x > 1$ can be represented as a sum of at most c_1 primes. This theorem can be reformulated as follows. There exist two positive numbers c_2, c_3 such that every integer $x > c_2$ can be represented as a sum of at most c_3 primes. Denote the least number c_3 (for which there exists a respective number c_2) by the symbol S.

The lower bound $S \geqq 3$ is trivial; recently, Romanov attempted to give an upper bound; he asserts that $S \leqq 1104$. However, in addition to a number of printing errors and miscalculations, his paper also has an essential shortcoming: Romanov presents an asymptotic formula on the distribution of primes in arithmetic progressions in the proof of which he probably made an error. For, although Romanov asserts that his formula can be derived by a slight modification of Landau's reasoning (*Vorlesungen über Zahlentheorie*, Vol. 2), we are still unable to decide whether the formula is true or false.

Fortunately, this doubtful formula can be bypassed. Indeed, in the present paper we are able to prove not only the inequality $S \leqq 1104$, but the even stronger inequality $S \leqq 71$. However, we emphasize explicitly that we are indebted to Romanov's work—despite its shortcomings—for one of the main ideas of this paper; in addition, we also use a method of one of Landau's papers on Schnirelmann's theorem and one of the results of Chinčin on a certain additive property of number sets.

*Translated by Prof. V. Dlab (Ottawa).

10.

Über das Waringsche Problem

Acta Arithmetica 1 (1936), 212–221
[Commentary on p. nnn]

In der Behandlung des Waringschen Problems hat Herr Vinogradov kürzlich bedeutende Fortschritte erzielt[1]. Er bewies:

Hauptsatz: *Es sei für $k > 2$, k ganz, $G(k)$ die kleinste Zahl, so dass sich jede hinreichend grosse Zahl als Summe von höchstens $G(k)$ positiven k-ten Potenzen darstellen lässt. Dann ist*

$$G(k) = O(k \log k), \tag{1}$$

genauer

$$G(k) \leq 6k \log k + (4 + \log 216) k. \tag{2}$$

Schon die Existenz eines endlichen $G(k)$ ist nicht trivial; sie wurde erstmalig von Hilbert bewiesen. Die erste brauchbare Abschätzung verdankt man Hardy und Littlewood, die, grob gesprochen,

$$\overline{\lim_{k=\infty}}\, k^{-1} 2^{2-k} G(k) \leq 1 \tag{3}$$

bewiesen[2]. Da trivialerweise

---

[1] „On the upper bound of $G(n)$ in Waring's Problem" [Bulletin de l'Académie de l' U. R. S. S., VII série, Classe des sciences mathématiques et naturelles, 1934, S. 1455 — 1469], russisch mit englischer Zusammenfassung; „Une nouvelle variante de la démonstration du théorème de Waring" [Comptes Rendus **200** (1935), S. 182 — 184]; „On Waring's Problem" [Annals of Mathematics **36** (1935), S. 395 — 405].

[2] „Some Problems of „Partitio Numerorum" ": I. „A new solution of Waring's Problem" [Göttinger Nachrichten (1920), S. 33 — 54]; II. „Proof that every large number is the sum of at most 21 biquadrates" [Mathematische Zeitschrift **9** (1921), S. 14 — 27]; IV. „The singular series in Waring's Problem, and the value of the number $G(k)$" [ebenda **12** (1922), S. 161 — 188]; VI. „Further researches in Waring's Problem" [ebenda **23** (1925), S. 1 — 37].

Alle in der vorliegenden Arbeit benutzten Hardy-Littlewoodschen Resultate, mit Ausnahme von (10), stehen auch bei Landau „Vorlesungen über Zahlentheorie", Bd. 1, Teil VI. Für (10) vergleiche man Partitio Numerorum IV.

(4) $$G(k) \geqq k+1,$$

ist die Bedeutung des Vinogradovschen Resultates ohne weiteres klar. Sein Beweis ist teilweise die konsequente Weiterführung Hardy und Littlewoodscher Ideen: teilweise beruht er auf der Einführung eines neues Lemmas [3], das anstelle der schwierigen Weylschen diophantischen Approximationen tritt und dadurch die Rechnungen wesentlich vereinfacht.

Es ist jedoch dem Scharfsinn von Herrn Vinogradov entgangen, dass seine kompliziertesten Rechnungen unnötig sind und sich durch einen einfachen Kunstgriff umgehen lassen. Das soll in der vorliegenden Arbeit gezeigt werden, und ich erhalte nicht nur (1), sondern auch

(5) $$G(k) \leqq 6k\log k + \left(4 + 3\log\left(3 + \frac{2}{k}\right)\right)k + 3,$$

was etwas besser ist als (2). Es ist nicht unmöglich diese Formel weiter zu verbessern, doch kann ich zur Zeit nicht einmal

(6) $$\overline{\lim_{k=\infty}}\, k^{-1}\log^{-1} k\, G(k) < 6$$

beweisen.

B e z e i c h n u n g e n. Es sei

$$k \geqq 3 \text{ ganz}, \; s = 4k, \; \rho = \frac{k}{k+1}, \; l = [k\log(3k^2+2k)] + 1,$$

also

(7) $$\left(1 - \frac{1}{k}\right)^l < \left(1 - \frac{1}{k}\right)^{k\log(3k^2+2k)} < e^{-\log(3k^2+2k)} = (3k^2+2k)^{-1}.$$

$N > 0$ sei die darzustellende Zahl. Wir setzen

$$P = N^{1/k}.$$

Es sei $0 \leqq \alpha \leqq 1$. Wir teilen diese Strecke in Intervalle ein, indem wir jedem Farey-Bruch $\frac{a}{q}$, wo

$$1 \leqq q < P^{k-\rho}, \; 1 \leqq a \leqq q, \; (a, q) = 1,$$

in bekannter Weise eine Umgebung zuordnen, so dass die Strecke

[3]) Lemma 10 dieser Arbeit.

$0 \leqq \alpha \leqq 1$ schlicht bedeckt wird. Wir teilen diese Intervalle in zwei Klassen:

1. Ist $1 \leqq q \leqq P^\rho$, so nennen wir das Intervall einen Major Arc. Bezeichnung: $\mathfrak{M}$.

2. Ist $P^\rho < q < P^{k-\rho}$, so nennen wir das Intervall einen Minor Arc. Bezeichnung: $\mathfrak{m}$.

In jedem Fall hat das Intervall die Form:

$$(8) \quad \alpha = \frac{a}{q} + \beta, \quad -\vartheta_1 q^{-1} P^{\rho-k} \leqq \beta \leqq \vartheta_2 q^{-1} P^{\rho-k}, \quad \frac{1}{2} \leqq \vartheta_1 \leqq 1, \quad \frac{1}{2} \leqq \vartheta_2 \leqq 1.$$

u durchlaufe alle Zahlen, die sich als Summe von höchstens l positiven k-ten Potenzen darstellen lassen.

Wir setzen:

$$T = T(\alpha) = \sum_{1 \leqq \mu \leqq P} e^{2\pi i \mu^k \alpha}, \quad I = I(\beta) = \int_0^P e^{2\pi i \beta v^k} dv,$$

$$R = R(\alpha) = \sum_{u \leqq \frac{1}{4} P^k} e^{2\pi i u \alpha}, \quad A_R = \sum_{u \leqq \frac{1}{4} P^k} 1,$$

$$S = S(\alpha) = \sum_{1 \leqq y \leqq P^{\rho/k}} \sum_{u \leqq \frac{1}{4} P^{k-\rho}} e^{2\pi i y^k \cdot u\alpha}, \quad A_s = \sum_{1 \leqq y \leqq P^{\rho/k}} \sum_{u \leqq \frac{1}{4} P^{k-\rho}} 1.$$

O — Abschätzungen gelten bei festem k gleichmässig in allen übrigen Variablen. o — Abschätzungen gelten bei festem k für $P \to \infty$.

$c_1, \ldots, c_6$ sind positive Konstanten, die nur von k abhängen.

Major Arcs.

Lemma 1 (Hardy — Littlewood): *Setzt man*

$$S_{a,q} = \sum_{\nu=1}^{q} e^{2\pi i \nu^k \frac{a}{q}},$$

so ist

$$(9) \quad S_{a,q} = O\left(q^{1-\frac{1}{k}}\right).$$

Lemma 2 (Hardy — Littlewood): *Setzt man für ganzes* n

$$\mathfrak{S}=\mathfrak{S}(n,k)=\sum_{q=1}^{\infty} q^{-s}\sum_{a=1}^{q} e^{-2\pi i n\frac{a}{q}} S_{a,q}^{s},$$

so konvergiert diese Reihe absolut nach Lemma 1 und es ist

$$\mathfrak{S}>c_1. \tag{10}$$

Lemma 3: *Es sei für* $\tau_1 \leqq \tau \leqq \tau_2$

$$0\leqq f(\tau)\leqq B_0,\ 0\leqq f'(\tau)\leqq B_1,\ 0\leqq f''(\tau).$$

Dann ist

$$\sum_{\tau_1\leqq\lambda\leqq\tau_2} e^{2\pi i f(\lambda)}=\int_{\tau_1}^{\tau_2} e^{2\pi i f(\tau)}\,d\tau+O(1+B_1+B_0B_1).$$

Beweis: Nach der Eulerschen Summenformel ist die Differenz von Summe und Integral

$$=O(1)+2\pi i\int_{\tau_1}^{\tau_2}\left(\tau-[\tau]-\frac{1}{2}\right)f'(\tau)\,e^{2\pi i f(\tau)}\,d\tau$$

$$=O(1)+2\pi i\int_{\tau_1}^{\tau_2}\left(\tau-[\tau]-\frac{1}{2}\right)f'(\tau)\{(2\pi(1+i)f(\tau))+1-(2\pi f(\tau)+1$$

$$-\cos(2\pi f(\tau)))-i(2\pi f(\tau)-\sin(2\pi f(\tau)))\}\,d\tau=O(1)+O(B_1(B_0+1))$$

nach dem zweiten Mittelwertsatz, da jeder der vier Summanden in der geschweiften Klammer nach Multiplikation mit einer Einheitswurzel zwischen 0 und $2\pi\sqrt{2}\,B_0+2$ liegt und monoton nicht fällt

Lemma 4: *Auf* $\mathfrak{M}$ *gilt*

$$T=q^{-1}S_{a,q}I+O(q+P^{2\rho-1}).$$

Beweis: *Es ist*

$$T=\sum_{\nu=1}^{q}\ \sum_{\frac{1-\nu}{q}\leqq\lambda\leqq\frac{P-\nu}{q}} e^{2\pi i(\nu+\lambda q)^k\left(\frac{a}{q}+\beta\right)}$$

$$=\sum_{\nu=1}^{q} e^{2\pi i\nu^k\frac{a}{q}}\sum_{\frac{1-\nu}{q}\leqq\lambda\leqq\frac{P-\nu}{q}} e^{2\pi i(\nu+\lambda q)^k\beta}.$$

Wendet man auf die innere Summe Lemma 3 mit

$$B_0 = P^k \mid \beta \mid \leqq P^k q^{-1} P^{\rho-k} \leqq q^{-1} P^\rho,$$

$$B_1 = k q P^{k-1} \mid \beta \mid \leqq k P^{\rho-1}$$

an, so folgt die Behauptung wegen

$$B_1 = O(1),\ B_0 B_1 = O(q^{-1} P^{2\rho-1}).$$

Lemma 5:

$$\sum_{\mathfrak{M}} \int_{\mathfrak{M}} \mid T^s - q^{-s} S_{a,q}^s I^s \mid d\alpha = O(P^{s-k-(1-\rho)}).$$

Beweis: Nach Lemma 4 ist auf $\mathfrak{M}$

$$T - q^{-1} S_{a,q} I = O(P^\rho). \tag{11}$$

Für $0 \leqq \mid \beta \mid \leqq P^{-k}$ ist nach Lemma 1

$$q^{-1} S_{a,q} I = O(q^{-\frac{1}{k}} P). \tag{12}$$

Für $P^{-k} \leqq \mid \beta \mid \leqq q^{-1} P^{\rho-k}$ ist

$$I = \int_0^P e^{2\pi i \beta v^k} dv = \mid \beta \mid^{-\frac{1}{k}} \int_0^{|\beta|^{\frac{1}{k}} P} e^{\pm 2\pi i v^k} dv = O\left(\mid \beta \mid^{-\frac{1}{k}}\right), \tag{13}$$

da $\int_0^\infty e^{2\pi i v^k} dv$ konvergiert.

Ferner ist wegen $q \leqq P^\rho$

$$P^\rho \leqq q^{-\frac{1}{k}} P \tag{14}$$

und für $\mid \beta \mid \leqq q^{-1} P^{\rho-k}$

$$P^\rho = P^{1-\rho/k} = (P^{\rho-k})^{-\frac{1}{k}} \leqq q^{-\frac{1}{k}} \mid \beta \mid^{-\frac{1}{k}}. \tag{15}$$

Nach (11), (12) und (14) ist also für $0 \leqq \mid \beta \mid \leqq P^{-k}$

$$\mid T^s - q^{-s} S_{a,q}^s I^s \mid = O\left(P^\rho q^{-\frac{s-1}{k}} P^{s-1}\right).$$

Proof. Let $\mathfrak{M}'$ be the set of $p + p'$ which are divisible by 30, and $M'(x)$ the number of these numbers $\leqq x$. Then

$$M'(30x) = M(x).$$

One has

$$(16) \qquad \left(\sum_{\substack{y=1 \\ y \equiv 0 \,(\mathrm{mod}\, 30)}}^{30x} A(y) \right)^2 \leqq M'(30x) \sum_{\substack{y=1 \\ y \equiv 0 \,(\mathrm{mod}\, 30)}}^{30x} A^2(y).$$

By Theorem 2, part 1, with $j = 30$, one has

$$\sum_{\substack{y=1 \\ y \equiv 0 \,(\mathrm{mod}\, 30)}}^{30x} A(y) = B(30x, 30) \sim \frac{1}{2\varphi(30)} \frac{(30x)^2}{\log^2 x} = \frac{225}{4} \frac{x^2}{\log^2 x}.$$

By (16) and Theorem 5 one therefore has

$$\varliminf_{x=\infty} \frac{M(x)}{x} = \varliminf_{x=\infty} \frac{M'(30x)}{x}$$

$$\geqq \frac{\varliminf_{x=\infty} \frac{\log^4 x}{x^4} \left(\sum_{\substack{y=1 \\ y \equiv 0 \,(\mathrm{mod}\, 30)}}^{30x} A(y) \right)^2}{\varlimsup_{x=\infty} \frac{\log^4 x}{x^3} \sum_{\substack{y=1 \\ y \equiv 0 \,(\mathrm{mod}\, 30)}}^{30x} A^2(y)}$$

$$\geqq \frac{\frac{225^2}{4^2}}{\frac{1}{24}\alpha A \cdot 30^3} = \frac{1}{\frac{16}{45} A\alpha} > \frac{1}{\left[\frac{16}{45} A\alpha\right] + 1}.$$

*

We shall now prove relation (13) with

$$A = 15\sigma e^{-2C} \frac{1}{(1 - e^{-\frac{1}{2}\Theta})^2} \beta$$

for every Θ with (1), (2). By Theorem 7, this then yields immediately Theorem 13 which was stated at the beginning of this first part.

By Theorem 9 (with n in place of i, and $m-1$ in place of m) one therefore has

$$\left|\varrho^{(m)}-\varrho^{(m-1)}L_m\right| \leqq \sum_{n=0}^{2m-2} \sigma_{2m-n+1}^{(m)} \sum_{\substack{\sum_1^{m-1} j_l = n \\ \sum_1^r j_l \leqq 2r \text{ for } 1 \leqq r < m-1}} \sigma_{j_1}^{(1)} \ldots \sigma_{j_{m-1}}^{(m-1)}.$$

Introducing

$$2m-n+1=j_m,$$

then one therefore has

$$\left|\varrho^{(m)}-\varrho^{(m-1)}L_m\right| \leqq \sum_{\substack{\sum_1^{m} j_l = 2m+1 \\ \sum_1^r j_l \leqq 2r \text{ for } 1 \leqq r < m-1}} \sigma_{j_1}^{(1)} \ldots \sigma_{j_m}^{(m)}$$

$$\leqq \sum_{\substack{\sum_1^{m} j_l = 2m+1 \\ \sum_1^r j_l \leqq 2r \text{ for } 1 \leqq r < m}} \sigma_{j_1}^{(1)} \ldots \sigma_{j_m}^{(m)};$$

because

$$\sum_1^{m-1} j_l = 2m+1-j_m \leqq 2(m-1).$$

By Theorem 8 and (24) one therefore has

$$\left|\varrho^{(m)}-\varrho^{(m-1)}L_m\right| \leqq \sum_{\substack{\sum_1^{m} j_l = 2m+1 \\ \sum_1^r j_l \leqq 2r \text{ for } 1 \leqq r < m}} \frac{\Theta^{j_1} \ldots \Theta^{j_m}}{j_1! \ldots j_m!}$$

$$= \frac{\Theta^{2m+1}}{(2m+1)!} T \leqq \frac{\Theta^{2m+1}}{(2m+1)!} m^{2m},$$

hence by (22)

$$\left| \frac{\varrho^{(m)}}{\prod_1^m L_i} - \frac{\varrho^{(m-1)}}{\prod_1^{m-1} L_i} \right| \leqq e^{m\Theta} \frac{\Theta^{2m+1}}{(2m+1)!} m^{2m}.$$

Thus

$$\mathfrak{S} = \mathfrak{S}(n, k) = \sum_{q=1}^{\infty} q^{-s} \sum_{a=1}^{q} e^{-2\pi i n \frac{a}{q}} S_{a,q}^{s},$$

so konvergiert diese Reihe absolut nach Lemma 1 *und es ist*

$$\mathfrak{S} > c_1. \tag{10}$$

Lemma 3: *Es sei für* $\tau_1 \leqq \tau \leqq \tau_2$

$$0 \leqq f(\tau) \leqq B_0, \; 0 \leqq f'(\tau) \leqq B_1, \; 0 \leqq f''(\tau).$$

Dann ist

$$\sum_{\tau_1 \leqq \lambda \leqq \tau_2} e^{2\pi i f(\lambda)} = \int_{\tau_1}^{\tau_2} e^{2\pi i f(\tau)} d\tau + O(1 + B_1 + B_0 B_1).$$

Beweis: Nach der Eulerschen Summenformel ist die Differenz von Summe und Integral

$$= O(1) + 2\pi i \int_{\tau_1}^{\tau_2} \left(\tau - [\tau] - \frac{1}{2}\right) f'(\tau) e^{2\pi i f(\tau)} d\tau$$

$$= O(1) + 2\pi i \int_{\tau_1}^{\tau_2} \left(\tau - [\tau] - \frac{1}{2}\right) f'(\tau) \{(2\pi(1+i) f(\tau)) + 1 - (2\pi f(\tau) + 1$$

$$- \cos(2\pi f(\tau))) - i(2\pi f(\tau) - \sin(2\pi f(\tau)))\} d\tau = O(1) + O(B_1(B_0 + 1))$$

nach dem zweiten Mittelwertsatz, da jeder der vier Summanden in der geschweiften Klammer nach Multiplikation mit einer Einheitswurzel zwischen 0 und $2\pi\sqrt{2}\,B_0 + 2$ liegt und monoton nicht fällt

Lemma 4: *Auf* $\mathfrak{M}$ *gilt*

$$T = q^{-1} S_{a,q} I + O(q + P^{2\rho - 1}).$$

Beweis: *Es ist*

$$T = \sum_{\nu=1}^{q} \sum_{\frac{1-\nu}{q} \leqq \lambda \leqq \frac{P-\nu}{q}} e^{2\pi i (\nu + \lambda q)^k \left(\frac{a}{q} + \beta\right)}$$

$$= \sum_{\nu=1}^{q} e^{2\pi i \nu^k \frac{a}{q}} \sum_{\frac{1-\nu}{q} \leqq \lambda \leqq \frac{P-\nu}{q}} e^{2\pi i (\nu + \lambda q)^k \beta}.$$

und nach (11), (13) und (15) ist für $P^{-k} \leqq |\beta| \leqq q^{-1} P^{\rho-k}$

$$|T^s - q^{-s} S_{a,q}^s I^s| = O\left(P^\rho q^{-\frac{s-1}{h}} |\beta|^{-\frac{s-1}{k}}\right).$$

Also

$$\int_{\mathfrak{M}} |T^s - q^{-s} S_{a,q}^s I^s| \, d\alpha = O\left(P^{-k+\rho+s-1} q^{-\frac{s-1}{k}}\right)$$

$$+ O\left(P^\rho q^{-\frac{s-1}{k}} \int_{P^{-k}}^{\infty} \beta^{-\frac{s-1}{k}} d\beta\right) = O\left(q^{-\frac{s-1}{k}} P^{s-k-(1-\rho)}\right)$$

$$+ O\left(q^{-\frac{s-1}{k}} P^{\rho-k\left(1-\frac{s-1}{k}\right)}\right) = O\left(q^{-\frac{s-1}{k}} P^{s-k-(1-\rho)}\right).$$

Hieraus folgt die Behauptung, da

$$\sum_{\mathfrak{M}} q^{-\frac{s-1}{k}} \leqq \sum_{q=1}^{\infty} q^{1-\frac{s-1}{k}} = \sum_{q=1}^{\infty} q^{-3+\frac{1}{k}} = O(1).$$

Lemma 6 (E. Landau[4]):

$$\int_{-\infty}^{\infty} e^{-2\pi i n \beta} I^s \, d\beta = \frac{\Gamma^s(1+1/k)}{\Gamma(s/k)} n^{s/k-1} \quad \textit{für } 0 < n \leqq P^k.$$

Lemma 7:

$$\int_{\mathfrak{M}} e^{-2\pi i n \beta} I^s \, d\beta = \frac{\Gamma^s(1+1/k)}{\Gamma(s/k)} n^{s/k-1} + O\left(P^{s-k-\rho\left(\frac{s}{k}-1\right)} q^{\frac{s}{k}-1}\right),$$

für $0 < n \leqq P^k$.

Beweis: Nach (13) ist

$$\int_{\beta \text{ nicht auf } \mathfrak{M}} |I|^s \, d\beta = O \int_{\frac{1}{2} q^{-1} P^{\rho-k}} \beta^{-\frac{s}{k}} d\beta = O\left((q^{-1} P^{\rho-k})^{1-\frac{s}{k}}\right)$$

$$= O\left(P^{s-k-\rho\left(\frac{s}{k}-1\right)} q^{\frac{s}{k}-1}\right).$$

[4]) „Über die neue Winogradoffsche Behandlung des Waringschen Problems" [Mathematische Zeitschrift 31 (1929), S. 319—338].

Lemma 8:

$$\sum_{\mathfrak{M}} \int_{\mathfrak{M}} e^{-2\pi i n \alpha} T^s \, d\alpha = \frac{\Gamma^s(1+1/k)}{\Gamma(s/k)} \mathfrak{S}(n) n^{s/k-1} + O(P^{s-k-(1-\rho)}),$$

für $0 < n \leqq P^k$, *also ist für* $\frac{1}{4} P^k \leqq n \leqq P^k$ *bei hinreichend grossem* P

$$\Re \sum_{\mathfrak{M}} \int_{\mathfrak{M}} e^{-2\pi i n \alpha} T^s \, d\alpha > c_2 P^{s-k}.$$

Beweis: Nach Lemma 5 ist die linke Seite

$$= O(P^{s-k-(1-\rho)}) + \sum_{\mathfrak{M}} q^{-s} S^s_{a,q} \int_{\mathfrak{M}} e^{-2\pi i n \alpha} I^s \, d\alpha$$

$$= O(P^{s-k-(1-\rho)}) + \sum_{\mathfrak{M}} q^{-s} S^s_{a,q} e^{-2\pi i n \frac{a}{q}} \int_{\mathfrak{M}} e^{-2\pi i n \beta} I^s \, d\beta$$

$$= O(P^{s-k-(1-\rho)}) + \frac{\Gamma^s(1+1/k)}{\Gamma(s/k)} n^{s/k-1} \sum_{\mathfrak{M}} q^{-s} S^s_{a,q} e^{-2\pi i n \frac{a}{q}}$$

$$+ O\left(\sum_{\mathfrak{M}} q^{-\frac{s}{k}} P^{s-k-\rho\left(\frac{s}{k}-1\right)} q^{\frac{s}{k}-1} \right) \quad \text{(Lemma 7 und 1}$$

$$= O(P^{s-k-(1-\rho)}) + \frac{\Gamma^s(1+1/k)}{\Gamma(s/k)} n^{\frac{s}{k}-1} \left(\mathfrak{S} + \sum_{q > P^\rho} O\left(q^{1-\frac{s}{k}}\right) \right)$$

$$+ O\left(P^{s-k-\rho\left(\frac{s}{k}-1\right)}\right) \sum_{\mathfrak{M}} q^{-1}$$

$$= O(P^{s-k-(1-\rho)}) + \frac{\Gamma^s(1+1/k)}{\Gamma(s/k)} \mathfrak{S} n^{s/k-1} + O\left(P^{s-k-\rho\left(\frac{s}{k}-2\right)}\right).$$

Minor Arcs.

Lemma 10 (Vinogradov): *Es sei*

$$\left| \alpha - \frac{a}{q} \right| \leqq q^{-2}, \quad (a, q) = 1, \quad q > 1.$$

ξ *durchlaufe* Ξ *ganze Zahlen eines Intervalls der Länge* X *und* η

durchlaufe Η *ganze Zahlen eines Intervalls der Länge Y.*

Dann ist

$$\left|\sum_{\xi,\eta} e^{2\pi i \xi\eta\alpha}\right|^2 = O\left(\Xi \mathrm{H} X Y \frac{\log q}{q}\left(1+\frac{q}{X}\right)\left(1+\frac{q}{Y}\right)\right);$$

also, falls $Y \leqq q \leqq X$,

$$\left|\sum_{\xi,\eta} e^{2\pi i \xi\eta\alpha}\right|^2 = O(\Xi \mathrm{H} X \log q).$$

Beweis:

$$\left|\sum_{\xi,\eta} e^{2\pi i \xi\eta\alpha}\right|^2 \leqq \sum_{\xi} 1 \sum_{x=M+1}^{M+X} \left|\sum_{\eta} e^{2\pi i x\eta\alpha}\right|^2$$

$$= \Xi \sum_{\eta}\sum_{\eta'}\sum_{x=M+1}^{M+X} e^{2\pi i x(\eta-\eta')\alpha} \leqq \Xi \sum_{\eta}\sum_{\eta'} \mathrm{Min}\left(X, \frac{1}{\{(\eta-\eta')\alpha\}}\right)$$

$$\leqq \Xi \mathrm{H} \sum_{y=1-Y}^{Y-1} \mathrm{Min}\left(X, \frac{1}{\{y\alpha\}}\right) = O\left(\Xi \mathrm{H}\left(\frac{Y}{q}+1\right)(X+q\log q)\right).$$

Lemma 11: *Auf* $\mathfrak{m}$ *ist*

$$|S(\alpha)|^2 = O(A_S P^{k-\rho} \log P).$$

Beweis: Lemma 10 mit

$$\Xi \mathrm{H} = A_S, \quad X = P^{k-\rho}, \quad Y = P^{\rho}.$$

Lemma 12:

$$\sum_{\mathfrak{m}} \int_{\mathfrak{m}} |T^s(\alpha) R^2(\alpha) S(\alpha)|\, d\alpha = O\left(P^s A_R A_S^{\frac{1}{2}} P^{\frac{1}{2}(k-\rho)} \log P\right).$$

Beweis:

$$\sum_{\mathfrak{m}} \int_{\mathfrak{m}} |T^s(\alpha) R^2(\alpha) S(\alpha)|\, d\alpha \leqq \underset{0\leqq\alpha\leqq 1}{\mathrm{Max}} |T^s| \underset{\alpha \text{ auf } \mathfrak{m}}{\mathrm{Max}} |S| \int_0^1 |R(\alpha)|^2 d\alpha$$

$$\leqq P^s O(A_S P^{k-\rho} \log P)^{\frac{1}{2}} A_R.$$

Lemma 13 (Hardy-Littlewood):

$$A_R \geqq c_3 P^{k\left(1-\left(1-\frac{1}{k}\right)^l\right)},$$

$$A_S \geqq c_4 P^{\frac{\rho}{k}+(k-\rho)\left(1-\left(1-\frac{1}{k}\right)^l\right)}.$$

Lemma 14:

$$P^s A_R A_S^{\frac{1}{2}} P^{\frac{1}{2}(k-\rho)} \log P = o\,(P^{s-k} A_R^2 A_S).$$

Beweis: Wegen

$$2k\left(1-\left(1-\frac{1}{k}\right)^l\right)+\frac{\rho}{k}+(k-\rho)\left(1-\left(1-\frac{1}{k}\right)^l\right)=3k-\rho+\frac{\rho}{k}$$

$$-(3k-\rho)\left(1-\frac{1}{k}\right)^l$$

$$> 3k-\rho+\frac{\rho}{k}-(3k-\rho)\,(3k^2+2k^{-1})^{-1} \quad \text{(nach (7))}$$

$$= 3k-\rho+\frac{\rho}{k}-\frac{1}{k+1}=3k-\rho$$

gilt

$$P^{3k-\rho} \log^2 P = o\,(A_R^2 A_S),$$

woraus die Behauptung folgt.

Lemma 15: *Für hinreichend grosses N ist*

$$\int_0^1 e^{-2\pi i N\alpha}\, T^s(\alpha)\, R^2(\alpha)\, S(\alpha)\, d\alpha > 0.$$

Beweis: Es genügt zu zeigen, dass das Integral von 0 verschieden ist. Es ist

$$\Re \sum_{\mathfrak{M}} \int_{\mathfrak{M}} e^{-2\pi i N\alpha}\, T^s(\alpha)\, R^2(\alpha)\, S(\alpha)\, d\alpha$$

$$= \Re \sum_{u_1 \leqq \frac{1}{4}P^k} \; \sum_{u_2 \leqq \frac{1}{4}P^k} \; \sum_{1 \leqq y \leqq P^{\rho/k}} \; \sum_{u_3 \leqq \frac{1}{4}P^{k-\rho}} \; \sum_{\mathfrak{M}}$$

$$\times \int_{\mathfrak{M}} e^{-2\pi i (N-u_1-u_2-y^k u_3)\alpha}\, T^s\, d\alpha$$

$$\geqq \sum_{u_1 \leqq \frac{1}{4} P^k} \sum_{u_2 \leqq \frac{1}{4} P^k} \sum_{1 \leqq y \leqq P^{\rho/k}} \sum_{u_3 \leqq \frac{1}{4} P^{k-\rho}} c_2 P^{s-k} \qquad \text{(Lemma 8)}$$

$$= c_2 A_R^2 A_S P^{s-k},$$

und aus Lemma 12 und 14 folgt die Behauptung.

Hauptsatz:

$$G(k) \leqq 4k + 3l \leqq k(3 \log (3k^2 + 2k) + 4) + 3.$$

Beweis: Nach Lemma 15 ist jedes grosse N in der Form

$$N = x_1^k + \ldots + x_s^k + u_1 + u_2 + y^k u_3$$

darstellbar.

(Eingegangen am 7. Juli 1935)

[*Cambridge*]

10t.

On Waring's Problem

Translation of
Acta Arithmetica 1 (1936), 212–221
[Commentary on p. 577]

Recently Vinogradov has made significant progress in the treatment of Waring's problem.[1] He proved:

Main Theorem. *For $k > 2$, k integral, let $G(k)$ be the smallest number such that every sufficiently large number can be represented as a sum of at most $G(k)$ positive k-th powers. Then*

$$G(k) = O(k \log k); \tag{1}$$

more precisely,

$$G(k) \leqq 6k \log k + (4 + \log 216)k. \tag{2}$$

The existence of a finite $G(k)$ is not trivial; it was first proved by Hilbert. The first useful bound is due to Hardy and Littlewood, who proved,[2] roughly speaking,

$$\overline{\lim_{k=\infty}}\, k^{-1}2^{2-k}G(k) \leqq 1. \tag{3}$$

[1] "On the upper bound of $G(n)$ in Waring's problem" [*Bulletin de l'Académie de l'U.R.S.S., VII série, Classe des sciences mathématiques et naturelles*, 1934, pp. 1455–1469], Russian with English summary; "Une nouvelle variante de la démonstracion du théorème de Waring" [*Comptes Rendus* **200** (1935), 182–184]; "On Waring's Problem" [*Annals of Mathematics* **36** (1935), 395–405].

[2] "Some problems of Partitio Numerorum": I. "A new solution of Waring's problem" [*Göttinger Nachrichten* (1920), pp. 33–54]; II. "Proof that every large number is the sum of at most 21 biquadrates" [*Mathematische Zeitschrift* **9** (1921), 14–27]; IV. "The singular series in Waring's problem and the value of the number $G(k)$" [ibid. **12** (1922), 161–188]; VI. "Further researches in Waring's Problem" [ibid. **23** (1925), 1–37].

With the exception of (10), all the results of Hardy-Littlewood which are used in the present article also appear in Landau, *Vorlesungen über Zahlentheorie*, Vol. I, part VI. For (10), cf. Partitio Numerorum IV.

Since trivially

$$(4) \qquad G(k) \geqq k+1,$$

the significance of Vinogradov's result is immediately clear. His proof is partly a consequent continuation of the ideas of Hardy and Littlewood; partly it rests on the introduction of a new lemma[3] which takes the place of the difficult diophantine approximations of Weyl and thereby simplifies the calculations considerably.

However, it escaped the ingenuity of Vinogradov that his complicated calculations are unnecessary and can be avoided by a simple trick. This will be demonstrated in the present article, and I obtain not only (1) but also

$$(5) \qquad G(k) \leqq 6k \log k + \left(4 + 3\log\left(3 + \frac{2}{k}\right)\right)k + 3,$$

which is slightly better than (2). It is not impossible to improve this formula still further, but I cannot even prove at present that

$$(6) \qquad \overline{\lim_{k=\infty}}\, k^{-1} \log^{-1} k G(k) < 6.$$

Notations. Let

$$k \geqq 3 \text{ be integral}, \quad s = 4k, \quad \varrho = \frac{k}{k+1}, \quad l = \left[k \log(3k^2 + 2k)\right] + 1,$$

hence

$$(7) \quad \left(1 - \frac{1}{k}\right)^{l} < \left(1 - \frac{1}{k}\right)^{k \log(3k^2+2k)} < e^{-\log(3k^2+2k)} = (3k^2 + 2k)^{-1}.$$

Let $N > 0$ be the number to be represented. We put

$$P = N^{\frac{1}{k}}.$$

Let $0 \leqq \alpha \leqq 1$. We divide this interval into subintervals by associating in a well-known fashion to each Farey fraction $\frac{a}{q}$, where

$$1 \leqq q \leqq P^{k-\varrho}, \quad 1 \leqq a \leqq q, \quad (a, q) = 1,$$

[3] Lemma 10 of this article.

a neighborhood such that the interval $0 \leqq \alpha \leqq 1$ if covered without repetition. We divide these subintervals into two classes:

1. If $1 \leqq q \leqq P^{\varrho}$, then we call the interval a major arc. Notation: $\mathfrak{M}$.
2. If $P^{\varrho} < q < P^{k-\varrho}$, then we call the interval a minor arc. Notation: $\mathfrak{m}$.

In every case the interval has the form:

$$(8)\quad \alpha = \frac{a}{q} + \beta, \; -\vartheta_1 q^{-1} P^{\varrho-k} \leqq \beta \leqq \vartheta_2 q^{-1} P^{\varrho-k}, \frac{1}{2} \leqq \vartheta_1 \leqq 1, \quad \frac{1}{2} \leqq \vartheta_2 \leqq 1.$$

Let u run over all numbers which can be represented as a sum of at most l positive kth powers.

We put:

$$T = T(\alpha) + \sum_{1 \leqq \mu \leqq P} e^{2\pi i \mu^k \alpha}, \qquad I = I(\beta) = \int_0^P e^{2\pi i \beta v^k}\, dv,$$

$$R = R(\alpha) = \sum_{u \leqq \frac{1}{4} P^k} e^{2\pi i u \alpha}, \qquad A_R = \sum_{u \leqq \frac{1}{4} P^k} 1,$$

$$S = S(\alpha) = \sum_{1 \leqq y \leqq P^{\varrho/k}} \; \sum_{u \leqq \frac{1}{4} P^{k-\varrho}} e^{2\pi i y^k u \alpha}, \qquad A_S = \sum_{1 \leqq y \leqq P^{\varrho/k}} \; \sum_{u \leqq \frac{1}{4} P^{k-\varrho}} 1.$$

O-estimates are, for k fixed, valid uniformly in all the other variables. o-estimates are, for k fixed, valid for $P \to \infty$.

$c_1, \ldots, c_6$ are positive constants which depend only on k.

Major Arcs

Lemma 1 (Hardy-Littlewood). *Put*

$$S_{a,q} = \sum_{\nu=1}^{q} e^{2\pi i \nu^k \frac{a}{q}}.$$

Then

$$(9)\qquad S_{a,q} = O\left(q^{1-\frac{1}{k}}\right).$$

Lemma 2 (Hardy-Littlewood). *Put, for integral n,*

$$\mathfrak{S} = \mathfrak{S}(n, k) = \sum_{q=1}^{\infty} q^{-s} \sum_{a=1}^{q} e^{-2\pi i n \frac{a}{q}} S_{a,q}^{s}.$$

Then by Lemma 1 *the series converges absolutely and one has*

$$\mathfrak{S} > c_1. \tag{10}$$

Lemma 3. *Assume that for* $\tau_1 \leqq \tau \leqq \tau_2$

$$0 \leqq f(\tau) \leqq B_0, \quad 0 \leqq f'(\tau) \leqq B_1, \quad 0 \leqq f''(\tau).$$

Then

$$\sum_{\tau_1 \leqq \lambda \leqq \tau_2} e^{2\pi i f(\lambda)} = \int_{\tau_1}^{\tau_2} e^{2\pi i f(\tau)}\, d\tau + O(1 + B_1 + B_0 B_1).$$

Proof. By Euler's sum formula the difference between the sum and the integral is

$$= O(1) + 2\pi i \int_{\tau_1}^{\tau_2} \left(\tau - [\tau] - \frac{1}{2}\right) f'(\tau) e^{2\pi i f(\tau)}\, d\tau$$

$$= O(1) + 2\pi i \int_{\tau_1}^{\tau_2} \left(\tau - [\tau] - \frac{1}{2}\right) f'(\tau)\{(2\pi(1 + i) f(\tau)) + 1$$

$$-(2\pi f(\tau) + 1 - \cos(2\pi f(\tau)) - i(2\pi f(\tau) - \sin(2\pi f(\tau))))\}\, d\tau$$

$$= O(1) + O(B_1(B_0 + 1))$$

by the second Mean Value Theorem, since each of the four summands in the braces lies, after multiplication with a root of unity, between 0 and $2\pi\sqrt{2}\, B_0 + 2$ and is monotonically nondecreasing.

Lemma 4. *On* $\mathfrak{M}$ *one has*

$$T = q^{-1} S_{a,q} I + O(q + P^{2\varrho - 1}).$$

Proof. One has

$$T = \sum_{\nu=1}^{q} \sum_{\frac{1-\nu}{q} \leqq \lambda \leqq \frac{P-\nu}{q}} e^{2\pi i(\nu + \lambda q)^k \left(\frac{a}{q} + \beta\right)}$$

$$= \sum_{\nu=1}^{q} e^{2\pi i \nu^k \frac{a}{q}} \sum_{\frac{1-\nu}{q} \leqq \lambda \leqq \frac{P-\nu}{q}} e^{2\pi i(\nu+\lambda q)^k \beta}.$$

If one applies Lemma 3 to the inner sum with

$$B_0 = P^k|\beta| \leqq P^k q^{-1} P^{\varrho-k} \leqq q^{-1} P^{\varrho},$$

$$B_1 = kqP^{k-1}|\beta| \leqq kP^{\varrho-1}$$

then the assertion follows because

$$B_1 = O(1), \quad B_0 B_1 = O(q^{-1}P^{2\varrho-1}).$$

***Lemma* 5.**

$$\sum_{\mathfrak{M}} \int_{\mathfrak{M}} |T^s - q^{-s} S_{a,q}^s I^s| \, d\alpha = O(P^{s-k-(1-\varrho)}).$$

Proof. By Lemma 4 one has on $\mathfrak{M}$

$$T - q^{-1} S_{a,q} I = O(P^{\varrho}). \tag{11}$$

For $0 \leqq |\beta| \leqq P^{-k}$ one has by Lemma 1

$$q^{-1} S_{a,q} I = O\left(q^{-\frac{1}{k}} P\right). \tag{12}$$

For $P^{-k} \leqq |\beta| \leqq q^{-1} P^{\varrho-k}$ one has

$$I = \int_0^P e^{2\pi i \beta v^k} \, dv = |\beta|^{-1/k} \int_0^{|\beta|^{1/k} P} e^{\pm 2\pi i v^k} \, dv = O\left(|\beta|^{-\frac{1}{k}}\right). \tag{13}$$

since $\int_0^\infty e^{2\pi i v^k} \, dv$ converges.

Moreover, since $q \leqq P^{\varrho}$, one has

$$P^{\varrho} \leqq q^{-\frac{1}{k}} P \tag{14}$$

and for $|\beta| \leqq q^{-1} P^{\varrho-k}$

$$P^{\varrho} = P^{1-\varrho/k} = \left(P^{\varrho-k}\right)^{-\frac{1}{k}} \leqq q^{-\frac{1}{k}}|\beta|^{-\frac{1}{k}}. \tag{15}$$

By (11), (12), and (14) one therefore has for $0 \leqq |\beta| \leqq P^{-k}$

$$|T^s - q^{-s}S^s_{a,q}I^s| = O\left(P^{\varrho}q^{-\frac{s-1}{k}}P^{s-1}\right).$$

and by (11), (13), and (15) one has for $P^{-k} \leqq |\beta| \leqq q^{-1}P^{\varrho-k}$

$$|T^s - q^{-s}S^s_{a,q}I^s| = O\left(P^{\varrho}q^{-\frac{s-1}{k}}|\beta|^{-\frac{s-1}{k}}\right).$$

Thus

$$\int_{\mathfrak{M}} |T^s - q^{-s}S^s_{a,q}I^s|\,d\alpha = O\left(P^{-k+\varrho+s-1}q^{-\frac{s-1}{k}}\right) + O\left(P^{\varrho}q^{-\frac{s-1}{k}}\int_{P^{-k}}^{\infty}\beta^{-\frac{s-1}{k}}\,d\beta\right)$$

$$= O\left(q^{-\frac{s-1}{k}}P^{s-k-(1-\varrho)}\right) + O\left(q^{-\frac{s-1}{k}}P^{\varrho-k\left(1-\frac{s-1}{k}\right)}\right)$$

$$= O\left(q^{-\frac{s-1}{k}}P^{s-k-(1-\varrho)}\right).$$

From this the assertion follows since

$$\sum_{\mathfrak{M}} q^{-\frac{s-1}{k}} \leqq \sum_{q=1}^{\infty} q^{1-\frac{s-1}{k}} = \sum_{q=1}^{\infty} q^{-3+\frac{1}{k}} = O(1).$$

Lemma 6 (E. Landau[4]).

$$\int_{-\infty}^{\infty} e^{-2\pi in\beta}I^s\,d\beta = \frac{\Gamma^s(1+1/k)}{\Gamma(s/k)}n^{s/k-1} \quad \textit{for } 0 < n \leqq P^k.$$

Lemma 7.

$$\int_{\mathfrak{M}} e^{-2\pi in\beta}I^s\,d\beta = \frac{\Gamma^s(1+1/k)}{\Gamma(s/k)}n^{s/k-1} + O\left(P^{s-k-\varrho\left(\frac{s}{k}-1\right)}q^{\frac{s}{k}-1}\right),$$

for $0 < n \leqq P^k$.

[4] "Über die neue Winogradoffsche Behandlung des Waringschen Problems" [*Mathematische Zeitschrift* **31** (1929), 319–338].

Proof. By (13) one has

$$\int_{\beta \text{ not on } \mathfrak{M}} |I|^s \, d\beta = O \int_{\frac{1}{2}q^{-1}P^{\varrho-k}} \beta^{\frac{-s}{k}} \, d\beta = O\left(\left(q^{-1}P^{\varrho-k}\right)^{1-\frac{s}{k}}\right)$$

$$= O\left(P^{s-k-\varrho\left(\frac{s}{k}-1\right)} q^{\frac{s}{k}-1}\right).$$

Lemma 8.

$$\sum_{\mathfrak{M}} \int_{\mathfrak{M}} e^{-2\pi i n \alpha} T^s \, d\alpha = \frac{\Gamma^s(1+1/k)}{\Gamma(s/k)} \mathfrak{S}(n) n^{s/k-1} + O(P^{s-k-(1-\varrho)}),$$

for $0 < n \leqq P^k$, *hence for* $\frac{1}{4}P^k \leqq n \leqq P^k$ *with sufficiently large* P *one has*

$$\Re \sum_{\mathfrak{M}} \int_{\mathfrak{M}} e^{-2\pi i n \alpha} T^s \, d\alpha > c_2 P^{s-k}.$$

Proof. By Lemma 5 the left side

$$= O(P^{s-k-(1-\varrho)}) + \sum_{\mathfrak{M}} q^{-s} S_{a,q}^s \int_{\mathfrak{M}} e^{-2\pi i n \alpha} I^s \, d\alpha$$

$$= O(P^{s-k-(1-\varrho)}) + \sum_{\mathfrak{M}} q^{-s} S_{a,q}^s e^{-2\pi i n \frac{a}{q}} \int_{\mathfrak{M}} e^{-2\pi i n \beta} I^s \, d\beta$$

$$= O(P^{s-k-(1-\varrho)}) + \frac{\Gamma^s(1+1/k)}{\Gamma(s/k)} n^{s/k-1} \sum_{\mathfrak{M}} q^{-s} S_{a,q}^s e^{-2\pi i n \frac{a}{q}}$$

$$+ O\left(\sum_{\mathfrak{M}} q^{-\frac{s}{k}} P^{s-k-\varrho\left(\frac{s}{k}-1\right)} q^{\frac{s}{k}-1}\right) \quad \text{(Lemma 7 and 1)}$$

$$= O(P^{s-k-(1-\varrho)}) + \frac{\Gamma^s(1+1/k)}{\Gamma(s/k)} n^{\frac{s}{k}-1} \left(\mathfrak{S} + \sum_{q > P^\varrho} O\left(q^{1-\frac{s}{k}}\right)\right)$$

$$+ O\left(P^{s-k-\varrho\left(\frac{s}{k}-1\right)}\right) \sum_{\mathfrak{M}} q^{-1}$$

$$= O(P^{s-k-(1-\varrho)}) + \frac{\Gamma^s(1+1/k)}{\Gamma(s/k)}\mathfrak{S}n^{s/k-1} + O\left(P^{s-k-\varrho\left(\frac{s}{k}-2\right)}\right).$$

Minor Arcs

Lemma 10 (Vinogradov). *Let*

$$\left|\alpha - \frac{a}{q}\right| \leqq q^{-2}, \quad (a,q) = 1, \quad q > 1.$$

Let ξ run through Ξ integers in an interval of length X and η run through H integers in an interval of length Y. Then one has

$$\left|\sum_{\xi,\eta} e^{2\pi i\xi\eta\alpha}\right|^2 = O\left(\Xi H XY\frac{\log q}{q}\left(1+\frac{q}{X}\right)\left(1+\frac{q}{Y}\right)\right);$$

hence, if $Y \leqq q \leqq X$,

$$\left|\sum_{\xi,\eta} e^{2\pi i\xi\eta\alpha}\right|^2 = O(\Xi H X \log q).$$

Proof.

$$\left|\sum_{\xi,\eta} e^{2\pi i\xi\eta\alpha}\right|^2 \leqq \sum_{\xi} 1 \sum_{x=M+1}^{M+X} \left|\sum_{\eta} e^{2\pi i x\eta\alpha}\right|^2$$

$$= \Xi\sum_{\eta}\sum_{\eta'}\sum_{x=M+1}^{M+X} e^{2\pi i x(\eta-\eta')\alpha}$$

$$\leqq \Xi\sum_{\eta}\sum_{\eta'} \operatorname{Min}\left(X, \frac{1}{\{(\eta-\eta')\alpha\}}\right)$$

$$\leqq \Xi H \sum_{y=1-Y}^{Y-1} \operatorname{Min}\left(X, \frac{1}{\{y\alpha\}}\right)$$

$$= O\left(\Xi H\left(\frac{Y}{q}+1\right)(X + q\log q)\right).$$

Lemma 11. *On* $\mathfrak{m}$ *one has*

$$|S(\alpha)|^2 = O(A_S P^{k-\varrho} \log P).$$

Proof. Lemma 10 with

$$\Xi H = A_S, \quad X = P^{k-\varrho}, \quad Y = P^{\varrho}.$$

Lemma 12.

$$\sum_{\mathfrak{m}} \int_{\mathfrak{m}} |T^s(\alpha) R^2(\alpha) S(\alpha)| \, d\alpha = O\left(P^s A_R A_S^{\frac{1}{2}} P^{\frac{1}{2}(k-\varrho)} \log P \right).$$

Proof.

$$\sum_{\mathfrak{m}} \int_{\mathfrak{m}} |T^s(\alpha) R^2(\alpha) S(\alpha)| \, d\alpha \leqq \max_{0 \leqq \alpha \leqq 1} |T^s| \max_{\alpha \text{ on } \mathfrak{m}} |S| \int_0^1 |R(\alpha)|^2 \, d\alpha$$

$$\leqq P^s O(A_S P^{k-\varrho} \log P)^{\frac{1}{2}} A_R.$$

Lemma 13 (Hardy-Littlewood).

$$A_R \geqq c_3 P^{k\left(1-\left(1-\frac{1}{k}\right)^l\right)}$$

$$A_S \geqq c_4 P^{\frac{\varrho}{k}+(k-\varrho)\left(1-\left(1-\frac{1}{k}\right)^l\right)}.$$

Lemma 14.

$$P^s A_R A_S^{\frac{1}{2}} P^{\frac{1}{2}(k-\varrho)} \log P = o(P^{s-k} A_R^2 A_S).$$

Proof. Since

$$2k\left(1 - \left(1 - \frac{1}{k}\right)^l\right) + \frac{\varrho}{k} + (k-\varrho)\left(1 - \left(1 - \frac{1}{k}\right)^l\right)$$

$$= 3k - \varrho + \frac{\varrho}{k} - (3k-\varrho)\left(1 - \frac{1}{k}\right)^l$$

$$> 3k - \varrho + \frac{\varrho}{k} - (3k-\varrho)(3k^2 + 2k^{-1})^{-1} \quad \text{(by (7))}$$

$$= 3k - \varrho + \frac{\varrho}{k} - \frac{1}{k+1} = 3k - \varrho,$$

one has

$$P^{3k-\varrho} \log^2 P = o\left(A_R^2 A_S\right),$$

from which the assertion follows.

Lemma 15. *For a sufficiently large N one has*

$$\int_0^1 e^{-2\pi i N\alpha} T^s(\alpha) R^2(\alpha) S(\alpha)\, d\alpha > 0.$$

Proof. It suffices to show that the integral is different from 0. One has

$$\Re \sum_{\mathfrak{M}} \int_{\mathfrak{M}} e^{-2\pi i N\alpha} T^s(\alpha) R^2(\alpha) S(\alpha)\, d\alpha$$

$$= \Re \sum_{u_1 \leqq \frac{1}{4}P^k} \sum_{u_2 \leqq \frac{1}{4}P^k} \sum_{1 \leqq y \leqq P^{\varrho/k}} \sum_{u_3 \leqq \frac{1}{4}P^{k-\varrho}} \sum_{\mathfrak{M}}$$

$$\times \int_{\mathfrak{M}} e^{-2\pi i (N - u_1 - u_2 - y^k u_3)\alpha} T^s\, d\alpha$$

$$\geqq \sum_{u_1 \leqq \frac{1}{4}P^k} \sum_{u_2 \leqq \frac{1}{4}P^k} \sum_{1 \leqq y \leqq P^{\varrho/k}} \sum_{u_3 \leqq \frac{1}{4}P^{k-\varrho}} c_2 P^{s-k} \quad \text{(Lemma 8)}$$

$$= c_2 A_R^2 A_S P^{s-k},$$

and from Lemmas 12 and 14 the assertion follows.

Main Theorem.

$$G(k) \leqq 4k + 3l \leqq k\left(3\log(3k^2 + 2k) + 4\right) + 3.$$

Proof. By Lemma 15 every large N is representable in the form

$$N = x_1^k + \cdots + x_s^k + u_1 + u_2 + y^k u_3.$$

(Received 7 July 1935)
[Cambridge]

11.

On the Zeros of Certain Dirichlet Series I

(With H. Davenport)*

Journal of the London Mathematical Society 11 (1936), 181–185
[Commentary on p. 588]

1. The function $\zeta(s, a)$ is defined for $0 < a \leqslant 1$ by

$$\zeta(s, a) = \sum_{n=0}^{\infty} (n+a)^{-s}.$$

For $a = 1$, $\zeta(s, a) = \zeta(s)$, and for $a = \frac{1}{2}$, $\zeta(s, a) = (2^s - 1)\zeta(s)$; hence, for these two values of a, $\zeta(s, a)$ has no zeros in the half-plane $\sigma \geqslant 1$. We give in this note a simple proof that, for any other rational value of a, $\zeta(s, a)$ has an infinity of zeros for $\sigma > 1$. As regards irrational values of a, we show that the result holds also for any transcendental a. We are unable to attack the problem at all if a is an algebraic irrational number.

For a positive definite quadratic form

$$Q(x, y) = ax^2 + bxy + cy^2$$

with integer coefficients and a fundamental discriminant $d = b^2 - 4ac$, the Epstein zeta-function $\zeta(s, Q)$ is defined by

$$\zeta(s, Q) = \sum_{\substack{x, y = -\infty \\ x, y \neq 0, 0}}^{\infty} Q(x, y)^{-s}.$$

Potter and Titchmarsh† have recently proved that $\zeta(s, Q)$ has an infinity of zeros on $\sigma = \frac{1}{2}$, so that the analogue of Hardy's result holds. Using the same method as for $\zeta(s, a)$ with rational a, we prove that, if the class-number $h(d)$ is even, then $\zeta(s, Q)$ has an infinity of zeros for $\sigma > 1$‡. The condition $h(d)$ even is satisfied unless $d = -4, -8$, or $-p$, p prime§. If $h(d) = 1$, then

$$\zeta(s, Q) = c\zeta(s) \prod_p \left\{1 - \left(\frac{d}{p}\right) p^{-s}\right\}^{-1},$$

and so has no zeros for $\sigma > 1$. It can be proved that, if $h(d)$ is odd but different from 1, then again $\zeta(s, Q)$ has an infinity of zeros for $\sigma > 1$, but the proof is more complicated and will be given in a second note.

2. Let $a = l/k$, $k > 2$, $(l, k) = 1$, $0 < l < k$. If χ runs through the Dirichlet characters mod k, we have

$$\zeta(s, a) = \frac{k^s}{\phi(k)} \sum_{\chi} \bar{\chi}(l) L(s, \chi).$$

---

* Received 11 January, 1936; read 16 January, 1936.

† *Proc. London Math. Soc.* (2), 39 (1935), 372–384.

‡ The fact that not all zeros of $\zeta(s, Q)$ lie on $\sigma = \frac{1}{2}$ was discovered in the case $Q = x^2 + 5y^2$ by Potter and Titchmarsh (*loc. cit.*).

§ Hecke, *Theorie der algebraischen Zahlen* (1923), Satz 132.

We denote by χ_0 the principal character, and by χ_1 an arbitrary real non-principal character, which will be fixed throughout. For any prime p, we define

$$a(p) = \begin{cases} 1 & \text{if } \chi_1(p) = 0 \text{ or } 1, \\ i & \text{if } \chi_1(p) = -1; \end{cases}$$

and, for $n = p_1^{\nu_1} \dots p_r^{\nu_r}$, we define $a(n) = a(p_1)^{\nu_1} \dots a(p_r)^{\nu_r}$. We write

$$M(s, \chi) = \sum_{n=1}^{\infty} a(n)\chi(n)n^{-s} = \prod_p \{1 - a(p)\chi(p)p^{-s}\}^{-1},$$

$$Z(s) = \frac{k^s}{a(l)\phi(k)} \sum_\chi \bar{\chi}(l) M(s, \chi)$$

$$= \frac{k^s}{a(l)} \sum_{\substack{n=1 \\ n \equiv l \ (\mathrm{mod}\ k)}}^{\infty} a(n)n^{-s}.$$

For $n \equiv l \pmod k$, we have $a(n) = \pm a(l)$; hence $Z(s)$ is real for $s > 1$, and it is easily seen that

$$\frac{1}{a(l)} \{M(s, \chi_0) + \chi_1(l) M(s, \chi_1)\} = 2\Re\left(\frac{M(s, \chi_0)}{a(l)}\right)$$

is also real for $s > 1$.

Lemma 1. *For $s > 1$,*

$$\log M(s, \chi_0) = -\tfrac{1}{2}(1+i)\log(s-1) + O(1),$$
$$\log M(s, \chi_1) = -\tfrac{1}{2}(1-i)\log(s-1) + O(1),$$

and, for $\chi \neq \chi_0, \chi_1$,

$$\log M(s, \chi) = O(1).$$

Proof. We have

$$\log M(s, \chi) = \sum_p a(p)\chi(p)p^{-s} + O(1)$$
$$= \tfrac{1}{2}\sum_p \{1 + \chi_1(p)\}\chi(p)p^{-s} + \tfrac{1}{2}i\sum_p \{1 - \chi_1(p)\}\chi(p)p^{-s} + O(1)$$
$$= \tfrac{1}{2}(1+i)\log L(s, \chi) + \tfrac{1}{2}(1-i)\log L(s, \chi\chi_1) + O(1),$$

whence the result.

Lemma 2. *For any $\delta > 0$, $Z(s)$ has a zero in $1 < s < 1 + \delta$.*

Proof. By the last part of Lemma 1,

$$Z(s) = \frac{2k^s}{\phi(k)} \Re\left(\frac{1}{a(l)} M(s, \chi_0)\right) + O(1)$$

for $s > 1$. By the first part of Lemma 1, we can, for any $H > 0$, find s_1 and

s_2 such that $1 < s_1 < s_2 < 1+\delta$ and

$$\mathfrak{R}\left(\frac{1}{a(l)} M(s_1, \chi_0)\right) > H,$$

$$\mathfrak{R}\left(\frac{1}{a(l)} M(s_2, \chi_0)\right) < -H.$$

Then, if H is sufficiently large,

$$Z(s_1) > 0 > Z(s_2).$$

Thus there is a value of s with $1 < s < 1+\delta$ for which $Z(s) = 0$. This proves the lemma.

THEOREM 1. *If a is rational and $a \neq \frac{1}{2}$ or 1, $\zeta(s, a)$ has an infinity of zeros for $\sigma > 1$.*

Proof. By Kronecker's theorem*, given $\epsilon > 0$ and $P > 2$, there exists a real number τ such that $|p^{-i\tau} - a(p)| < \epsilon$ for all $p \leqslant P$. Hence, for any $\epsilon > 0$ and $\delta > 0$, there is a τ such that

$$|\zeta(s+\tau i, l/k) - a(l)\, Z(s)| < \epsilon$$

for $\sigma > 1+\delta$.

Let $s_1 > 1$ be a zero of $Z(s)$. For any $\eta > 0$ there exists an η_1 with $0 < \eta_1 < \eta$, $\eta_1 < s_1 - 1$, such that $Z(s) \neq 0$ for $|s - s_1| = \eta_1$. Take

$$\epsilon = \min_{|s-s_1|=\eta_1} |Z(s)|,$$

and $\delta < s_1 - \eta_1 - 1$; then the conditions of Rouché's theorem are satisfied on $|s - s_1| = \eta_1$. Hence $\zeta(s, l/k)$ has a zero in the circle $|s - s_1 - \tau i| < \eta$. This proves the theorem.

3. THEOREM 2. *For transcendental a, $\zeta(s, a)$ has an infinity of zeros for $\sigma > 1$.*

Proof. The numbers $\log(n+a)$ are linearly independent. For a relation

$$\sum_{n=0}^{N} c_n \log(n+a) = 0$$

with integers c_n not all zero implies that

$$\prod_{n=0}^{N} (a+n)^{c_n} = 1,$$

and this cannot be an identity in a since it is not satisfied by $a = -n$ if $c_n \neq 0$.

* Titchmarsh, *The zeta-function of Riemann* (1930), 11.

For any $\delta > 0$ we choose m such that

$$\sum_{0}^{m} (n+a)^{-1-\delta} > \sum_{m+1}^{\infty} (n+a)^{-1-\delta},$$

and let $a(n) = 1$ for $n \leqslant m$ and -1 for $n > m$. Then

$$Z(s) = \sum_{0}^{\infty} a(n)(n+a)^{-s}$$

has a zero for $1 < s < 1+\delta$. Applying Kronecker's theorem as before, we find that $\zeta(s, a)$ has a zero with $1 < \sigma < 1+\delta$.

4. Let Q be a positive definite quadratic form with a fundamental discriminant d for which $h(d)$ is even, and so $d \leqslant -15$. It is well known* that the various non-equivalent quadratic forms of discriminant d correspond (1, 1) to the classes $\mathfrak{K}$ of ideals in the quadratic field $P(\sqrt{d})$ in such a way that the number of representations of a positive integer n by the quadratic form is precisely twice the number of integer ideals of norm n in the corresponding class. Hence, if $\mathfrak{K}_1$ is the class corresponding to Q,

$$\zeta(s, Q) = 2 \sum_{\mathfrak{a} \text{ in } \mathfrak{K}_1} N\mathfrak{a}^{-s}.$$

Let χ run through the characters of the group of ideal classes, and let χ_0 denote the principal character. Then

$$\zeta(s, Q) = \frac{2}{h(d)} \sum_{\chi} \bar{\chi}(\mathfrak{K}_1) L(s, \chi),$$

where

$$L(s, \chi) = \sum_{\mathfrak{a}} \chi(\mathfrak{a}) N\mathfrak{a}^{-s}$$
$$= \prod_{\mathfrak{p}} \{1-\chi(\mathfrak{p}) N\mathfrak{p}^{-s}\}^{-1},$$

for $\sigma > 1$, where $\mathfrak{a}$ runs through all integer ideals other than 0, and $\mathfrak{p}$ through all prime ideals.

It is known that, for $\chi \neq \chi_0$, $L(s, \chi)$ is regular at $s=1$, and $L(1, \chi) \neq 0$, also that $L(s, \chi_0)$ has a simple pole at $s = 1$.

Since the order of the class-group is even, there exists† a real non-principal character χ_1.

Theorem 3. *$\zeta(s, Q)$ has an infinity of zeros for $\sigma > 1$.*

The proof is identical with that of § 3, except for one point in connection with the application of Kronecker's theorem. The numbers $N\mathfrak{p}$ are not

---

* Hecke, *Theorie der algebraischen Zahlen* (1923), § 53.

† The number of characters, being the order of the class-group, is even; hence the number of real characters is even.

all different, but the same $N\mathfrak{p}$ can arise only from two different $\mathfrak{p}$'s, and the product $\mathfrak{p}_1\mathfrak{p}_2$ of two such $\mathfrak{p}$'s is a rational prime. Hence $\chi_1(\mathfrak{p}_1)\chi_1(\mathfrak{p}_2) = 1$, *i.e.* $a(\mathfrak{p}_1) = a(\mathfrak{p}_2)$. Thus the application of Kronecker's theorem to the logarithms of the set* consisting of the different values assumed by $N\mathfrak{p}$ is sufficient.

Trinity College,
Cambridge.

12.

On the Zeros of Certain Dirichlet Series II

(With H. Davenport) *

Journal of the London Mathematical Society 11 (1936), 307–312
[Commentary on p. 588]

1. In our previous paper† we proved that, if

$$Q(x, y) = ax^2 + bxy + cy^2$$

is a positive definite quadratic form, with integer coefficients and a fundamental discriminant $d = b^2 - 4ac$ such that the class-number $h(d)$ is even, then the function

$$\zeta(s, Q) = \sum_{\substack{x, y = -\infty \\ x, y \neq 0, 0}}^{\infty} Q(x, y)^{-s}$$

has an infinity of zeros for $\sigma > 1$. In this paper we shall prove that this also holds if $h(d)$ is odd and greater than 1. The proof, though based on the same idea, is rather more complicated, since the auxiliary arithmetical function $a(p)$ which we use has to be defined by induction.

2. We have, for $\sigma > 1$, the identity‡

$$\zeta(s, Q) = \frac{2}{h(d)} \sum_{\chi} \bar{\chi}(\mathfrak{K}) \, L(s, \chi),$$

where $\mathfrak{K}$ is the ideal-class in the field $\mathrm{P}(\sqrt{d})$ corresponding to the quadratic form Q, and χ runs through the characters of the class-group. Here

$$L(s, \chi) = \sum_{\mathfrak{a}} \chi(\mathfrak{a}) \, (N\mathfrak{a})^{-s}$$

$$= \prod_{\mathfrak{p}} \{1 - \chi(\mathfrak{p}) \, (N\mathfrak{p})^{-s}\}^{-1},$$

where $\mathfrak{a}$ runs through all non-zero integer ideals, and $\mathfrak{p}$ through all prime ideals of the field $\mathrm{P}(\sqrt{d})$.

---

* Received 28 May, 1936; read 18 June, 1936.
† *Journal London Math. Soc.*, 11 (1936), 181–185.
‡ *Loc. cit.*, § 4.

Since $h(d)$ is odd and different from 1, $-d$ is a prime*. The law of decomposition of the rational primes into prime ideals in the field $\mathrm{P}(\sqrt{d})$ is:

Primes ϖ with $(d/\varpi)=-1$ are themselves prime ideals in the field.

Primes π with $(d/\pi)=1$ are products of two different conjugate prime ideals in the field.

The prime $|d|$ is the square of the principal prime ideal $(\sqrt{d})$.

We divide the primes π into two classes, denoting by q those whose prime ideal factors are principal and by p those for which they are non-principal. Then we have

$$L(s,\chi)=(1-|d|^{-s})^{-1}\prod_{\varpi}(1-\varpi^{-2s})^{-1}\prod_{q}(1-q^{-s})^{-2}$$

$$\times\prod_{\substack{p\\ p=\mathfrak{p}_1\mathfrak{p}_2}}\{1-\chi(\mathfrak{p}_1)p^{-s}\}^{-1}\{1-\chi(\mathfrak{p}_2)p^{-s}\}^{-1}.$$

Hence

$$\zeta(s,Q)=\frac{2}{h(d)}(1-|d|^{-s})^{-1}\prod_{\varpi}(1-\varpi^{-2s})^{-1}\prod_{q}(1-q^{-s})^{-2}F(s),$$

where
$$F(s)=\sum_{\chi}\bar{\chi}(\mathfrak{K})\prod_{p}\{1-2\mathfrak{R}\chi(p)p^{-s}+p^{-2s}\}^{-1},$$

in which $\chi(p)$ denotes either of the two conjugate complex numbers $\chi(\mathfrak{p}_1)$, $\chi(\mathfrak{p}_2)$.

We write $h(d)=2N+1$, and denote by $\chi_1,\dots,\chi_N$ a set of N non-principal characters no two of which are reciprocal†. Then

$$F(s)=\prod_{p}(1-p^{-s})^{-2}+2\sum_{\nu=1}^{N}\mathfrak{R}\chi_\nu(\mathfrak{K})\prod_{p}\{1-2\mathfrak{R}\chi_\nu(p)p^{-s}+p^{-2s}\}^{-1}. \tag{1}$$

3. Lemma 1. *Let $\mathfrak{G}$ be an Abelian group of $h=2N+1$ elements; let $\gamma_1,\dots,\gamma_N$ be a set of N elements of $\mathfrak{G}$ none of which is the unit element and no two of which are reciprocal, and let $\chi_1,\dots,\chi_N$ be a set of N characters of $\mathfrak{G}$ none of which is the principal character and no two of which are reciprocal. Then*

* Hecke, *Theorie der algebraischen Zahlen* (1923), Satz 132.

† Since the group of characters is of odd order, there is no character of order 2, *i.e.* no real non-principal character.

the determinant

$$|1-\mathfrak{B}\chi_k(\gamma_j)|$$

is different from zero.

Proof. We have to prove that, if

$$\sum_{j=1}^{N} z_j\{1-\mathfrak{B}\chi_k(\gamma_j)\}=0 \tag{2}$$

for $k=1, 2, \ldots, N$, then $z_1=z_2=\ldots=z_N=0$. It follows from (2) that

$$\sum_{j=1}^{N} z_j\{2-\psi(\gamma_j)-\psi(\gamma_j^{-1})\}=0 \tag{3}$$

for every character ψ of $\mathfrak{G}$. Multiplying (3) by $\psi(\gamma_l^{-1})$ and summing over all characters ψ, we obtain

$$z_l=0.$$

4. We denote by $\mathfrak{K}_1, \mathfrak{K}_1^{-1}, \ldots, \mathfrak{K}_N, \mathfrak{K}_N^{-1}$ the ideal classes, other than the principal class, of the field $P(\sqrt{d})$. The two prime ideal factors of one of our primes p of § 2 lie in one pair $\mathfrak{K}_j, \mathfrak{K}_j^{-1}$ of these classes, and we divide the primes p into sets $\mathfrak{L}_1, \ldots, \mathfrak{L}_N$, placing in $\mathfrak{L}_j$ those whose prime ideal factors lie in $\mathfrak{K}_j, \mathfrak{K}_j^{-1}$. It was proved by de la Vallée-Poussin* that the number of primes p in $\mathfrak{L}_j$ not exceeding x is asymptotically

$$\frac{1}{h(d)}\frac{x}{\log x}$$

as $x\to\infty$.

It will be convenient to order the primes p by denoting the primes in the set $\mathfrak{L}_j$, taken in increasing order, by

$$p_j,\ p_{j+N},\ p_{j+2N},\ \ldots.$$

It follows from the result just stated that

$$p_n\sim\frac{h(d)}{N}\,n\log n \tag{4}$$

as $n\to\infty$.

* "Recherches analytiques sur la théorie des nombres premiers (troisième partie)", *Annales de la Société scientif. de Bruxelles*, 21A (1897), 1–13. The theorem is proved also by Landau, "Über Ideale und Primideale in Idealklassen", *Math. Zeitschrift*, 2 (1918), 52–154, Satz 85. We require only the case $\mathfrak{f}=\mathfrak{o}$.

5. For any set of numbers $a(p)$ with $|a(p)| = 1$ we define a function $F_a(s)$ in analogy with (1) by

$$F_a(s) = \prod_p \{1 - a(p)p^{-s}\}^{-2} \tag{5}$$

$$+2 \sum_{\nu=1}^{N} \mathfrak{K}\chi_\nu(\mathfrak{K}) \prod_p \{1 - 2\mathfrak{K}\chi_\nu(p)a(p)p^{-s} + a^2(p)p^{-2s}\}^{-1}.$$

LEMMA 2. *For any given $\delta > 0$ there is a set of numbers $a(p)$ with $|a(p)| = 1$ such that $F_a(s)$ has a zero in $1 < s < 1+\delta$.*

Proof. We shall show that we can find a set of numbers $a(p)$ and a value of s such that

$$\prod_p \{1 - a(p)p^{-s}\}^{-2} = -2N\mathfrak{K}\chi_\nu(\mathfrak{K}) \prod_p \{1 - 2\mathfrak{K}\chi_\nu(p)a(p)p^{-s} + a^2(p)p^{-2s}\}^{-1}$$

for $\nu = 1, 2, \ldots, N$. If we take logarithms, it is sufficient to have

$$\sum_p \sum_{m=1}^{\infty} \frac{1}{m} \{a(p)p^{-s}\}^m = c_1(\nu) + \sum_p \sum_{m=1}^{\infty} \frac{1}{m} \{a(p)p^{-s}\}^m \mathfrak{K}\{\chi_\nu^m(p)\},$$

where

$$c_1(\nu) = \tfrac{1}{2}\pi i + \tfrac{1}{2} \log \{2N\mathfrak{K}\chi_\nu(\mathfrak{K})\}.$$

This may be written in the form

$$\sum_p \{1 - \mathfrak{K}\chi_\nu(p)\} a(p)p^{-s} = c_1(\nu) + \sum_p c_2\{s, \nu, p, a(p)\} p^{-2s}, \tag{6}$$

$c_2\{s, \nu, p, a(p)\}$ being a given function which satisfies

$$|c_2\{s, \nu, p, a(p)\}| < c_3,$$

where c_3 is an absolute constant.

By Lemma 1 we can transform the N equations (6) into

$$\sum_{p \text{ in } \mathfrak{L}_\nu} a(p)p^{-s} = c_4(\nu) + \sum_p c_5\{s, \nu, p, a(p)\} p^{-2s}$$

or

$$\sum_{\substack{n=1 \\ n \equiv \nu(N)}}^{\infty} a(p_n)p_n^{-s} = c_4(\nu) + \sum_{n=1}^{\infty} c_5\{s, \nu, p_n, a(p_n)\} p_n^{-2s} \tag{7}$$

for $\nu = 1, 2, \ldots, N$ with

$$|c_4(\nu)| < c_6, \quad |c_5\{s, \nu, p_n, a(p_n)\}| < c_6,$$

c_6 depending only on d.

Without loss of generality we can suppose that $\delta \leqslant \frac{1}{2}$. Then it can be deduced from (4) that there exists an $n_0 > N$ such that, for $n > n_0$,

$$\sum_{m=n-N}^{n-1} p_m^{-2} \leqslant \frac{1}{2c_6} p_n^{-1-\delta} \tag{8}$$

and

$$\sum_{\substack{m=n+N \\ m\equiv n \pmod N}}^{\infty} p_m^{-1-\delta} \geqslant 3p_n^{-1}. \tag{9}$$

It also follows from (4) that we can now choose s so near to 1 that, for $1 \leqslant \nu \leqslant N$,

$$\tfrac{1}{2} \sum_{\substack{m>n_0 \\ m\equiv \nu \pmod N}} p_m^{-s} > \sum_{\substack{m\leqslant n_0 \\ m\equiv \nu \pmod N}} p_m^{-1} + c_6\left(1+\frac{\pi^2}{6}\right). \tag{10}$$

Write*

$$S_n = \sum_{\substack{m\leqslant n \\ m\equiv n(N)}} a(p_m)\, p_m^{-s} - c_4(n) - \sum_{m<n} c_5\{s, n, p_m, a(p_m)\} p_m^{-2s}. \tag{11}$$

For $m \leqslant n_0$ we define $a(p_m) = 1$. For $m > n_0$ we define $a(p_m)$ by induction as

$$a(p_m) = -\frac{S_{m-N}}{|S_{m-N}|} \tag{12}$$

unless $S_{m-N} = 0$, in which case we take $a(p_m) = 1$. We shall prove that with this definition

$$|S_n| \leqslant \tfrac{1}{2} \sum_{\substack{m>n \\ m\equiv n \pmod N}} p_m^{-s}. \tag{13}$$

Then (7) will be satisfied, and the lemma will be established.

Equations (10), (11) show that (13) is true for $n \leqslant n_0$. Hence we may suppose that $n > n_0$ and that

$$|S_{n-N}| \leqslant \tfrac{1}{2} \sum_{\substack{m>n-N \\ m\equiv n \pmod N}} p_m^{-s}. \tag{14}$$

We have

$$S_n = S_{n-N} + a(p_n)\, p_n^{-s} - \sum_{m=n-N}^{n-1} c_5\{s, n, p_m, a(p_m)\} p_m^{-2s}.$$

---

* We define $c_4(n)$ for $n > N$ by $c_4(n) = c_4(n-N)$, and similarly for $c_5\{s, n, p, a(p)\}$.

By (12),

$$|S_{n-N}+a(p_n)p_n^{-s}| = \big||S_{n-N}|-p_n^{-s}\big|$$

$$\leqslant \max (p_n^{-s}, |S_{n-N}|-p_n^{-s}).$$

Hence
$$|S_n| \leqslant \max (p_n^{-s}, |S_{n-N}|-p_n^{-s})+c_6 \sum_{m=n-N}^{n-1} p_m^{-2}$$

$$\leqslant \max (p_n^{-s}, |S_{n-N}|-p_n^{-s})+\tfrac{1}{2}p_n^{-s} \quad \text{[by (8)]}$$

$$\leqslant \max (\tfrac{3}{2}p_n^{-s}, \tfrac{1}{2} \sum_{\substack{m>n \\ m\equiv n \pmod N}} p_m^{-s}) \quad \text{[by (14)]};$$

and this, together with (9), proves (13).

6. Theorem. *If d is a negative fundamental discriminant and $h(d)$ is odd and different from 1, then $\zeta(s, Q)$ has an infinity of zeros for $\sigma > 1$.*

Proof. The theorem follows from Lemma 2 by the same application of Kronecker's theorem as was made in our previous paper. Given $\epsilon > 0$ and $x > 0$ there exists a real number τ such that

$$|p^{-i\tau}-a(p)| < \epsilon$$

for all $p < x$. Since all the products on the right of (5) are uniformly convergent for $\sigma \geqslant 1+\rho > 1$, it follows that, for any $\epsilon > 0$ and $\rho > 0$, we can find a real number τ such that

$$|F(s+\tau i)-F_a(s)| < \epsilon$$

for $\sigma \geqslant 1+\rho$.

Let $s_1 > 1$ be a zero of $F_a(s)$. For any $\eta > 0$ there exists an η_1 with $0 < \eta_1 < \eta$ and $\eta_1 < s_1-1$ such that $F_a(s) \neq 0$ for $|s-s_1| = \eta_1$. Take

$$\epsilon = \min_{|s-s_1|=\eta_1} |F_a(s)|,$$

and $\rho = s_1-\eta_1-1$, then the conditions of Rouché's theorem are satisfied by $F_a(s)$ and $F(s+\tau i)-F_a(s)$ for the circle $|s-s_1| = \eta_1$. Hence $F(s)$ has a zero in the circle $|s-s_1-\tau i| < \eta$. This proves the theorem.

Trinity College,
Cambridge.

13.

On Waring's Problem for Fourth Powers

(With H. Davenport)

Proceedings of the London Mathematical Society 41 (1936), 143–150
[Commentary on p. 577]

[Received 25 January, 1936.—Read 20 February, 1936.]

Introduction.

The object of this paper is to give a proof that every large positive integer is representable as the sum of seventeen fourth powers. This result has also been proved independently by Dr. Estermann†.

The fact that every large number is representable as the sum of nineteen fourth powers was established by Hardy and Littlewood‡. We make in this paper a straightforward application of the Hardy-Littlewood method as modified recently by Vinogradov§.

It is well known that there are an infinity of numbers not representable as the sum of less than sixteen fourth powers.

Notation.

Let N be any sufficiently large integer, and let $P = \frac{1}{2}N^{\frac{1}{4}}$. Let

$$T(a, P) = \sum_{P<n<2P} e(n^4 a) \quad [e(x) = e^{2\pi i x}],$$

$$T_i(a) = T(a, 2^{-i} P^{(\frac{3}{4})^i}) \quad (i = 0, 1, 2, 3),$$

$$Q(a) = T_1(a)\, T_2(a)\, T_3^{\,2}(a) = \sum_n r_4'(n)\, e(na),$$

$$R(a) = T_0(a)\, Q(a) = \sum_n r_5'(n)\, e(na),$$

$$T_0^{\,9}(a)\, Q^2(a) = \sum_n r_{17}'(n)\, e(na).$$

---

† See *Proc. London Math. Soc.* (2), 41 (1936), 126–142. The two proofs were obtained at about the same time.

‡ "Further researches in Waring's problem", *Math. Zeitschrift*, 23 (1925), 1–37.

§ See, for example, *Annals of Math.*, 36 (1935), 395–405. For a simplified account see Heilbronn, *Acta Arithmetica*, 1 (1935), Part 2.

We divide the interval $0 \leqslant \alpha \leqslant 1$ in the usual way into Farey arcs belonging to all rational points a/q with $1 \leqslant q \leqslant P^3$, $0 \leqslant a < q$, $(a, q) = 1$. We divide these arcs into major arcs $\mathfrak{M}$ $(1 \leqslant q \leqslant P)$ and minor arcs $\mathfrak{m}$ $(P < q \leqslant P^3)$. In either case the arc has the form

$$\alpha = \frac{a}{q} + \beta, \quad -\vartheta_1 q^{-1} P^{-3} \leqslant \beta \leqslant \vartheta_2 q^{-1} P^{-3},$$

where
$$\tfrac{1}{2} \leqslant \vartheta_i \leqslant 1.$$

It is convenient to divide each major arc into three parts, as follows:

(I) $$|\beta| \leqslant \tfrac{1}{2} P^{-4},$$

(II) $$\tfrac{1}{2} P^{-4} \leqslant |\beta| \leqslant \tfrac{1}{2} q^{-\frac{4}{5}} P^{-\frac{16}{5}},$$

(III) $$|\beta| \geqslant \tfrac{1}{2} q^{-\frac{4}{5}} P^{-\frac{16}{5}}.$$

This is possible, since

$$0 < \tfrac{1}{2} P^{-4} \leqslant \tfrac{1}{2} q^{-\frac{4}{5}} P^{-\frac{16}{5}} \leqslant \vartheta_i q^{-1} P^{-3}.$$

We use the classical notation:

$$S_{a,q} = \sum_{v=1}^{q} e\left(v^4 \frac{a}{q}\right),$$

$$\mathfrak{S}_9(n) = \sum_{q=1}^{\infty} q^{-9} \sum_{\substack{a=1 \\ (a,q)=1}}^{q} e\left(-n \frac{a}{q}\right) S_{a,q}^9.$$

Let
$$\psi_{a,q}(\alpha) = \tfrac{1}{4} \frac{S_{a,q}}{q} \sum_{P^4 < j < (2P)^4} \frac{\Gamma(j+\frac{1}{4})}{j!} e(j\beta).$$

We use ϵ to denote an arbitrarily small positive number, and $c_1, c_2, \ldots$ to denote positive absolute constants.

In order to avoid the repetition of many well-known arguments, we make frequent references to Gelbcke† and Landau‡.

Minor arcs.

LEMMA 1. $$r_5'(n) = O(P^{\epsilon}).$$

Proof. If $n = x_0^4 + x_1^4 + x_2^4 + x_3^4 + x_3'^4$, where

$$2^{-v} P^{(\frac{3}{4})^v} < x_v < 2^{1-v} P^{(\frac{3}{4})^v},$$

---

† *Math. Annalen*, 105 (1931), 637–652. Our definitions are not identical with those of Gelbcke, but results for which a reference to Gelbcke is given follow by the same arguments as those used there.

‡ *Vorlesungen über Zahlentheorie* (1927), 1.

the values of x_0, x_1, x_2 are unique. For, if v is the first of the suffixes 0, 1, 2 for which $x_v \neq x_v{}^*$, then

$$|x_v{}^4 - x_v^{*4}| > 4\,.\,16^{-v} P^{3(\frac{3}{4})^v} = 4(2^{1-(v+1)} P^{(\frac{3}{4})^{v+1}})^4 \geqslant x_{v+1}^4 + \ldots + x_3'^4.$$

Also, the number of representations of $n - x_0{}^4 - x_1{}^4 - x_2{}^4$ as the sum of two fourth powers is $O(P^\epsilon)$.

LEMMA 2. $\sum_{\mathfrak{m}} \int_{\mathfrak{m}} |T_0{}^7(\alpha)\, R^2(\alpha)|\, d\alpha = O\{P^{5-\frac{1}{32}+\epsilon} Q^2(0)\}$.

Proof. By Gelbcke, Satz 10,

$$T_0(\alpha) = O(P^{\frac{7}{8}+\epsilon}).$$

Hence
$$\sum_{\mathfrak{m}} \int_{\mathfrak{m}} |T_0{}^7(\alpha)\, R^2(\alpha)|\, d\alpha = O\left(P^{7\cdot\frac{7}{8}+7\epsilon} \int_0^1 |R(\alpha)|^2 d\alpha\right)$$

$$= O\left(P^{6+\frac{1}{8}+7\epsilon} \sum_n [r_5'(n)]^2\right)$$

$$= O\left(P^{6+\frac{1}{8}+8\epsilon} \sum_n r_5'(n)\right)$$

$$= O\{P^{7+\frac{1}{8}+8\epsilon} Q(0)\}.$$

Also
$$Q(0) > c_1 P^{\frac{3}{4}+(\frac{3}{4})^2+2(\frac{3}{4})^3} = c_1 P^{2+\frac{1}{8}+\frac{1}{32}};$$

hence the result.

Major arcs.

LEMMA 3. *For all a, q, and β, provided that $|\beta| \leqslant \frac{1}{2}$, we have*

$$\psi_{a,q}(\alpha) = O\{q^{-\frac{1}{4}} \min(P, |\beta|^{-\frac{1}{4}})\}.$$

Proof. Gelbcke, Satz 8, with the classical result

$$S_{a\,q} = O(q^{\frac{3}{4}})$$

(Landau, Satz 315), in place of Gelbcke's Satz 5.

LEMMA 4. *For all a, q, and β, provided that $|\beta| \leqslant \frac{1}{2}$, we have*

$$T(\alpha, P) - \psi_{a,q}(\alpha) = O\{q^{1-\frac{1}{4}+\epsilon}(1 + |\beta|\, P^4)\}.$$

Proof. Gelbcke, Satz 12, with the inequality

$$\sum_{h=0}^{m} e\left(h^4 \frac{a}{q}\right) = O(q^{\frac{3}{4}+\epsilon})$$

for $0 \leqslant m \leqslant q$, in place of Gelbcke's Satz 5. A proof of this inequality, based on well-known arguments, will be published shortly†.

Lemma 5. *If $q \leqslant P$ and $|\beta| \leqslant P^{-4}$, then*

$$T(a, P) = O\{q^{-\frac{1}{4}+\epsilon}\, T(0, P)\}.$$

Proof. By Lemmas 3, 4,

$$T(a, P) = O(q^{-\frac{1}{4}} P) + O(q^{1-\frac{1}{4}+\epsilon}) = O\{q^{-\frac{1}{4}+\epsilon}\, T(0, P)\}.$$

Lemma 6. *On $\mathfrak{M}$, provided that $q \leqslant P^{\frac{1}{4}}$, we have*

$$Q^2(\alpha) = O\{q^{-1+\epsilon}\, Q^2(0)\}.$$

Proof. Since

$$q^{-1} P^{-3} < (2^{-1} P^{\frac{3}{4}})^{-4} < (2^{-2} P^{(\frac{3}{4})^2})^{-4},$$

and

$$q \leqslant P^{\frac{1}{4}} < P^{(\frac{3}{4})^2} < P^{\frac{3}{4}},$$

we have, by Lemma 5,

$$T_i(\alpha) = O\{q^{-\frac{1}{4}+\epsilon}\, T_i(0)\}$$

for $i = 1, 2$, which establishes Lemma 6.

Lemma 7. *On* (I),

$$T_0^{\,9}(\alpha) - \psi_{a,q}^9(\alpha) = O(q^{-1-\frac{1}{4}+\epsilon} P^8).$$

Proof. Lemmas 3 and 4, since $q^{1-\frac{1}{4}} \leqslant q^{-\frac{1}{4}} P$.

Lemma 8. *On* (II),

$$T_0^{\,9}(\alpha) - \psi_{a,q}^9(\alpha) = O(q^{-1-\frac{1}{4}+\epsilon} P^4 |\beta|^{-1}).$$

Proof. Lemmas 3 and 4, since on (II)

$$q^{1-\frac{1}{4}} |\beta| P^4 \leqslant q^{-\frac{1}{4}} |\beta|^{-\frac{1}{4}}.$$

Lemma 9. *On* (III),

$$T_0^{\,9}(\alpha) - \psi_{a,q}^9(\alpha) = O(q^{-\frac{9}{8}} P^{4+\frac{1}{2}+\epsilon} |\beta|^{-\frac{9}{8}}).$$

Proof. By Gelbcke, Satz 14,

$$T_0(\alpha) = O(q^{-\frac{1}{8}} P^{\frac{1}{2}+\epsilon} |\beta|^{-\frac{1}{8}}),$$

and, by Lemma 3,

$$\psi_{a,q}(\alpha) = O(q^{-\frac{1}{4}} |\beta|^{-\frac{1}{4}}) = O(q^{-\frac{1}{8}} P^{\frac{1}{2}+\epsilon} |\beta|^{-\frac{1}{8}}).$$

† For primes q, an equivalent inequality was given by Hardy and Littlewood, *Quart. J. of Math.*, 48 (1920), 277, foot-note.

LEMMA 10.

$$\sum_{\mathfrak{M}}' \int_{\mathfrak{M}} |T_0^9(\alpha)-\psi_{a,q}^9(\alpha)|\,|Q(\alpha)|^2\,d\alpha = O\{P^{5-\frac{1}{10}+\epsilon}\,Q^2(0)\},$$

where the summation is over the major arcs with $q \leqslant P^{\frac{1}{4}}$.

Proof. It is sufficient to prove that

$$\int_{\mathfrak{M}} |T_0^9(\alpha)-\psi_{a,q}^9(\alpha)|\,d\alpha = O(q^{-1}\,P^{5-\frac{1}{10}+\epsilon}),$$

since then, by Lemma 6, the sum in question is

$$O\Big(\sum_{\mathfrak{M}}' q^{-1}\,P^{5-\frac{1}{10}+\epsilon}\,q^{-1+\epsilon}\,Q(0)^2\Big) = O\Big(\sum_{q\leqslant P^{\frac{1}{4}}} q.q^{-2+\epsilon}\,P^{5-\frac{1}{10}+\epsilon}\,Q(0)^2\Big).$$

By Lemma 7,

$$\int_I |T_0^9-\psi^9|\,d\alpha = O(q^{-1-\frac{1}{4}+\epsilon}\,P^8\,P^{-4}) = O(q^{-1}\,P^{5-\frac{1}{10}}).$$

By Lemma 8,

$$\int_{II} |T_0^9-\psi^9|\,d\alpha = O\Big(q^{-1-\frac{1}{4}+\epsilon}\,P^4 \int_{\frac{1}{2}P^{-4}}^{\frac{1}{2}} \beta^{-1}\,d\beta\Big)$$
$$= O(q^{-1}\,P^{5-\frac{1}{10}}).$$

By Lemma 9,

$$\int_{III} |T_0^9-\psi^9|\,d\alpha = O\Big(q^{-\frac{9}{8}}\,P^{4+\frac{1}{2}+\epsilon} \int_{\frac{1}{2}q^{-\frac{1}{2}}P^{-\frac{16}{5}}}^{\infty} \beta^{-\frac{9}{8}}\,d\beta\Big)$$
$$= O(q^{-\frac{9}{8}}\,P^{4+\frac{1}{2}+\epsilon}\,q^{\frac{1}{16}}\,P^{\frac{2}{5}})$$
$$= O(q^{-1}\,P^{5-\frac{1}{10}+\epsilon}).$$

LEMMA 11. *If* $\bar{\mathfrak{M}}_{a,q}$ *denotes the part of the interval* (0, 1) *not belonging to* $\mathfrak{M}_{a,q}$, *then*

$$\sum_{q\leqslant P^{\frac{1}{4}}} \sum_{\substack{a=1\\(a,q)=1}}^{q} \int_{\bar{\mathfrak{M}}_{a,q}} |\psi_{a,q}(\alpha)|^9\,|Q(\alpha)|^2\,d\alpha = O\{P^{5-\frac{1}{4}}\,Q^2(0)\}.$$

Proof. By Lemma 3,

$$\int_{\bar{\mathfrak{M}}_{a,q}} |\psi_{a,q}(\alpha)|^9\,d\alpha = O\Big(\int_{\frac{1}{4}q^{-1}P^{-3}}^{\infty} q^{-\frac{9}{4}}\,\beta^{-\frac{9}{4}}\,d\beta\Big)$$
$$= O(q^{-\frac{9}{4}+\frac{5}{4}}\,P^{\frac{15}{4}})$$
$$= O(q^{-1}\,P^{4-\frac{1}{4}}).$$

Hence

$$\sum_{q \leqslant P^{\frac12}} \sum_{\substack{a=1 \\ (a,q)=1}}^{q} \int_{\mathfrak{M}_{a,q}} |\psi_{a,q}(a)|^9 |Q(a)|^2 \, da = O\Big(\sum_{q \leqslant P^{\frac12}} q \cdot q^{-1} P^{4-\frac14} Q^2(0)\Big),$$

giving the result.

LEMMA 12.

$$\sum_{\mathfrak{M}}{}'' \int_{\mathfrak{M}} |T_0{}^7(a) R^2(a)| \, da = O\{P^{5-\frac{1}{32}+\epsilon} Q^2(0)\},$$

where the summation is over the major arcs with $q > P^{\frac12}$.

Proof. By Lemmas 3 and 4 we have, for $P^{\frac12} < q \leqslant P$,

$$T_0(a) = O(q^{-\frac14} P) + O\{q^{\frac34+\epsilon}(1+q^{-1}P)\}$$
$$= O(q^{-\frac14+\epsilon} P) = O(P^{\frac78+\epsilon}).$$

The result now follows exactly as in the proof of Lemma 2.

The singular series.

LEMMA 13.

$$\int_0^1 e(-na)\,\psi_{a,q}^9(a)\, da = \left(\frac{S_{a,q}}{q}\right)^9 e\left(-\frac{na}{q}\right) F(n),$$

where, for $10P^4 < n < 100P^4$,

$$c_2 P^5 < F(n) < c_3 P^5.$$

Proof. The integral in question is

$$(\tfrac14)^9 \left(\frac{S_{a,q}}{q}\right)^9 e\left(-\frac{na}{q}\right) \sum_{\substack{j_1+\ldots+j_9=n \\ P^4<j_i<(2P)^4}} \frac{\Gamma(j_1+\frac14)\ldots\Gamma(j_9+\frac14)}{j_1!\ldots j_9!}.$$

Since
$$c_4 j^{-\frac34} j! < \Gamma(j+\tfrac14) < c_5 j^{-\frac34} j!,$$

we have
$$c_6 (P^{-3})^9 \sum_{\substack{j_1+\ldots+j_9=n \\ P^4<j_i<(2P)^4}} 1 < F(n) < c_7 (P^{-3})^9 \sum_{\substack{j_1+\ldots+j_9=n \\ P^4<j_i<(2P)^4}} 1.$$

Hence the result, since $10P^4 < n < 100P^4$, and so

$$c_8 (P^4)^8 < \sum_{\substack{j_1+\ldots+j_9=n \\ P^4<j_i<(2P)^4}} 1 < c_9 (P^4)^8.$$

LEMMA 14. $\mathfrak{S}_9(n) \geqslant 0$ *for all* n, *and* $\mathfrak{S}_9(n) > c_{10}$ *provided that* n *is congruent to one of* 1, 2, ..., 9 (mod 16).

Proof. In Landau's notation,

$$\mathfrak{S}_9(n) = \prod_p \chi_p,$$

where
$$\chi_p = \sum_{\nu=0}^{\infty} p^{-9\nu} \sum_{\substack{a=1 \\ p \nmid a}}^{p^\nu - 1} e\left(-\frac{na}{p^\nu}\right) (S_{a,p^\nu})^9,$$

the product being absolutely convergent. For $p > c_{11}$,

$$\chi_p > 1 - p^{-\frac{5}{4}}$$

(Landau, Satz 324).

For $p > 2$, we have, by (349) of Landau†,

$$\chi_p \geqslant \begin{cases} p^{-8} N(p, 0) & \text{if } p^4 | n, \\ p^{-8} N(p, n) & \text{if } p^4 \nmid n, \end{cases}$$

where, by Satz 301 (2), $N(p, n) \geqslant 1$ and $N(p, 0) \geqslant 1$ since

$$9 \geqslant r + 1 = (4, p-1) + 1.$$

For $p = 2$, provided that $16 \nmid n$, we have, by the same equation,

$$\chi_2 = 16^{-8} N(16, n).$$

Now, if n is congruent to one of 1, 2, ..., 9 (mod 16), $N(16, n) \geqslant 1$, and so $\chi_2 \geqslant 16^{-8}$.

By equation (349) of Landau, $\chi_p \geqslant 0$ for all p and n, and if n is congruent to one of 1, 2, ..., 9 (mod 16),

$$\mathfrak{S}_9(n) > 16^{-8} \prod_{p \leqslant c_{11}} p^{-8} \prod_{p > c_{11}} (1 - p^{-\frac{5}{4}}) = c_{10}.$$

Proof of the theorem.

THEOREM. $r'_{17}(N) > 0$ *for* $N > c_{12}$.

Proof. By Lemmas 2, 10, 11, 12

$$r'_{17}(N) = \int_0^1 e(-N\alpha)\, T_0^{\,9}(\alpha)\, Q^2(\alpha)\, d\alpha$$

$$= \sum_{\mathfrak{M}}{}' \int_0^1 e(-N\alpha)\, \psi_{a,q}^9(\alpha)\, Q^2(\alpha)\, d\alpha + O\{P^{5-\frac{1}{32}+\epsilon} Q^2(0)\}.$$

---

† $N(p^l, m)$ is the number of solutions of the congruence $x_1^4 + \ldots + x_9^4 \equiv m \pmod{p^l}$ with not all of $x_1, \ldots, x_9$ divisible by p.

The sum on the right is

$$\sum_{n_1}\sum_{n_2} r_4'(n_1)\, r_4'(n_2) \sum_{\mathfrak{M}}' \int_0^1 e\{-(N-n_1-n_2)\,\alpha\}\,\psi_{a,q}^9(\alpha)\,d\alpha$$

$$=\sum_{n_1}\sum_{n_2} r_4'(n_1)\, r_4'(n_2)\, F(N-n_1-n_2) \sum_{q\leqslant P^{\frac12}} \sum_{\substack{a=1\\(a,q)=1}}^{q} (q^{-1}S_{a,q})^9\, e\left(-(N-n_1-n_2)\frac{a}{q}\right)$$

by Lemma 13. Now

$$\sum_{q\leqslant P^{\frac12}} \sum_{\substack{a=1\\(a,q)=1}}^{q} (q^{-1}S_{a,q})^9\, e\left(-\frac{na}{q}\right) = \mathfrak{S}_9(n)+O\left(\sum_{q>P^{\frac12}} q(q^{-\frac14})^9\right)$$

$$=\mathfrak{S}_9(n)+O(P^{-\frac18}).$$

The condition $10P^4 < N-n_1-n_2 < 100P^4$ of Lemma 13 is satisfied, if $r_4'(n_1)>0$, $r_4'(n_2)>0$. Hence we have

$$r_{17}'(N)=\sum_{n_1}\sum_{n_2} r_4'(n_1)\, r_4'(n_2)\, F(N-n_1-n_2)\,\mathfrak{S}_9(N-n_1-n_2)+O\{P^{5-\frac{1}{32}+\epsilon}\, Q^2(0)\}.$$

We define t to be 0 if N is congruent to one of 1, 2, ..., 8 (mod 16), and to be 1 if N is congruent to one of 9, 10, ..., 16 (mod 16). If $n_1\equiv n_2\equiv 4t$ (mod 16), $N-n_1-n_2$ is congruent to one of 1, 2, ..., 8 (mod 16). Also, clearly,

$$\sum_{n_1\equiv 4t\,(16)} r_4'(n_1) > c_{13}\sum_{n_1} r_4'(n_1) = c_{13}\, Q(0).$$

Hence

$$r_{17}'(N) > c_2\, P^5\, c_{10}\left(\sum_{n_1\equiv 4t} r_4'(n_1)\right)^2 + O\{P^{5-\frac{1}{32}+\epsilon}\, Q^2(0)\}$$

$$> c_2\, P^5\, c_{10}\, c_{13}^2\, Q^2(0) + O\{P^{5-\frac{1}{32}+\epsilon}\, Q^2(0)\}$$

$$> 0.$$

Trinity College,
Cambridge.

14.

On an Exponential Sum

(With H. Davenport)

Proceedings of the London Mathematical Society 41 (1936), 449–453
[Commentary on p. 578]

[Received 3 April, 1936.—Read 23 April, 1936.]

Introduction.

Let q and k be positive integers with $q > 1$, $k > 2$. For $(a, q) = 1$, we write

$$S_{a,q} = \sum_{x=1}^{q} e_q(ax^k) \quad [e_q(t) = e^{2\pi i t/q}].$$

Our object in this note is to prove the inequality

$$\sum_{x=1}^{m} e_q(ax^k) = \frac{m}{q} S_{a,q} + O(q^{3/4+\epsilon}),$$

for every $\epsilon > 0$*. This inequality has a certain interest in connection with the "major arcs" in Waring's problem, and in a recent paper† we have made use of it in the case $k = 4$.

Lemmas.

For any integer b, we write

$$S(a, b, q) = \sum_{x=1}^{q} e_q(ax^k + bx).$$

LEMMA 1. *If* $p \nmid a$, *then*

$$S(a, b, p) = O(p^{3/4}).$$

* The constant implied in the symbol O depends only on k and ϵ.

† "On Waring's problem for fourth powers", *Proc. London Math. Soc.* (2), 41 (1936), 143–150.

Proof (Mordell*). We have the identity:

$$(1)\quad \sum_{u=0}^{p-1}\sum_{v=0}^{p-1}|S(u,v,p)|^4$$

$$=\sum_{u=0}^{p-1}\sum_{v=0}^{p-1}\sum_{x_1=0}^{p-1}\cdots\sum_{x_4=0}^{p-1} e_p\Big(u(x_1^k+x_2^k-x_3^k-x_4^k)+v(x_1+x_2-x_3-x_4)\Big)$$

$$=\sum_{\substack{x_1=0\\ x_1+x_2\equiv x_3+x_4(p)\\ x_1^k+x_2^k\equiv x_3^k+x_4^k(p)}}^{p-1}\cdots\sum_{x_4=0}^{p-1} p^2=p^2\sum_{w=0}^{p-1}\sum_{t=0}^{p-1}R^2(w,t),$$

where $R(w,t)$ denotes the number of solutions of the congruences

$$x+y\equiv w \pmod p,$$

$$x^k+y^k\equiv t \pmod p.$$

This pair of congruences has at most k solutions, unless $w\equiv t\equiv 0$, in which case it has at most p solutions†. Hence the right-hand side of (1) is $O(p^4)$. Now for any $\lambda\not\equiv 0 \pmod p$ we have $S(a,b,p)=S(a\lambda^k,b\lambda,p)$. Hence

$$|S(a,b,p)|^4=\frac{1}{p-1}\sum_{\lambda=1}^{p-1}|S(a\lambda^k,b\lambda,p)|^4\leqslant\frac{k}{p-1}\sum_{u=0}^{p-1}\sum_{v=0}^{p-1}|S(u,v,p)|^4=O(p^3),$$

which establishes the lemma.

Lemma 2. *If* $\nu>1$, $p\nmid a$, $p^\nu\nmid b^4$, *then*

$$S(a,b,p^\nu)=O(p^{\frac{3}{4}\nu}).$$

Proof. Let $b=p^\beta b'$, where $p\nmid b'$, and write $\tau=[\frac{1}{2}(\nu+1)]$; then we have $\beta<\frac{1}{4}\nu$ and

$$\beta\leqslant\tfrac{1}{4}(\nu-1)<\tfrac{1}{2}(\nu-1)\leqslant\nu-\tau.$$

In the definition of $S(a,b,p^\nu)$ replace x by $p^\tau y+z$. Since $2\tau\geqslant\nu$, we obtain

$$S(a,b,p^\nu)=\sum_{z=1}^{p^\tau} e_{p^\nu}(az^k+bz)\sum_{y=0}^{p^{\nu-\tau}-1} e_{p^{\nu-\tau}}\Big((kaz^{k-1}+b)\,y\Big)$$

$$=p^{\nu-\tau}\sum_{\substack{z=1\\ kaz^{k-1}+b\equiv 0(p^{\nu-\tau})}}^{p^\tau} e_{p^\nu}(az^k+bz).$$

---

* *Quart. J. of Math.*, 3 (1932), 161–167.

† The congruence $x^k+(w-x)^k\equiv t \pmod p$ can only be an identity in x if $w\equiv t\equiv 0 \pmod p$, or if $k\equiv 0 \pmod p$, and in the latter case the number of solutions does not exceed $p<k$.

Hence $$|S(a, b, p^\nu)| \leqslant p^{\nu-\tau} \frac{p^\tau}{p^{\nu-\tau}} N,$$

where N is the number of solutions of the congruence

$$kaz^{k-1}+b \equiv 0 \pmod{p^{\nu-\tau}}. \tag{2}$$

Now let $k=p^\kappa k'$, where $p \nmid k'$. Clearly $0 \leqslant \kappa < k-1$. Since $p \nmid a$, and $\beta < \nu-\tau$, the congruence (2) can have solutions only if $\beta = \kappa+(k-1)\rho$, where $\rho \geqslant 0$ is an integer, and in this case (2) is the same as

$$k' a(z/p^\rho)^{k-1}+b' \equiv 0 \pmod{p^{\nu-\tau-\beta}}.$$

This congruence* has not more than $2(k-1)$ solutions (mod $p^{\nu-\tau-\beta}$); hence

$$N = O\left(\frac{p^{\nu-\tau}}{p^{\nu-\tau-\beta+\rho}}\right) = O(p^{\beta-\rho}).$$

Hence $$S(a, b, p^\nu) = O(p^{\tau+\beta-\rho}).$$

If $\kappa = 0$, we have

$$\tau+\beta-\rho = \tau \leqslant \tfrac{1}{2}(\nu+1) \leqslant \tfrac{3}{4}\nu, \quad \text{if} \quad \rho = 0,$$

$$\tau+\beta-\rho \leqslant \tau+\beta-1 \leqslant \tfrac{1}{2}(\nu+1)+\tfrac{1}{4}(\nu-1)-1 < \tfrac{3}{4}\nu \quad \text{if} \quad \rho > 0.$$

If $\kappa > 0$, so that $p \mid k$, we have

$$S(a, b, p^\nu) = O(p^{\tau+\beta}) = O(p^{\frac{1}{2}(\nu+1)+\frac{1}{4}(\nu-1)}) = O(p^{\frac{3}{4}\nu}).$$

LEMMA 3. *If* $(a, q) = 1$ *and* $b \neq 0$, *then* $S(a, b, q) = O\left(q^{\frac{3}{4}+\epsilon}(q, b)\right)$.

Proof. If $(q_1, q_2) = 1$, then

$$S(a, b, q_1 q_2) = \sum_{x=1}^{q_1} \sum_{y=1}^{q_2} e_{q_1 q_2}\left(a(xq_2+yq_1)^k+b(xq_2+yq_1)\right)$$
$$= S(aq_2^{k-1}, b, q_1)\, S(aq_1^{k-1}, b, q_2).$$

By Lemmas 1 and 2, if $p^\beta \| b$ and $\beta < \frac{1}{4}\nu$,

$$S(a, b, p^\nu) = O(p^{\frac{3}{4}\nu}) = O\left(p^{\frac{3}{4}\nu}(p^\nu, b)\right).$$

Also, if $\beta \geqslant \frac{1}{4}\nu$,

$$S(a, b, p^\nu) = O(p^\nu) = O\left(p^{\frac{3}{4}\nu}(p^\nu, b)\right).$$

---

* The congruence $x^l \equiv c \pmod{p^\lambda}$, where $p \nmid c$, has at most l solutions, unless $p = 2$, in which case it has at most $2l$ solutions.

The result follows, on noting that

$$\prod_{p|q} O(1) = O(q^{\epsilon}),$$

and that $(q_1 q_2, b) = (q_1, b)(q_2, b)$ if $(q_1, q_2) = 1$.

LEMMA 4. *Let $f(x)$ be any polynomial with integral coefficients, and let*

$$\sum_{x=1}^{q} e_q\big(f(x)+nx\big) = F(n).$$

Then

$$\sum_{x=1}^{m} e_q\big(f(x)\big) = \frac{m}{q}\, F(0) + O\Big(\sum_{n=1}^{q-1} \frac{1}{n}\,\big(|F(n)|+|F(-n)|\big)\Big) + O(\log q).$$

Proof. Without loss of generality, we can suppose that $1 \leqslant m \leqslant q$. It is well known that, if t is not an integer,

$$\{t\} = t-[t]-\tfrac{1}{2} = -\frac{1}{2\pi i} \sum_{n=-(q-1)}^{q-1}{}' \frac{1}{n}\, e^{2\pi i n t} + O\Big(\frac{1}{q\|t\|}\Big), \tag{3}$$

where the summation excludes $n = 0$, and $\|t\|$ denotes the distance of t from the nearest integer. Now the expression

$$\Big[\frac{x-\frac{1}{2}}{q}\Big] - \Big[\frac{x-m-\frac{1}{2}}{q}\Big] = \frac{m}{q} - \Big\{\frac{x-\frac{1}{2}}{q}\Big\} + \Big\{\frac{x-m-\frac{1}{2}}{q}\Big\}$$

has the value 1 for $1 \leqslant x \leqslant m$ and 0 for $m+1 \leqslant x \leqslant q$. Hence

$$\sum_{x=1}^{n} e_q\big(f(x)\big) = \frac{m}{q}\, F(0) - \sum_{x=1}^{q} \Big\{\frac{x-\frac{1}{2}}{q}\Big\}\, e_q\big(f(x)\big) + \sum_{x=1}^{q} \Big\{\frac{x-m-\frac{1}{2}}{q}\Big\}\, e_q\big(f(x)\big).$$

Now if h denotes either of the numbers $\frac{1}{2}$, $m+\frac{1}{2}$, we have, by (3),

$$\sum_{x=1}^{q} \Big\{\frac{x-h}{q}\Big\}\, e_q\big(f(x)\big) = -\frac{1}{2\pi i} \sum_{n=-(q-1)}^{q-1}{}' \frac{1}{n}\, e_q(-nh)\, F(n) + O\Bigg(\sum_{x=1}^{q} \frac{1}{q\Big\|\frac{x-h}{q}\Big\|}\Bigg)$$

$$= O\Big(\sum_{n=1}^{q-1} \frac{1}{n}\,\big(|F(n)|+|F(-n)|\big)\Big) + O(\log q).$$

This establishes the lemma.

Proof of the inequality.

THEOREM. *With the notation of the introduction,*

$$\sum_{x=1}^{m} e_q(ax^k) = \frac{m}{q}\, S_{a,q} + O(q^{\frac{1}{2}+\epsilon}).$$

Proof. By Lemma 4 with $f(x) = ax^k$, it suffices to prove that

$$\sum_{n=1}^{q-1} \frac{1}{n} \left(|S(a, n, q)| + |S(a, -n, q)| \right) = O(q^{\frac{1}{2}+\epsilon}).$$

By Lemma 3, the sum on the left is

$$O\left(\sum_{n=1}^{q-1} \frac{1}{n}\, q^{\frac{1}{2}+\epsilon}(q, n) \right) = O\left(q^{\frac{1}{2}+\epsilon} \sum_{d \mid q} d \sum_{m=1}^{(q-1)/d} \frac{1}{md} \right)$$

$$= O\left(q^{\frac{1}{2}+\epsilon} \log q \sum_{d \mid q} 1 \right) = O(q^{\frac{1}{2}+2\epsilon}).$$

Trinity College,
Cambridge.

15.

On Waring's Problem: Two Cubes and One Square

(With H. Davenport)

Proceedings of the London Mathematical Society 43 (1937), 73–104
[Commentary on p. 578]

[Received 12 November, 1936.—Read 12 November, 1936.]

Introduction.

The object of this paper is to give proofs of the following results:

THEOREM 1. *Almost all positive integers are representable as*

$$x^3+y^3+z^2, \tag{1}$$

where x, y, z are positive integers.

THEOREM 2. *Almost all positive integers not congruent to* 3 (mod 9) *are representable in the form*

$$x^3+y^3+6z^2, \tag{2}$$

and almost all positive integers not congruent to 6 (mod 9) *are representable in the form*

$$x^3+y^3+12z^2. \tag{3}$$

Consequently almost all positive integers are representable in one of the two forms

$$x^3+y^3+(z+1)^3-(z-1)^3, \quad x^3+y^3+(z+2)^3-(z-2)^3,$$

and so we have

THEOREM 3. *Almost all positive integers are representable as the sum of four integer cubes (positive or negative).*

The phrase "almost all positive integers" has its usual meaning in connection with Waring's problem, namely that the number of positive integers not exceeding ξ which do not have the property in question is

$o(\xi)$ as $\xi \to \infty$. As regards the numbers (mod 9) excluded in Theorem 2, it is immediately seen that they are not representable in the relevant form, since a cube is congruent to 0 or 1 or -1 (mod 9) and $6z^2 \equiv 0$ or 6 (mod 9) and $12z^2 \equiv 0$ or 3 (mod 9).

The work in this paper is a fairly straightforward application of the Hardy-Littlewood method, except as regards the singular series, where problems new in this connection arise. The singular series is essentially a partial sum of the Dirichlet series

$$M(s) = \prod_{p \equiv 1 \pmod 3} \left(1 + \left(\frac{n}{p}\right)\{2 + \chi(4n) + \bar{\chi}(4n)\} p^{-s}\right)$$

at $s = 1$, where χ, $\bar{\chi}$ are the non-principal cubic characters (mod p). The function $M(s)$ can be expressed in terms of the zeta-functions of certain algebraic fields, with a factor which is absolutely convergent at $s = 1$. It appears to be difficult to obtain a positive lower bound for our partial sum by completing it to the infinite series $M(1)$ without making use of some unproved hypothesis concerning the zeros of these zeta-functions†. We avoid this difficulty by proving that, for our purpose, the partial sum can be replaced by the first factors of a "singular product", for which a sufficiently good lower bound can be obtained.

Notation.

P is any sufficiently large positive integer. W is 1 or 6 or 12 [according as we are considering the form (1) or (2) or (3)]. All small Latin letters (whether with or without suffixes) except c, e, and i denote positive integers, and p denotes a prime throughout. $c_1, c_2, \ldots$ denote positive absolute constants. We use the abbreviations

$$e(\alpha) = e^{2\pi i\alpha}, \quad e_q(\alpha) = e^{2\pi i\alpha/q}.$$

Let
$$T(\alpha) = T(\alpha, P) = \sum_{x=1}^{P} e(\alpha x^3),$$

$$U(\alpha) = U(\alpha, P, W) = \sum_{x=1}^{P^{\frac{3}{2}}} e(\alpha W x^2),$$

$$V(\alpha) = T^2(\alpha)\, U(\alpha) = \sum_n r(n)\, e(\alpha n),$$

† See the Appendix.

so that, for $n \leqslant P^3$, $r(n)$ is the number of representations of n in the form (1) or (2) or (3) according to the value of W. Let

$$S_{a,q}^{(3)} = \sum_{x=1}^{q} e_q(ax^3), \qquad S_{a,q}^{(2)} = \sum_{x=1}^{q} e_q(ax^2),$$

$$I^{(3)}(\beta) = \tfrac{1}{3} \sum_{n=1}^{P^3} n^{-\frac{2}{3}} e(\beta n), \qquad I^{(2)}(\beta) = \tfrac{1}{2} \sum_{n=1}^{P^3} n^{-\frac{1}{2}} e(\beta n),$$

$$T^*(\alpha, a, q) = q^{-1} S_{a,q}^{(3)} I^{(3)}\left(\alpha - \frac{a}{q}\right),$$

$$U^*(\alpha, a, q) = q^{-1} S_{Wa,q}^{(2)} I^{(2)}\left(W\left(\alpha - \frac{a}{q}\right)\right),$$

$$V^*(\alpha, a, q) = T^{*2}(\alpha, a, q)\, U^*(\alpha, a, q),$$

$$A(n, q) = q^{-3} \sum_{\substack{a=1 \\ (a,q)=1}}^{q} (S_{a,q}^{(3)})^2 (S_{Wa,q}^{(2)})\, e_q(-an).$$

Let $\delta = \frac{1}{100}$, $R = [P^{1+\delta}]$, $X = \frac{1}{2} \log R$. Let

$$\mathfrak{S}(R, n) = \sum_{q=1}^{R} A(n, q),$$

$$\mathfrak{S}'(R, n) = \prod_{p \leqslant X} \left(\sum_{\substack{\nu=0 \\ p^\nu \leqslant X}}^{\infty} A(n, p^\nu) \right).$$

We divide the interval $0 \leqslant \alpha \leqslant 1$ into "Farey arcs" surrounding all rational points a/q with $(a, q) = 1$, $a \leqslant q \leqslant P^{2-\delta}$, the arc surrounding a/q extending as far as the mediants $(a+a_1)/(q+q_1)$, $(a+a_2)/(q+q_2)$ of a/q and the neighbouring Farey fractions a_1/q_1, a_2/q_2 of order $P^{2-\delta}$. The interval $(0, 1)$ is covered exactly once by these arcs, when we make the obvious convention about the arc surrounding $1/1$. The points of the Farey arc surrounding a/q are of the form $\alpha = a/q + \beta$ with

$$-\vartheta_1 q^{-1} P^{-2+\delta} \leqslant \beta \leqslant \vartheta_2 q^{-1} P^{-2+\delta},$$

where $\frac{1}{2} \leqslant \vartheta_1 \leqslant 1$, $\frac{1}{2} \leqslant \vartheta_2 \leqslant 1$.

If $q \leqslant P^{1+\delta}$ we call the arc surrounding a/q a major arc and denote it by $\mathfrak{M}_{a,q}$, and if $P^{1+\delta} < q \leqslant P^{2-\delta}$ we call the arc a minor arc and denote it by $\mathfrak{m}_{a,q}$.

We divide each major arc into two parts:

(I) $$|\beta| \leqslant P^{-3(1-\delta)},$$

(II) $$|\beta| \geqslant P^{-3(1-\delta)}.$$

It may happen that the whole of an arc belongs to (I).

Throughout the paper the constants implied by the use of the symbol O depend on ϵ, and on ϵ only.

The approximation of $T(\alpha)$, $U(\alpha)$.

Let

$$S^{(2)}(a, b, q) = \sum_{x=1}^{q} e_q(ax^2+bx), \qquad S^{(3)}(a, b, q) = \sum_{x=1}^{q} e_q(ax^3+bx).$$

LEMMA 1. *If* $(a, q) = 1$,

$$S^{(2)}(aW, b, q) = O(q^{\frac{1}{2}}).$$

Proof. We have

$$|S^{(2)}(Wa, b, q)|^2 = \sum_{x=1}^{q} \sum_{y=1}^{q} e_q\Big(Wa(y^2-x^2)+b(y-x)\Big)$$

$$= \sum_{x=1}^{q} \sum_{z=1}^{q} e_q\Big(Wa(z^2+2zx)+bz\Big),$$

on writing $y \equiv x+z \pmod q$. The sum over x is zero unless $2Wz \equiv 0 \pmod q$. Hence

$$|S^{(2)}(Wa, b, q)|^2 \leqslant q \sum_{\substack{z=1 \\ q|2Wz}}^{q} 1 \leqslant 2Wq.$$

LEMMA 2. *If* $(a, q) = 1$,

$$S^{(3)}(a, b, q) = O\Big(q^{\frac{2}{3}+\epsilon}(q, b)\Big).$$

Proof. If $(q_1, q_2) = 1$, we have

$$S^{(3)}(a, b, q_1 q_2) = \sum_{x_1=1}^{q_1} \sum_{x_2=1}^{q_2} e_{q_1 q_2}\Big(a(x_1 q_2+x_2 q_1)^3+b(x_1 q_2+x_2 q_1)\Big)$$

$$= S^{(3)}(aq_2{}^2, b, q_1)\, S^{(3)}(aq_1{}^2, b, q_2).$$

Hence it is sufficient to prove that for any prime $p \nmid a$ and any positive integer ν,

$$S^{(3)}(a, b, p^\nu) = O\Big(p^{\frac{2}{3}\nu}(p^\nu, b)\Big), \tag{4}$$

since then

$$S^{(3)}(a, b, q) = q^{\frac{2}{3}}(q, b) \prod_{p|q} O(1) = O\Big(q^{\frac{2}{3}+\epsilon}(q, b)\Big).$$

For $\nu = 1$, (4) is a special case of a result due to Mordell†. We therefore suppose that $\nu \geqslant 2$.

---

† *Quart. J. of Math.*, 3 (1932), 161–167. Actually the exponent $\frac{2}{3}$ can be replaced by $\frac{5}{8}$; for this see Davenport, *Journal für Math.*, 169 (1933), 158–176, Theorem 1.

Let $\tau = [\frac{1}{2}(\nu+1)]$, so that $2\tau \geqslant \nu$ and $\tau \leqslant \frac{2}{3}\nu$. Then, replacing x by $p^\tau y+z$, we obtain

$$S^{(3)}(a, b, p^\nu) = \sum_{z=1}^{p^\tau} e_{p^\nu}(az^3+bz) \sum_{y=1}^{p^{\nu-\tau}} e_{p^{\nu-\tau}}\big((3az^2+b)y\big)$$

$$= p^{\nu-\tau} \sum_{\substack{z=1 \\ 3az^2+b\equiv 0(p^{\nu-\tau})}}^{p^\tau} e_{p^\nu}(az^3+bz),$$

so that
$$|S^{(3)}(a, b, p^\nu)| \leqslant p^{\nu-\tau} \frac{p^\tau}{p^{\nu-\tau}} N,$$

where N is the number of solutions of the congruence

$$3az^2+b \equiv 0 \pmod{p^{\nu-\tau}}. \tag{5}$$

Suppose first that $p \neq 3$. Let $b = p^\mu b'$, where $p \nmid b'$ and $\mu \geqslant 0$. If $\mu \geqslant \nu-\tau$, we have

$$S^{(3)}(a, b, p^\nu) = O(p^\nu) = O(p^{\frac{1}{3}\nu} p^{\nu-\tau}) = O\big(p^{\frac{2}{3}\nu}(p^\nu, b)\big).$$

Hence we may suppose that $\mu < \nu-\tau$. In order that (5) may have any solutions at all it is necessary that $\mu = 2\mu_1$, and then $z = p^{\mu_1}z'$ and

$$3az'^2+b' \equiv 0 \pmod{p^{\nu-\tau-2\mu_1}}, \qquad z' \leqslant p^{\nu-\tau-\mu_1}. \tag{6}$$

The number of values of z' satisfying (6) is $O(p^{\mu_1})$. Hence $N = O(p^{\mu_1})$ and

$$S^{(3)}(a, b, p^\nu) = O(p^{\tau+\mu_1}) = O(p^{\tau+\mu}) = O\big(p^{\frac{2}{3}\nu}(p^\nu, b)\big).$$

Suppose now that $p = 3$. Let $b = 3^\mu b'$, where $3 \nmid b'$. If $\mu \geqslant \nu-\tau-1$, we have

$$S^{(3)}(a, b, 3^\nu) = O(3^\nu) = O(3^{\nu-1}) = O(3^{\frac{1}{3}\nu} 3^{\nu-\tau-1}) = O\big(3^{\frac{2}{3}\nu}(3^\nu, b)\big).$$

Hence we may suppose that $\mu < \nu-\tau-1$. Then $\mu = 2\mu_1+1$, and $z = 3^{\mu_1}z'$ and

$$az'^2+b' \equiv 0 \pmod{3^{\nu-\tau-\mu}}, \qquad z' \leqslant 3^{\nu-\tau-\mu_1}.$$

The rest of the proof is practically as before.

Lemma 3. *Let $f(x)$ be any polynomial with integral coefficients, and let*

$$F(n) = \sum_{x=1}^{q} e_q\big(f(x)+nx\big).$$

Then

$$\sum_{x=1}^{m} e_q\big(f(x)\big) = \frac{m}{q} F(0) + O\Big(\sum_{n=1}^{q-1} \frac{1}{n}\big(|F(n)|+|F(-n)|\big)\Big) + O(\log q).$$

This is Lemma 4 of our paper "On an exponential sum", *Proc. London Math. Soc.* (2), 41 (1936), 449–453.

LEMMA 4. *If* $(a, q) = 1$,

$$\sum_{x=1}^{m} e_q(ax^3) = \frac{m}{q} S^{(3)}_{a,q} + O(q^{\frac{3}{4}+\epsilon}), \tag{7}$$

$$\sum_{x=1}^{m} e_q(Wax^2) = \frac{m}{q} S^{(2)}_{Wa,q} + O(q^{\frac{1}{2}+\epsilon}). \tag{8}$$

Proof. These results follow from Lemma 3 on taking $f(x)$ to be ax^3 or Wax^2 and using Lemmas 1, 2. In the case of (7), for example, we obtain

$$\sum_{x=1}^{m} e_q(ax^3) = \frac{m}{q} S^{(3)}_{a,q} + O\Big(\sum_{n=1}^{q-1} \frac{1}{n} q^{\frac{3}{4}+\epsilon}(q, n)\Big) + O(\log q)$$

$$= \frac{m}{q} S^{(3)}_{a,q} + O\Big(q^{\frac{3}{4}+\epsilon} \sum_{d \mid q} d \sum_{m \leqslant (q-1)/d} \frac{1}{md}\Big) + O(\log q)$$

$$= \frac{m}{q} S^{(3)}_{a,q} + O\big(q^{\frac{3}{4}+\epsilon} d(q) \log q\big) + O(\log q)$$

$$= \frac{m}{q} S^{(3)}_{a,q} + O(q^{\frac{3}{4}+2\epsilon}).$$

LEMMA 5. *If* $(a, q) = 1$, *and* $\alpha = a/q + \beta$, *then*

$$T(\alpha) = T^*(\alpha, a, q) + O\big(q^{\frac{3}{4}+\epsilon}(1 + P^3|\beta|)\big), \tag{9}$$

$$U(\alpha) = U^*(\alpha, a, q) + O\big(q^{\frac{1}{2}+\epsilon}(1 + P^3|\beta|)\big). \tag{10}$$

Proof. The proofs are almost the same in the two cases, so we prove only (9). Let

$$S_0 = 0, \quad S_n = \sum_{x \leqslant n^{\frac{1}{3}}} e_q(ax^3).$$

Then, by partial summation,

$$T(\alpha) = \sum_{x=1}^{P} e_q(ax^3)\, e(\beta x^3)$$

$$= \sum_{n=1}^{P^3} (S_n - S_{n-1})\, e(\beta n)$$

$$= \sum_{n=1}^{P^3-1} S_n\Big(e(\beta n) - e\big(\beta(n+1)\big)\Big) + S_{P^3} e(\beta P^3).$$

By Lemma 4,

$$S_n = \frac{n^{\frac{1}{3}}}{q} S^{(3)}_{a,q} + O(q^{\frac{2}{3}+\epsilon}).$$

Hence

$$T(\alpha) = q^{-1} S^{(3)}_{a,q} \sum_{n=1}^{P^3-1} n^{\frac{1}{3}}\Big(e(\beta n) - e\big(\beta(n+1)\big)\Big)$$

$$+ q^{-1} S^{(3)}_{a,q} P e(\beta P^3) + O\Big(q^{\frac{2}{3}+\epsilon} \sum_{n=1}^{P^3-1} |1-e(\beta)|\Big) + O(q^{\frac{2}{3}+\epsilon})$$

$$= q^{-1} S^{(3)}_{a,q} \sum_{n=1}^{P^3} \Big(n^{\frac{1}{3}} - (n-1)^{\frac{1}{3}}\Big) e(\beta n) + O\Big(q^{\frac{2}{3}+\epsilon}(1+P^3|\beta|)\Big).$$

Now

$$n^{\frac{1}{3}} - (n-1)^{\frac{1}{3}} = \tfrac{1}{3} n^{-\frac{2}{3}} + O(n^{-\frac{5}{3}}).$$

Hence

$$T(\alpha) = q^{-1} S^{(3)}_{a,q} I^{(3)}(\beta) + O\Big(q^{\frac{2}{3}+\epsilon}(1+P^3|\beta|)\Big)$$

$$= T^*(\alpha, a, q) + O\Big(q^{\frac{2}{3}+\epsilon}(1+P^3|\beta|)\Big).$$

LEMMA 6. *If* $(a, q) = 1$,

$$|S^{(2)}_{Wa,q}| < c_1 q^{\frac{1}{2}}, \tag{11}$$

$$|S^{(3)}_{a,q}| < c_2 q^{\frac{2}{3}}. \tag{12}$$

Proof. The first result is a special case of Lemma 1. For the second, see Landau, *Vorlesungen über Zahlentheorie*, 1, Satz 315. This book will in future be referred to simply as Landau.

LEMMA 7. *If* $|\beta| \leqslant \frac{1}{2}$, *we have*

$$I^{(3)}(\beta) = O(|\beta|^{-\frac{1}{3}}),$$

$$I^{(2)}(W\beta) = O(|\beta|^{-\frac{1}{2}}).$$

Proof. The proofs are similar in the two cases, so we prove the result for $I^{(3)}(\beta)$ only. If $|\beta| \leqslant P^{-3}$ the result is obvious, since $|I^{(3)}(\beta)| \leqslant P$. If $|\beta| \geqslant P^{-3}$, we have

$$I^{(3)}(\beta) = \tfrac{1}{3} \sum_{n \leqslant |\beta|^{-1}} n^{-\frac{2}{3}} e(\beta n) + \tfrac{1}{3} \sum_{|\beta|^{-1} < n \leqslant P^3} n^{-\frac{2}{3}} e(\beta n)$$

$$= O(|\beta|^{-\frac{1}{3}}) + O\Big(|\beta|^{\frac{2}{3}} |1-e(\beta)|^{-1}\Big),$$

by applying Abel's lemma to the second sum. Hence the result.

Minor arcs.

LEMMA 8. $$\sum_{\mathfrak{m}}\int_{\mathfrak{m}}|V(\alpha)|^2 d\alpha = O(P^{4-\delta+\epsilon}).$$

Proof. By (10), (11), and the fact that $|I^{(2)}(W\beta)| \leqslant P^{\frac{3}{2}}$, we have

$$U(\alpha) = O(q^{-\frac{1}{2}}P^{\frac{3}{2}}) + O\left(q^{\frac{1}{2}+\epsilon}(1+P^3|\beta|)\right)$$
$$= O(q^{-\frac{1}{2}}P^{\frac{3}{2}}) + O\left(q^{\frac{1}{2}+\epsilon}(1+P^3q^{-1}P^{-2+\delta})\right).$$

For a minor arc, $P^{1+\delta} < q \leqslant P^{2-\delta}$. Hence, on a minor arc,

$$U(\alpha) = O(P^{1-\frac{1}{2}\delta}) + O(P^{1-\frac{1}{2}\delta+2\epsilon}) + O(P^{1+\delta-(\frac{1}{2}-\epsilon)(1+\delta)}) = O(P^{1-\frac{1}{2}\delta+2\epsilon}).$$

Hence it is sufficient to prove that

$$\sum_{\mathfrak{m}}\int_{\mathfrak{m}}|T(\alpha)|^4 d\alpha = O(P^{2+\epsilon}).$$

Now $$\int_0^1 |T(\alpha)|^4 d\alpha = \sum_{n=1}^{2P^3}\left(r_2(n)\right)^2,$$

where $r_2(n)$ is the number of representations of n as $x_1^3+x_2^3$ with $x_1 \leqslant P$, $x_2 \leqslant P$. Since $r_2(n) = O(n^\epsilon) = O(P^{3\epsilon})$, we have

$$\int_0^1 |T(\alpha)|^4 d\alpha = O\left(P^{3\epsilon}\sum_{n=1}^{2P^3} r_2(n)\right) = O(P^{2+3\epsilon}).$$

Major arcs.

LEMMA 9. *On $\mathfrak{M}_{a,q}$, if $\alpha = a/q+\beta$, $\beta \neq 0$, we have*

$$T(\alpha) = O(P^{\frac{1}{2}(1+\delta)+\epsilon}q^{-\frac{1}{2}-\epsilon}|\beta|^{-\frac{1}{2}-\epsilon}).$$

Proof. On $\mathfrak{M}_{a,q}$ we have

$$q \leqslant P^{1+\delta}, \quad |\beta| \leqslant q^{-1}P^{-2+\delta}.$$

There exist a_1, q_1, with $(a_1, q_1) = 1$, such that

$$q_1 \leqslant \frac{2}{q|\beta|},$$

$$\left|\alpha - \frac{a_1}{q_1}\right| \leqslant \frac{q|\beta|}{2q_1}. \tag{13}$$

Let $\beta_1 = \alpha - a_1/q_1$. Then

$$|\beta_1| \leqslant \tfrac{1}{2}q_1^{-1}P^{-2+\delta}. \tag{14}$$

It is impossible that $q_1 \leqslant P^{2-\delta}$, for then, by (14), α would be a point of the Farey arc surrounding a_1/q_1, which is only possible if $q_1 = q$, $a_1 = a$, $\beta_1 = \beta$, and this is contrary to (13). Hence

$$q_1 > P^{2-\delta}. \tag{15}$$

Let
$$S_0 = 0, \quad S_n = \sum_{x \leqslant n^{\frac13}} e_{q_1}(a_1 x^3).$$

As in the proof of Lemma 5,

$$T(\alpha) = \sum_{n=1}^{P^3-1} S_n \Big(e(\beta_1 n) - e\big(\beta_1(n+1)\big)\Big) + S_{P^3} e(\beta_1 P^3).$$

By Landau, Satz 267, we have

$$S_n = O\Big(P^{\epsilon} q_1^{\epsilon}(P^{\frac34} + P q_1^{-\frac14} + P^{\frac14} q_1^{\frac14})\Big),$$

for $n \leqslant P^3$. By (15),

$$P^{\frac34} \leqslant P^{\frac14(1+\delta)} q_1^{\frac14}, \quad P q_1^{-\frac14} \leqslant P^{\frac14(1+\delta)} q_1^{\frac14}.$$

Hence
$$T(\alpha) = O\Big(P^{\epsilon} q_1^{\epsilon} P^{\frac14(1+\delta)} q_1^{\frac14}(1 + P^3|\beta_1|)\Big).$$

By (14),
$$|\beta_1| \leqslant \tfrac12 q_1^{-1} P^{-2+\delta} < P^{-3}.$$

Hence
$$\begin{aligned} T(\alpha) &= O(P^{\frac14(1+\delta)+\epsilon} q_1^{\frac14+\epsilon}) \\ &= O(P^{\frac14(1+\delta)+\epsilon} q^{-\frac14-\epsilon} |\beta|^{-\frac14-\epsilon}). \end{aligned}$$

Lemma 10. *On the part I of a major arc* $\mathfrak{M}_{a,q}$,

$$V(\alpha) - V^*(\alpha, a, q) = O(P^{\frac52+7\delta+\epsilon} q^{-\frac13}).$$

Proof. By Lemma 5, we have, on I,

$$T - T^* = O(P^{3\delta} q^{\frac12+\epsilon}), \tag{16}$$

$$U - U^* = O(P^{3\delta} q^{\frac12+\epsilon}). \tag{17}$$

By Lemma 6, noting that $|I^{(3)}(\beta)| \leqslant P$ and $|I^{(2)}(W\beta)| \leqslant P^{\frac32}$, we have

$$T^* = O(P q^{-\frac13}), \tag{18}$$

$$U^* = O(P^{\frac32} q^{-\frac12}). \tag{19}$$

By (16), (18),

$$T + T^* = O(P^{1+4\delta} q^{-\frac13+\epsilon}). \tag{20}$$

By (17, (19),

$$U = O(P^{\frac{3}{2}} q^{-\frac{1}{2}+\epsilon}). \tag{21}$$

By (16), (17), (18), (20), (21),

$$\begin{aligned} V - V^* &= (T - T^*)(T + T^*)U + T^{*2}(U - U^*) \\ &= O(P^{3\delta} q^{\frac{2}{3}+\epsilon} P^{1+4\delta} q^{-\frac{1}{3}+\epsilon} P^{\frac{3}{2}} q^{-\frac{1}{2}+\epsilon}) + O(P^2 q^{-\frac{2}{3}} P^{3\delta} q^{\frac{1}{2}+\epsilon}) \\ &= O(P^{\frac{5}{2}+7\delta} q^{-\frac{1}{6}+3\epsilon}) + O(P^{2+3\delta} q^{-\frac{1}{6}+\epsilon}) \\ &= O(P^{\frac{5}{2}+7\delta} q^{-\frac{1}{6}+3\epsilon}). \end{aligned}$$

LEMMA 11. *On the part II of a major arc* $\mathfrak{M}_{a,q}$,

$$V(\alpha) - V^*(\alpha, a, q) = O(P^{\frac{1}{2}(1+\delta)+\epsilon} q^{-1} |\beta|^{-1}) + O(P^{\frac{7}{2}+\frac{1}{2}\delta+\epsilon} |\beta|^{\frac{1}{2}}).$$

Proof, By Lemma 5,

$$U - U^* = O(P^3 q^{\frac{1}{2}+\epsilon} |\beta|). \tag{22}$$

By Lemma 9,

$$T = O(P^{\frac{1}{4}(1+\delta)+\epsilon} q^{-\frac{1}{4}} |\beta|^{-\frac{1}{4}-\epsilon}). \tag{23}$$

By Lemmas 6, 7,

$$T^* = O(q^{-\frac{1}{3}} |\beta|^{-\frac{1}{3}}), \tag{24}$$

$$U^* = O(q^{-\frac{1}{2}} |\beta|^{-\frac{1}{2}}). \tag{25}$$

Since

$$\begin{aligned} q^{-\frac{1}{3}} |\beta|^{-\frac{1}{3}} &\leqslant q^{-\frac{1}{3}} |\beta|^{-\frac{1}{3}} (P^{3(1-\delta)} |\beta|)^{\frac{1}{12}} \\ &= P^{\frac{1}{4}(1-\delta)} q^{-\frac{1}{3}} |\beta|^{-\frac{1}{4}} \leqslant P^{\frac{1}{4}(1+\delta)+\epsilon} q^{-\frac{1}{4}} |\beta|^{-\frac{1}{4}-\epsilon}, \end{aligned}$$

we have, from (23), (24),

$$|T| + |T^*| = O(P^{\frac{1}{4}(1+\delta)+\epsilon} q^{-\frac{1}{4}} |\beta|^{-\frac{1}{4}-\epsilon}). \tag{26}$$

By (22), (25), (26),

$$\begin{aligned} |V - V^*| &\leqslant (|T| + |T^*|)^2 (|U^*| + |U - U^*|) \\ &= O\left(P^{\frac{1}{2}(1+\delta)+2\epsilon} q^{-\frac{1}{2}} |\beta|^{-\frac{1}{2}-2\epsilon} (q^{-\frac{1}{2}} |\beta|^{-\frac{1}{2}} + P^3 q^{\frac{1}{2}+\epsilon} |\beta|)\right) \\ &= O(P^{\frac{1}{2}(1+\delta)+2\epsilon} q^{-1} |\beta|^{-1-2\epsilon}) + O(P^{\frac{7}{2}+\frac{1}{2}\delta+2\epsilon} q^{\epsilon} |\beta|^{\frac{1}{2}-2\epsilon}). \end{aligned}$$

Hence we have the result, since $|\beta|^{-\epsilon} = O(P^{3\epsilon})$ and $q^{\epsilon} = O(P^{2\epsilon})$.

LEMMA 12.

$$\sum_{\mathfrak{M}} \int_{\mathfrak{M}} |V(\alpha) - V^*(\alpha, a, q)|^2 d\alpha = O(P^{4-\delta}).$$

Proof. The contribution of the parts I is, by Lemma 10,

$$O\left(\sum_{q=1}^{P^{1+\delta}} \sum_{a=1}^{q} P^{5+14\delta+2\epsilon} q^{-\frac{1}{3}} P^{-3(1-\delta)}\right) = O(P^{2+17\delta+2\epsilon} \sum_{q=1}^{P^{1+\delta}} q^{\frac{2}{3}})$$

$$= O(P^{2+17\delta+2\epsilon+\frac{5}{3}(1+\delta)}) = O(P^{4-\delta}),$$

since $\delta = \frac{1}{100}$ and ϵ is arbitrarily small.

By Lemma 11, the contribution of the part II of a single major arc $\mathfrak{M}_{a,q}$ is

$$O\left(P^{1+\delta+2\epsilon} q^{-2} \int_{P^{-3(1-\delta)}}^{\infty} \beta^{-2} d\beta\right) + O\left(P^{7+\delta+2\epsilon} \int_{0}^{q^{-1}P^{-2+\delta}} \beta\, d\beta\right)$$

$$= O(P^{1+\delta+2\epsilon+3(1-\delta)} q^{-2}) + O(P^{7+\delta+2\epsilon} q^{-2} P^{-4+2\delta})$$

$$= O(P^{4-2\delta+2\epsilon} q^{-2}).$$

Hence the total contribution of the parts II is

$$O\left(P^{4-2\delta+2\epsilon} \sum_{q=1}^{P^{1+\delta}} q \cdot q^{-2}\right) = O(P^{4-2\delta+2\epsilon} \log P)$$

$$= O(P^{4-\delta}).$$

Some further lemmas.

Let $\|\xi\|$ denote the distance of a real number ξ from the nearest integer, so that $0 \leqslant \|\xi\| \leqslant \frac{1}{2}$.

LEMMA 13. *If $q \leqslant R$, and Q is a positive integer, and $q \nmid Q$, then*†

$$\sum_{\substack{a=1 \\ (a,q)=1}}^{q} \min\left(R, \frac{1}{\|Qa/q\| - 1/q}\right) = O(R \log R).$$

Proof. Put $d = (Q, q)$, $Q = Q'd$, $q = q'd$, so that $(Q', q') = 1$ and $q' > 1$. Then the sum in question does not exceed

$$\sum_{\substack{a=1 \\ q' \nmid a}}^{q} \min\left(R, \frac{1}{\|Q'a/q'\| - 1/dq'}\right) = d \sum_{a=1}^{q'-1} \min\left(R, \frac{1}{\|Q'a/q'\| - 1/dq'}\right)$$

$$= d \sum_{n=1}^{q'-1} \min\left(R, \frac{1}{\|n/q'\| - 1/dq'}\right)$$

$$\leqslant 2d \sum_{n \leqslant \frac{1}{2}q'} \min\left(R, \frac{q'}{n - 1/d}\right).$$

† $\min(R, 1/0) = R$.

If $d = 1$, this is

$$2R+2 \sum_{2 \leqslant n \leqslant \frac{1}{2}q'} \frac{q'}{n-1} = O(R)+O(q' \log q') = O(R \log R).$$

If $d > 1$, it does not exceed

$$2d \sum_{n \leqslant \frac{1}{2}q'} \frac{q'}{n-\frac{1}{2}} = O(dq' \log q') = O(R \log R).$$

LEMMA 14.

$$\sum_{q=1}^{R} q^{-1} \sum_{\substack{a=1 \\ (a,\, q)=1}}^{q} \left(\sum_{Q=1}^{R} \sum_{\substack{A=1 \\ (A,\, Q)=1 \\ A,\, Q \neq a,\, q}}^{Q} \min\left(R, \frac{1}{|A-Qa/q|-1/q}\right)\right)^2 = O(R^4 \log^2 R).$$

Proof. We note that, under the conditions of summation,

$$\left| A - \frac{Qa}{q} \right| \geqslant \frac{1}{q}.$$

For $q = 1$ the inner sum is

$$\sum_{Q=2}^{R} \sum_{\substack{A=1 \\ (A,\, Q)=1}}^{Q} \min\left(R, \frac{1}{|A-Q|-1}\right) \leqslant \sum_{Q=1}^{R} \left(2R+2 \sum_{n=1}^{Q} \frac{1}{n}\right)$$

$$= O(R^2).$$

Hence the contribution of $q = 1$ to the sum of the lemma is $O(R^4)$.

For $q \geqslant 2$, $q \mid Q$, we have, on writing $Q/q = Q'$,

$$\sum_{\substack{A=1 \\ (A,\, Q)=1 \\ A \neq Q'a}}^{Q} \min\left(R, \frac{1}{|A-Q'a|-1/q}\right) \leqslant 2 \sum_{n=1}^{Q} \frac{1}{n-\frac{1}{2}} = O(\log R).$$

For $q \geqslant 2$, $q \nmid Q$, we have

$$\sum_{\substack{A=1 \\ (A,\, Q)=1}}^{Q} \min\left(R, \frac{1}{|A-Qa/q|-1/q}\right)$$

$$\leqslant 2 \min\left(R, \frac{1}{\|Qa/q\|-1/q}\right) + 2 \sum_{n=1}^{Q} \frac{1}{n-1/q}$$

$$\leqslant 2 \min\left(R, \frac{1}{\|Qa/q\|-1/q}\right) + O(\log R).$$

Hence we have (for $q \geqslant 2$)

$$(27) \quad \sum_{Q=1}^{R} \sum_{\substack{A=1 \\ (A,Q)=1 \\ A,Q \neq a,q}}^{Q} \min\left(R, \frac{1}{|A-Qa/q|-1/q}\right)$$

$$\leqslant 2 \sum_{\substack{Q=1 \\ q \nmid Q}}^{R} \min\left(R, \frac{1}{\|Qa/q\|-1/q}\right) + O(R \log R).$$

In particular, the sum on the left of (27) is $O(R^2)$.

In the sum of the lemma (but with the lower limit 2 instead of 1 for q) we substitute for one of the inner sums the upper bound $O(R^2)$, and for the other the upper bound given by (27). We obtain

$$O\left(R^2 \sum_{q=2}^{R} q^{-1} \sum_{\substack{a=1 \\ (a,q)=1}}^{q} \sum_{\substack{Q=1 \\ q \nmid Q}}^{R} \min\left(R, \frac{1}{\|Qa/q\|-1/q}\right)\right)$$

$$+O\left(R^2 \sum_{q=2}^{R} q^{-1} \sum_{\substack{a=1 \\ (a,q)=1}}^{q} R \log R\right).$$

The last term is $O(R^4 \log R)$. By Lemma 13, the first sum is

$$O\left(R^2 \sum_{q=2}^{R} q^{-1} \sum_{Q=1}^{R} R \log R\right) = O(R^4 \log^2 R).$$

This establishes Lemma 14.

Lemma 15.

$$\int_0^1 \left| \sum_{Q=1}^{R} \sum_{\substack{A=1 \\ (A,Q)=1}}^{Q}{}' V^*(\alpha, A, Q) \right|^2 d\alpha = O(P^{4-\frac{1}{3}}),$$

where the dash on the sign of summation means that if α is on a major arc $\mathfrak{M}_{a,q}$, the term $Q=q$, $A=a$ is to be omitted from the sum.

Proof. We make a Farey dissection of order R, and denote the typical Farey arc of this dissection by $\mathrm{M}_{a,q}$. Each point a/q is also the centre of a major arc $\mathfrak{M}_{a,q}$ of the former Farey dissection, and conversely, and $\mathfrak{M}_{a,q}$ is entirely inside $\mathrm{M}_{a,q}$.

The integral of the lemma is

$$\sum_{q=1}^{R} \sum_{\substack{a=1 \\ (a,q)=1}}^{q} \int_{\mathrm{M}_{a,q}-\mathfrak{M}_{a,q}} \left| \sum_{Q=1}^{R} \sum_{\substack{A=1 \\ (A,Q)=1}}^{Q} V^*(\alpha, A, Q) \right|^2 d\alpha$$

$$+ \sum_{q=1}^{R} \sum_{\substack{a=1 \\ (a,q)=1}}^{q} \int_{\mathfrak{M}_{a,q}} \left| \sum_{Q=1}^{R} \sum_{\substack{A=1 \\ (A,Q)=1 \\ A,Q \neq a,q}}^{Q} V^*(\alpha, A, Q) \right|^2 d\alpha.$$

The first sum does not exceed

$$2\sum_{q=1}^{R}\sum_{\substack{a=1\\(a,q)=1}}^{q}\int_{\mathrm{M}_{a,q}-\mathfrak{M}_{a,q}}|V^*(\alpha,a,q)|^2 d\alpha$$

$$+2\sum_{q=1}^{R^*}\sum_{\substack{a=1\\(a,q)=1}}^{q}\int_{\mathrm{M}_{a,q}-\mathfrak{M}_{a,q}}\Bigg|\sum_{Q=1}^{R}\sum_{\substack{A=1\\(A,Q)=1\\A,Q\neq a,q}}^{Q}V^*(\alpha,A,Q)\Bigg|^2 d\alpha.$$

Hence the integral of the lemma does not exceed

$$2\sum_{q=1}^{R}\sum_{\substack{a=1\\(a,q)=1}}^{q}\int_{\mathrm{M}_{a,q}-\mathfrak{N}_{a,q}}|V^*(\alpha,a,q)|^2 d\alpha$$

$$+2\sum_{q=1}^{R}\sum_{\substack{a=1\\(a,q)=1}}^{q}\int_{\mathrm{M}_{a,q}}\Bigg|\sum_{Q=1}^{R}\sum_{\substack{A=1\\(A,Q)=1\\A,Q\neq a,q}}^{Q}V^*(\alpha,A,Q)\Bigg|^2 d\alpha$$

$$=2\Sigma_1+2\Sigma_2,$$

say.

By Lemmas 6, 7,

$$V^*(\alpha,a,q)=O(q^{-\frac{7}{6}}|\alpha-a/q|^{-\frac{7}{6}}). \tag{28}$$

Hence

$$\Sigma_1=O\left(\sum_{q=1}^{R}q\int_{\frac{1}{2}q^{-1}P^{-2+\delta}}^{\infty}q^{-\frac{7}{3}}\beta^{-\frac{7}{3}}d\beta\right)$$

$$=O\left(\sum_{q=1}^{R}q^{1-\frac{7}{3}}(q^{-1}P^{-2+\delta})^{-\frac{4}{3}}\right)$$

$$=O\left(P^{\frac{8}{3}-\frac{4}{3}\delta}\sum_{q=1}^{R}1\right)=O(P^{4-\frac{1}{3}}).$$

In order to apply (28) in Σ_2, we must first obtain a lower bound for $|\alpha-A/Q|$ when α is a point of the arc $\mathrm{M}_{a,q}$ and $Q,A\neq q,a$. First, we have

$$\left|\alpha-\frac{A}{Q}\right|\geqslant\left|\frac{a}{q}-\frac{A}{Q}\right|-\left|\alpha-\frac{a}{q}\right|$$

$$\geqslant\left|\frac{a}{q}-\frac{A}{Q}\right|-q^{-1}R^{-1}\geqslant\left|\frac{a}{q}-\frac{A}{Q}\right|-q^{-1}Q^{-1},$$

since, on $\mathrm{M}_{a,q}$, $|\alpha-a/q|\leqslant q^{-1}R^{-1}$. Secondly, $\mathrm{M}_{A,Q}$ contains all points of the interval

$$\left(\frac{A}{Q}-\tfrac{1}{2}Q^{-1}R^{-1},\ \frac{A}{Q}+\tfrac{1}{2}Q^{-1}R^{-1}\right),$$

and α does not lie on $\mathbf{M}_{A,Q}$. Hence

$$|\alpha-A/Q|\geqslant\tfrac{1}{2}Q^{-1}R^{-1}.$$

Thus we have

$$Q\left|\alpha-\frac{A}{Q}\right|\geqslant\tfrac{1}{2}\max\left(\left|A-\frac{Qa}{q}\right|-\frac{1}{q},\ \frac{1}{R}\right). \tag{29}$$

Now, for any ξ, η with $\xi\geqslant 0$, $\eta>0$, we have

$$\bigl(\max(\xi,\eta)\bigr)^{-\frac{7}{8}}\leqslant\eta^{-\frac{1}{8}}\min(\xi^{-1},\eta^{-1}). \tag{30}$$

By (28), (29), (30),

$$V^*(\alpha,A,Q)=O\left(R^{\frac{1}{8}}\min\left(R,\frac{1}{|A-Qa/q|-1/q}\right)\right).$$

Substituting this in Σ_2 we have

$$\Sigma_2=O\left(\sum_{q=1}^{R}\sum_{\substack{a=1\\(a,q)=1}}^{q}q^{-1}R^{-1}\left(\sum_{Q=1}^{R}\sum_{\substack{A=1\\(A,Q)=1\\A,Q\neq a,q}}^{Q}R^{\frac{1}{8}}\min\left(R,\frac{1}{|A-Qa/q|-1/q}\right)\right)^2\right)$$

$$=O(R^{-1+\frac{1}{3}}R^4\log^2R)=O(P^{4-\frac{1}{3}}),$$

by Lemma 14.

Lemma 16.

$$\int_0^1\left|V(\alpha)-\sum_{Q=1}^{R}\sum_{\substack{A=1\\(A,Q)=1}}^{Q}V^*(\alpha,A,Q)\right|^2d\alpha=O(P^{4-\delta+\epsilon}).$$

Proof. The integral is

$$\sum_{\mathfrak{m}}\int_{\mathfrak{m}}\left|V(\alpha)-\sum_{Q=1}^{R}\sum_{\substack{A=1\\(A,Q)=1}}^{Q}V^*(\alpha,A,Q)\right|^2d\alpha$$

$$+\sum_{a,q}\int_{\mathfrak{M}_{a,q}}\left|V(\alpha)-\sum_{Q=1}^{R}\sum_{\substack{A=1\\(A,Q)=1}}^{Q}V^*(\alpha,A,Q)\right|^2d\alpha$$

$$\leqslant 2\sum_{\mathfrak{m}}\int_{\mathfrak{m}}|V(\alpha)|^2d\alpha+2\sum_{\mathfrak{m}}\int_{\mathfrak{m}}\left|\sum_{Q=1}^{R}\sum_{\substack{A=1\\(A,Q)=1}}^{Q}V^*(\alpha,A,Q)\right|^2d\alpha$$

$$+2\sum_{a,q}\int_{\mathfrak{M}_{a,q}}|V(\alpha)-V^*(\alpha,a,q)|^2d\alpha+2\sum_{a,q}\int_{\mathfrak{M}_{a,q}}\left|\sum_{Q=1}^{R}\sum_{\substack{A=1\\(A,Q)=1\\A,Q\neq a,q}}^{Q}V^*(\alpha,A,Q)\right|^2d\alpha.$$

The first of these sums is $O(P^{4-\delta+\epsilon})$ by Lemma 8. The third sum is $O(P^{4-\delta})$ by Lemma 12. The second and fourth sums together give exactly

$$\int_0^1 \left| \sum_{Q=1}^{R} \sum_{\substack{A=1 \\ (A,Q)=1}}^{Q}{}' V^*(\alpha, A, Q) \right|^2 d\alpha,$$

where the dash has the same meaning as in Lemma 15, and by Lemma 15 this is $O(P^{4-\frac{1}{3}})$. Hence we have the result.

LEMMA 17.

$$\sum_{n=W+2}^{P^3} \Big(r(n) - \psi(n)\mathfrak{S}(R, n) \Big)^2 = O(P^{4-\delta+\epsilon}), \tag{31}$$

where, for $W+2 \leqslant n \leqslant P^3$,

$$c_3 n^{\frac{1}{3}} < \psi(n) < c_4 n^{\frac{1}{3}}. \tag{32}$$

Proof. We have

$$\Big(I^{(3)}(\alpha - A/Q) \Big)^2 I^{(2)}\Big(W(\alpha - A/Q) \Big)$$

$$= \tfrac{1}{18} \sum_{n_1=1}^{P^3} \sum_{n_2=1}^{P^3} \sum_{n_3=1}^{P^3} n_1^{-\frac{2}{3}} n_2^{-\frac{2}{3}} n_3^{-\frac{1}{2}} e\Big((\alpha - A/Q)(n_1 + n_2 + W n_3) \Big)$$

$$= \sum_{n=W+2}^{(W+2)P^3} \psi(n) e\Big((\alpha - A/Q) n \Big),$$

where

$$\psi(n) = \tfrac{1}{18} \sum_{\substack{n_1=1 \\ n_1+n_2+Wn_3=n}}^{P^3} \sum_{n_2=1}^{P^3} \sum_{n_3=1}^{P^3} n_1^{-\frac{2}{3}} n_2^{-\frac{2}{3}} n_3^{-\frac{1}{2}}. \tag{33}$$

Hence

$$V^*(\alpha, A, Q) = \sum_{n=W+2}^{(W+2)P^3} \psi(n) e(n\alpha) Q^{-3} (S^{(3)}_{A,Q})^2 (S^{(2)}_{WA,Q}) e_Q(-An).$$

Summing over A, Q, we have

$$\sum_{Q=1}^{R} \sum_{\substack{A=1 \\ (A,Q)=1}}^{Q} V^*(\alpha, A, Q) = \sum_{n=W+2}^{(W+2)P^3} \psi(n) \mathfrak{S}(R, n) e(n\alpha).$$

Substituting this in Lemma 16, we obtain (31), on omitting the terms with $n > P^3$.

If $n \leqslant P^3$, the upper limits for n_1, n_2, n_3 in (33) can be omitted. If $n \geqslant W+2$, the number of representations of n as $n_1+n_2+Wn_3$ is greater than $c_5 n^2$, and in each representation

$$n_1^{-\frac{2}{3}} n_2^{-\frac{2}{3}} n_3^{-\frac{1}{2}} \geqslant n^{-\frac{2}{3}-\frac{2}{3}-\frac{1}{2}} = n^{-2+\frac{1}{6}}.$$

This proves one half of (32). For the other half we observe that

$$\sum_{n_1+n_2=m} n_1^{-\frac{2}{3}} n_2^{-\frac{2}{3}} \leqslant 2 \sum_{n_1 \leqslant \frac{1}{2}m} n_1^{-\frac{2}{3}} (\tfrac{1}{2}m)^{-\frac{2}{3}} < c_6 m^{-\frac{1}{3}}$$

and

$$\sum_{m+Wn_3=n} m^{-\frac{1}{3}} n_3^{-\frac{1}{2}} \leqslant \sum_{m \leqslant \frac{1}{2}n} m^{-\frac{1}{3}} (n/2W)^{-\frac{1}{2}} + \sum_{n_3 \leqslant n/2W} (\tfrac{1}{2}n)^{-\frac{1}{3}} n_3^{-\frac{1}{2}} < c_7 n^{\frac{1}{6}}.$$

Lemmas for the singular series.

LEMMA 18. $$|A(n, q)| < c_8 q^{-\frac{1}{6}}.$$

Proof. By Lemma 6 and the definition of $A(n, q)$, we have

$$|A(n, q)| \leqslant q^{-3} \sum_{a=1}^{q} c_1 q^{\frac{1}{2}} (c_2 q^{\frac{2}{3}})^2 = c_8 q^{-\frac{1}{6}}.$$

LEMMA 19. *If* $(q_1, q_2) = 1$,

$$A(n, q_1 q_2) = A(n, q_1) A(n, q_2).$$

Proof. We have

$$A(n, q_1 q_2) = (q_1 q_2)^{-3} \sum_{\substack{a_1=1 \\ (a_1, q_1)=1}}^{q_1} \sum_{\substack{a_2=1 \\ (a_2, q_2)=1}}^{q_2} e_{q_1 q_2}\big(-n(a_1 q_2 + a_2 q_1)\big) \tag{34}$$

$$\times (S^{(3)}_{a_1 q_2 + a_2 q_1,\, q_1 q_2})^2 S^{(2)}_{W(a_1 q_2 + a_2 q_1),\, q_1 q_2}.$$

Now

$$S^{(3)}_{a_1 q_2 + a_2 q_1,\, q_1 q_2} = \sum_{x=1}^{q_1 q_2} e_{q_1 q_2}\big((a_1 q_2 + a_2 q_1) x^3\big)$$

$$= \sum_{x=1}^{q_1 q_2} e_{q_1}(a_1 x^3)\, e_{q_2}(a_2 x^3)$$

$$= \sum_{x_1=1}^{q_1} \sum_{x_2=1}^{q_2} e_{q_1}(a_1 x_1^3)\, e_{q_2}(a_2 x_2^3)$$

$$= S^{(3)}_{a_1, q_1} S^{(3)}_{a_2, q_2}.$$

Similarly $$S^{(2)}_{W(a_1q_2+a_2q_1),\, q_1q_2} = S^{(2)}_{Wa_1,\, q_1} S^{(2)}_{Wa_2,\, q_2}.$$

Substituting in (34), we obtain the result.

Lemma 20. *Let p be a prime $\neq 2$ and $p \nmid a$. Then*

$$S^{(2)}_{a,\, p^\nu} = \begin{cases} p^{\frac{1}{2}\nu} & \text{if } \nu \text{ is even;} \\ (a/p)\, S^{(2)}_{1,\, p}\, p^{\frac{1}{2}(\nu-1)} & \text{if } \nu \text{ is odd;} \end{cases}$$

and

$$(S^{(2)}_{1,\, p})^2 = \left(\frac{-1}{p}\right) p.$$

Proof. By Landau, Satz 312, if $\nu > 2$,

$$S^{(2)}_{a,\, p^\nu} = p\, S^{(2)}_{a,\, p^{\nu-2}}.$$

Also, by Landau, Satz 307,

$$S^{(2)}_{a,\, p^2} = p.$$

By Landau, Satz 310,

$$S^{(2)}_{a,\, p} = \left(\frac{a}{p}\right) \tau,$$

where τ is independent of a, so that $\tau = S^{(2)}_{1,\, p}$. The last of the results enunciated follows from Landau, Satz 211.

Lemma 21. *Let p be a prime $\neq 3$ and $p \nmid a$. Then*

$$S^{(3)}_{a,\, p^\nu} = \begin{cases} p^{\frac{2}{3}\nu} & \text{if } \nu \equiv 0 \pmod 3; \\ 0 & \text{if } \nu \equiv 1 \pmod 3,\ p \equiv 2 \pmod 3; \\ S^{(3)}_{a,\, p}\, p^{\frac{2}{3}(\nu-1)} & \text{if } \nu \equiv 1 \pmod 3,\ p \equiv 1 \pmod 3; \\ p^{\frac{1}{3}(2\nu-1)} & \text{if } \nu \equiv 2 \pmod 3. \end{cases}$$

Also, if $p \equiv 1 \pmod 3$,

$$S^{(3)}_{a,\, p} = \chi(a)\, \tau(\bar{\chi}) + \bar{\chi}(a)\, \tau(\chi),$$

where χ, $\bar{\chi}$ are the non-principal cubic characters (mod p),

$$\tau(\chi) = \sum_{x=1}^{p} \chi(x)\, e_p(x),$$

and

$$|\tau(\chi)| = |\tau(\bar{\chi})| = p^{\frac{1}{2}}.$$

Proof. By Landau, Satz 312, if $\nu > 3$,

$$S^{(3)}_{a,\, p^\nu} = p^2\, S^{(3)}_{a,\, p^{\nu-3}}.$$

Also, by Landau, Satz 307,

$$S^{(3)}_{a,\,p^2}=p, \quad S^{(3)}_{a,\,p^3}=p^2.$$

If $p\equiv 2 \pmod 3$, the congruence $x^3\equiv y \pmod p$ has exactly one solution for given y, hence

$$S^{(3)}_{a,\,p}=\sum_{x=1}^{p} e_p(ax^3)=\sum_{y=1}^{p} e_p(ay)=0.$$

If $p\equiv 1 \pmod 3$, then, by Landau, Satz 310,

$$S^{(3)}_{a,\,p}=\chi(a)\,\tau(\bar{\chi})+\bar{\chi}(a)\,\tau(\chi),$$

where χ, χ are the non-principal cubic characters $\pmod p$.

By Landau, Satz 308,

$$|\tau(\chi)|=|\tau(\bar{\chi})|=p^{\frac{1}{2}}.$$

This establishes all the assertions of Lemma 21.

Lemma 22. *If* $p|n$, $p\neq 2, 3$, *and* $\nu\leqslant 5$, *then*

$$|A(n, p^\nu)|\leqslant 4p^{-1}.$$

Proof. We consider each value of ν separately, and apply Lemmas 20, 21.

(1) If $p\equiv 2 \pmod 3$, then $A(n, p)=0$. If $p\equiv 1 \pmod 3$, then

$$A(n, p)=p^{-3}\sum_{a=1}^{p-1}\left(\frac{Wa}{p}\right) S^{(2)}_{1,\,p}\big(\chi(a)\,\tau(\bar{\chi})+\bar{\chi}(a)\,\tau(\chi)\big)^2$$

$$=0,$$

since none of the characters

$$\left(\frac{a}{p}\right)\chi^2(a),\quad \left(\frac{a}{p}\right)\chi\bar{\chi}(a)=\left(\frac{a}{p}\right),\quad \left(\frac{a}{p}\right)\bar{\chi}^2(a)$$

is a principal character.

(2) We have

$$|A(n, p^2)|=\left|p^{-6}\sum_{\substack{a=1\\ p\nmid a}}^{p^2} e_{p^2}(-na)\,p^3\right|\leqslant p^{-1}.$$

(3) We have

$$A(n, p^3) = p^{-9} \sum_{\substack{a=1\\ p \nmid a}}^{p^3} e_{p^3}(-na) p^4 \left(\frac{Wa}{p}\right) p\, S^{(2)}_{1,p}.$$

This is zero if $p^3 | n$. It is also zero if $n = pn'$, where $p \nmid n'$, for

$$\sum_{\substack{a=1\\ p \nmid a}}^{p^3} e_{p^2}(-n'a)\left(\frac{a}{p}\right) = p \sum_{a_1=1}^{p-1} \sum_{a_2=1}^{p} \left(\frac{a_1}{p}\right) e_{p^2}\big(-n'(a_1+pa_2)\big) = 0.$$

Hence we are left with the case $n = p^2 n''$, where $p \nmid n''$. In this case

$$|A(n, p^3)| = \left| p^{-4} S^{(2)}_{1,p}\, p^2 \sum_{a=1}^{p-1} \left(\frac{Wa}{p}\right) e_p(-n''a) \right|$$

$$= \left| p^{-2} S^{(2)}_{1,p} \left(\frac{-Wn''}{p}\right) S^{(2)}_{1,p} \right| = p^{-1}.$$

(4) If $p \equiv 2 \pmod 3$, $A(n, p^4) = 0$. If $p \equiv 1 \pmod 3$,

$$|A(n, p^4)| = \left| p^{-12} \sum_{\substack{a=1\\ p \nmid a}}^{p^4} e_{p^4}(-na) p^4 p^2 (S^{(3)}_{a,p})^2 \right| \leqslant 4p^{-1}.$$

(5) We have

$$|A(n, p^5)| = \left| p^{-15} \sum_{\substack{a=1\\ p \nmid a}}^{p^5} e_{p^5}(-na) p^6 p^2 S^{(2)}_{1,p} \left(\frac{Wa}{p}\right) \right| < p^{-1}.$$

Lemma 23. *If $p \neq 2, 3$, and $p \nmid n$, then, for $\nu > 1$,*

$$A(n, p^\nu) = 0.$$

Proof. By Lemmas 20, 21, if $a \equiv a' \pmod p$, $p \nmid a$,

$$S^{(3)}_{a,p^\nu} = S^{(3)}_{a',p^\nu}, \quad S^{(2)}_{Wa,p^\nu} = S^{(2)}_{Wa',p^\nu}.$$

Hence

$$A(n, p^\nu) = p^{-3\nu} \sum_{a_1=1}^{p-1} \sum_{a_2=1}^{p^{\nu-1}} e_{p^\nu}\big(-n(a_1+pa_2)\big) (S^{(3)}_{a_1,p^\nu})^2 S^{(2)}_{Wa_1,p^\nu} = 0.$$

Lemma 24. *If $p \neq 2, 3$, and $p \nmid n$, then*

$$A(n, p) = \begin{cases} 0 & \text{if } p \equiv 2 \pmod 3; \\ \left(\frac{Wn}{p}\right)\{2+\chi(4n)+\bar\chi(4n)\}p^{-1} & \text{if } p \equiv 1 \pmod 3; \end{cases}$$

where $\chi, \bar\chi$ are the non-principal cubic characters (mod p).

Proof. The first result follows at once from Lemma 21. For the second part we have†, if $p \nmid a$,

$$(S_{a,p}^{(3)})^2 = \sum_{x=1}^{p} \sum_{y=1}^{p} e_p\left(a(x^3+y^3)\right)$$

$$= \sum_{x=1}^{p} \sum_{y=1}^{p} e_p\left(a\left((x+y)^3+(x-y)^3\right)\right)$$

$$= \sum_{x=1}^{p} \sum_{y=1}^{p} e_p(2ax^3+6axy^2)$$

$$= p + S_{1,p}^{(2)} \sum_{x=1}^{p-1} \left(\frac{6ax}{p}\right) e_p(2ax^3).$$

Using this and Lemma 20, we have

$$A(n, p) = p^{-3}\left(\frac{W}{p}\right) S_{1,p}^{(2)} \sum_{a=1}^{p-1} \left(\frac{a}{p}\right) p\, e_p(-na)$$

$$+p^{-3}\left(\frac{W}{p}\right) (S_{1,p}^{(2)})^2 \sum_{a=1}^{p-1} \left(\frac{a}{p}\right) e_p(-na) \sum_{x=1}^{p-1} \left(\frac{6ax}{p}\right) e_p(2ax^3)$$

$$= p^{-2}(S_{1,p}^{(2)})^2 \left(\frac{-nW}{p}\right) + p^{-3}(S_{1,p}^{(2)})^2 \left(\frac{6W}{p}\right) \sum_{a=1}^{p} \sum_{x=1}^{p-1} \left(\frac{x}{p}\right) e_p\left(a(2x^3-n)\right)$$

$$-p^{-3}(S_{1,p}^{(2)})^2 \left(\frac{6W}{p}\right) \sum_{x=1}^{p-1} \left(\frac{x}{p}\right)$$

$$= p^{-2}(S_{1,p}^{(2)})^2 \left(\frac{-nW}{p}\right) + p^{-2}(S_{1,p}^{(2)})^2 \left(\frac{3W}{p}\right) \sum_{2x^3 \equiv n \pmod p} \left(\frac{2x}{p}\right).$$

By Lemma 20, $$(S_{1,p}^{(2)})^2 = \left(\frac{-1}{p}\right) p.$$

Also, if $2x^3 \equiv n \pmod p$, then $(2x/p) = (n/p)$. Hence

$$A(n, p) = p^{-1}\left(\frac{nW}{p}\right) + p^{-1}\left(\frac{-3W}{p}\right)\left(\frac{n}{p}\right) \sum_{2x^3 \equiv n \pmod p} 1.$$

Since $(-3/p) = 1$ for $p \equiv 1 \pmod 3$, and since the number of solutions of $2x^3 \equiv n \pmod p$ is $1+\chi(4n)+\bar{\chi}(4n)$, we have the result stated.

The transition to the singular product.

For any positive integer q let $q^* = \prod_{p|q} p$, and let $\nu(q)$ denote the number of different prime factors of q.

† A proof can, of course, be obtained by using the last part of Lemma 21, but then we require the relations between the Gaussian sums for the characters of order 6.

Lemma 25. *For any quadratfrei m, and any n,*

$$\sum_{\substack{q \geqslant m \\ q^* = m}} |A(n, q)| \leqslant c_9^{\nu(m)} m^{-1}.$$

In particular, for any q and n,

$$|A(n, q)| \leqslant \frac{c_9^{\nu(q)}}{q^*}.$$

Proof. If $m = 1$, the only value for q is $q = 1$, and the result is obviously true. Now suppose that $m = p_1 \ldots p_s$, where $p_1, \ldots, p_s$ are different primes. The possible values for q are $q = p_1^{l_1} \ldots p_s^{l_s}$. By Lemma 19,

$$\sum_{\substack{q \geqslant m \\ q^* = m}} |A(n, q)| = \prod_{j=1}^{s} \left(\sum_{l=1}^{\infty} |A(n, p_j^{\,l})| \right).$$

If $p = 2$ or 3, then, by Lemma 18,

$$\sum_{l=1}^{\infty} |A(n, p^l)| < \sum_{l=1}^{\infty} c_8 p^{-\frac{1}{8}l} < c_{10} p^{-1}.$$

If $p > 3$, $p | n$, then, by Lemmas 18, 22,

$$\sum_{l=1}^{\infty} |A(n, p^l)| < \sum_{l=1}^{5} 4p^{-1} + \sum_{l=6}^{\infty} c_8 p^{-\frac{1}{8}l} < c_{11} p^{-1}.$$

If $p > 3$, $p \nmid n$, then, by Lemmas 23, 24,

$$\sum_{l=1}^{\infty} |A(n, p^l)| < c_{12} p^{-1}.$$

This establishes the result.

Lemma 26. *If $q_1 \neq q_2$, then, for all n_1, n_2,*

$$\left| \sum_{n=n_1}^{n_2} A(n, q_1) A(n, q_2) \right| \leqslant \frac{q_1 q_2}{q_1^* q_2^*} c_9^{\nu(q_1)} c_9^{\nu(q_2)}.$$

Proof. We have

$$A(n, q_1) A(n, q_2)$$

$$= (q_1 q_2)^{-3} \sum_{\substack{a_1=1 \\ (a_1, q_1)=1}}^{q_1} \sum_{\substack{a_2=1 \\ (a_2, q_2)=1}}^{q_2} (S^{(3)}_{a_1, q_1})^2 S^{(2)}_{Wa_1, q_1} (S^{(3)}_{a_2, q_2})^2 S^{(2)}_{Wa_2, q_2}$$

$$\times e_{q_1 q_2}\big(-n(a_1 q_2 + a_2 q_1)\big).$$

Now $a_1q_2+a_2q_1\not\equiv 0 \pmod{q_1q_2}$, for this would imply $q_1|q_2$ and $q_2|q_1$, contrary to hypothesis. Hence

$$\sum_{n=n_1+1}^{n_1+q_1q_2} A(n, q_1q_2)=0$$

for all n_1. Therefore

$$\left|\sum_{n=n_1}^{n_2} A(n, q_1)A(n, q_2)\right| \leqslant \sum_{n=1}^{q_1q_2} |A(n, q_1)A(n, q_2)|$$

$$\leqslant \frac{q_1q_2}{q_1{}^*q_2{}^*} c_9^{\nu(q_1)} c_9^{\nu(q_2)},$$

by Lemma 25.

Lemma 27. $\displaystyle\sum_{q=1}^{R} \frac{q}{q^*} c_9^{\nu(q)} = O(R^{1+\epsilon})$ *for any* $\epsilon>0$.

Proof. For any $\sigma>0$ we have

$$\sum_{q=1}^{\infty} \frac{1}{q^*} c_9^{\nu(q)} q^{-\sigma} = \prod_p \left\{1+\frac{c_9}{p}\sum_{l=1}^{\infty} p^{-l\sigma}\right\}$$

$$= \prod_p \left\{1+\frac{c_9}{p(p^\sigma-1)}\right\},$$

the product (and therefore the series) being convergent. Hence, for any $\epsilon>0$,

$$\sum_{q=1}^{R} \frac{1}{q^*} c_9^{\nu(q)} q^{-\epsilon} = O(1),$$

and so

$$\sum_{q=1}^{R} \frac{q}{q^*} c_9^{\nu(q)} = O(R^{1+\epsilon}).$$

Lemma 28. *For any* n,

$$\sum_{q>X} \left(A(n, q)\right)^2 < c_{13}X^{-\frac{1}{4}}.$$

Proof. We have

$$\sum_{q>X}\left(A(n, q)\right)^2 = \sum_{\substack{q>X\\ q^*<X}} \left(A(n, q)\right)^2 + \sum_{\substack{q>X\\ q^*\geqslant X}} \left(A(n, q)\right)^2.$$

By Lemmas 18, 25,

$$\sum_{\substack{q>X\\ q^*<X}} \left(A(n, q)\right)^2 < c_8X^{-\frac{1}{2}} \sum_{\substack{q>X\\ q^*<X}} |A(n, q)| \leqslant c_8X^{-\frac{1}{2}} \sum_{m<X} c_9^{\nu(m)} m^{-1}.$$

Now
$$\sum_{m<X} c_9^{\nu(m)} m^{-1} = O(X^{\epsilon})$$

for any $\epsilon > 0$. Hence

$$\sum_{\substack{q>X \\ q^*<X}} \left(A(n, q)\right)^2 < c_{14} X^{-\frac{1}{2}}.$$

By Lemma 25,

$$\sum_{\substack{q>X \\ q^*\geqslant X}} \left(A(n, q)\right)^2 \leqslant \sum_{m\geqslant X} \left(\sum_{\substack{q\geqslant m \\ q^*=m}} |A(n, q)| \right)^2$$
$$\leqslant \sum_{m\geqslant X} (c_9^{\nu(m)} m^{-1})^2$$
$$= O(X^{-1+\epsilon})$$

for any $\epsilon > 0$. This proves Lemma 28.

LEMMA 29.

$$\sum_{n=1}^{P^3} \left(\mathfrak{S}(R, n) - \mathfrak{S}'(R, n)\right)^2 = O\left(P^3 (\log P)^{-\frac{1}{2}}\right).$$

Proof. For sufficiently large P we have [denoting by $\pi(X)$ the number of primes not exceeding X]

$$X^{\pi(X)} < X^{2X/\log X} = e^{2X} = R.$$

Hence we can write

$$\mathfrak{S}(R, n) - \mathfrak{S}'(R, n) = \sum_{q=1}^{R} \theta(q) A(n, q),$$

where $\theta(q)$ is 0 or 1, and $\theta(q) = 0$ for $q \leqslant X$. We have, therefore,

$$\sum_{n=1}^{P^3} \left(\sum_{q=1}^{R} \theta(q) A(n, q) \right)^2$$
$$= \sum_{n=1}^{P^3} \sum_{q=1}^{R} \theta(q) \left(A(n, q)\right)^2 + \sum_{n=1}^{P^3} \sum_{q_1=1}^{R} \sum_{\substack{q_2=1 \\ q_1\neq q_2}}^{R} \theta(q_1)\theta(q_2) A(n, q_1) A(n, q_2)$$
$$= \Sigma_1 + \Sigma_2,$$

say. By Lemma 28,

$$\Sigma_1 \leqslant \sum_{n=1}^{P^3} \sum_{q>X} \left(A(n, q)\right)^2 < c_{13} P^3 X^{-\frac{1}{2}} = O\left(P^3 (\log P)^{-\frac{1}{2}}\right).$$

By Lemmas 26, 27,

$$\Sigma_2 = O\left(\sum_{q_1=1}^{R}\sum_{q_2=1}^{R}\frac{q_1 q_2}{q_1{}^* q_2{}^*}c_9^{\nu(q_1)}c_9^{\nu(q_2)}\right)$$

$$= O\left(\sum_{q=1}^{R}\frac{q}{q^*}c_9^{\nu(q)}\right)^2$$

$$= O(R^{2(1+\epsilon)}) = O\left(P^3(\log P)^{-\frac{1}{7}}\right).$$

The singular product.

Lemma 30. *There exist* c_{15}, c_{16}, c_{17} *such that*

$$\prod_{c_{15}\leqslant p\leqslant X}\left(\sum_{\substack{\nu=0\\ p^\nu\leqslant X}}^{\infty} A(n, p^\nu)\right) > c_{16}(\log\log P)^{-c_{17}}.$$

Proof. By Lemmas 18, 22, if $p>3$ and $p|n$,

$$\sum_{\substack{\nu=0\\ p^\nu\leqslant X}}^{\infty} A(n, p^\nu) \geqslant 1-\frac{20}{p}-\sum_{\nu=6}^{\infty} c_8 p^{-\frac{1}{6}\nu} > 1-\frac{c_{18}}{p}.$$

By Lemmas 23, 24, if $p>3$ and $p \nmid n$,

$$\sum_{\substack{\nu=0\\ p^\nu\leqslant X}}^{\infty} A(n, p^\nu) \geqslant 1-\frac{4}{p}.$$

Hence, if $c_{15}>2c_{18}$,

$$\prod_{c_{15}\leqslant p\leqslant X}\left(\sum_{\substack{\nu=0\\ p^\nu\leqslant X}}^{\infty} A(n, p^\nu)\right) > \prod_{2c_{18}\leqslant p\leqslant X}\left(1-\frac{c_{18}}{p}\right)$$

$$> c_{16}(\log X)^{-c_{17}}$$

$$> c_{16}(\log\log P)^{-c_{17}}.$$

This establishes Lemma 30.

We introduce the (temporary) notation

$$S'^{(3)}_{a,\, p^\nu} = \sum_{\substack{x=1\\ p\nmid x}}^{p^\nu} e_{p^\nu}(ax^3),$$

$$B(n, p^\nu) = p^{-3\nu}\sum_{\substack{a=1\\ p\nmid a}}^{p^\nu} S'^{(3)}_{a,\, p^\nu}\, S^{(3)}_{a,\, p^\nu}\, S^{(2)}_{Wa,\, p^\nu}\, e_{p^\nu}(-na).$$

LEMMA 31.

$$\sum_{\nu=1}^{l} A(n, p^\nu) \geqslant -\frac{1}{p} + \sum_{\nu=1}^{l} B(n, p^\nu).$$

Proof. We have

$$1 + \sum_{\nu=1}^{l} A(n, p^\nu) = \sum_{\nu=0}^{l} p^{-3\nu} \sum_{\substack{a=1 \\ p \nmid a}}^{p^\nu} (S^{(3)}_{a, p^\nu})^2 S^{(2)}_{Wa, p^\nu} e_{p^\nu}(-na)$$

$$= p^{-3l} \sum_{\nu=0}^{l} \sum_{\substack{a=1 \\ p \nmid a}}^{p^\nu} (S^{(3)}_{ap^{l-\nu}, p^l})^2 S^{(2)}_{Wap^{l-\nu}, p^l} e_{p^l}(-nap^{l-\nu})$$

$$= p^{-3l} \sum_{a=1}^{p^l} (S^{(3)}_{a, p^l})^2 S^{(2)}_{Wa, p^l} e_{p^l}(-na)$$

$$= p^{-2l} N(n, p^l),$$

where $N(n, p^l)$ is the number of solutions of the congruence

$$x^3 + y^3 + Wz^2 \equiv n \pmod{p^l}.$$

By exactly the same argument we find that

$$1 - \frac{1}{p} + \sum_{\nu=1}^{l} B(n, p^\nu) = p^{-2l} N'(n, p^l),$$

where $N'(n, p^l)$ is the number of solutions of

$$x^3 + y^3 + Wz^2 \equiv n \pmod{p^l}, \quad p \nmid x.$$

The truth of the result enunciated is now obvious.

LEMMA 32. *For* $p \neq 3$,

$$\sum_{\nu=0}^{l} A(n, p^\nu) > c_{19}.$$

Proof. If $p \neq 3$, we have, for $\nu > 1$, $p \nmid a$,

$$S'^{(3)}_{a, p^\nu} = \sum_{\substack{x=1 \\ p \nmid x}}^{p^\nu} e_{p^\nu}(ax^3) = \sum_{y=1}^{p} \sum_{\substack{z=1 \\ p \nmid z}}^{p^{\nu-1}} e_{p^\nu}\left(a(p^{\nu-1}y + z)^3\right)$$

$$= \sum_{y=1}^{p} \sum_{\substack{z=1 \\ p \nmid z}}^{p^{\nu-1}} e_{p^\nu}(az^3)\, e_p(3ayz^2)$$

$$= 0.$$

Hence, by Lemma 31,

$$\sum_{\nu=0} A(n, p^\nu) \geqslant 1 - p^{-1} + B(n, p).$$

For $p \equiv 2 \pmod 3$, $S^{(3)}_{a,\,p} = 0$ by Lemma 21, hence, in this case,

$$B(n, p) = 0,$$

and the result is established.

For $p \equiv 1 \pmod 3$, we observe that

$$S'^{(3)}_{a,\,p} = S^{(3)}_{a,\,p} - 1.$$

Hence

$$|B(n, p)| = \left| A(n, p) - p^{-3} \sum_{a=1}^{p-1} S^{(3)}_{a,\,p} S^{(2)}_{Wa,\,p} e_p(-na) \right|$$

$$\leqslant |A(n, p)| + 2p^{-2}(p-1).$$

By part (1) of the proof of Lemma 22, $A(n, p) = 0$ if $p \mid n$. By Lemma 24, $|A(n, p)| \leqslant 4p^{-1}$ if $p \nmid n$. Hence

$$|B(n, p)| \leqslant 4p^{-1} + 2p^{-1} - 2p^{-2},$$

and

$$\sum_{\nu=0}^{l} A(n, p^{\nu}) \geqslant 1 - p^{-1} - |B(n, p)|$$

$$\geqslant 1 - 7p^{-1} + 2p^{-2}$$

$$\geqslant 1 - \tfrac{7}{7} + \tfrac{2}{49}$$

$$> c_{19}.$$

LEMMA 33. *For* $l \geqslant 2$,

$$\sum_{\nu=0}^{l} A(n, 3^{\nu}) \geqslant \tfrac{2}{3} + B(n, 9).$$

Proof. The argument at the beginning of the proof of the previous Lemma shows that if $3 \nmid a$, $S'^{(3)}_{a,\,3^{\nu}} = 0$ for $\nu > 2$. Also, if $3 \nmid a$,

$$S^{(3)}_{a,\,3} = 1 + e_3(a) + e_3(8a) = 0.$$

Hence $B(n, 3) = 0$. The result now follows from Lemma 31.

LEMMA 34. *If* $W = 1$, *then, for all* n,

$$\tfrac{2}{3} + B(n, 9) \geqslant \tfrac{4}{9}.$$

Proof. Let $\zeta = e_9(1)$. Then if $3 \nmid a$,

$$S'^{(3)}_{a,\,9} = 3(\zeta^a + \zeta^{-a}),$$

$$S^{(3)}_{a,\,9} = 3(1 + \zeta^a + \zeta^{-a}),$$

$$S^{(2)}_{a,\,9} = 3 + 2(\zeta^a + \zeta^{4a} + \zeta^{7a}) = 3. \tag{35}$$

Hence $$B(n, 9) = \tfrac{1}{27} \sum_{\substack{a=1 \\ 3 \nmid a}}^{9} \zeta^{-na}(2+\zeta^{a}+\zeta^{-a}+\zeta^{2a}+\zeta^{-2a}).$$

It is easily seen that

$$(36) \qquad \sum_{\substack{a=1 \\ 3 \nmid a}}^{9} \zeta^{-ma} = \begin{cases} 0 & \text{if } 3 \nmid m, \\ -3 & \text{if } 3 \mid m, \quad 9 \nmid m, \\ 6 & \text{if } 9 \mid m. \end{cases}$$

Now of the numbers $n, n, n-1, n+1, n-2, n+2$, exactly two are divisible by 3. Hence

$$B(n, 9) \geqslant \frac{-6}{27} = -\tfrac{2}{3}+\tfrac{4}{9}.$$

LEMMA 35. *If* $W = 6$ *or* 12, *and* $n \not\equiv -W \pmod 9$, *then*

$$\tfrac{2}{3}+B(n, 9) \geqslant \tfrac{1}{9}.$$

Proof. Suppose that $W = 6$. Then instead of (35) we have

$$S^{(2)}_{Wa, 9} = 3+6\zeta^{6a}.$$

Hence

$$B(n, 9) = \tfrac{1}{27} \sum_{\substack{a=1 \\ 3 \nmid a}}^{9} \zeta^{-na}(2+\zeta^{a}+\zeta^{-a}+\zeta^{2a}+\zeta^{-2a})(1+2\zeta^{6a}).$$

Suppose first that $n \not\equiv 0 \pmod 3$. By the first part of (36),

$$B(n, 9) = \tfrac{1}{27} \sum_{\substack{a=1 \\ 3 \nmid a}}^{9} \zeta^{-na}(\zeta^{a}+3\zeta^{-a}+\zeta^{2a}+3\zeta^{-2a}+2\zeta^{4a}+2\zeta^{-4a}).$$

Since $$\zeta^{a}+\zeta^{-2a}+\zeta^{4a} = \zeta^{-a}+\zeta^{2a}+\zeta^{-4a} = 0$$

for $3 \nmid a$, we have

$$B(n, 9) = \tfrac{1}{27} \sum_{\substack{a=1 \\ 3 \nmid a}}^{9} \zeta^{-na}(2\zeta^{-a}+2\zeta^{-2a}+\zeta^{4a}+\zeta^{-4a}).$$

Of the numbers $n+1, n+1, n+2, n+2, n-4, n+4$, not all can be divisible by 3. Hence

$$B(n, 9) \geqslant -\tfrac{15}{27} = -\tfrac{2}{3}+\tfrac{1}{9}.$$

Suppose next that $n \equiv 0 \pmod 3$. By the first part of (36),

$$B(n, 9) = \tfrac{2}{27} \sum_{\substack{a=1 \\ 3 \nmid a}}^{9} \zeta^{-na}(1+2\zeta^{6a}).$$

Hence
$$B(6,\ 9) = \tfrac{2}{27}(-3+12) > 0,$$
and
$$B(9,\ 9) = \tfrac{2}{27}(6-6) = 0.$$

Thus the lemma is established in the case $W = 6$. For the case $W = 12$, we observe that, if we write $B_W(n, 9)$ to indicate the dependence of $B(n, 9)$ on W, then
$$B_{12}(n,\ 9) = B_6(n',\ 9)$$
if $n \equiv -n' \pmod 9$.

Lemma 36. *If* $l \geqslant 2$, *the inequality*
$$\sum_{\nu=0}^{l} A(n,\ 3^\nu) > c_{20}$$
holds (a) for all n *if* $W = 1$, *(b) for all* n *not congruent to* $-W \pmod 9$ *if* $W = 6$ *or* 12.

Proof. Lemmas 33, 34, 35.

Lemma 37. *The inequality*
$$\mathfrak{S}'(R,\ n) > c_{21}(\log\log P)^{-c_{17}}$$
holds (a) for all n *if* $W = 1$, *(b) for all* n *not congruent to* $-W \pmod 9$ *if* $W = 6$ *or* 12.

Proof. By Lemmas 30, 32, 36,
$$\mathfrak{S}'(R,\ n) > c_{20}\Big(\prod_{p<c_{15}} c_{19}\Big)\, c_{16}(\log\log P)^{-c_{17}}$$
$$= c_{21}(\log\log P)^{-c_{17}}.$$

Proof of Theorems 1, 2.

By Lemma 17,
$$\sum_{n=W+2}^{P^3} \Big(r(n)-\psi(n)\,\mathfrak{S}(R,\ n)\Big)^2 = O(P^{4-\delta+\epsilon}).$$

By Lemma 29 and the inequality (32),
$$\sum_{n=W+2}^{P^3} \Big(\psi(n)\,\mathfrak{S}(R,\ n)-\psi(n)\,\mathfrak{S}'(R,\ n)\Big)^2 = O\Big(P^4(\log P)^{-\frac{1}{2}}\Big).$$

Hence
$$\sum_{n=W+2}^{P^3} \Big(r(n)-\psi(n)\,\mathfrak{S}'(R,\ n)\Big)^2 = O\Big(P^4(\log P)^{-\frac{1}{4}}\Big).$$

Let a dash attached to the sign of summation denote (a) if $W=1$, no restriction on n, (b) if $W=6$ or 12 the restriction $n \not\equiv -W \pmod 9$. Then

$$\sum_{\substack{n=W+2 \\ r(n)=0}}^{P^3}{}' \Big(\psi(n)\,\mathfrak{S}'(R,n)\Big)^2 = O\Big(P^4(\log P)^{-\frac{1}{4}}\Big).$$

By Lemma 37 and the inequality (32) this implies

$$\sum_{\substack{n=W+2 \\ r(n)=0}}^{P^3}{}' n^{\frac{1}{3}} = O\Big(P^4(\log P)^{-\frac{1}{4}}(\log\log P)^{2c_{17}}\Big)$$

$$= O\Big(P^4(\log P)^{-\frac{1}{8}}\Big).$$

Let $Z=[P^3(\log P)^{-\frac{1}{11}}]$. Then

$$\sum_{\substack{n=1 \\ r(n)=0}}^{P^3}{}' 1 \leqslant Z + \sum_{\substack{n=Z \\ r(n)=0}}^{P^3}{}' 1$$

$$\leqslant Z + Z^{-\frac{1}{3}} \sum_{\substack{n=Z \\ r(n)=0}}^{P^3}{}' n^{\frac{1}{3}}$$

$$= O(Z) + O\Big(Z^{-\frac{1}{3}}P^4(\log P)^{-\frac{1}{8}}\Big)$$

$$= O\Big(P^3(\log P)^{-\frac{1}{11}}\Big).$$

This proves Theorems 1 and 2.

Appendix. The infinite singular series.

We give here the connection (referred to in the Introduction) between the infinite singular series and certain L-series in algebraic fields. For simplicity we suppose that $W=1$ and that $4n$ is neither a square nor a cube.

By Lemmas 19, 23, 24 we have

$$\sum_{q=1}^{\infty} A(n,q)\,q^{1-s} = \phi(s)\,M(s), \tag{37}$$

where $M(s)$ is as defined in the Introduction and

$$\phi(s) = \prod_{p\mid 6n}\left\{\sum_{\nu=0}^{\infty} A(n,p^{\nu})\right\}.$$

We denote by n_3 the least positive integer congruent to 0, 1, or $-1 \pmod 9$ for which $27.4n/n_3$ is a cube, and by n_2 the least positive integer congruent to 0 or 1 (mod 4) for which $4n/n_2$ is a square.

We introduce the four algebraic fields

$$K_2 = \mathbf{P}(\sqrt{-3}), \quad K_4 = K_2(\sqrt{n_2}), \quad K_6 = K_2(\sqrt[3]{n_3}), \quad K_{12} = K_2(\sqrt{n_2}, \sqrt[3]{n_3}),$$

and we denote by $\zeta_r(s)$ the Dedekind ζ-function of K_r. Then it follows from the general theory of relatively Abelian fields that

$$\zeta_4(s) = \zeta_2(s)\, L_A(s), \tag{38}$$

$$\zeta_6(s) = \zeta_2(s)\, L_B^2(s), \tag{39}$$

$$\zeta_{12}(s) = \zeta_2(s)\, L_A(s)\, L_B^2(s)\, L_C^2(s), \tag{40}$$

where $L_A(s)$, $L_B(s)$, $L_C(s)$ are Abelian L-series in K_2. As such they are integral functions of s, and do not vanish at $s = 1$.

By (38), (39), (40),

$$\left(\frac{\zeta_{12}(s)\, \zeta_4(s)}{\zeta_6(s)\, \zeta_2(s)}\right)^{\frac{1}{2}} = L_A(s)\, L_C(s). \tag{41}$$

If we express the functions $\zeta_r(s)$ in rational terms by using the laws of decomposition of primes into prime ideals in metabelian fields, we find that

$$(42) \quad L_A(s)\, L_C(s) = (1-\gamma_2 3^{-s})^{-1} (1-\gamma_2 \gamma_3 3^{-s})^{-1}$$

$$\times \prod_{p \equiv 1 \pmod 3} \left(1-\left(\frac{n_2}{p}\right) p^{-s}\right)^{-2} \left(1-\left(\frac{n_2}{p}\right) \chi_p(n_3) p^{-s}\right)^{-1} \left(1-\left(\frac{n_2}{p}\right) \bar{\chi}_p(n_3) p^{-s}\right)^{-1}$$

$$\times \prod_{\substack{p \equiv 2 \pmod 3 \\ p \nmid n_2}} (1-p^{-2s})^{-2} \prod_{\substack{p \equiv 2 \pmod 3 \\ p \nmid n_2 \\ p \mid n_3}} (1-p^{-2s}),$$

where

$$\gamma_2 = \begin{cases} \left(\dfrac{n_2}{3}\right) & \text{if } 3 \nmid n_2, \\[2ex] \left(\dfrac{-\frac{1}{3}n_2}{3}\right) & \text{if } 3 \mid n_2; \end{cases}$$

$$\gamma_3 = \begin{cases} 1 & \text{if } n_3 \equiv \pm 1 \pmod 9, \\ 0 & \text{if } n_3 \equiv 0 \pmod 9). \end{cases}$$

Hence

$$M(s) = g_1(s)\, g_2(s)\, L_A(s)\, L_C(s), \tag{43}$$

where we collect together in $g_1(s)$ the product over the prime divisors of $3n$. Then $g_2(s)$ is a product which is absolutely convergent for $\sigma \geqslant \frac{1}{2} + \epsilon$ for any $\epsilon > 0$, uniformly in n.

Hence

$$\sum_{q=1}^{\infty} A(n, q) = \phi(1) g_1(1) g_2(1) L_A(1) L_C(1).$$

By using (41) it is possible to express $L_A(1)\ L_C(1)$ in terms of the class-numbers and regulators of the fields K_2, K_4, K_6, K_{12}.

Using the recent work of Siegel† to obtain a lower bound for $L_A(1)$, it is possible to prove that

$$\sum_{q=1}^{\infty} A(n, q) > c(\epsilon) n^{-\epsilon}.$$

But only if we assume some hypothesis of the nature of the Riemann hypothesis about the zeros of $L_A(s)$ and $L_C(s)$ does it seem possible to prove the same about $\sum_{q=1}^{R} A(n, q)$.

Trinity College,
Cambridge.

† *Acta Arithmetica*, 1 (1936), 83–86.

16.

Note on a Result in the Additive Theory of Numbers

(With H. Davenport)

Proceedings of the London Mathematical Society 43 (1937), 142–151
[Commentary on p. 579]

[Received 1 December, 1936.—Read 10 December, 1936.]

1. It has been proved by Romanoff† that the positive integers which are representable as the sum of a prime and a k-th power have positive density. The object of this note is to give a proof that almost all‡ positive integers are representable as $p+x^k$.

The proof can easily be modified to show that, if k is odd, almost all positive integers are representable in one of the two forms $p+x^k$, $2p+x^k$, with $p \equiv 1 \pmod 4$. It follows from this that almost all positive integers are representable as $x^2+y^2+z^k$.

2. Let P be a large positive integer,

$$R = [\log P]^k, \quad X = [\tfrac{1}{4}\log P], \quad \xi = e^{-\sqrt{\log P}}.$$

Throughout the paper, all small Latin letters except c, e, i, s denote positive integers, and p always denotes a prime. $c_1, c_2, \ldots$ denote positive numbers depending only on k. The constants implied by the symbol O depend only on k and ϵ.

We divide the interval $0 \leqslant \alpha \leqslant 1$ in the usual way into Farey arcs, corresponding to all Farey fractions a/q with

$$q \leqslant P^{k-\frac{1}{2}}, \quad a \leqslant q, \quad (a, q) = 1.$$

We divide these arcs into major arcs $\mathfrak{M}$ $(q \leqslant R)$ and minor arcs $\mathfrak{m}$ $(R < q \leqslant P^{k-\frac{1}{2}})$. In either case an arc has the form

$$\alpha = a/q + \beta, \quad -\vartheta_1 q^{-1} P^{-k+\frac{1}{2}} \leqslant \beta \leqslant \vartheta_2 q^{-1} P^{-k+\frac{1}{2}},$$

where $\frac{1}{2} \leqslant \vartheta_1 \leqslant 1$, $\frac{1}{2} \leqslant \vartheta_2 \leqslant 1$.

---

† *Math. Annalen*, 109 (1934), 668–678.

‡ *I.e.*, the number of non-representable numbers not exceeding η is $o(\eta)$ as $\eta \to \infty$.

It is convenient to divide each major arc into two parts, as follows:

$\mathfrak{M}_1$: $$|\beta| \leqslant P^{-k} R^2,$$

$\mathfrak{M}_2$: $$|\beta| \geqslant P^{-k} R^2.$$

We use the abbreviations

$$e(\eta) = e^{2\pi i \eta}, \quad e_q(\eta) = e(\eta/q).$$

Let $$T(\alpha) = \sum_{x=1}^{P} e(\alpha x^k), \quad U(\alpha) = \sum_{p \leqslant P^k} e(\alpha p),$$

$$T(\alpha)\, U(\alpha) = \sum_n r(n)\, e(n\alpha),$$

so that, for $n \leqslant P^k$, $r(n)$ is the number of representations of n as $p + x^k$. Let

$$S_{a,q} = \sum_{x=1}^{q} e_q(ax^k),$$

$$I^{(k)}(\beta) = \frac{1}{k} \sum_{n=1}^{P^k} n^{-1+(1/k)} e(\beta n), \qquad I(\beta) = \sum_{n=2}^{P^k} \frac{1}{\log n}\, e(\beta n),$$

$$T^*(\alpha, a, q) = q^{-1} S_{a,q}\, I^{(k)}(\alpha - a/q), \quad U^*(\alpha, a, q) = \frac{\mu(q)}{\phi(q)}\, I(\alpha - a/q),$$

$$A(n, q) = \frac{\mu(q)}{q\phi(q)} \sum_{\substack{a=1 \\ (a,q)=1}}^{q} S_{a,q}\, e_q(-na),$$

$$\mathfrak{S}(n) = \sum_{q=1}^{R} A(n, q),$$

$$\mathfrak{S}'(n) = \prod_{p \leqslant X} \Big(1 + A(n, p)\Big),$$

$$\pi(x) = \sum_{p \leqslant x} 1, \quad \pi(x;\, q, l) = \sum_{\substack{p \leqslant x \\ p \equiv l \ (\mathrm{mod}\ q)}} 1,$$

$$\mathrm{li}(x) = \int_2^x \frac{d\eta}{\log \eta}.$$

3. Lemma 1. *If $q \leqslant R$, $(l, q) = 1$, $n \leqslant P^k$, then*

$$\pi(n;\, q, l) = \frac{1}{\phi(q)}\, \mathrm{li}(n) + O(P^k \xi^{c_1}).$$

Proof. Let $\sigma(q)$ be the greatest real zero of any Dirichlet L-function to modulus q. It was proved by Page† that

$$\pi(n;\, q,\, l) = \frac{1}{\phi(q)}\,\mathrm{li}(n) + O(ne^{-c_2\sqrt{\log n}}) + O\left(\frac{n^{\sigma(q)}}{\phi(q)\log n}\right).$$

Hence it suffices to prove that

$$\sigma(q) \leqslant 1 - C(\epsilon)\,q^{-\epsilon} \quad [C(\epsilon) > 0], \tag{1}$$

for then we have, taking $\epsilon = (2k)^{-1}$,

$$n^{\sigma(q)} \leqslant P^{k(1-C(\epsilon)\,q^{-\epsilon})} \leqslant P^k\, P^{-kC(\epsilon)\,R^{-\epsilon}} \leqslant P^k\, e^{-kC(\epsilon)(\log P)} = P^k\, \xi^{c_3}.$$

For L-functions formed with complex characters, (1) was proved by Gronwall‡.

For L-functions formed with real non-principal characters, it is sufficient to consider those arising from primitive characters mod q. If χ is such a character, there exists an integer $\Delta = \pm q$ such that $\chi(n) = (\Delta/n)$ (Kronecker symbol). The fact that the greatest real zero of $L(s, \chi) = L_\Delta(s)$ satisfies (1) is an easy consequence of Siegel's result§

$$L_\Delta(1) = \sum_{n=1}^{\infty} \left(\frac{\Delta}{n}\right) n^{-1} > C_1(\epsilon)\,|\Delta|^{-\epsilon} \quad [C_1(\epsilon) > 0],$$

and the elementary inequality

$$L_\Delta'(s) = O(\log^2 |\Delta|) \quad \text{for} \quad s \geqslant 1 - (\log|\Delta|)^{-1}.$$

Lemma 2. *If $q \leqslant R$, $(a, q) = 1$, $\alpha = a/q + \beta$, then*

$$U(\alpha) - U^*(\alpha, a, q) = O\left(P^k\, \xi^{c_4}(1 + P^k|\beta|)\right).$$

Proof. We have, for $n \leqslant P^k$,

$$S_n = \sum_{p \leqslant n} e_q(ap) = \sum_{\substack{l=1 \\ (l,q)=1}}^{q} \pi(n;\, q,\, l)\, e_q(al) + O\left(d(q)\right)$$

$$= \frac{1}{\phi(q)}\,\mathrm{li}(n) \sum_{\substack{l=1 \\ (l,q)=1}}^{q} e_q(al) + O(q\, P^k\, \xi^{c_1})$$

$$= \frac{\mu(q)}{\phi(q)}\,\mathrm{li}(n) + O(P^k\, \xi^{c_4}),$$

† *Proc. London Math. Soc.* (2), 39 (1935), 116–141, Theorem 1.

‡ *Rend. di Palermo*, 35 (1913), 145–159, end of § 1.

§ *Acta Arithmetica*, 1 (1936), 83–86.

by Lemma 1. Hence

$$U(\alpha) = \sum_{n=2}^{P^k} (S_n - S_{n-1})\, e(\beta n)$$

$$= \sum_{n=2}^{P^k} S_n\Big(e(\beta n) - e\big(\beta(n+1)\big)\Big) + S_{P^k} e\big(\beta(P^k+1)\big)$$

$$= \frac{\mu(q)}{\phi(q)} \Big\{ \sum_{n=2}^{P^k} \mathrm{li}(n)\Big(e(\beta n) - e\big(\beta(n+1)\big)\Big) + \mathrm{li}(P^k)\, e\big(\beta(P^k+1)\big)\Big\}$$

$$+ O\Big(P^k \xi^{c_4}\Big\{1 + \sum_{n=2}^{P^k} |1 - e(\beta)|\Big\}\Big)$$

$$= \frac{\mu(q)}{\phi(q)} \sum_{n=3}^{P^k} \{\mathrm{li}(n) - \mathrm{li}(n-1)\}\, e(\beta n) + O\big(P^k \xi^{c_4}(1 + P^k|\beta|)\big),$$

whence the result, since

$$\mathrm{li}(n) - \mathrm{li}(n-1) = \frac{1}{\log n} + O\Big(\frac{1}{n \log n}\Big).$$

LEMMA 3. *If* $(a, q) = 1$, *and* $\alpha = a/q + \beta$, *then*

$$T(\alpha) - T^*(\alpha, a, q) = O\big(q(1 + P^k|\beta|)\big).$$

Proof. Let

$$S_n = \sum_{x^k \leqslant n} e_q(ax^k).$$

Clearly

$$S_n = \frac{n^{1/k}}{q} S_{a,q} + O(q).$$

The result now follows by the usual partial summation, as in the proof of the previous lemma.

LEMMA 4. *If* $(a, q) = 1$, $\alpha = a/q + \beta$, $|\beta| \leqslant \frac{1}{2}$, *then*

$$U^*(\alpha, a, q) = O\Big(\frac{1}{\phi(q)} \min\Big(\frac{P^k}{\log P}, |\beta|^{-1}\Big)\Big), \tag{2}$$

$$T^*(\alpha, a, q) = O\big(q^{-1/k} \min(P, |\beta|^{-1/k})\big). \tag{3}$$

Proof. Since†

$$S_{a,q} = O(q^{1-(1/k)}),$$

† Landau, *Vorlesungen über Zahlentheorie*, 1, Satz 315.

and

$$I(\beta)=O\left(\frac{P^k}{\log P}\right), \qquad I^{(k)}(\beta)=O(P),$$

it suffices to prove that

$$I(\beta)=O(|\beta|^{-1}), \quad I^{(k)}(\beta)=O(|\beta|^{-1/k}).$$

These inequalities follow at once on dividing the sums into $n \leqslant |\beta|^{-1}$ and $n > |\beta|^{-1}$ and applying Abel's lemma to the second parts.

4. Lemma 5. *On* $\mathfrak{m}$, *if* $q \geqslant P^{\frac{1}{2}}$, $T(\alpha)=O(P^{1-c_5})$.

Proof. Let S_n be as in the proof of Lemma 3. Writing $K=2^{k-1}$ we have†

$$|S_n|^{Kk}=O\left(n^{\epsilon} q^{\epsilon}(n^{K-1}+n^K q^{-k}+n^{K-k} q^k)\right).$$

Hence, for $n \leqslant P^k$,

$$S_n=O(P^{1-c_5}).$$

The result now follows by the usual partial summation (as in the proof of Lemma 2) when we observe that

$$P^k|\beta| \leqslant P^k q^{-1} P^{-k+\frac{1}{2}} \leqslant 1.$$

Lemma 6. *On* $\mathfrak{m}$, *if* $q < P^{\frac{1}{2}}$, *then* $T(\alpha)=O(PR^{-1/k})$.

Proof. By Lemma 3 and (3),

$$T(\alpha)=O(P^{\frac{1}{2}})+O(PR^{-1/k})=O(PR^{-1/k}).$$

Lemma 7. *On* $\mathfrak{M}_2$, $T(\alpha)=O(PR^{-2/k})$.

Proof. By Lemma 3 and (3),

$$T(\alpha)=O\left(R(1+P^{\frac{1}{2}})\right)+O\left(P(R^2)^{-1/k}\right)=O(PR^{-2/k}).$$

Lemma 8. $\left(\sum_{\mathfrak{m}}\int_{\mathfrak{m}}+\sum_{\mathfrak{M}_2}\int_{\mathfrak{M}_2}\right)|T(\alpha)U(\alpha)|^2\,d\alpha=O\left(\frac{P^{k+2}}{\log P}R^{-2/k}\right).$

† Landau, *loc. cit.*, Satz 267.

Proof. By Lemmas 5, 6, 7 we have, on $\mathfrak{m}$ and $\mathfrak{M}_2$,

$$T(\alpha) = O(PR^{-1/k}).$$

Also
$$\int_0^1 |U(\alpha)|^2 d\alpha = \pi(P^k) = O\left(\frac{P^k}{\log P}\right).$$

LEMMA 9. *On* $\mathfrak{M}_1$,

$$T(\alpha)\,U(\alpha) - T^*(\alpha, a, q)\,U^*(\alpha, a, q) = O(P^{k+1}\,\xi^{c_6}).$$

Proof. By Lemmas 2, 3, 4,

$$\begin{aligned} |TU - T^*U^*| &\leqslant |U-U^*|\,|T| + |T-T^*|\,|U^*| \\ &= O(P^k\,\xi^{c_4}\,R^2\,P) + O(qR^2\,P^k) \\ &= O(P^{k+1}\,\xi^{c_6}). \end{aligned}$$

LEMMA 10.

$$\sum_{\mathfrak{M}_1}\int_{\mathfrak{M}_1} |T(\alpha)\,U(\alpha) - T^*(\alpha, a, q)\,U^*(\alpha, a, q)|^2 d\alpha = O(P^{k+2}\,\xi^{c_7}).$$

Proof. By Lemma 9, the expression on the left is

$$\begin{aligned} O\left(\sum_{q=1}^{R}\ \sum_{\substack{a=1\\(a,q)=1}}^{q} P^{-k}\,R^2\,P^{2k+2}\,\xi^{2c_6}\right) &= O(P^{k+2}\,R^4\,\xi^{2c_6}) \\ &= O(P^{k+2}\,\xi^{c_7}). \end{aligned}$$

LEMMA 11.

$$\int_0^1 \left|\sum_{q=1}^{R}\ {\sum_{\substack{a=1\\(a,q)=1}}^{q}}{}' \ T^*(\alpha, a, q)\,U^*(\alpha, a, q)\right|^2 d\alpha = O(P^{k+2}\,R^{-4/k}),$$

where the dash denotes that, if α *is on one of the arcs* $\mathfrak{M}_1$, *the term* a, q *corresponding to this arc is to be omitted from the sum.*

Proof. By Cauchy's inequality,

$$\left|\sum_q {\sum_a}' T^*U^*\right|^2 \leqslant \left(\sum_q {\sum_a}' 1\right)\left(\sum_q {\sum_a}' |T^*U^*|^2\right).$$

Hence the integral in question does not exceed

$$R^2 \sum_{q=1}^{R} \sum_{\substack{a=1 \\ (a,q)=1}}^{q} \left(\int_0^1 - \int_{\mathfrak{M}_1}\right) | T^*(\alpha, a, q)\, U^*(\alpha, a, q)|^2 d\alpha$$

$$= O\left(R^2 \sum_{q=1}^{R} \sum_{\substack{a=1 \\ (a,q)=1}}^{q} \int_{P^{-k}R^2}^{\infty} \frac{1}{\phi^2(q)}\, q^{-2/k} \beta^{-2-(2/k)}\, d\beta \right)$$

$$= O\left(R^2 \sum_{q=1}^{R} \frac{1}{\phi(q)}\, q^{-2/k} P^{k+2} (R^2)^{-1-(2/k)} \right)$$

$$= O(P^{k+2} R^{-4/k}).$$

Lemma 12.

$$\int_0^1 \left| T(\alpha)\, U(\alpha) - \sum_{q=1}^{R} \sum_{\substack{a=1 \\ (a,q)=1}}^{q} T^*(\alpha, a, q)\, U^*(\alpha, a, q) \right|^2 d\alpha = O\left(P^{k+2} (\log P)^{-3} \right).$$

Proof. Lemmas 8, 10, 11.

Lemma 13.

$$\sum_{q=1}^{R} \sum_{\substack{a=1 \\ (a,q)=1}}^{q} T^*(\alpha, a, q)\, U^*(\alpha, a, q) = \sum_n \mathfrak{S}(n)\, \psi(n)\, e(n\alpha),$$

where, for $3 \leqslant n \leqslant P^k$,

$$\frac{c_8 n^{1/k}}{\log n} < \psi(n) < \frac{c_9 n^{1/k}}{\log n}. \tag{4}$$

Proof. The sum on the left is

$$\frac{1}{k} \sum_{q=1}^{R} \sum_{\substack{a=1 \\ (a,q)=1}}^{q} q^{-1} S_{a,q} \frac{\mu(q)}{\phi(q)} \sum_{n_1=1}^{P^k} n_1^{-1+(1/k)} \sum_{n_2=2}^{P^k} \frac{1}{\log n_2}\, e\left((n_1+n_2)(\alpha - a/q) \right)$$

$$= \sum_n \mathfrak{S}(n)\, \psi(n)\, e(n\alpha),$$

where
$$\psi(n) = \frac{1}{k} \sum_{\substack{n_1=1 \\ n_1+n_2=n}}^{P^k} \sum_{n_2=2}^{P^k} n_1^{-1+(1/k)} (\log n_2)^{-1}.$$

For $3 \leqslant n \leqslant P^k$, $\psi(n)$ obviously satisfies the inequalities (4).

7. Lemma 14.

$$-\frac{(k, p-1)-1}{p-1} \leqslant A(n, p) \leqslant \frac{1}{p-1},$$

and $$|A(n, q)| \leqslant \frac{k^{\nu(q)}}{\phi(q)},$$

where $\nu(q)$ *denotes the number of different prime factors of* q.

Proof. We have

$$\begin{aligned} A(n, p) &= \frac{\mu(p)}{\phi(p)} p^{-1} \sum_{a=1}^{p-1} S_{a,p} e_p(-na) \\ &= -\frac{1}{p(p-1)} \sum_{a=1}^{p-1} \sum_{x=1}^{p} e_p\left(a(x^k-n)\right) \\ &= -\frac{1}{p(p-1)}(-p+pN), \end{aligned}$$

where N is the number of solutions of the congruence $x^k \equiv n \pmod{p}$. Since $0 \leqslant N \leqslant (k, p-1)$, the first part of the lemma is established.

It is easily proved that, if $(q_1, q_2) = 1$, then

$$A(n, q_1 q_2) = A(n, q_1) A(n, q_2),$$

and this gives the second part of the lemma, if we note that $A(n, q) = 0$ unless q is quadratfrei.

LEMMA 15.

$$\sum_{n=1}^{P^k} \left(\mathfrak{S}(n) - \mathfrak{S}'(n)\right)^2 = O\left(P^k (\log P)^{-1+\epsilon}\right).$$

Proof. We have

$$\prod_{p \leqslant X} p < X^{\pi(X)} < e^{2X} \leqslant P^{\frac{1}{4}}.$$

Hence

$$\mathfrak{S}(n) - \mathfrak{S}'(n) = \sum_{X < q < P^{\frac{1}{4}}} \theta(q) A(n, q),$$

where $\theta(q)$ is 0, 1, or -1, and is independent of n. Therefore

$$\begin{aligned} \sum_{n=1}^{P^k} \left(\mathfrak{S}(n) - \mathfrak{S}'(n)\right)^2 &= \sum_{n=1}^{P^k} \sum_{X < q < P^{\frac{1}{4}}} \theta(q)^2 A(n, q)^2 \\ &\quad + 2 \sum_{X < q_1 < q_2 < P^{\frac{1}{4}}} \theta(q_1)\theta(q_2) \sum_{n=1}^{P^k} A(n, q_1) A(n, q_2) \\ &= \Sigma_1 + \Sigma_2, \end{aligned}$$

say. By Lemma 14,

$$\begin{aligned} \sum_{X < q < P^{\frac{1}{4}}} \theta(q)^2 A(n, q)^2 &\leqslant \sum_{q=X}^{\infty} \frac{k^{2\nu(q)}}{\phi^2(q)} \\ &= O(X^{-1+\epsilon}). \end{aligned}$$

Hence

$$\Sigma_1 = O\Big(P^k(\log P)^{-1+\epsilon}\Big).$$

To deal with Σ_2 we observe that, if $q_1 \neq q_2$ and $(a_1, q_1) = (a_2, q_2) = 1$, then $a_1 q_2 + a_2 q_1 \not\equiv 0 \pmod{q_1 q_2}$. Hence $A(n, q_1)A(n, q_2)$ is periodic in n with period $q_1 q_2$ and average value zero.

Hence, using Lemma 14, we have

$$\left|\sum_{n=1}^{P^k} A(n, q_1)A(n, q_2)\right| \leqslant \sum_{n=1}^{q_1 q_2} |A(n, q_1)\, A(n, q_2)|$$

$$\leqslant \frac{q_1 q_2 k^{\nu(q_1)} k^{\nu(q_2)}}{\phi(q_1)\phi(q_2)} = O(q_1^{\epsilon} q_2^{\epsilon}).$$

Thus

$$\Sigma_2 = O\Big(\sum_{X<q_1<q_2<P^{\frac{1}{2}}} q_1^{\epsilon} q_2^{\epsilon}\Big) = O(P^{1+\epsilon}),$$

and this completes the proof of Lemma 15.

LEMMA 16. $\mathfrak{S}'(n) > c_{10}(\log\log P)^{-k}$.

Proof. By Lemma 14,

$$\mathfrak{S}'(n) = \prod_{p\leqslant 2k}\Big(1 + A(n, p)\Big) \prod_{2k<p\leqslant X}\Big(1 + A(n, p)\Big)$$

$$\geqslant \prod_{p\leqslant 2k} \frac{1}{p-1} \prod_{2k<p\leqslant X}\Big(1 - \frac{k}{p-1}\Big)$$

$$> c_{11} \prod_{p\leqslant X} (1-p^{-1})^k$$

$$> c_{10}(\log\log P)^{-k}.$$

8. THEOREM. *Almost all positive integers are representable in the form* $p + x^k$.

Proof. By Lemmas 12, 13,

$$\sum_{n=3}^{P^k}\Big(r(n) - \mathfrak{S}(n)\psi(n)\Big)^2 = O\Big(P^{k+2}(\log P)^{-3}\Big).$$

By Lemma 15 and (4),

$$\sum_{n=3}^{P^k}\Big(\mathfrak{S}(n)\psi(n) - \mathfrak{S}'(n)\psi(n)\Big)^2 = O\Big(P^{k+2}(\log P)^{-3+\epsilon}\Big).$$

Hence

$$\sum_{n=3}^{P^k}\Big(r(n)-\mathfrak{S}'(n)\psi(n)\Big)^2 = O\Big(P^{k+2}(\log P)^{-3+\epsilon}\Big).$$

Thus, by Lemma 16 and (4),

$$\sum_{\substack{n=3\\ r(n)=0}}^{P^k}\frac{n^{2/k}}{(\log n)^2}(\log\log P)^{-2k} = O\Big(P^{k+2}(\log P)^{-3+\epsilon}\Big),$$

and so

$$\sum_{\substack{n=3\\ r(n)=0}}^{P^k} n^{2/k} = O\Big(P^{k+2}(\log P)^{-1+2\epsilon}\Big).$$

Let N denote the number of numbers $n \leqslant P^k$ for which $r(n)=0$. Then

$$\frac{1}{1+2/k}N^{1+2/k} \leqslant \sum_{n=1}^{N} n^{2/k} \leqslant \sum_{\substack{n=3\\ r(n)=0}}^{P^k} n^{2/k} = O\Big(P^{k+2}(\log P)^{-1+2\epsilon}\Big),$$

$$N = O\Big(P^k(\log P)^{-\frac{(1-2\epsilon)k}{k+2}}\Big) = o(P^k).$$

Trinity College,
Cambridge.

17.

On Real Characters

Acta Arithmetica 2 (1937), 212–213
[Commentary on p. 598]

Let $\chi(n)$ denote a real non-principal character. We define

$$S_1(x) = \sum_{n \leq x} \chi(n), \quad S_m(x) = \sum_{n \leq x} S_{m-1}(n) \qquad \text{for } m \geq 2.$$

S. Chowla [1] considered the hypothesis that

$$S_m(x) \geq 0 \qquad \text{for } x \geq 1, \tag{1}$$

if $m \geq m_0(\chi)$.

It will be shewn in this paper that this is not the case for all real characters χ.

Put

$$\frac{\zeta(2s)}{\zeta(s)} = \sum_{n=1}^{\infty} \lambda(n) n^{-s} \qquad \text{for } s > 1$$

and

$$f(y) = \sum_{n=1}^{\infty} \lambda(n) e^{-ny} \qquad \text{for } y > 0.$$

Then

$$\Gamma(s) \frac{\zeta(2s)}{\zeta(s)} = \int_0^{\infty} y^{s-1} f(y)\, dy$$

[1]) Acta Arithmetica, Vol. 1, p. 113.

for $s>1$, and the integral tends to 0 as $s\to 1$. Hence, $f(y)$ not being identically 0,

$$f(y)<0$$

for some suitably chosen $y>0$.

We choose an integer $a>0$ such that

$$f(y)+2\sum_{n=a+1}^{\infty} e^{-ny}=f(y)+2\frac{e^{-ay}}{e^{y}-1}<0 \tag{2}$$

and an integer d (positive or negative) such that

$$\left(\frac{d}{p}\right)=-1$$

for all primes $p\leqq a$. Here $\left(\frac{d}{p}\right)$ denotes the Legendre-Kronecker symbol. Then

$$\chi(n)=\left(\frac{d}{n}\right)=\lambda(n) \qquad \text{for } n\leqq a,$$

and by (2)

$$\sum_{n=1}^{\infty}\chi(n)\,e^{-ny}\leqq f(y)+2\sum_{n=a+1}^{\infty} e^{-ny}<0. \tag{3}$$

As

$$\sum_{n=1}^{\infty}\chi(n)\,e^{-ny}=(1-e^{-y})^{m}\sum_{n=1}^{\infty} S_m(n)\,e^{-ny},$$

(3) shews that (1) is not true for any value of m.

(Received 26 September, 1936.)

[Cambridge]

18.

On an Inequality in the Elementary Theory of Numbers

Proceedings of the Cambridge Philosophical Society 33 (1937), 207–209
[Commentary on p. 564]

[Received 11 February.—Read 8 March, 1937.]

Let $a_1, a_2, \ldots, a_n$ be a set of n positive integers. Then it is easily seen that the set of positive integers not divisible by any a_ν has a density, i.e. that if $N_n(z)$ is the number of such integers not exceeding z, then $z^{-1}N_n(z)$ tends to a limit when $z \to \infty$; and that

$$\lim_{z=\infty} z^{-1} N_n(z) = A_n, \tag{1}$$

where $$A_n = 1 - \sum_{\nu_1=1}^{n} \frac{1}{a_{\nu_1}} + \sum_{\nu_1=2}^{n} \sum_{\nu_2=1}^{\nu_1-1} \frac{1}{[a_{\nu_1}, a_{\nu_2}]} - \ldots + (-1)^n \frac{1}{[a_1, \ldots, a_n]},$$

and where $[u_1, \ldots, u_\mu]$ denotes the least positive common multiple of the positive integers $u_1, \ldots, u_\mu$.

The object of this paper is to prove the inequality

$$A_n \geqslant \prod_{\nu=1}^{n} \left(1 - \frac{1}{a_\nu}\right), \tag{2}$$

and to draw several conclusions from it.

(2) is true for $n = 1$. Hence it is sufficient to assume that (2) holds for n and to prove it for $n+1$.

We divide all positive integers into four classes:

I. Integers divisible by a_{n+1}, and by at least one a_ν $(1 \leqslant \nu \leqslant n)$.

II. Integers not divisible by a_{n+1}, but by at least one a_ν $(1 \leqslant \nu \leqslant n)$.

III. Integers divisible by a_{n+1}, but not by any a_ν $(1 \leqslant \nu \leqslant n)$.

IV. Integers not divisible by a_{n+1}, and not by any a_ν $(1 \leqslant \nu \leqslant n)$.

Again, it can be easily seen that each of these classes has a density; these densities we denote by $\delta_1, \ldots, \delta_4$ respectively.

We have the trivial relations

$$\delta_1 + \delta_3 = \frac{1}{a_{n+1}}, \tag{3}$$

$$\delta_2 + \delta_4 = 1 - \frac{1}{a_{n+1}}. \tag{4}$$

By the inductive hypothesis, we have

$$\delta_3 + \delta_4 = A_n \geqslant \prod_{\nu=1}^{n} \left(1 - \frac{1}{a_\nu}\right). \tag{5}$$

Further, if v is an integer belonging to one of the classes I or II, then $a_{n+1}v$ belongs to I. Hence the density of I is at least $1/a_{n+1}$ times the density of the combined classes I and II, and we have

$$\delta_1 \geqslant \frac{1}{a_{n+1}}(\delta_1+\delta_2), \tag{6}$$

that is

$$(a_{n+1}-1)\,\delta_1 \geqslant \delta_2.$$

Applying (3) and (4), we obtain from this

$$(a_{n+1}-1)\left(\frac{1}{a_{n+1}}-\delta_3\right) \geqslant 1-\frac{1}{a_{n+1}}-\delta_4,$$

$$\delta_4 \geqslant (a_{n+1}-1)\,\delta_3,$$

$$a_{n+1}\delta_4 \geqslant (a_{n+1}-1)\,(\delta_3+\delta_4),$$

$$\delta_4 \geqslant \left(1-\frac{1}{a_{n+1}}\right)(\delta_3+\delta_4).$$

Since

$$\delta_4 = A_{n+1},$$

we have from (5)

$$A_{n+1} \geqslant \prod_{\nu=1}^{n+1}\left(1-\frac{1}{a_\nu}\right),$$

and our inequality is established.

Actually we can prove (2) in a more general form. If $\xi_1, \ldots, \xi_n$ are n real numbers with $0 \leqslant \xi_\nu \leqslant 1$, we define

$$A_n(\xi_1,\ldots,\xi_n) = 1-\sum_{\nu_1=1}^{n}\frac{\xi_{\nu_1}}{a_{\nu_1}}+\sum_{\nu_1=2}^{n}\sum_{\nu_2=1}^{\nu_1-1}\frac{\xi_{\nu_1}\xi_{\nu_2}}{[a_{\nu_1},a_{\nu_2}]}-\ldots+(-1)^n\frac{\xi_1\ldots\xi_n}{[a_1,\ldots,a_n]}.$$

Then we have

$$A_n(\xi_1,\ldots,\xi_n) \geqslant \prod_{\nu=1}^{n}\left(1-\frac{\xi_\nu}{a_\nu}\right). \tag{7}$$

For the function

$$A_n(\xi_1,\ldots\xi_n)-\prod_{\nu=1}^{n}\left(1-\frac{\xi_\nu}{a_\nu}\right)$$

is linear in each of the variables $\xi_1, \ldots, \xi_n$; hence it assumes its minimum at one of the vertices of the unit-cube $0 \leqslant \xi_\nu \leqslant 1$, and it is sufficient to show that (7) holds if ξ_ν is restricted to the values 0 and 1. But in this case (7) is identical with (2) applied to the subset a_ν $(1 \leqslant \nu \leqslant n,\ \xi_\nu = 1)$.

We shall give one application of (7).

Let b and m be integers, $b>1$, $m>0$, $(b, m) = 1$, and let $f(m)$ denote the order of b (mod m). Then*

$$\sum_{\substack{m=1\\(m,b)=1}}^{\infty}\frac{\mu(m)}{mf(m)} \geqslant \prod_{p \nmid b}\left(1-\frac{1}{pf(p)}\right)>0, \tag{8}$$

where p runs through all primes not dividing b.

* If f is the smallest positive integer satisfying the congruence $b^f \equiv 1 \pmod{m}$, then f is called the order of b (mod m).

It was proved by Romanoff* that

$$\sum_{\substack{m=1\\(m,b)=1}}^{\infty} \frac{1}{mf(m)}$$

converges; hence the sum and the product in (8) are absolutely convergent and the product is positive.

Let p_ν ($\nu = 1, 2, \ldots$) run through all primes not dividing b taken in increasing order.

Then we put
$$\xi_\nu = \frac{1}{p_\nu}, \quad a_\nu = f(p_\nu).$$

Since
$$f(uv) = [f(u), f(v)]$$

for $(u, v) = 1$, we have for $n > 0$

$$A_n(\xi_1, \ldots, \xi_n) = \sum_{\substack{m=1\\(m,b)=1}}^{\infty}{}' \frac{\mu(m)}{mf(m)},$$

where the dash indicates that the sum is extended only over those values of m which are not divisible by $p_{n+1}, p_{n+2}, \ldots$.

Hence we have by (7)

$$\sum_{\substack{m=1\\(m,b)=1}}^{\infty}{}' \frac{\mu(m)}{mf(m)} \geqslant \prod_{\nu=1}^{n}\left(1-\frac{1}{p_\nu f(p_\nu)}\right) > \prod_{p}\left(1-\frac{1}{pf(p)}\right).$$

This proves (8), since

$$\left|\sum_{\substack{m=1\\(m,b)=1}}^{\infty}{}' \frac{\mu(m)}{mf(m)} - \sum_{\substack{m=1\\(m,b)=1}}^{\infty} \frac{\mu(m)}{mf(m)}\right| \leqslant \sum_{m=p_{n+1}}^{\infty} \frac{1}{mf(m)} = v(1),$$

as $n \to \infty$.

* *Math. Ann.* 109 (1934), 668–78. An upper bound for the sum was obtained by Landau, *Acta Arithm.* 1 (1935), 43–62 and by Erdös and Turan, *Bull. de l'inst. de math. et méc. à l'univ. Koulycheff de Tomsk*, 1 (1935), 144–7.

19.

On Dirichlet Series Which Satisfy a Certain Functional Equation

The Quarterly Journal of Mathematics, Oxford Series, 9 (1938), 194–195
[Commentary on p. 599]

WE consider a Dirichlet series

$$\phi(s) = \sum_{n=1}^{\infty} a_n n^{-s} \tag{1}$$

which satisfies the following four conditions:

(i) the series (1) has a half-plane of convergence, so that

$$f_\mu(\gamma) = \sum_{n=1}^{\infty} a_n e^{-n^\mu \gamma}$$

is absolutely convergent for $\gamma > 0$, $\mu > 0$;

(ii) $\phi(s)$ is regular for all finite values of s apart from a finite number of poles;

(iii) we can find a positive integer k and positive numbers A and λ such that

$$\Phi(s) = A^{-s}\Gamma^k(s/\lambda)\phi(s)$$

satisfies

$$\Phi(s) = \Phi(1-s); \tag{2}$$

(iv)

$$\phi(s) = O(|t|^P) \tag{3}$$

uniformly for $\sigma \geqslant \frac{1}{2}$, $|t| \to \infty$, where the constant P depends on the function only.*

By $\Phi_0(s)$ we denote the rational function which vanishes at infinity and makes $\Phi(s)-\Phi_0(s)$ an integral function; $x_1,\ldots, x_k$ are positive variables; we denote their sum by $S(x)$ and their product by $N(x)$; μ is an abbreviation for λ/k.

Then we have for all values of s

$$\Phi(s) = \Phi_0(s) + \int_{N(x)\geqslant 1} \cdots \int \{N(x)^{s/\lambda} + N(x)^{(1-s)/\lambda}\} f_\mu\{A^\mu S(x)\} \frac{dx_1}{x_1} \cdots \frac{dx_k}{x_k}. \tag{4}$$

The proof of (4) is easy. It follows from (3) that the difference of the two sides of (4) is regular for all s and bounded if σ is bounded. Hence it suffices to show that the difference tends to zero as σ tends to $+\infty$ and $-\infty$. But (2) shows that all the three terms in (4) are

* This condition can be replaced by a weaker one by the use of an argument of Phragmén-Lindelöf.

even functions of $s-\frac{1}{2}$. Hence we need investigate the case $\sigma \to +\infty$ only.

For sufficiently large positive σ the series in (1) is absolutely convergent. Hence we have for $n > 0$

$$A^{-s/k}\Gamma(s/\lambda)n^{-s/k} = \int_0^\infty x^{s/\lambda}e^{-A^\mu n^\mu x}\frac{dx}{x},$$

$$A^{-s}\Gamma^k(s/\lambda)n^{-s} = \int_{N(x)>0}\cdots\int N(x)^{s/\lambda}e^{-A^\mu n^\mu S(x)}\frac{dx_1}{x_1}\cdots\frac{dx_k}{x_k},$$

$$\Phi(s) = \int_{N(x)>0}\cdots\int N(x)^{s/\lambda}f_\mu\{A^\mu S(x)\}\frac{dx_1}{x_1}\cdots\frac{dx_k}{x_k}.$$

But the two integrals

$$\int_{0\leqslant N(x)\leqslant 1}\cdots\int N(x)^{s/\lambda}f_\mu\{A^\mu S(x)\}\frac{dx_1}{x_1}\cdots\frac{dx_k}{x_k}$$

and

$$\int_{N(x)>1}\cdots\int N(x)^{(1-s)/\lambda}f_\mu\{A^\mu S(x)\}\frac{dx_1}{x_1}\cdots\frac{dx_k}{x_k}$$

tend to zero as $\sigma \to \infty$. So does $\Phi_0(s)$, and (4) is proved.

We mention one application of (4).

Let d_1, d_2, d_3 be discriminants of quadratic fields and let $L_{d_i}(s)$ denote the series

$$\sum_{n=1}^\infty (d_i/n)n^{-s}.$$

Suppose further that $d_1 d_2 d_3$ is a perfect square. Then

$$\phi(s) = \zeta(s)\prod_{i=1}^3 L_{d_i}(s)$$

has non-negative coefficients and satisfies our conditions (i)–(iv) with

$$k = \tfrac{1}{2}\Big(5 + \sum_{i=1}^3 \operatorname{sign} d_i\Big),$$
$$\lambda = \tfrac{1}{2}k,$$
$$A = 2^{4-k}\pi^2\prod_{i=1}^3 |d_i|^{-\frac{1}{2}};$$

and (4) is the identity from which Siegel* derived his important result

$$\log L_d(1) = o(\log|d|).$$

* *Acta Arithmetica*, 1, 83–6.

20.

On Euclid's Algorithm in Real Quadratic Fields

Proceedings of the Cambridge Philosophical Society 34 (1938), 521–526
[Commentary on p. 586]

[Received 5 August.—Read 17 October, 1938.]

The object of this paper is to complete the proof of the

THEOREM. *Let* $P(\sqrt{d})$ *be the quadratic field of discriminant* $d > 0$. *Then Euclid's algorithm does not hold in* $P(\sqrt{d})$ *if* d *is sufficiently large.*

This was proved for even discriminants by Berg† in a surprisingly simple way.

The problem was further investigated by Behrbohm and Rédei‡, who proved the theorem in several special cases. In particular, they showed that unique factorization can only hold in $P(\sqrt{d})$ for odd d if d is either a prime or a product of two different primes $\equiv 3 \pmod 4$. Since unique factorization follows from Euclid's algorithm, we can restrict ourselves to these two cases.

For the first of these cases (d prime), the theorem was proved by Erdös and Ko§ in a recent paper. The method employed there is applied in the present paper to both cases, so that no knowledge of the former paper is required here.

Notation. All Latin letters except e denote positive rational integers; p, q and r always denote primes.

The discriminant d of our field $P(\sqrt{d})$ is either of the form

$$d = q_1, \quad q_1 \equiv 1 \pmod 4$$

or of the form

$$d = q_1 q_2, \quad q_1 < q_2, \quad q_1 \equiv q_2 \equiv 3 \pmod 4.$$

The symbol O involves absolute constants only.

Let r_1, r_2, r_3 be the three smallest primes which satisfy

$$\left(\frac{d}{r_1}\right) = \left(\frac{d}{r_2}\right) = \left(\frac{d}{r_3}\right) = -1, \quad r_1 < r_2 < r_3.$$

LEMMA 1. *Let* r_1 *and* r_2 *divide* a_1 *and* a_2 *respectively, each to an odd power, where*

$$a_1 + a_2 = d.$$

Then Euclid's algorithm does not hold in $P(\sqrt{d})$ *if* a_1 *or* a_2 *is a quadratic residue* mod d *in the naive sense*||.

† *Kungl. Fysiogr. Sällskapeti Lund Förhandlingar*, 5 (1935), 5.

‡ *Journal für Math.* 174 (1936), 192–205.

§ *Journal London Math. Soc.* 13 (1938), 3–8.

|| We say that u is a quadratic residue mod v in the naive sense, if the congruence

$$y^2 \equiv u \pmod v$$

has a solution.

Proof (indirect). Let us assume that Euclid's algorithm holds in $P(\sqrt{d})$ and that a_2 is a quadratic residue mod d.

Then we can find, for every integer z, an integer α in the field such that

$$z \equiv \alpha \pmod{\sqrt{d}},$$

$$|N(\alpha)| < |N(\sqrt{d})| = d.$$

The conjugate congruence $\quad z \equiv \alpha' \pmod{\sqrt{d}}$

is also valid; hence we have $\quad z^2 \equiv N(\alpha) \pmod{\sqrt{d}}$.

Since z^2 and $N(\alpha)$ are rational integers, and since d is quadratfrei, it follows that

$$z^2 \equiv N(\alpha) \pmod{d}.$$

We now choose z as a solution of the congruence

$$z^2 \equiv a_2 \pmod{d}.$$

Then we have an integer α satisfying

$$N(\alpha) \equiv a_2 \pmod{d},$$

$$|N(\alpha)| < d.$$

Hence, either $\quad N(\alpha) = a_2$ or $N(\alpha) = a_2 - d = -a_1$.

But neither a_2 nor $-a_1$ can be the norm of a number of $P(\sqrt{d})$, since they are each divisible by a prime to an odd multiplicity for which d is a quadratic non-residue.

This contradiction proves the Lemma.

LEMMA 2. *If $d \nmid m$, then*

$$\sum_{\substack{n=1 \\ n \equiv 1\,(m)}}^{x} \left(\frac{d}{n}\right) = O\{m^{\frac{1}{2}} d^{\frac{1}{2}} \log(md)\}.$$

Proof. It is well known that, for any non-principal character χ mod k,

$$\sum_{n=1}^{x} \chi(n) = O(k^{\frac{1}{2}} \log k).$$

Let ψ be any character mod m. Then $\psi(n)\,(d/n)$ is a character mod (md), and is not principal since $d \nmid m$.

Hence, if ψ runs through all characters mod m,

$$\sum_{\substack{n=1 \\ n \equiv 1\,(\mathrm{mod}\ m)}}^{x} \left(\frac{d}{n}\right) = \frac{1}{\phi(m)} \sum_{\psi} \sum_{n=1}^{x} \psi(n) \left(\frac{d}{n}\right)$$

$$= \frac{1}{\phi(m)} \sum_{\psi} O\{m^{\frac{1}{2}} d^{\frac{1}{2}} \log(md)\}$$

$$= O\{m^{\frac{1}{2}} d^{\frac{1}{2}} \log(md)\}.$$

LEMMA 3. $$r_3 = O(d^{\frac{1}{2}-\epsilon})$$

for a suitably chosen absolute constant $\epsilon > 0$.

Proof†. We choose two absolute constants $\delta > 0$, $\epsilon > 0$, such that

$$\tfrac{1}{2}e^{-\frac{1}{2}+7\delta} + \epsilon = \tfrac{1}{3}, \quad \tfrac{1}{2} - 6\delta > 0.$$

We put

$$m = \prod_{\substack{e=1 \\ r_e < \delta^{-1}}}^{3} r_e \prod_{\substack{q \mid d \\ q < \delta^{-1}}} q$$

(so that m is bounded), and $\quad x = [d^{\frac{1}{2}} \log^2 d]$.

We assume at once that d is large enough to satisfy the following conditions:

$$d > m, \tag{1}$$

$$x^{e^{-\frac{1}{2}+6\delta}} < d^{\frac{1}{2}e^{-\frac{1}{2}+7\delta}}. \tag{2}$$

Further we may assume that r_3 is so large that

$$\sum_{r_3 \leqslant p \leqslant x} \frac{1}{p} \leqslant \log_+\left(\frac{\log x}{\log r_3}\right) + \delta, \tag{3}$$

where $\log_+(\eta)$ denotes max $(\log \eta, 0)$.

(1) shows that Lemma 2 is applicable. So we have

$$\sum_{\substack{n=1 \\ n \equiv 1\,(m)}}^{x} \left(\frac{d}{n}\right) = O\{m^{\frac{1}{2}} d^{\frac{1}{2}} \log(md)\}$$

$$= O(d^{\frac{1}{2}} \log d). \tag{4}$$

On the other hand, $(d/n) = 1$ unless n is divisible by q_1, q_2, r_1, r_2 or a prime $\geqslant r_3$.

Hence

$$\sum_{\substack{n=1 \\ n \equiv 1\,(m)}}^{x} \left(\frac{d}{n}\right) \geqslant \sum_{\substack{n=1 \\ n \equiv 1\,(m)}}^{x} 1 - 2\sum_{p}{}^{*} \sum_{\substack{n=1 \\ n \equiv 1\,(m) \\ n \equiv 0\,(p)}}^{x} 1,$$

where the star indicates that the summation is extended over all primes between r_3 and x, both inclusive, and over those of the primes q_1, q_2, r_1, r_2 which do not divide m and which are therefore greater than or equal to δ^{-1}.

Since

$$\sum_{\substack{n=1 \\ n \equiv 1\,(m) \\ n \equiv 0\,(p)}}^{x} 1 \leqslant \frac{x}{mp} + 1,$$

we have

$$\sum_{\substack{n=1 \\ n \equiv 1\,(m)}}^{x} \left(\frac{d}{n}\right) \geqslant \frac{x}{m} - 2\left\{4\left(\frac{\delta x}{m} + 1\right) + \sum_{r_3 \leqslant p \leqslant x}\left(\frac{x}{mp} + 1\right)\right\}$$

$$\geqslant \frac{x}{m}\left\{1 - 8\delta - 2\sum_{r_3 \leqslant p \leqslant x} \frac{1}{p}\right\} - 8 - \pi(x)$$

$$\geqslant \frac{x}{m}\left\{1 - 8\delta - 2\log_+\left(\frac{\log x}{\log r_3}\right) - 2\delta\right\} - 8 - \pi(x),$$

by (3).

† Vinogradov, *Trans. Amer. Math. Soc.* 29 (1927), 218–26 proved $r_1 = O(d^{\frac{1}{2}e^{-\frac{1}{2}+\epsilon}})$. His proof is easily generalized to obtain Lemma 3. See also Erdös and Ko, *loc. cit.* Lemma 3.

Combining this inequality with (4), we have

$$\frac{x}{m}\left\{1-10\delta-2\log_+\left(\frac{\log x}{\log r_3}\right)\right\} \leqslant O\{\pi(x)+d^{\frac{1}{2}}\log d\},$$

i.e. dividing by x/m, $\quad 1-10\delta-2\log_+\left(\frac{\log x}{\log r_3}\right) \leqslant O\left(\frac{1}{\log d}\right).$

Hence, for sufficiently large d,

$$1-11\delta-2\log_+\left(\frac{\log x}{\log r_3}\right) \leqslant 0,$$

$$\log_+\left(\frac{\log x}{\log r_3}\right) \geqslant \tfrac{1}{2}-6\delta > 0,$$

$$\log\frac{\log x}{\log r_3} \geqslant \tfrac{1}{2}-6\delta,$$

$$\begin{aligned} r_3 &\leqslant x^{e^{-\frac{1}{2}+6\delta}} \\ &\leqslant d^{\frac{1}{2}e^{-\frac{1}{2}+7\delta}} \qquad \text{(by (2))} \\ &= d^{\frac{1}{3}-\epsilon}. \end{aligned}$$

Lemma 4. *If $r_1 \geqslant 5$, $s < r_1$, we can find a prime p_0 not dividing s, which satisfies the inequalities*

$$p_0 < r_1, \tag{5}$$

$$p_0 \leqslant \log d, \textit{ if } d \textit{ is sufficiently large.} \tag{6}$$

Proof. Let p_0 be the least prime not dividing s. If $s \leqslant 2$, $p_0 = s+1 < r_1$; if $s > 2$, no prime-divisor of $s-1$ exceeds s, hence $p_0 \leqslant s-1 < r_1$. This proves (5).

Further, by the prime number theorem, for large d

$$\begin{aligned} \sum_{p \leqslant \log d} \log p &= \log d + o(\log d) \\ &> \log r_3 \quad \text{(by Lemma 3)} \\ &> \log r_1 \\ &> \log s. \end{aligned}$$

This proves (6).

Lemma 5. *If d is sufficiently large, then we can find two numbers s and t with the following properties:*

$$d = sr_2r_3 + tr_1; \tag{7}$$

$$\left(\frac{d}{s}\right) \geqslant 0; \tag{8}$$

either $\quad (s, r_2r_3) = 1 \quad (9)$

or $\quad (s, r_2^3r_3) = r_2^2; \quad (10)$

either $\quad (t, r_1) = 1 \quad (11)$

or $\quad (t, r_1^3) = r_1^2. \quad (12)$

Proof. We note, first, that (8) and (9) are automatically satisfied if every prime factor of s is less than r_1, in particular, if $s < r_1$.

We distinguish five cases.

First case. $r_1 \geqslant 5$. We choose d so large that Lemma 4 becomes applicable, and that

$$r_3^3(1+\log d) < d. \tag{13}$$

This is possible by Lemma 3.

We choose s as the smallest positive root of the congruence

$$r_2 r_3 s \equiv d \pmod{r_1}.$$

Then

$$s < r_1, \quad s r_2 r_3 < r_3^3 < d$$

and we have a representation $\quad d = s r_2 r_3 + t r_1$

which satisfies all conditions of our lemma with the possible exception of (11) and (12).

If $r_1 \mid t$, we denote by n the smallest positive root of the congruence

$$s + n r_1 \equiv 0 \pmod{p_0},$$

where p_0 is the prime in Lemma 4, and we consider the representation

$$d = \frac{s + n r_1}{p_0} p_0 r_2 r_3 + (t - n r_2 r_3) r_1,$$

where (11) is now satisfied since $\quad n < p_0 < r_1$.

Further

$$\frac{s + n r_1}{p_0} < \frac{(n+1) r_1}{p_0} \leqslant r_1,$$

so that (8) and (9) are satisfied and

$$\begin{aligned} \frac{s + n r_1}{p_0} p_0 r_2 r_3 &< r_1 p_0 r_2 r_3 \\ &< r_3^3 \log d \\ &< d \end{aligned}$$

by (13).

Second case. $r_1 = 3$. We have a representation

$$d = s r_2 r_3 + 3t$$

with $s \leqslant 2$, which satisfies (8) and (9). If $3 \mid t$ we replace it by

$$d = 4 s r_2 r_3 + 3(t - s r_2 r_3),$$

so that (11) is satisfied and we have, for large d,

$$t - s r_2 r_3 > 0.$$

Third case. $r_1 = 2$, $r_2 > 3$. We have a representation

$$d = r_2 r_3 + 2t,$$

and also

$$d = 3 r_2 r_3 + 2(t - r_2 r_3)$$

and either t or $t - r_2 r_3$ is odd.

Fourth case. $r_1 = 2, r_2 = 3, r_3 \equiv 1 \pmod 4$. If we write

$$d = 3r_3 + 2t,$$

then $t \equiv 1 \pmod 2$, since $d \equiv 1 \pmod 4$ and $3r_3 \equiv 3 \pmod 4$.

Fifth case. $r_1 = 2, r_2 = 3, r_3 \equiv -1 \pmod 4$. We have a representation

$$d = 3r_3 + 4t,$$

from which we deduce

$$d = 15r_3 + 4(t - 3r_3),$$
$$d = 27r_3 + 4(t - 6r_3),$$
$$d = 135r_3 + 4(t - 33r_3).$$

All these four relations satisfy (8). The first two satisfy (9) and the last two satisfy (10). Finally one of them satisfies (12) since the numbers t, $t - 3r_3$, $t - 6r_3$, $t - 33r_3$ run through a complete residue system mod 4, and therefore one of them must be even but not divisible by 4.

Lemma 6. *If*

$$a_1 + a_2 = d,$$
$$\left(\frac{d}{a_1}\right) \geqslant 0,$$

then either a_1 or a_2 is a quadratic residue mod d *in the naive sense*†.

Proof. If $d = q_1$, it follows that $(a_1, q_1) = 1$,

$$\left(\frac{a_1}{q_1}\right) = 1,$$

and a_1 is a quadratic residue mod d in the ordinary sense.

If $d = q_1 q_2$, we note first that

$$\left(\frac{d}{a_1}\right) = \left(\frac{d}{a_2}\right) \geqslant 0;$$

hence, reducing the Kronecker symbols to Legendre symbols,

$$\left(\frac{a_1}{q_1}\right)\left(\frac{a_1}{q_2}\right) = \left(\frac{a_2}{q_1}\right)\left(\frac{a_2}{q_2}\right) \geqslant 0. \tag{14}$$

Since $q_1 \equiv q_2 \equiv 3 \pmod 4$, we have

$$\left(\frac{a_1}{q_1}\right) = -\left(\frac{a_2}{q_1}\right), \quad \left(\frac{a_1}{q_2}\right) = -\left(\frac{a_2}{q_2}\right).$$

Since $d \nmid a_1$, not all four Legendre symbols are zero. Hence at least one of them must be positive, say

$$\left(\frac{a_i}{q_j}\right) = 1,$$

and therefore by (14)

$$\left(\frac{a_i}{q_k}\right) \geqslant 0 \text{ for } k = 1 \text{ and } 2.$$

Hence a_i is a quadratic residue mod d in the naive sense.

Proof of the main theorem. Lemmas 1, 5 and 6.

† It may, of course, happen that a_1 and a_2 are both residues.

21.

Edmund Landau

(With G. H. Hardy *)

Journal of the London Mathematical Society 13 (1938), 302–310

Edmund Landau, an honorary member of this Society since 1924, was born in Berlin on 14 February 1877, and died there on 19 February 1938. He was the son of Professor Leopold Landau, a well-known gynaecologist. After passing through the "French Gymnasium" in Berlin, he entered the University of Berlin as a student of mathematics, and remained there, apart from two short intervals in Munich and Paris, until 1909. His favourite teacher was Frobenius, who lectured on algebra and the theory of numbers. Landau worked through these lectures very thoroughly, and used his notes of them throughout his life. He took his doctor's degree in 1899†, and obtained the "venia legendi", or right to give lectures, in 1901.

In 1909 Landau succeeded Minkowski as ordinary professor in Göttingen. The University of Göttingen was then in its mathematical prime; Klein and Hilbert were Landau's colleagues, and young mathematicians came to Göttingen for inspiration from every country. After the war the University recovered its position quickly, so that Landau had always ample opportunities of training able pupils, and often had a decisive influence on their careers.

In 1933 the political situation forced him to resign his chair. He retired to Berlin, but still lectured occasionally outside Germany. In 1935 he came to Cambridge as Rouse Ball Lecturer, and gave the lectures which he developed later into a Cambridge Tract‡. He continued to take an active interest in mathematics, and his last lectures were in Brussels in November 1937, only a few months before his sudden death.

Landau's first and most abiding interest was the analytic theory of numbers, and in particular the theory of the distribution of primes and prime ideals. In his "doctor-dissertation" (**1**) he gave a new proof of the identity

$$\sum_{1}^{\infty} \frac{\mu(n)}{n} = 0 \tag{1}$$

(conjectured by Euler and first proved by von Mangoldt), and this was

---

* We have to thank Dr. Alfred Brauer for information concerning the facts of Landau's career.

† His thesis was **1** in the list of papers on p. 310.

‡ G in the list of books on p. 307.

the first of a long series of papers on the zeta-function and the theory of primes.

The "prime number theorem"

$$\pi(x) \sim \frac{x}{\log x} \tag{2}$$

was first proved by Hadamard and de la Vallée-Poussin in 1896. De la Vallée-Poussin went further, and proved that

$$\pi(x) = \int_2^x \frac{dt}{\log t} + O\{xe^{-A\sqrt{(\log x)}}\} = \operatorname{li} x + O\{xe^{-A\sqrt{(\log x)}}\}, \tag{3}$$

for a certain positive A. The proofs of both Hadamard and de la Vallée-Poussin depended upon Hadamard's theory of integral functions, and in particular on the fact that $\zeta(s)$, apart from a simple pole at $s = 1$, is regular all over the complex plane.

In 1903 Landau found (in **2**) a new proof of the prime number theorem which does not depend upon the general theory of Hadamard. For this proof we need know only that $\zeta(s)$ can be continued "a little way over" the line $\sigma = 1$; we do not need the functional equation, the Weierstrass product, and the other machinery used in the earlier proofs. On the other hand we do not obtain quite so precise a formula as (3).

This discovery of Landau's was very important, since it permitted a decisive step in the theory of the prime ideals of an algebraic field κ. This theory depends upon the properties of the Dedekind zeta-function

$$\zeta_\kappa(s) = \Sigma \frac{1}{(N\mathfrak{a})^s}, \tag{4}$$

where $\mathfrak{a}$ runs through the integer ideals of κ (except 0), and $N\mathfrak{a}$ is the norm of $\mathfrak{a}$. It was not proved until much later (by Hecke in 1917) that $\zeta_\kappa(s)$ can be continued all over the plane, but Landau had no difficulty in showing that it has properties like those of $\zeta(s)$ used in his proof of the prime number theorem. He thus obtained the "prime ideal theorem": if $\pi_\kappa(x)$ is the number of prime ideals of κ whose norm is less than x, then

$$\pi_\kappa(x) \sim \frac{x}{\log x}. \tag{5}$$

Later*, using Hecke's discoveries, he proved the formula for $\pi_\kappa(x)$ corresponding to (3).

* C, § 20.

The logic of prime number theory has developed a good deal since 1903 and even since 1917, and Landau kept fully in touch with all these developments. Thus his paper **12** contains the shortest and most direct proof of the prime number theorem known today (a proof based on the ideas of Wiener). He was also intensely interested in the logical relations between different propositions in the theory. Thus he first proved (in **6** and **10**) that (1) and (2) are "equivalent", that each can be deduced from the other by "elementary" reasoning, although there is no "elementary" proof of either*. It was Landau who first enabled experts to classify the theorems of prime number theory according to their "depths".

Landau's second big discovery was in an entirely different direction. Picard's theorem states that an integral function, not a constant, assumes all values with at most one exception. Picard deduced his theorem in 1879 from the properties of the "modular function", and it was not until 1896 that Borel found the first elementary proof.

In 1904 Landau, studying Borel's proof, made a most important, and then very unexpected, extension (**3**). If a_0 and a_1 are given, $a_1 \neq 0$, and

$$f(x) = a_0 + a_1 x + \dots$$

is regular at the origin, then there is a number $\Omega = \Omega(a_0, a_1) > 0$, depending on a_0 and a_1 only, such that $f(x)$, if regular in the circle $|x| < \Omega$, must assume one of the values 0 and 1 somewhere in the circle. It is obvious that this theorem includes Picard's theorem.

A few weeks later Schottky developed Landau's theorem further. Suppose that $a_0 \neq 0$, $a_0 \neq 1$, and $0 < \vartheta < 1$. Then there is a number $\Phi = \Phi(a_0, \vartheta)$ with the following property: if $f(x)$ is regular, and never 0 or 1, for $|x| < 1$, and $f(0) = a_0$, then

$$|f(x)| < \Phi(a_0, \vartheta)$$

for $|x| < \vartheta$. Landau's theorem is a simple corollary†.

Schottky's theorem was imperfect in one important respect, since his function Φ was unbounded near $a_0 = 0$ and $a_0 = 1$. Landau, in **13**,

* The proof that (2) implies (1) is given in A, § 156, but that of the converse implication is later.

† See B (ed. ii), p. 103.

removed this imperfection. Suppose that $a > 0$, $0 < \vartheta < 1$. Then there is a number $\Psi = \Psi(a, \vartheta)$ with the property: if $f(x)$ is regular, and never 0 or 1, for $|x| < 1$, and $|f(0)| \leqslant a$, then

$$|f(x)| < \Psi(a, \vartheta)$$

for $|x| < \vartheta$. In **13** he makes an important application to the theory of $\zeta(s)$ and $\zeta_\kappa(s)$.

These theorems have inspired a great amount of later work. Carathéodory, for example, found the "best" Ω in terms of the modular function. The elementary proofs have also been transformed by the discovery of "Bloch's theorem", and are developed in this way in the second edition of Landau's *Ergebnisse* (B).

This theorem and the prime ideal theorem were probably the most striking of Landau's original discoveries. We state a few more with the minimum of comment.

(1) Every large positive integer is a sum of at most 8 positive integral cubes (**9**).

This 8 is the only number which has resisted the later analytic attacks on Waring's problem.

(2) Every positive definite polynomial with rational coefficients can be represented as a sum of 8 squares of polynomials with rational coefficients. In particular, every positive definite quadratic with rational coefficients is a sum of 5 squares of rational linear functions; and 5 is the best possible number (**5**).

Mordell has since proved the corresponding theorem for quadratics with *integral* coefficients.

(3) If $f(x) \sim x$ when $x \to \infty$, and $xf'(x)$ increases with x, then $f'(x) \to 1$ (**8**, 218).

This theorem (which contains the kernel of a differencing process used by de la Vallée-Poussin and others in the analytic theory of numbers) is, perhaps, the first genuine example of a "O-Tauberian" theorem.

(4) If $f(x) = a_0 + a_1 x + a_2 x^2 + \ldots$ is regular, and $|f(x)| < 1$, for $|x| < 1$, then

$$|a_0 + a_1 + \ldots + a_n| \leqslant 1 + \left(\frac{1}{2}\right)^2 + \left(\frac{1.3}{2.4}\right)^2 + \ldots + \left(\frac{1.3\ldots 2n-1}{2.4\ldots 2n}\right)^2.$$

There is equality, for every n, with an appropriate $f(x)$ depending on n (**11**).

(5) A Dirichlet's series $\Sigma a_n e^{-\lambda_n s}$, with non-negative coefficients, has a singularity at the real point of its line of convergence (**4**).

This had been proved before for power-series by Vivanti and Pringsheim, but their method of proof cannot be extended to the general case.

(6) If $N(\sigma_0, T)$ is the number of zeros of $\zeta(s)$ in the domain $\sigma \geqslant \sigma_0 > \frac{1}{2}$, $|t| \leqslant T$, then

$$N(\sigma_0, T) = o(T).$$

This was proved, first with O and then as stated, by Bohr and Landau in their joint papers **14** and **15**. It is known that $N(\frac{1}{2}, T)$ is of order $T \log T$, so that most of the zeros of $\zeta(s)$ lie very near $\sigma = \frac{1}{2}$. This was the first successful attempt to show that the Riemann hypothesis is at any rate "approximately" true. Carlson proved later that $o(T)$ may be replaced by $O(T^a)$, where $0 < a < 1$, and Titchmarsh and Ingham have since improved Carlson's value of a.

Landau also published a very large number of new, shorter, and simpler proofs of known theorems. We mention only his well-known proof of Weierstrass's approximation Theorem (**7**). This depends upon the singular integral

$$\int_{-\frac{1}{2}}^{\frac{1}{2}} \{1-(u-x)^2\}^n f(u)\, du,$$

and was perhaps the first proof in which the approximating functions are "visibly" polynomials. It is reproduced, in a more general form, in Hobson's book (vol. ii, ed. 2, 459–461).

Landau wrote over 250 papers, but it is possible that he will be remembered first for his books, of which he wrote seven.

A. *Handbuch der Lehre von der Verteilung der Primzahlen* (Leipzig and Berlin, Teubner, 1909: 2 vols., 961 pp.).

B. *Darstellung und Begründung einiger neuerer Ergebnisse der Funktionentheorie* (Berlin, Springer, 1916; second edition, 1929: 122 pp.).

C. *Einführung in die elementare und analytische Theorie der algebraischen Zahlen und Ideale* (Leipzig and Berlin, Teubner, 1918; second edition, 1927: 147 pp.).

D. *Vorlesungen über Zahlentheorie* (Leipzig, Hirzel, 1927: 3 vols., 1009 pp.).

E. *Grundlagen der Analysis* (Leipzig, Akademische Verlagsgesellschaft, 1930: 134 pp.).

F. *Einführung in die Differentialrechnung und Integralrechnung* (Groningen, Noordhoff, 1934: 368 pp.).

G. *Über einige neuere Fortschritte der additiven Zahlentheorie* (Cambridge Tracts in Mathematics, No. 35, 1937: 94 pp.).

Of these books, E and F are elementary, and we say nothing about them, interesting and individual as they are. All the rest are works of first-rate importance and high distinction.

Landau was the complete master of a most individual style, which it is easy to caricature (as some of his pupils sometimes did in an amusing way*), but whose merits are rare indeed. It has two variations, the "old Landau style", best illustrated by the *Handbuch*, which sweeps on majestically without regard to space, and the "new Landau style" of his post-war days, in which there is an incessant striving for compression. Each of these styles is a model of its kind. There are no mistakes—for Landau took endless trouble, and was one of the most accurate thinkers of his day—no ambiguities, and no omissions; the reader has no skeletons to fill, but is given every detail of every proof. He may, indeed, sometimes wish that a little more had been left to his imagination, since half the truth is often easier to picture vividly than the whole of it, and the very completeness of Landau's presentation sometimes makes it difficult to grasp the "main idea". But Landau would not, or could not, think or write vaguely, and a reader has to read as precisely and conscientiously as Landau wrote. If he will do so, and if he will then compare Landau's discussion of a theorem with those of other writers, he will be astonished to find how often Landau has given him the shortest, the simplest, and in the long run the most illuminating proof.

The *Handbuch* was probably the most *important* book he wrote. In it the analytic theory of numbers is presented for the first time, not as a collection of a few beautiful scattered theorems, but as a systematic science. The book transformed the subject, hitherto the hunting ground of a few

* For example in a mock *Festschrift* written on the occasion of his declining an invitation to leave Göttingen for another university.

adventurous heroes, into one of the most fruitful fields of research of the last thirty years. Almost everything in it has been superseded, and that is the greatest tribute to the book.

Landau would not publish a second edition of the *Handbuch* (which must necessarily have been a new book), but preferred to incorporate the results of later researches in his *Vorlesungen*, which is no doubt his *greatest* book. This remarkable work is complete in itself; he does not assume (as he had done in the *Handbuch*) even a little knowledge of number-theory or algebra. It stretches from the very beginning to the limits of knowledge, in 1927, of the "additive", "analytic", and "geometric" theories. Thus part 6 (vol. i, pp. 235–269) carries the solution of Waring's problem to where it stood before Vinogradov's recent work. Part 12 (vol. iii, pp. 201–328) contains practically everything then known about "Fermat's last theorem", and the rest of the book is conceived on the same scale. In spite of this enormous programme, Landau never deviates an inch from his ideal of absolute completeness. For example, he never refers to his *Algebraische Zahlen*, but proves from the beginning everything he needs.

The richness of content of the book, and the power of condensation it shows, are astonishing. Thus the classical theorems about decompositions into two, three, and four squares are proved in twenty-eight pages (vol. i, pp. 97–125). And Landau can find room (vol.i, pp. 153–171) for four different evaluations of Gauss's sums.

The *Vorlesungen* is not only Landau's finest book but also, in spite of the great difficulty and complexity of some of the subject matter, the most agreeably written. The style here is the rather informal style of his lectures, which he was persuaded by his friends to leave unchanged.

The *Algebraische Zahlen* gives a short and self-contained account (pp. 1–54) of the theory of algebraic numbers and ideals, intended as an introduction to the proofs of the prime ideal theorem and its refinements which occupy the remainder of the book. He does not go so deeply into the algebraic theory as, for example, Hecke, being content with what is required for his applications.

The *Ergebnisse* is probably Landau's most *beautiful* book. It contains a collection of elegant, significant, and entertaining theorems of modern function theory: Hadamard's and Fabry's "gap theorems", Fatou's theorem, the most striking "Tauberian" theorems, Bloch's theorem, the Picard-Landau group of theorems, and the fundamental theorems concerning "schlicht" functions. It is one of the most attractive little volumes in recent mathematical literature, and the most effective answer to any one who suggests that Landau's mathematics was dull.

Finally, his last work, the Cambridge Tract originating from his Rouse Ball lecture, gives an account of Vinogradov's "Waring" and Schnirelmann's "Goldbach" theorems, and of a group of half solved "elementary" problems of additive number theory which open a new field of research for young and unprejudiced mathematicians. There is a review of this tract by Mr. Ingham in the current volume of the *Mathematical Gazette*, to which we should have little to add.

Landau was certainly one of the hardest workers of our times. His working day often began at 7 a.m. and continued, with short intervals, until midnight. He loved lecturing, more perhaps even than he realized himself; and a lecture from Landau was a very serious thing, since he expected his students to work in the spirit in which he worked himself, and would never tolerate the tiniest rough end or the slightest compromise with the truth. His enforced retirement must have been a terrible blow to him; it was quite pathetic to see his delight when he found himself again in front of a blackboard in Cambridge, and his sorrow when his opportunity came to an end.

No one was ever more passionately devoted to mathematics than Landau, and there was something rather surprisingly impersonal, in a man of such strong personality, in his devotion. Everybody prefers to do things himself, and Landau was no exception; but most of us are at bottom a little jealous of progress by others, while Landau seemed singularly free from such unworthy emotions. He would insist on his own rights, even a little pedantically, but he would insist in the same spirit and with the same rigour on the rights of others.

This was all part of his passion for order in the world of mathematics. He could not stand untidiness in his chosen territory, blunders, obscurity, or vagueness, unproved assertions or half substantiated claims. If X had proved something, it was up to X to print his proof, and until that happened the something was nothing to Landau. And the man who did his job incompetently, who spoilt Landau's world, received no mercy; that was the unpardonable sin in Landau's eyes, to make a mathematical mess where there had been order before.

Landau received many honours in his lifetime. He was a member of the Academies of Berlin, Göttingen, Halle, Leningrad, and Rome; but no honour seemed to please him quite so much as his election to honorary membership of this society, and he came specially from Germany to attend our sixtieth anniversary dinner. This was natural, since there was no country where his reputation stood quite so high as in England, and none where his work has borne more fruit,

References.

This list contains only papers referred to in the notice.

1. " Neuer Beweis der Gleichung $\Sigma\,\mu(n)/n = 0$ ", *Inaugural-Dissertation* (Berlin, 1899).
2. " Neuer Beweis des Primzahlsatzes und Beweis des Primidealsatzes ", *Math. Annalen*, 56 (1903), 645–670.
3. " Über eine Verallgemeinerung des Picardschen Satzes ", *Berliner Sitzungsberichte* (1904), 1118–1133.
4. " Über einen Satz von Tschebyschef ", *Math. Annalen*, 61 (1905), 527–550.
5. " Über die Darstellung definiter Funktionen durch Quadrate ", *Math. Annalen*, 62 (1906), 272–285.
6. " Über den Zusammenhang einiger neuerer Sätze der analytischen Zahlentheorie ", *Wiener Sitzungsberichte*, 115, 2a (1906), 589–632.
7. " Über die Approximation einer stetigen Funktion durch eine ganze rationale Funktion ". *Rend. di Palermo*, 25 (1908), 337–345.
8. " Beiträge zur analytischen Zahlentheorie ", *Rend. di Palermo*, 26 (1908), 169–302.
9. " Über eine Anwendung der Primzahltheorie auf das Waringsche Problem in der elementaren Zahlentheorie ", *Math. Annalen*, 66 (1909), 102–105.
10. " Über die Äquivalenz zweier Hauptsätze der analytischen Zahlentheorie ", *Wiener Sitzungsberichte*, 120, 2a (1911), 973–988.
11. " Abschätzung der Koeffizientensumme einer Potenzreihe ", *Archiv d. Math. u. Phys*, (3), 24 (1916), 250–260.
12. " Über den Wienerschen neuen Weg zum Primzahlsatz. ", *Berliner Sitzungsberichte* (1932), 514–521.

With Harald Bohr.

13. " Über das Verhalten von $\zeta(s)$ und $\zeta_\kappa(s)$ in der Nähe der Geraden $\sigma = 1$ ", *Göttinger Nachrichten* (1910), 303–330.
14. " Ein Satz über Dirichletsche Reihen mit Anwendungen auf die ζ-Funktion und die L-Funktionen ", *Rend. di Palermo*, 37 (1914), 269–272.
15. " Sur les zéros de la fonction $\zeta(s)$ de Riemann ", *Comptes rendus*, 158 (1914), 106–110.

22.

On Indefinite Quadratic Forms in Five Variables

(With H. Davenport) *

Journal of the London Mathematical Society 21 (1946), 185–193
[Commentary on p. 580]

1. A well-known theorem, due to Meyer†, tells us that any indefinite quadratic form in five variables, with rational coefficients and non-zero determinant, represents zero non-trivially. It has been conjectured‡ that such a form with *real* coefficients necessarily assumes values that are arbitrarily small numerically. We shall prove that this is the case when the quadratic form is of the type

$$Q(x_1, \ldots, x_5) = \lambda_1 x_1^2 + \ldots + \lambda_5 x_5^2.$$

The corresponding result with 9 in place of 5 was proved by Chowla§, who deduced it from a theorem of Jarník and Walfisz|| on lattice-points in a five-dimensional ellipsoid.

* Received 18 March, 1946; read 21 March, 1946.

† *Vierteljahrsschrift der naturforschenden Ges. in Zürich*, 29 (1884), 209–222. A proof is given in Dickson, *Modern elementary theory of numbers*, Ch. 11.

‡ See, for example, Oppenheim, *Annals of Math.* (2), 32 (1931), 272.

§ *Journal London Math. Soc.*, 9 (1934), 162–163.

|| *Math. Zeitschrift*, 32 (1930), 152–160.

We use a modification of the Hardy-Littlewood method, and prove the following:—

THEOREM*. *Let $\lambda_1, \ldots, \lambda_5$ be real numbers, not all of the same sign, and none of them zero, such that one at least of the ratios λ_r/λ_s is irrational. Then there exist arbitrarily large integers P such that the inequalities*

$$1 \leqslant x_1 \leqslant P, \quad \ldots, \quad 1 \leqslant x_5 \leqslant P, \qquad |Q(x_1, \ldots, x_5)| < 1$$

have more than γP^3 solutions, where $\gamma = \gamma(\lambda_1, \ldots, \lambda_5) > 0$.

To deduce the corresponding result for the inequality $|Q(x_1, \ldots, x_5)| < \delta$ for any $\delta > 0$, we need only apply the Theorem to the quadratic form Q/δ.

That $\lambda_1 x_1^2 + \lambda_2 x_2^2$ need not assume values which are arbitrarily small numerically is easily proved by taking $-\lambda_1/\lambda_2$ to be a positive irrational number whose square root has a continued fraction development with bounded partial quotients. Whether a similar form in three or four variables necessarily assumes arbitrarily small values remains an open question.

2. Throughout this paper, small Latin letters other than e and i denote integers, and q, q_0, q_1, q_2 are restricted to positive integers. Small Greek letters denote real numbers. ϵ denotes an arbitrarily small positive number, and $\gamma_1, \gamma_2, \ldots$ are positive numbers depending only on $\epsilon, \lambda_1, \ldots, \lambda_5$. The constants implied by the symbol O also depend only on $\epsilon, \lambda_1, \ldots, \lambda_5$.

We assume, without loss of generality, that λ_1/λ_2 is irrational.

For any positive integer P we define

$$S(P) = \sum_{x=1}^{P} e(\alpha x^2), \quad I(P) = \int_0^P e(\alpha \xi^2)\, d\xi,$$

where $e(\zeta) = e^{2\pi i \zeta}$. Later (in § 6), P will be restricted to a particular sequence of values. We suppose throughout that P is large.

3. We begin with some well-known lemmas.

LEMMA 1. *For $R \geqslant 1$ and any α there exist a, q with*

$$(a, q) = 1, \quad q \leqslant R, \quad |\alpha - a/q| \leqslant R^{-1} q^{-1}.$$

For a proof, see Landau, *Vorlesungen über Zahlentheorie*, **1**, Satz 158.

---

* The same method, together with the use of an inequality due to Hua (*Quarterly J. of Math.*, 9 (1938), 199–202), proves the corresponding theorem for $\lambda_1 x_1^k + \ldots + \lambda_s x_s^k$, where $s = 2^k + 1$.

LEMMA 2. *If $r(m)$ denotes the number of representations of m as the sum of two integral squares, then $r(m)=O(m^\epsilon)$.*

For a proof, see Landau, **1**, Satz 262.

LEMMA 3. *If $(a, q)=1$ and*

$$A(m)=A_{a,q}(m)=\sum_{x=1}^{m} e(ax^2/q),$$

then
$$A(m)=mq^{-1}A(q)+O\left((q\log q)^{\frac{1}{2}}\right),$$

and
$$A(q)=O\left((q\log q)^{\frac{1}{2}}\right).$$

Proof. We assume first that $1\leqslant m\leqslant \frac{1}{2}q$. Then

$$|A(m)|^2=\sum_{x=1}^{m}\sum_{y=1}^{m} e\left(a(x^2-y^2)/q\right)$$

$$=m+2\Re\sum_{1\leqslant y<x\leqslant m} e\left(a(x^2-y^2)/q\right).$$

Hence, putting $x=y+z$,

$$|A(m)|^2=m+2\Re\sum_{z=1}^{m-1}\sum_{y=1}^{m-z} e\left(a(2yz+z^2)/q\right)$$

$$\leqslant m+2\sum_{z=1}^{m-1}\left|\sum_{y=1}^{m-z} e(2ayz/q)\right|$$

$$\leqslant m+4\sum_{z=1}^{m-1}|1-e(2az/q)|^{-1}$$

$$\leqslant m+4\sum_{n=1}^{q-1}|1-e(n/q)|^{-1}$$

$$=O(q)+\sum_{1\leqslant n\leqslant \frac{1}{2}q} O(qn^{-1})$$

$$=O(q\log q).$$

Since $e(ax^2/q)$ is an even function of x with period q, we have, for $1\leqslant m\leqslant q$,

$$A(m)=O\left((q\log q)^{\frac{1}{2}}\right).$$

Lastly, if $m>q$,

$$A(m)=\left[\frac{m}{q}\right]A(q)+A\left(m-\left[\frac{m}{q}\right]q\right)$$

$$=\frac{m}{q}A(q)+O\left((q\log q)^{\frac{1}{2}}\right).$$

Lemma 4. *For any* η,

$$\int_{-\infty}^{\infty} e(\eta a)\left(\frac{\sin \pi a}{\pi a}\right)^2 da = \max(0,\ 1-|\eta|).$$

Proof. The integral is

$$2\int_0^{\infty} \frac{\cos 2\pi\eta a \sin^2 \pi a}{\pi^2 a^2}\, da = \frac{2}{\pi}\int_0^{\infty} \frac{\cos 2\eta a \sin^2 a}{a^2}\, da$$

$$= \frac{1}{\pi}\int_0^{\infty} \frac{\sin^2(\eta+1)a+\sin^2(\eta-1)a-2\sin^2 \eta a}{a^2}\, da$$

$$= \tfrac{1}{2}\{|\eta+1|+|\eta-1|-2|\eta|\},$$

whence the result.

4. We now establish some straightforward estimates for $S(a)$ and $S(a)-I(a)$.

Lemma 5. *If* $a = O(P^{-\frac{3}{2}})$, *then*

$$S(a)-I(a) = O(1).$$

Proof. By Euler's summation formula, we have

$$S(a) = \sum_{x=1}^{P} e(ax^2) = -\tfrac{1}{2}+\tfrac{1}{2}e(aP^2)+I(a)+\int_0^P (\xi-[\xi]-\tfrac{1}{2})\frac{d}{d\xi}e(a\xi^2)\,d\xi.$$

Let $\Phi(\xi) = \int_0^{\xi} (\eta-[\eta]-\tfrac{1}{2})\,d\eta$. Then

$$\int_0^P (\xi-[\xi]-\tfrac{1}{2})\frac{d}{d\xi}e(a\xi^2)\,d\xi = \left[\Phi(\xi)\frac{d}{d\xi}e(a\xi^2)\right]_0^P - \int_0^P \Phi(\xi)\frac{d^2}{d\xi^2}e(a\xi^2)\,d\xi.$$

Since $\Phi(\xi) = O(1)$ and, for $0 \leqslant \xi \leqslant P$,

$$\frac{d}{d\xi}e(a\xi^2) = 4\pi i a\xi e(a\xi^2) = O(P|a|) = O(1)$$

and

$$\frac{d^2}{d\xi^2}e(a\xi^2) = 4\pi i a e(a\xi^2)+(4\pi i a\xi)^2 e(a\xi^2) = O(|a|+P^2a^2) = O(P^{-1}),$$

we have
$$S(a) = I(a)+O(1).$$

Lemma 6. *If* $(a,\ q) = 1,\ a = a/q+\beta,\ \beta = O(P^{-1}q^{-1})$, *then*

$$S(a) = O(Pq^{-\frac{1}{2}+\epsilon}+q^{\frac{1}{2}+\epsilon}).$$

Proof. If $A(m)$ has the meaning assigned to it in Lemma 3, we have

$$S(\alpha) = \sum_{x=1}^{P} e(ax^2/q)\, e(\beta x^2) = \sum_{x=1}^{P} \big(A(x) - A(x-1)\big)\, e(\beta x^2)$$

$$= \sum_{x=1}^{P} A(x) \Big\{ e(\beta x^2) - e\big(\beta(x+1)^2\big) \Big\} + A(P)\, e\big(\beta(P+1)^2\big).$$

Similarly,

$$q^{-1} A(q)\, S(\beta) = q^{-1} A(q) \sum_{x=1}^{P} x \Big\{ e(\beta x^2) - e\big(\beta(x+1)^2\big) \Big\} + q^{-1} A(q)\, P e\big(\beta(P+1)^2\big).$$

By Lemma 3, $A(x) - xq^{-1} A(q) = O(q^{\frac{1}{2}+\epsilon})$. Hence, by subtraction,

$$S(\alpha) - q^{-1} A(q)\, S(\beta) = O\Big(q^{\frac{1}{2}+\epsilon} \Big\{ 1 + \sum_{x=1}^{P} \Big| e(\beta x^2) - e\big(\beta(x+1)^2\big) \Big| \Big\}\Big)$$

$$= O(q^{\frac{1}{2}+\epsilon} \{1 + P^2 |\beta|\})$$

$$= O(Pq^{-\frac{1}{2}+\epsilon} + q^{\frac{1}{2}+\epsilon}).$$

Finally, $S(\beta) = O(P)$ trivially, so that $q^{-1} A(q)\, S(\beta) = O(Pq^{-\frac{1}{2}+\epsilon})$.

LEMMA 7. *If* $\alpha = O(P^{-1})$ *and* $\alpha \neq 0$, *then*

$$S(\alpha) = O(|\alpha|^{-\frac{1}{2}-\epsilon}).$$

Proof. Plainly $2|\alpha|^{-1} > 1$. Hence, by Lemma 1, there exist a, q with $(a, q) = 1$ such that

$$q \leqslant 2|\alpha|^{-1}, \quad |\alpha - a/q| \leqslant \tfrac{1}{2}|\alpha| q^{-1}.$$

Clearly $a \neq 0$. Hence

$$\frac{1}{q} \leqslant \frac{|a|}{q} \leqslant |\alpha| + \tfrac{1}{2}|\alpha| q^{-1} \leqslant \tfrac{3}{2}|\alpha|.$$

The conditions of Lemma 6 are satisfied, since

$$\beta = \alpha - a/q = O(|\alpha| q^{-1}) = O(P^{-1} q^{-1}).$$

Hence

$$S(\alpha) = O(Pq^{-\frac{1}{2}+\epsilon} + q^{\frac{1}{2}+\epsilon}) = O(|\alpha|^{-1} |\alpha|^{\frac{1}{2}-\epsilon} + |\alpha|^{-\frac{1}{2}-\epsilon}),$$

as stated.

LEMMA 8. *For* $\mu \geqslant 0$ *and* $1 \leqslant r \leqslant 5$,

$$\int_{\mu}^{\infty} |S(\lambda_r \alpha)|^4 \left(\frac{\sin \pi\alpha}{\pi\alpha}\right)^2 d\alpha = O\left(\frac{P^{2+\epsilon}}{\mu+1}\right).$$

Proof. We have

$$\int_m^{m+1} |S(a)|^4 da = \int_0^1 \left| \sum_{x=1}^{P} \sum_{y=1}^{P} e\left(a(x^2+y^2)\right) \right|^2 da$$

$$\leqslant \sum_{t=1}^{2P^2} r^2(t) = O(P^{2+\epsilon})$$

by Lemma 2. Hence, in particular,

$$(1) \qquad \int_0^1 |S(a)|^4 \left(\frac{\sin \pi a/\lambda_r}{\pi a/\lambda_r}\right)^2 da = O(P^{2+\epsilon}).$$

Further, if $m \geqslant 1$,

$$\int_m^{m+1} |S(a)|^4 \left(\frac{\sin \pi a/\lambda_r}{\pi a/\lambda_r}\right)^2 da \leqslant \frac{\lambda_r^2}{(\pi m)^2} \int_m^{m+1} |S(a)|^4 da,$$

whence

$$(2) \qquad \int_m^{\infty} |S(a)|^4 \left(\frac{\sin \pi a/\lambda_r}{\pi a/\lambda_r}\right)^2 da = O\left(\frac{P^{2+\epsilon}}{m}\right).$$

The result follows from (1) and (2).

5. We now come to the main part of the proof.

LEMMA 9.

$$\sum_{x_1=1}^{P} \dots \sum_{x_5=1}^{P} \max\left(0, 1-|Q(x_1, \dots, x_5)|\right) = \int_{-\infty}^{\infty} S(\lambda_1 a) \dots S(\lambda_5 a) \left(\frac{\sin \pi a}{\pi a}\right)^2 da,$$

$$\int_0^P \dots \int_0^P \max\left(0, 1-|Q(\xi_1, \dots, \xi_5)|\right) d\xi_1 \dots d\xi_5$$

$$= \int_{-\infty}^{\infty} I(\lambda_1 a) \dots I(\lambda_5 a) \left(\frac{\sin \pi a}{\pi a}\right)^2 da.$$

Proof. Lemma 4.

LEMMA 10. $\int_0^P \dots \int_0^P \max\left(0, 1-|Q(\xi_1, \dots, \xi_5)|\right) d\xi_1 \dots d\xi_5 > \gamma_1 P^3.$

Proof. If we write $\xi_r = |\lambda_r|^{-\frac{1}{2}} \eta_r^{\frac{1}{2}}$, the integral becomes

$$2^{-5} |\lambda_1 \dots \lambda_5|^{-\frac{1}{2}} \int_0^{P^2|\lambda_1|} \dots \int_0^{P^2|\lambda_5|} \max(0, 1-|\pm\eta_1 \pm \dots \pm \eta_5|) \frac{d\eta_1 \dots d\eta_5}{(\eta_1 \dots \eta_5)^{\frac{1}{2}}},$$

where the sign prefixed to η_r is that of λ_r. Without loss of generality we can suppose, in the proof of this lemma, that λ_1 is positive and λ_5 negative.

Let

$$\gamma_2 = \tfrac{1}{3}\min(|\lambda_1|, \ldots, |\lambda_5|).$$

We restrict $\eta_1, \ldots, \eta_4$ to the intervals

$$\gamma_2 P^2 \leqslant \eta_1 \leqslant 2\gamma_2 P^2, \quad 0 \leqslant \eta_r \leqslant \tfrac{1}{4}\gamma_2 P^2 \quad (r = 2, 3, 4).$$

For each set of values of $\eta_1, \ldots, \eta_4$ we restrict η_5 to the interval

$$\eta_1 \pm \eta_2 \pm \eta_3 \pm \eta_4 - \tfrac{1}{2} \leqslant \eta_5 \leqslant \eta_1 \pm \eta_2 \pm \eta_3 \pm \eta_4 + \tfrac{1}{2}.$$

In this interval we have

$$\eta_5 \geqslant \gamma_2 P^2 - \tfrac{3}{4}\gamma_2 P^2 - \tfrac{1}{2} > \gamma_3 P^2,$$

$$\eta_5 \leqslant 2\gamma_2 P^2 + \tfrac{3}{4}\gamma_2 P^2 + \tfrac{1}{2} < 3\gamma_2 P^2 \leqslant P^2|\lambda_5|.$$

Thus the integral of the lemma is greater than

$$2^{-5}|\lambda_1 \ldots \lambda_5|^{-\frac{1}{2}} \int_{\gamma_2 P^2}^{2\gamma_2 P^2} \int_0^{\frac{1}{4}\gamma_2 P^2} \int_0^{\frac{1}{4}\gamma_2 P^2} \int_0^{\frac{1}{4}\gamma_2 P^2} \tfrac{1}{2} \frac{d\eta_1 \ldots d\eta_4}{(\eta_1 \ldots \eta_4 P^2 |\lambda_5|)^{\frac{1}{2}}} = \gamma_1 P^3.$$

LEMMA 11.

$$\int_{-P^{-1}}^{P^{-1}} S(\lambda_1 \alpha) \ldots S(\lambda_5 \alpha) \left(\frac{\sin \pi\alpha}{\pi\alpha}\right)^2 d\alpha = \int_{-\infty}^{\infty} I(\lambda_1 \alpha) \ldots I(\lambda_5 \alpha) \left(\frac{\sin \pi\alpha}{\pi\alpha}\right)^2 d\alpha + O(P^{\frac{5}{3}}).$$

Proof. For $|\alpha| \leqslant P^{-\frac{7}{3}}$ we have, by Lemma 5,

$$S(\lambda_r \alpha) - I(\lambda_r \alpha) = O(1),$$

and (trivially) $I(\lambda_r \alpha) = O(P)$. Hence

$$\prod_{r=1}^{5} S(\lambda_r \alpha) - \prod_{r=1}^{5} I(\lambda_r \alpha) = O(P^4),$$

and so

$$\int_{-P^{-\frac{7}{3}}}^{P^{-\frac{7}{3}}} \left| \prod_{r=1}^{5} S(\lambda_r \alpha) - \prod_{r=1}^{5} I(\lambda_r \alpha) \right| \left(\frac{\sin \pi\alpha}{\pi\alpha}\right)^2 d\alpha = O(P^{\frac{5}{3}}).$$

Further, for $P^{-\frac{7}{3}} \leqslant |\alpha| \leqslant P^{-1}$ we have, by Lemma 7,

$$S(\lambda_r \alpha) = O(|\alpha|^{-\frac{3}{7}-\epsilon}).$$

Hence

$$\int_{P^{-\frac{7}{3}} \leqslant |\alpha| \leqslant P^{-1}} \left| \prod_{r=1}^{5} S(\lambda_r \alpha) \right| \left(\frac{\sin \pi\alpha}{\pi\alpha}\right)^2 d\alpha = O\left(\int_{P^{-\frac{7}{3}}}^{\infty} \alpha^{-5(\frac{3}{7}+\epsilon)} d\alpha\right) = O(P^{\frac{8}{3}+\frac{35}{3}\epsilon}).$$

Finally, for any $\alpha \neq 0$,

$$I(\alpha) = \frac{1}{2|\alpha|^{\frac{1}{2}}} \int_0^{P^2|\alpha|} e(\pm\eta)\, \eta^{-\frac{1}{2}}\, d\eta = O(|\alpha|^{-\frac{1}{2}}).$$

Hence

$$\int_{|a|\geqslant P^{-\frac{3}{2}}}\left|\prod_{r=1}^{5} I(\lambda_r a)\right|\left(\frac{\sin \pi a}{\pi a}\right)^2 da = O\left(\int_{P^{-\frac{3}{2}}}^{\infty} a^{-\frac{5}{2}} da\right) = O(P^{\frac{9}{4}}).$$

The Lemma follows from these three estimates.

Lemma 12. *For any positive absolute constant* ρ, *we have*

$$\int_{|a|\geqslant P^{\rho}} S(\lambda_1 a)\dots S(\lambda_5 a)\left(\frac{\sin \pi a}{\pi a}\right)^2 da = O(P^{3-\rho+\epsilon}).$$

Proof. The result follows from Lemma 8, since $S(\lambda_5 a) = O(P)$ and

$$|S(\lambda_1 a)\dots S(\lambda_4 a)| \leqslant \sum_{r=1}^{4} |S(\lambda_r a)|^4.$$

6. We now restrict P to a certain sequence of positive integral values. By Lemma 1, since λ_1/λ_2 is irrational, there exists an infinity of pairs a_0, q_0, for which

$$(a_0, q_0) = 1, \quad \left|\frac{\lambda_1}{\lambda_2} - \frac{a_0}{q_0}\right| \leqslant \frac{1}{q_0^2}.$$

We take $P = q_0^2$ for any such pair, and have, therefore,

$$\frac{\lambda_1}{\lambda_2} - \frac{a_0}{q_0} = O(P^{-1}). \tag{3}$$

Lemma 13. *If* ρ *is an absolute constant with* $0 < \rho < \frac{1}{10}$, *and* $P^{-1} \leqslant |a| \leqslant P^{\rho}$, *then*

$$\min\left(|S(\lambda_1 a)|, \ |S(\lambda_2 a)|\right) = O(P^{1-\rho+\epsilon}).$$

Proof. By Lemma 1 there exist a_1, q_1, a_2, q_2 satisfying

$$\begin{cases} (a_1, q_1) = 1, \quad q_1 \leqslant P, \quad |\lambda_1 a - a_1/q_1| \leqslant P^{-1} q_1^{-1}, \\ (a_2, q_2) = 1, \quad q_2 \leqslant P, \quad |\lambda_2 a - a_2/q_2| \leqslant P^{-1} q_2^{-1}. \end{cases} \tag{4}$$

If $q_1 \geqslant P^{2\rho}$ or $q_2 \geqslant P^{2\rho}$, the desired result follows from Lemma 6. Also if $a_1 = 0$ or $a_2 = 0$ the result follows from Lemma 7, since then $a = O(P^{-1})$ by (4), and $|a|^{-\frac{1}{2}-\epsilon} = O(P^{\frac{1}{2}+\epsilon})$.

We may therefore suppose, in the remainder of the proof of this lemma, that

$$a_1 \neq 0, \quad a_2 \neq 0, \quad q_1 < P^{2\rho}, \quad q_2 < P^{2\rho}. \tag{5}$$

We observe that

$$a_2 = O(q_2|\alpha| + P^{-1}) = O(P^{3\rho}). \tag{6}$$

By (4),

$$\alpha = \lambda_1^{-1} a_1 q_1^{-1}\big(1+O(P^{-1})\big) = \lambda_2^{-1} a_2 q_2^{-1}\big(1+O(P^{-1})\big).$$

Hence $$\frac{a_1 q_2}{a_2 q_1} = \frac{\lambda_1}{\lambda_2}\big(1+O(P^{-1})\big) = \frac{\lambda_1}{\lambda_2} + O(P^{-1}),$$

whence, by (3),

$$\frac{a_1 q_2}{a_2 q_1} - \frac{a_0}{q_0} = O(P^{-1}).$$

It is impossible that the difference on the left should be zero, for this would imply $q_0 | a_2 q_1$, whereas $q_0 = P^{\frac{1}{2}}$ and $a_2 q_1 = O(P^{5\rho})$ by (5) and (6). Hence

$$\frac{1}{|a_2 q_1 q_0|} \leqslant \left|\frac{a_1 q_2}{a_2 q_1} - \frac{a_0}{q_0}\right| = O(P^{-1}).$$

But this also is impossible, since $a_2 q_1 q_0 = O(P^{5\rho+\frac{1}{2}}) = o(P)$. This completes the proof of the lemma.

LEMMA 14.

$$\int_{P^{-1}\leqslant|\alpha|\leqslant P^{\rho}} S(\lambda_1 \alpha) \ldots S(\lambda_5 \alpha)\left(\frac{\sin \pi\alpha}{\pi\alpha}\right)^2 d\alpha = O(P^{3-\rho+\epsilon}).$$

Proof. The absolute value of the integrand does not exceed

$$\min\big(|S(\lambda_1\alpha)|,\ |S(\lambda_2\alpha)|\big)\left(\sum_{r=1}^{5} |S(\lambda_r \alpha)|^4\right)\left(\frac{\sin \pi\alpha}{\pi\alpha}\right)^2,$$

and the result follows from Lemma 13 and Lemma 8 (with $\mu = 0$).

7. *Proof of the Theorem.* By Lemmas 9, 10, 11, 12, 14,

$$\sum_{x_1=1}^{P} \ldots \sum_{x_5=1}^{P} \max\big(0,\ 1-|Q(x_1, \ldots, x_5)|\big) > \gamma_1 P^3 + O(P^{3-\rho+\epsilon}).$$

Hence the number of solutions of the inequalities of the Theorem is greater than γP^3.

University College,
London.

23.

On the Minimum of a Bilinear Form

(With H. Davenport)

The Quarterly Journal of Mathematics, Oxford Series, 18 (1947), 107–121
[Commentary on p. 589]

1. In this paper we study the minimum of a factorizable bilinear form

$$B(x,y,z,t) = (\alpha x+\beta y)(\gamma z+\delta t), \tag{1}$$

where $\alpha, \beta, \gamma, \delta$ are real, and x, y, z, t take all integral values, subject to

$$xt-yz = \pm 1. \tag{2}$$

We suppose that $$\Delta = \alpha\delta-\beta\gamma \neq 0,$$

and we suppose also that α/β and γ/δ are irrational, so that B does not represent zero.

Two bilinear forms will be called equivalent if one can be transformed into the other by one of the two following substitutions:

$$\text{(i)} \quad \begin{cases} x = px'+qy', & y = rx'+sy', \\ z = pz'+qt', & t = rz'+st', \end{cases}$$

$$\text{(ii)} \quad \begin{cases} x = pz'+qt', & y = rz'+st', \\ z = px'+qy', & t = rx'+sy', \end{cases}$$

where p, q, r, s are integers and $ps-qr = \pm 1$. Associated with the bilinear form is the indefinite quadratic form

$$Q(x,y) = (\alpha x+\beta y)(\gamma x+\delta y),$$

and such a quadratic form has associated with it the two equivalent bilinear forms

$$(\alpha x+\beta y)(\gamma z+\delta t), \qquad (\gamma x+\delta y)(\alpha z+\beta t).$$

If two bilinear forms are equivalent, so are the corresponding quadratic forms, and conversely.

The minimum of an indefinite binary quadratic form was studied in detail by Markoff in his classical memoir.* The minimum† $M(Q)$ of $|Q|$ satisfies

$$M(Q) \leqslant \frac{1}{\sqrt{5}}|\Delta|$$

* Markoff, *Math. Annalen*, 15 (1879), 381–407; 17 (1880), 379–400. Accounts of Markoff's work are given in Bachmann, *Arithmetik der quadratischen Formen*, II, Kapitel 4, and in Dickson, *Studies in the Theory of Numbers*, chapter 7.

† We use 'minimum' to mean 'lower bound'.

for all Q, and equality occurs if and only if Q is equivalent to a multiple of

$$x^2+xy-y^2. \tag{3}$$

For all forms other than these,

$$M(Q) \leqslant \frac{1}{\sqrt{8}}|\Delta|,$$

and equality occurs if and only if Q is equivalent to a multiple of

$$x^2-2y^2. \tag{4}$$

For all forms other than these,

$$M(Q) \leqslant \sqrt{\left(\frac{25}{221}\right)}|\Delta|, \tag{5}$$

and so on. The sequence of special forms continues indefinitely, the nth form Q_n having

$$M(Q_n) = \sqrt{\left(\frac{u_n^2}{9u_n^2-4}\right)}|\Delta|,$$

where u_n takes the sequence of values

$$1, 2, 5, 13, 29, 34, \ldots,$$

the 'Markoff numbers'. For all forms not equivalent to a multiple of one of these,

$$M(Q) \leqslant \tfrac{1}{3}|\Delta|.$$

We shall prove results for the possible values of $M(B)$, the minimum of $|B|$, which, though similar to those above as far as the first two minima are concerned, show a fundamental difference when we come to the third minimum. We establish four theorems.

THEOREM 1. *For all forms B,*

$$M(B) \leqslant \frac{3-\sqrt{5}}{2\sqrt{5}}|\Delta|, \tag{6}$$

and equality occurs if and only if B is equivalent to a multiple of

$$B_1 = \left(x+\frac{1+\sqrt{5}}{2}y\right)\left(z+\frac{1-\sqrt{5}}{2}t\right), \tag{7}$$

in which case the minimum is attained.

THEOREM 2. *For all forms other than those specified in Theorem 1,*

$$M(B) \leqslant \frac{2-\sqrt{2}}{4}|\Delta|, \tag{8}$$

and equality occurs if and only if B is equivalent to a multiple of

$$B_2 = (x-\sqrt{2}\,y)(z+\sqrt{2}\,t), \tag{9}$$

in which case the minimum is attained.

THEOREM 3. *For all forms other than those specified in Theorems* 1 *and* 2,

$$M(B) \leqslant \frac{\sqrt{2}-1}{3}|\Delta|, \tag{10}$$

and equality occurs if and only if B is equivalent to a multiple of

$$B_3 = (x-\sqrt{2}\,y)\{z+(3-\sqrt{2})t\}, \tag{11}$$

*in which case the minimum is attained.**

THEOREM 4. *For any $\delta > 0$ there exists a set of forms, no one of which is equivalent to a multiple of another, for which*

$$M(B) > \left(\frac{\sqrt{2}-1}{3}-\delta\right)|\Delta|,$$

and the set has the cardinal number of the continuum.

2. Throughout the paper, small Latin letters denote rational integers.

LEMMA 1. *The particular forms B_1 and B_2 have*

$$M(B_1) = \frac{3-\sqrt{5}}{2}, \qquad M(B_2) = \sqrt{2}-1,$$

and these minima are attained.

Proof. We have

$$\begin{aligned} B_1(x,y,z,t) &= xz-yt+\tfrac{1}{2}(1+\sqrt{5})yz+\tfrac{1}{2}(1-\sqrt{5})xt \\ &= xz-yt+yz\pm\tfrac{1}{2}(1-\sqrt{5}), \end{aligned}$$

by (2). Hence

$$|B_1| \geqslant \min_n |n\pm\tfrac{1}{2}(1-\sqrt{5})| = \frac{3-\sqrt{5}}{2},$$

since $0 < \frac{1}{2}(3-\sqrt{5}) < \frac{1}{2}$. On the other hand

$$B_1(1,-1,0,1) = \tfrac{1}{2}(3-\sqrt{5}).$$

Similarly, $\qquad B_2(x,y,z,t) = xz-2yt\pm\sqrt{2},$

so that $|B_2| \geqslant \sqrt{2}-1$, and equality occurs for $x = y = z = 1$, $t = 0$.

* It may be noted here that, whereas the minima of B_1 and B_2 are attained infinitely often, that of B_3 is not.

LEMMA 2. *For the particular form B_3, we have*

$$M(B_3) = \sqrt{2}-1,$$

and the minimum is attained.

Proof. We have to show that

$$|(x-\sqrt{2}\,y)\{z+(3-\sqrt{2})t\}| \geqslant \sqrt{2}-1 \tag{12}$$

for all x, y, z, t satisfying (2). Equality obviously occurs when $x = y = z = 1$, $t = 0$.

It is clear that (12) holds when $y = 0$, and when $t = 0$; we can therefore suppose, without loss of generality, that

$$y \geqslant 1, \qquad t \geqslant 1.$$

We write (12) in the form

$$\left|\left(\frac{x}{y}-\sqrt{2}\right)\left(\frac{z}{t}+3-\sqrt{2}\right)\right| \geqslant \frac{\sqrt{2}-1}{yt}. \tag{13}$$

If $z/t \geqslant -\frac{3}{2}$, we have

$$\frac{z}{t}+3-\sqrt{2} \geqslant \left|\frac{z}{t}+\sqrt{2}\right|,$$

and (13) follows from the result proved for B_2 in the preceding lemma. A similar argument holds if $x/y \leqslant -\frac{3}{2}$, since then

$$\sqrt{2}-\frac{x}{y} \geqslant \left|3+\frac{x}{y}+\sqrt{2}\right|,$$

and

$$\left|\left(3+\frac{z}{t}-\sqrt{2}\right)\left(3+\frac{x}{y}+\sqrt{2}\right)\right| \geqslant \frac{\sqrt{2}-1}{yt},$$

by considering $B_2(z+3t, t, x+3y, y)$.

We may therefore suppose that

$$\frac{x}{y} > -\frac{3}{2} > \frac{z}{t}, \qquad \frac{x}{y}-\frac{z}{t} = \frac{1}{yt}.$$

Since

$$\frac{x}{y}+\frac{3}{2} \geqslant \frac{1}{2y}, \qquad -\frac{3}{2}-\frac{z}{t} \geqslant \frac{1}{2t}, \tag{14}$$

it follows that

$$\frac{1}{yt} \geqslant \frac{1}{2y}+\frac{1}{2t},$$

i.e. $2 \geqslant y+t$, which implies $y = t = 1$, and therefore, by the equality in (14), $x = -1$, $z = -2$. For these values,

$$|B_3| = |(-1-\sqrt{2})\{-2+(3-\sqrt{2})\}| = 1 > \sqrt{2}-1.$$

LEMMA 3. *If B is not equivalent to a multiple of B_1 or B_2, then the quadratic form Q corresponding to B satisfies*

$$M(Q) \leqslant \sqrt{\left(\frac{25}{221}\right)}|\Delta| = \frac{|\Delta|}{2{\cdot}973...}. \tag{15}$$

Proof. This follows at once from the results of Markoff which were stated in § 1.

LEMMA 4. *Suppose Q assumes a value satisfying*

$$|Q(p,q)| = \frac{|\Delta|}{\mu} \tag{16}$$

for some p, q with $(p,q) = 1$. Then the corresponding B is equivalent to a multiple of

$$(x+\beta y)\{z+(\mu+\beta)t\}, \tag{17}$$

where β satisfies

$$0 \leqslant \beta+\tfrac{1}{2}\mu \leqslant \tfrac{1}{2}. \tag{18}$$

Proof. Plainly Q is equivalent to

$$\pm\frac{\Delta}{\mu}(x+\beta' y)(x+\delta' y),$$

for some β', δ'. Hence B is equivalent to

$$\frac{\Delta}{\mu}(x+\beta' y)(z+\delta' t),$$

and by comparison of determinants,

$$\beta'-\delta' = \pm\mu.$$

Hence B is equivalent to a multiple of

$$(x+\beta' y)\{z+(\mu+\beta')t\}.$$

By the further substitution

$$x = x'+ny', \qquad y = y', \qquad z = z'+nt', \qquad t = t',$$

we can replace β' by $\beta'+n$, and can therefore ensure that

$$|\beta'+\tfrac{1}{2}\mu| \leqslant \tfrac{1}{2}.$$

If $\beta'+\frac{1}{2}\mu < 0$, we apply the substitution $x = z'$, $y = -t'$, $z = x'$, $t = -y'$ and take $\beta = -\mu-\beta'$.

LEMMA 5. *If $\frac{5}{2} < \mu \leqslant 3$, there is a unique positive number $\eta = \eta(\mu)$ such that*

$$\left(2\mu-5-\frac{2\mu}{\eta}\right)\left(2+\frac{\mu}{\eta}\right) = \frac{\mu}{\eta}. \tag{19}$$

Also $\quad \eta(3) = 3(1+\sqrt{2})$, *and* $\quad \eta(\mu) > \eta(3)$ *for* $\quad \mu < 3$.

Proof. The condition (19) is the same as

$$\eta^2(2\mu-5)+\eta(\mu^2-5\mu)-\mu^2 = 0, \tag{20}$$

and this equation has obviously a unique positive root. When $\mu = 3$, it redúces to

$$\eta^2-6\eta-9 = 0,$$

and the positive root is then $3(1+\sqrt{2})$.

If we denote the left-hand side of (20) by $\phi(\eta,\mu)$, we have

$$\begin{aligned}\frac{\partial}{\partial\mu}\phi\{\eta(3),\mu\} &= 2\eta^2(3)+(2\mu-5)\eta(3)-2\mu\\ &\geqslant 2\{\eta^2(3)-\mu\} > 0.\end{aligned}$$

Since $\phi\{\eta(3),3\} = 0$, it follows that

$$\phi\{\eta(3),\mu\} < 0$$

for $\mu < 3$, and so $\eta(\mu) > \eta(3)$ for $\mu < 3$.

Lemma 6. *If* $\mu \geqslant 3$, *there is a unique positive number* $\eta = \eta(\mu)$ *such that*

$$\left(\mu-1+\frac{\mu}{\eta}\right)\left(1-\frac{2\mu}{\eta}\right) = \frac{\mu}{\eta}. \tag{21}$$

Also $\quad \eta(3) = 3(1+\sqrt{2})$, *and* $\eta > \eta(3)$ *for* $\mu > 3$.

Proof. The condition (21) is the same as

$$\eta^2(\mu-1)+\eta(2\mu-2\mu^2)-2\mu^2 = 0, \tag{22}$$

and this equation has obviously a unique positive root. When $\mu = 3$, it reduces to $2\eta^2-12\eta-18 = 0$, whence $\eta(3) = 3(1+\sqrt{2})$.

If we denote the left-hand side of (22) by $\psi(\eta,\mu)$, we have

$$\begin{aligned}\frac{\partial}{\partial\mu}\psi\{\eta(3),\mu\} &= \eta^2(3)+(2-4\mu)\eta(3)-4\mu\\ &< \eta(3)\{\eta(3)+2-4\mu\} < 0,\end{aligned}$$

since $\eta(3) < 8 < 4\mu-2$. It follows that

$$\psi\{\eta(3),\mu\} < 0$$

for $\mu > 3$, and so $\eta > \eta(3)$.

Lemma 7. *If* $2{\cdot}96 \leqslant \mu \leqslant 3{\cdot}14$, *and* η *has the significance given to it in Lemmas* 5 *and* 6, *then*

$$(1+\tfrac{1}{2}\mu)|3-\mu| < \frac{\mu}{\eta}, \tag{23}$$

$$|\tfrac{1}{2}\mu-\tfrac{3}{2}| < \frac{\mu}{\eta}. \tag{24}$$

Proof. Since $1+\frac{1}{2}\mu > \frac{1}{2}$, (23) implies

$$\tfrac{1}{2}|\mu-3| < \frac{\mu}{\eta},$$

which is (24).

To prove (23), we first show that $\eta < 8$. With the notation already adopted,

$$\begin{aligned}\phi(8,\mu) &= 8^2(2\mu-5)+8(\mu^2-5\mu)-\mu^2\\ &= 7\mu^2+88\mu-320\\ &= 7(3-\mu)^2-130(3-\mu)+7 > 0\end{aligned}$$

for $2{\cdot}96 \leqslant \mu \leqslant 3$. Similarly

$$\begin{aligned}\psi(8,\mu) &= 8^2(\mu-1)+8(2\mu-2\mu^2)-2\mu^2\\ &= -18\mu^2+80\mu-64\\ &= -18(\mu-3)^2-28(\mu-3)+14 > 0\end{aligned}$$

for $3 \leqslant \mu \leqslant 3{\cdot}14$. This proves $\eta < 8$.

Finally,

$$(1+\tfrac{1}{2}\mu)|3-\mu| \leqslant \left(\frac{1}{\mu}+\frac{1}{2}\right)\mu(\cdot 14) \leqslant \frac{2{\cdot}48}{2{\cdot}96}(\cdot 14)\mu < \frac{\mu}{8} < \frac{\mu}{\eta}.$$

Lemma 8. *If* $|Q(p,q)| = |\Delta|/\mu$ *for some* p, q *with* $(p,q) = 1$, *then*

$$M(B) \leqslant \frac{|\Delta|}{\eta} \quad \textit{if} \quad 2{\cdot}96 \leqslant \mu \leqslant 3{\cdot}14,$$

and

$$M(B) < \frac{|\Delta|}{\eta(3)} \quad \textit{if} \quad \mu = 3,$$

unless B *is equivalent to a multiple of* B_3.

Proof. By Lemma 4, it will suffice to consider the form

$$(x+\beta y)\{z+(\mu+\beta)t\},$$

where β satisfies

$$0 \leqslant \beta+\tfrac{1}{2}\mu \leqslant \tfrac{1}{2}.$$

We have, by (24),

$$\beta \leqslant \tfrac{1}{2}-\tfrac{1}{2}\mu < -1+\frac{\mu}{\eta},$$

and

$$\mu+\beta \leqslant \tfrac{1}{2}+\tfrac{1}{2}\mu < 2+\frac{\mu}{\eta}.$$

If $\beta > -1-\mu/\eta$, then

$$|(1+\beta.1)\{1+(\beta+\mu).0\}| < \frac{\mu}{\eta} = \frac{|\Delta|}{\eta}.$$

If $\mu+\beta > 2-\mu/\eta$, then

$$|(1+\beta.0)\{2+(\beta+\mu)(-1)\}| < \frac{\mu}{\eta} = \frac{|\Delta|}{\eta}.$$

Hence we need only consider values of β which satisfy

$$-\tfrac{1}{2}\mu \leqslant \beta \leqslant \min\left(-1-\frac{\mu}{\eta},\ 2-\mu-\frac{\mu}{\eta}\right). \tag{25}$$

We consider the following value of B:

$$F(\beta) = (1-\beta)\{3-2(\mu+\beta)\}.$$

$F(\beta)$ is zero when $\beta = 1$ and when $\beta = \tfrac{3}{2}-\mu$, hence

$$F'(\tfrac{5}{4}-\tfrac{1}{2}\mu) = 0.$$

Since $\tfrac{5}{4}-\tfrac{1}{2}\mu > -1$, $F(\beta)$ decreases steadily for $\beta < -1$. Also

$$|F(-\tfrac{1}{2}\mu)| = (1+\tfrac{1}{2}\mu)|3-\mu| < \frac{\mu}{\eta}$$

by (23). If $\mu \leqslant 3$,

$$\left|F\left(-1-\frac{\mu}{\eta}\right)\right| = \left(2+\frac{\mu}{\eta}\right)\left(2\mu-5-\frac{2\mu}{\eta}\right) = \frac{\mu}{\eta}$$

by (19). If $\mu \geqslant 3$,

$$\left|F\left(2-\mu-\frac{\mu}{\eta}\right)\right| = \left(1-\frac{2\mu}{\eta}\right)\left(\mu-1+\frac{\mu}{\eta}\right) = \frac{\mu}{\eta}$$

by (21). As $-1-\mu/\eta < -1$, and, for $\mu \geqslant 3$, $2-\mu-\mu/\eta < -1$, the inequality

$$|F(\beta)| \leqslant \frac{\mu}{\eta}$$

is satisfied throughout the interval (25). This proves the first assertion of Lemma 8. For the second, we observe that, when $\mu = 3$,

$$|F(\beta)| < \frac{\mu}{\eta}$$

unless $\beta = -1-\dfrac{\mu}{\eta} = -1-\dfrac{3}{\eta(3)} = -1-\dfrac{1}{\sqrt{2}+1} = -\sqrt{2}.$

LEMMA 9. *If* $|Q(p,q)| = |\Delta|/\mu$ *for some* p, q *with* $(p,q) = 1$, *and* $\mu \geqslant 3{\cdot}14$, *then*

$$M(B) < \frac{|\Delta|}{7{\cdot}3} < \frac{|\Delta|}{3(1+\sqrt{2})}.$$

Proof. Again it suffices to consider the form

$$(x+\beta y)\{z+(\mu+\beta)t\},$$

where $|\Delta| = \mu$. If the result were false, we should have

$$|m+\beta| \geqslant \frac{\mu}{7{\cdot}3} \quad \text{and} \quad |n+(\mu+\beta)| \geqslant \frac{\mu}{7{\cdot}3}$$

for all m and n. In particular,

$$\frac{2\mu}{7\cdot3} \leqslant 1. \tag{26}$$

Also there would exist k and l such that

$$|\beta-k-\tfrac{1}{2}| < \frac{1}{2}-\frac{\mu}{7\cdot3} \quad \text{and} \quad |(\mu+\beta)-l-\tfrac{1}{2}| < \frac{1}{2}-\frac{\mu}{7\cdot3}.$$

Hence
$$|\mu-(l-k)| < 1-\frac{2\mu}{7\cdot3}.$$

If $l-k \leqslant 3$, then
$$\mu < 3+1-\frac{2\mu}{7\cdot3},$$

whence
$$\mu < 4\frac{7\cdot3}{9\cdot3} < 3\cdot14,$$

contrary to hypothesis. If $l-k \geqslant 4$, then

$$\mu > 4-\left(1-\frac{2\mu}{7\cdot3}\right),$$

whence
$$\mu > 3\frac{7\cdot3}{5\cdot3} > \frac{7\cdot3}{2},$$

contrary to (26).

3. *Proof of Theorems* 1, 2, 3. If B is equivalent to a multiple of B_1 or B_2, then, by Lemma 1, $M(B)$ satisfies either (6) with equality or (8) with equality, and the minimum is attained. If this is not the case, then, by Lemma 3, the quadratic form Q corresponding to B satisfies (15). Thus there exist p, q with $(p,q) = 1$ such that

$$|Q(p,q)| = \frac{|\Delta|}{\mu}, \qquad \mu > 2\cdot96.$$

If $2\cdot96 < \mu \leqslant 3\cdot14$, then, by Lemmas 5, 6, 8,

$$M(B) < \frac{|\Delta|}{\eta(3)} = \frac{|\Delta|}{3(1+\sqrt{2})}$$

unless $\mu = 3$ and B is equivalent to a multiple of B_3, in which case

$$M(B) = \frac{|\Delta|}{3(1+\sqrt{2})} \tag{27}$$

by Lemma 2, and the minimum is attained. Finally, (27) holds also if $\mu > 3\cdot14$, by Lemma 9.

4. Let ω be any real number satisfying $\omega > 10$, and let

$$\Omega = (1+\sqrt{2})^{\omega}.$$

We use $\delta_1, \delta_2, \ldots$ to denote positive numbers depending only on ω which tend to zero as $\omega \to +\infty$. We shall also use the notation

$$L = O_{\omega}(M),$$

where L, M depend on ω and other variables, and $M > 0$, to mean that

$$|L| < (1+\delta_1)^{\omega} M$$

for some δ_1. If $L = O(M)$, then $L = O_{\omega}(M)$.

We define N_i for all i by

$$N_i = [i\omega] \text{ for } i \geqslant 0, \qquad N_i = -N_{-i}+2 \text{ for } i < 0. \tag{28}$$

We define a_n for all n by

$$a_{N_i} = a_{N_i-1} = 1 \text{ for all } i, \quad \text{and} \quad a_n = 2 \text{ for all other } n. \tag{29}$$

We define the numbers p_n, q_n for all n by

$$p_0 = 1, \quad q_0 = 0; \qquad p_{-1} = 0, \quad q_{-1} = 1 \tag{30}$$

and by the recurrence relations

$$p_n = a_n p_{n-1} + p_{n-2}, \qquad q_n = a_n q_{n-1} + q_{n-2}. \tag{31}$$

We put

$$\theta_n = a_{n+1} + \frac{1}{a_{n+2}+}\,\frac{1}{a_{n+3}+}\cdots \tag{32}$$

for all n, and in particular we write $\theta = \theta_0$. We shall be concerned with the bilinear form

$$B_{\theta} = (x-\theta y)\{z+(3-\theta)t\}.$$

LEMMA 10. *If $\omega \neq \omega'$, then B_{θ} is not equivalent to a multiple of $B_{\theta'}$.*

Proof. If B_{θ} is equivalent to a multiple of $B_{\theta'}$, then the corresponding irrational numbers θ, θ' are equivalent.* In this case, there exists k such that $a_n = a'_{n+k}$ for all sufficiently large n. Since, obviously,

$$\frac{1}{M}\sum_{n=1}^{M}(2-a_n) \to \frac{2}{\omega} \quad \text{as} \quad M \to \infty,$$

the corresponding numbers ω and ω' must be equal.

* Two real numbers ρ and σ are equivalent, if integers a, b, c, d can be found such that $ad-bc = \pm 1$, $\rho = (a\sigma+b)/(c\sigma+d)$.

LEMMA 11. *For $n \geqslant 2$, we have*

$$a_{-n} = a_{n+1}, \tag{33}$$

$$q_{-n} = (-1)^{n-1} q_{n-1}, \tag{34}$$

$$p_{-n} + 3q_{-n} = (-1)^{n-1} p_{n-1}. \tag{35}$$

Proof. If $-n = N_i$, where $i < 0$, then $n+1 = -N_i+1 = N_{-i}-1$, by (28). Consequently, also, if $-n = N_i-1$ then $n+1 = N_{-i}$. In these cases, $a_{-n} = a_{n+1} = 1$. For all other values of n satisfying $n \geqslant 2$, $a_{-n} = a_{n+1} = 2$. This proves (33).

For (34) and (35), we note that, by (28), (29),

$$a_2 = 2, \quad a_1 = 2, \quad a_0 = 1, \quad a_{-1} = 1, \quad a_{-2} = 2.$$

Hence, by (30) and (31),

$$p_2 = 5, \quad p_1 = 2, \quad p_0 = 1, \quad p_{-1} = 0, \quad p_{-2} = 1, \quad p_{-3} = -1,$$

$$q_2 = 2, \quad q_1 = 1, \quad q_0 = 0, \quad q_{-1} = 1, \quad q_{-2} = -1, \quad q_{-3} = 2.$$

Thus (34) and (35) are valid for $n = 2$ and $n = 3$. But, if they are valid for n and $n-1$, they are also valid for $n+1$, since, on using (31) and (33),

$$q_{-n-1} = q_{-n+1} - a_{-n+1} q_{-n} = (-1)^n (q_{n-2} + a_n q_{n-1}) = (-1)^n q_n,$$

$$\begin{aligned} p_{-n-1} + 3q_{-n-1} &= (p_{-n+1} + 3q_{-n+1}) - a_{-n+1}(p_{-n} + 3q_{-n}) \\ &= (-1)^n (p_{n-2} + a_n p_{n-1}) = (-1)^n p_n. \end{aligned}$$

LEMMA 12. *For all n,*

$$p_n q_{n-1} - p_{n-1} q_n = (-1)^n, \tag{36}$$

$$(p_n - \theta q_n)\theta_n = -(p_{n-1} - \theta q_{n-1}), \tag{37}$$

$$(p_n - \theta q_n)(q_{n-1} + \theta_n q_n) = (-1)^n. \tag{38}$$

Proof. Now (36) is true for $n = 0$ by (30), and follows for $n > 0$ and $n < 0$ by induction, using (31). Similarly (37) is true for $n = 0$, and follows by induction, since

$$\frac{p_{n+1} - \theta q_{n+1}}{p_n - \theta q_n} = a_{n+1} + \frac{p_{n-1} - \theta q_{n-1}}{p_n - \theta q_n}$$

and

$$\left(-\frac{1}{\theta_{n+1}}\right) = a_{n+1} + (-\theta_n).$$

Finally (38) follows, since, by (37) and (36),

$$(p_n - \theta q_n)(q_{n-1} + \theta_n q_n) = (p_n - \theta q_n) q_{n-1} - (p_{n-1} - \theta q_{n-1}) q_n = (-1)^n.$$

LEMMA 13. *For any i, if $m = N_i$ and $m' = N_{i-1}$, we have*

$$(1+\delta_2)^{-\omega} < \Omega^{-1}\left|\frac{p_{m'}-\theta q_{m'}}{p_m-\theta q_m}\right| < (1+\delta_2)^{\omega}.$$

Proof. By (37), $$\left|\frac{p_{m'}-\theta q_{m'}}{p_m-\theta q_m}\right| = \prod_{r=m'+1}^{m} \theta_r.$$

The number of factors in the product is

$$m-m' = N_i - N_{i-1} = \omega+O(1).$$

If $r \leqslant m-k$, where $k \geqslant 3$, then

$$a_{r+1} = a_{r+2} = \ldots = a_{r+k-2} = 2,$$

and so, by (32), θ_r is a continued fraction whose first $k-2$ elements are all 2. Hence, for every such r,

$$(1+\eta_k)^{-1} < \frac{\theta_r}{1+\sqrt{2}} < 1+\eta_k,$$

where $\eta_k \to 0$ as $k \to \infty$. For other values of r, we have $1 < \theta_r < 3$ by (32). We have now

$$(1+\eta_k)^{-(m-m'-k)} < \frac{\prod \theta_r}{(1+\sqrt{2})^{m-m'-k}} < (1+\eta_k)^{m-m'-k}3^k,$$

and the result follows by taking, for example, $k = [\sqrt{\omega}]$.

LEMMA 14. *For any i,*

$$\theta_{N_i} = 1+\sqrt{2}+O(\Omega^{-2}).$$

Proof. We have, on writing N for N_i,

$$\theta_N = a_{N+1} + \frac{1}{a_{N+2}+}\ldots = 2+\frac{1}{2+}\,\frac{1}{2+}\ldots,$$

where the number of 2's is $j = \omega+O(1)$. Hence

$$\theta_N = 1+\sqrt{2}+O(Q_j^{-2}),$$

where Q_j is the denominator of the jth convergent to the infinite continued fraction for $1+\sqrt{2}$. Since

$$Q_j = \frac{(1+\sqrt{2})^{j+1}-(1-\sqrt{2})^{j+1}}{2\sqrt{2}} > C\Omega,$$

where C is an absolute constant, the result follows.

LEMMA 15. *If $n = N_i$, then*

$$\frac{p_{n-1}+(3-\theta)q_{n-1}}{p_n+(3-\theta)q_n} = 2-\sqrt{2}+O(\Omega^{-2})$$

for all i.

Proof. For $i = 0$, the result reduces to

$$3-\theta = 2-\sqrt{2}+O(\Omega^{-2}),$$

which is a consequence of Lemma 14.

If $i < 0$ we have, by (34), (35), (37),

$$\frac{p_{n-1}+(3-\theta)q_{n-1}}{p_n+(3-\theta)q_n} = -\frac{p_{-n}-\theta q_{-n}}{p_{-n-1}-\theta q_{-n-1}} = \frac{1}{\theta_{-n}}.$$

Now $-n = -N_i = N_{-i}-2$. Writing $m = N_{-i}$, we have

$$\theta_{-n} = \theta_{m-2} = a_{m-1}+\frac{1}{a_m+1/\theta_m} = 1+\frac{1}{1+1/\theta_m},$$

and

$$\theta_m = 1+\sqrt{2}+O(\Omega^{-2})$$

by Lemma 14. Hence the result, in this case, since

$$\left\{1+\frac{1}{1+(\sqrt{2}+1)^{-1}}\right\}^{-1} = 2-\sqrt{2}.$$

If $i > 0$, we write the fraction with which we are concerned as

$$\frac{q_{n-1}}{q_n}\left\{\frac{(p_{n-1}/q_{n-1})-\theta+3}{(p_n/q_n)-\theta+3}\right\}. \tag{39}$$

By (38), since $\theta_n > 1$,

$$\left|\frac{p_{n-1}}{q_{n-1}}-\theta\right| < \frac{1}{q_{n-1}^2}, \qquad \left|\frac{p_n}{q_n}-\theta\right| < \frac{1}{q_n^2}.$$

Now $q_n > q_{n-1} \geqslant q_{N_1-1} > q_{N_1-2} = Q_{N_1-2}$. As in the proof of the preceding lemma, $Q_{N_1-2} > C\Omega$. Hence the second factor in (39) is $1+O(\Omega^{-2})$, and it suffices to prove that

$$\frac{q_{n-1}}{q_n} = 2-\sqrt{2}+O(\Omega^{-2}). \tag{40}$$

By (31),

$$\frac{q_n}{q_{n-1}} = a_n+\frac{1}{a_{n-1}+}\,\frac{1}{a_{n-2}+}\cdots\frac{1}{a_2}$$

$$= 1+\frac{1}{1+}\,\frac{1}{2+}\,\frac{1}{2+}\cdots,$$

where the number of 2's is $\omega+O(1)$. By the argument used in the preceding lemma, this suffices to prove (40).

5. *Proof of Theorem* 4. By Lemma 10, it suffices to prove that for all x, y, z, t satisfying $xt-yz = \pm 1$,

$$|B_\theta(x,y,z,t)| > \sqrt{2}-1-\eta(\omega), \tag{41}$$

where $\eta(\omega) \to 0$ as $\omega \to +\infty$.

By Lemma 13, for given x, y, z, t we can find a value of i such that, when $m = N_i$,

$$(1+\delta_2)^{-\omega}\Omega^{-1} < \left|\frac{x-\theta y}{x+(3-\theta)y}\right| \frac{1}{(p_m-\theta q_m)^2} < (1+\delta_2)^{\omega}\Omega. \qquad (42)$$

(It suffices to take the least value of i for which the first half of the inequality holds.) We make the integral unimodular substitution

$$x = p_m X + p_{m-1} Y, \qquad y = q_m X + q_{m-1} Y,$$

$$z = p_m Z + p_{m-1} T, \qquad t = q_m Z + q_{m-1} T.$$

Then

$$x-\theta y = (p_m-\theta q_m)(X-\theta_m Y),$$

$$z+(3-\theta)t = \{p_m+(3-\theta)q_m\}(Z+\theta'_m T),$$

where

$$\theta'_m = \frac{p_{m-1}+(3-\theta)q_{m-1}}{p_m+(3-\theta)q_m}.$$

By Lemmas 14 and 15,

$$\theta_m = 1+\sqrt{2}+O(\Omega^{-2}), \qquad \theta'_m = 2-\sqrt{2}+O(\Omega^{-2}). \qquad (43)$$

Hence

$$B_\theta(x, y, z, t) = \Lambda(X-\theta_m Y)(Z+\theta'_m T),$$

and a comparison of determinants shows that

$$|\Lambda| = 1+O(\Omega^{-2}). \qquad (44)$$

We can obviously assume that

$$|(X-\theta_m Y)(Z+\theta'_m T)| < \sqrt{2}-1, \qquad (45)$$

since otherwise (41) is satisfied.

We have also

$$x+(3-\theta)y = \{p_m+(3-\theta)q_m\}(X+\theta'_m Y);$$

hence (42) can be written

$$(1+\delta_2)^{-\omega}\Omega^{-1} < \frac{1}{|\Lambda|}\left|\frac{X-\theta_m Y}{X+\theta'_m Y}\right| < (1+\delta_2)^{\omega}\Omega.$$

By (44), this can be replaced by

$$(1+\delta_3)^{-\omega}\Omega^{-1} < \left|\frac{X-\theta_m Y}{X+\theta'_m Y}\right| < (1+\delta_3)^{\omega}\Omega. \qquad (46)$$

We define ν by

$$|(X-\theta_m Y)(X+\theta'_m Y)|^{\frac{1}{2}} = \nu. \qquad (47)$$

Then (46) and (47) imply

$$(1+\delta_3)^{-\omega}\Omega^{-\frac{1}{2}}\nu < |X-\theta_m Y| < (1+\delta_3)^{\omega}\Omega^{\frac{1}{2}}\nu, \qquad (48)$$

$$(1+\delta_3)^{-\omega}\Omega^{-\frac{1}{2}}\nu < |X+\theta'_m Y| < (1+\delta_3)^{\omega}\Omega^{\frac{1}{2}}\nu. \qquad (49)$$

From (45)

$$Z+\theta'_m T = O_\omega(\Omega^{\frac{1}{2}}\nu^{-1}). \qquad (50)$$

Also, since

$$(X-\theta_m Y)(Z+\theta'_m T)-(X+\theta'_m Y)(Z-\theta_m T)$$
$$= (\theta_m+\theta'_m)(XT-YZ) = O(1),$$

we have $$(X+\theta'_m Y)(Z-\theta_m T) = O(1),$$

and so $$Z-\theta_m T = O_\omega(\Omega^{\frac{1}{2}}\nu^{-1}). \qquad (51)$$

By (48), (49), (50), (51),

$$X = O_\omega(\Omega^{\frac{1}{2}}\nu), \quad Y = O_\omega(\Omega^{\frac{1}{2}}\nu), \quad Z = O_\omega(\Omega^{\frac{1}{2}}\nu^{-1}), \quad T = O_\omega(\Omega^{\frac{1}{2}}\nu^{-1}),$$

and so $$XZ,\ XT,\ YZ,\ YT \text{ are all } O_\omega(\Omega). \qquad (52)$$

By (43), the coefficients of the bilinear form

$$(X-\theta_m Y)(Z+\theta'_m T)$$

differ from the corresponding coefficients in the form

$$\{X-(1+\sqrt{2})Y\}\{Z+(2-\sqrt{2})T\} = B_3(X-Y, Y, Z-T, T)$$

by amounts which are all $O(\Omega^{-2})$. Hence, by (52),

$$(X-\theta_m Y)(Z+\theta'_m T)-B_3 = O_\omega(\Omega^{-2}.\Omega) = O_\omega(\Omega^{-1}).$$

Since $|B_3| \geqslant \sqrt{2}-1$ by Lemma 2, and

$$(1+\delta_4)^\omega(1+\sqrt{2})^{-\omega} \to 0 \quad \text{as} \quad \omega \to \infty,$$

this proves (41), and so proves Theorem 4.

[Received 15 August, 1946] [London and Bristol]

24.

Asymmetric Inequalities for Non-homogeneous Linear Forms

(With H. Davenport)*

Journal of the London Mathematical Society 22 (1947), 53–61
[Commentary on p. 590]

1. Let α, β, γ, δ be real numbers with $\Delta = \alpha\delta - \beta\gamma \neq 0$, and let ξ, η denote the linear forms

$$\xi = \alpha u + \beta v, \quad \eta = \gamma u + \delta v.$$

Minkowski proved that for any real numbers λ, μ there exist integers u, v for which

$$|(\xi+\lambda)(\eta+\mu)| \leqslant \tfrac{1}{4}|\Delta|.$$

It is plain from the case $\xi = u$, $\eta = v$, $\lambda = \mu = \frac{1}{2}$, that this inequality is the best possible of its kind.†

Our object in this note is to find what inequality is valid for the product $(\xi+\lambda)(\eta+\mu)$ when the integral variables u, v are restricted to satisfy *either*

(A) $$\xi+\lambda \geqslant 0, \quad \eta+\mu \geqslant 0,$$

---

* Received 26 September, 1946; read 14 November, 1946.

† For various improved forms of Minkowski's inequality, see Davenport, "Non-homogeneous binary quadratic forms", *Proc. Kon. Nederl. Akad. Wetensch.*, 49 (1946), 815–821.

or

$$(\xi+\lambda)(\eta+\mu) \geqslant 0. \tag{B}$$

Our results are very simple; we find that in each case the constant $\frac{1}{4}$ of Minkowski's theorem must be replaced by 1. More precisely, we prove the two theorems:

THEOREM 1. *There exist integers u, v such that* (A) *holds, and a fortiori* (B) *holds, and such that*

$$(\xi+\lambda)(\eta+\mu) < |\Delta|.$$

THEOREM 2. *For any $\kappa < 1$ there exist α, β, γ, δ, λ, μ such that the inequality*

$$(\xi+\lambda)(\eta+\mu) < \kappa|\Delta|$$

has no solution satisfying (B), *and a fortiori no solution satisfying* (A).

Theorem 1 is not a deep theorem ; we shall prove it by the well-known method of "standardizing" a lattice relative to a lattice point at which the minimum of the product is attained, or almost attained. Theorem 2 is not so straightforward, but we give examples of two kinds of linear forms, for each of which the conclusion holds. Our reason for giving two examples lies in another method of approach to the problem, which relates it to the values assumed by the homogeneous product $\xi\eta$. It is well known that there exist integers u_0, v_0 for which*

$$(u_0, v_0) = 1, \quad \xi_0\eta_0 = \nu|\Delta|, \quad 0 \leqslant \nu < 1. \tag{1}$$

If we require only (B) to be satisfied, we can prove a more precise result than that of Theorem 1, provided (1) holds with a value of ν which is neither slightly less than $\frac{1}{2}$ nor slightly less than 1. The result is:

THEOREM 3. *There exist integers u, v satisfying* (B) *and satisfying*

$$(\xi+\lambda)(\eta+\mu) < \kappa(\nu)|\Delta|,$$

where $\kappa(\nu) = \frac{1}{2}+\nu$ for $0 \leqslant \nu < \frac{1}{2}$ and $\kappa(\nu) = \nu$ for $\frac{1}{2} \leqslant \nu < 1$.

Our two examples (Lemmas 1 and 2) show that in neither of the two exceptional neighbourhoods can the inequality be substantially improved.

---

* This follows on taking $-u/v$ to be a suitable convergent to the continued fraction for either β/α or δ/γ. For proofs based on the geometry of numbers, see Segre, *Duke Math. J.*, 12 (1945), 337–365 and Mahler, *ibid.*, 367–371.

We give a third example (Lemma 3) to show that the conclusion of Theorem 3 would be false if (B) were replaced by (A).

2. *Proof of Theorem* 1. Let M denote the lower bound of $(\xi+\lambda)(\eta+\mu)$ for integral u, v which satisfy (A). If $M=0$, there is nothing to prove, so we may suppose $M>0$. For any $\epsilon_0>0$ there exist integers u_1, v_1 such that

$$\xi_1+\lambda>0, \quad \eta_1+\mu>0, \quad (\xi_1+\lambda)(\eta_1+\mu)=\frac{M}{1-\epsilon}, \quad 0\leqslant\epsilon<\epsilon_0.$$

Put
$$x=\frac{\xi-\xi_1}{\xi_1+\lambda}, \quad y=\frac{\eta-\eta_1}{\eta_1+\mu};$$

then x, y are linear forms in the integral variables $u-u_1, v-v_1$ of determinant

$$\frac{1-\epsilon}{M}|\Delta|. \tag{2}$$

Also
$$x+1=\frac{\xi+\lambda}{\xi_1+\lambda}, \quad y+1=\frac{\eta+\mu}{\eta_1+\mu},$$

and for all integers u, v for which $x+1\geqslant 0$, $y+1\geqslant 0$ we have

$$(x+1)(y+1)\geqslant 1-\epsilon, \tag{3}$$

by the definition of M. Thus the points in the x, y plane which correspond to integral u, v form a lattice, whose determinant is given by (2), such that every lattice point with $x+1\geqslant 0$, $y+1\geqslant 0$ satisfies (3).

It follows that there is no lattice point other than the origin O in the square

$$|x|\leqslant 1, \quad |y|\leqslant 1. \tag{4}$$

For, by considering the image in O if necessary, we can suppose that $x+y\leqslant 0$, and by (3) we have

$$\{\tfrac{1}{2}(x+y)+1\}^2-\{\tfrac{1}{2}(x-y)\}^2\geqslant 1-\epsilon.$$

Thus $(x-y)^2\leqslant 4\epsilon$ and $\frac{1}{2}(x+y)+1\geqslant 1-\epsilon$. Hence any lattice point in (4) satisfies

$$|x+y|\leqslant 2\epsilon, \quad |x-y|\leqslant 2\sqrt{\epsilon}. \tag{5}$$

But then, if this lattice point is not O, a suitable integral multiple of the point will satisfy (4) and not (5), which is impossible.

By Minkowski's fundamental theorem, we have

$$\frac{1-\epsilon}{M}|\Delta| \geqslant 1,$$

and so $M < |\Delta|$ unless, possibly, $M = |\Delta|$ and $\epsilon = 0$. But in this case there would be a lattice point on the boundary of the square, and such a point, or its image in O, would violate (3). Hence $M < |\Delta|$, and Theorem 1 is established.

3. *Proof of Theorem* 3. By applying a suitable integral unimodular transformation to the variables u, v we can suppose that (1) is satisfied by $u_0 = 1$, $v_0 = 0$; so that

$$\alpha\gamma = \nu|\Delta|. \tag{6}$$

We determine real numbers u_0, v_0 to satisfy

$$\alpha u_0 + \beta v_0 + \lambda = 0, \quad \gamma u_0 + \delta v_0 + \mu = 0.$$

Then the assertion of Theorem 3 is that we can satisfy

$$0 \leqslant (\alpha u + \beta v)(\gamma u + \delta v) < \kappa(\nu)|\Delta| \tag{7}$$

by real numbers u, v subject to

$$u \equiv u_0 \pmod 1, \qquad v \equiv v_0 \pmod 1. \tag{8}$$

Case 1. *Suppose* $\nu = 0$. By (6) we can suppose, without loss of generality, that $\gamma = 0$, and $\alpha \neq 0$ since $\alpha\delta - \beta\gamma \neq 0$. We choose first v, subject to (8), to satisfy $|v| \leqslant \frac{1}{2}$; and then choose u, subject to (8), to satisfy

$$0 \leqslant \left| u + \frac{\beta}{\alpha} v \right| < 1, \quad \left(u + \frac{\beta}{\alpha} v \right) \alpha\delta v \geqslant 0.$$

Now
$$0 \leqslant (\alpha u + \beta v)\delta v < \tfrac{1}{2}|\alpha\delta| = \tfrac{1}{2}|\Delta|.$$

Case 2. *Suppose* $0 < \nu < \frac{1}{2}$. By (6), we can write the inequality (7), which is to be proved, as

$$0 \leqslant (u + \theta v)(u + \phi v) < \kappa(\nu)/\nu, \tag{9}$$

where
$$|\theta - \phi| = |\Delta|/(\alpha\gamma) = 1/\nu.$$

We choose v, subject to (8), to satisfy $|v| \leqslant \frac{1}{2}$, and can suppose, without loss of generality, that

$$0 \leqslant (\theta - \phi)v \leqslant 1/(2\nu). \tag{10}$$

Then we choose u, subject to (8), to satisfy $0 \leqslant u+\phi v < 1$. We now have*

$$0 \leqslant (u+\theta v)(u+\phi v) < u+\theta v = u+\phi v+(\theta-\phi)v < 1+1/(2\nu) = \kappa(\nu)/\nu.$$

Case 3. *Suppose* $\frac{1}{2} \leqslant \nu < 1$. As in Case 2, it suffices to prove that we can satisfy (9), subject to (8). We choose v as before, and have (10). We now choose u, subject to (8), to satisfy

$$0 \leqslant u+\tfrac{1}{2}(\theta+\phi)v < 1.$$

If $u+\phi v \geqslant 0$, then $u+\theta v \geqslant 0$ by (10), and

$$0 \leqslant (u+\theta v)(u+\phi v) \leqslant (u+\tfrac{1}{2}(\theta+\phi)v)^2 < 1.$$

If $u+\phi v < 0$, then $u-1+\phi v < 0$, and

$$u-1+\theta v = (u+\phi v)-1+(\theta-\phi)v < -1+1/(2\nu) \leqslant 0.$$

hus

$$0 \leqslant (u-1+\theta v)(u-1+\phi v) \leqslant \left(1-\{u+\tfrac{1}{2}(\theta+\phi)v\}\right)^2 \leqslant 1.$$

Equality cannot occur in this application of the inequality of the arithmetic and geometric means, since this would require $u-1+\theta v = u-1+\phi v$, $v=0$, contrary to $u+\phi v < 0$, $u+\frac{1}{2}(\theta+\phi)v \geqslant 0$. Thus in both cases we can satisfy (9), subject to (8), with $\kappa(\nu) = \nu$. This completes the proof of Theorem 3.

4. Lemma 1. *Let N be any positive integer. The inequality*

$$0 \leqslant -(u+\tfrac{1}{2})^2+N(N+1)v^2 < 2N-\tfrac{1}{4}$$

has no solution in integers u, v.

Proof†. Let $x = 2u+1$, $y = v$, and

(11) $$X = x-2\sqrt{\{N(N+1)\}}y, \quad X' = x+2\sqrt{\{N(N+1)\}}y.$$

The assertion of the lemma is that the inequality

(12) $$-(8N-1) < XX' \leqslant 0$$

* It will be observed that in this case we have really proved that we can satisfy the inequality of Theorem 3 with condition (A).

† The result can also be deduced from an argument of Tchebicheff (see Mathews. *Theory of Numbers*, 269).

has no solution in integers x, y with x odd. Suppose, if possible, that (12) has such a solution.

Let $\vartheta = 2N+1+2\sqrt{\{N(N+1)\}}$. With an obvious notation, $\vartheta\vartheta' = 1$; and ϑX, $\vartheta' X'$ is also a solution of (12) with an odd value of x. Hence there exists a solution of (12) for which

$$\frac{1}{\vartheta} < \left|\frac{X}{X'}\right| \leqslant \vartheta.$$

For this solution,

$$|X|^2 \leqslant (8N-1)\vartheta, \quad |X'|^2 < (8N-1)\vartheta.$$

Hence, by (11),

$$|y| = \frac{|X'-X|}{4\sqrt{\{N(N+1)\}}} < \frac{\sqrt{\{(8N-1)\vartheta\}}}{2\sqrt{\{N(N+1)\}}} < \frac{\sqrt{\{(8N-1)2(2N+1)\}}}{2\sqrt{\{N(N+1)\}}} < 2\sqrt{2}.$$

It follows that $y = 0$ or ± 1 or ± 2. $y = 0$ is plainly not possible in (12), since x is odd. If $y = \pm 1$ or ± 2, (12) becomes

$$-(8N-1) < x^2 - 4N(N+1) \leqslant 0 \quad \text{or} \quad -(8N-1) < x^2 - 16N(N+1) \leqslant 0,$$

which imply

$$(2N-1)^2 < x^2 < (2N+1)^2 \quad \text{or} \quad (4N+1)^2 < x^2 < (4N+2)^2,$$

and so there is no solution with x odd. This proves the lemma.

5. Lemma 2. *Let θ, θ' be the roots of $\theta^2 - N\theta - N = 0$, where N is a positive integer. Then the inequality*

$$(13) \qquad -\left(N - \frac{1}{N+4}\right) < \left(u - \theta v - \frac{\theta+2}{N+4}\right)\left(u - \theta' v - \frac{\theta'+2}{N+4}\right) \leqslant 0$$

has no solution in integers u, v.

Proof. Since $(U-\theta V)(U-\theta' V) = U^2 - NUV - NV^2$, the product in (13) is

$$\left(u - \frac{2}{N+4}\right)^2 - N\left(u - \frac{2}{N+4}\right)\left(v + \frac{1}{N+4}\right) - N\left(v + \frac{1}{N+4}\right)^2$$

$$= u^2 - Nuv - Nv^2 - u + 1/(N+4).$$

Hence it suffices to prove that the inequality

$$-N < u^2 - Nuv - Nv^2 - u < 0$$

has no solution in integers u, v. We suppose that such a solution exists; by changing the signs of u and v if necessary, we have

$$-N < u^2-Nuv-Nv^2 \pm u < 0, \quad v > 0 \tag{14}$$

for a certain sign $+$ or $-$. We can suppose also that $u > 0$, for the expression in (14) is

$$\{u-\tfrac{1}{2}(Nv\mp 1)\}^2-\{\tfrac{1}{2}(Nv\mp 1)\}^2-Nv^2$$

and is unaltered if u is replaced by u', where $u+u' = Nv\mp 1$, and one at least of u, u' is positive.

Among all solutions of (14) with $u > 0$ we select that one for which u is least. We have

$$u(u-Nv\pm 1) > N(v^2-1) \geqslant 0,$$

whence $u > Nv\mp 1$ and so $u \geqslant Nv$. Since (14) is not satisfied when $u = Nv$, we have

$$u > Nv. \tag{15}$$

Also
$$u(u-Nv\pm 1) < Nv^2 < uv,$$

whence $u < (N+1)v\mp 1$, $u \leqslant (N+1)v$. If $u = (N+1)v$, (14) becomes $-N < v^2 \pm (N+1)v < 0$, or $0 < v(N+1-v) < N$, which is impossible. Hence

$$u < (N+1)v. \tag{16}$$

Let
$$U = u-Nv\pm 1, \quad V = -u+(N+1)v\mp 1.$$

Then $U \geqslant 0$ and $V \geqslant 0$ by (15) and (16), and

$$u^2-Nuv-Nv^2\pm u = U^2-NUV-NV^2\mp U.$$

If $U = 0$ or $V = 0$ we get a contradiction in (14), and so $U > 0$, $V > 0$, and $U < u$, which contradicts the definition of u, v. Hence (14) has no solution, and the lemma is proved.

6. *Proof of Theorem* 2. As explained in §1, we give two proofs of Theorem 2. For the first proof, based on Lemma 1, we take

$$\alpha u+\beta v+\lambda = u+\sqrt{\{N(N+1)\}}v+\tfrac{1}{2}, \quad \gamma u+\delta v+\mu = -u+\sqrt{\{N(N+1)\}}v-\tfrac{1}{2}.$$

Then the inequality

$$0 \leqslant (\alpha u+\beta v+\lambda)(\gamma u+\delta v+\mu) < 2N-\tfrac{1}{4}$$

has no solution. Since $|\Delta| = 2\sqrt{\{N(N+1)\}}$ for these forms, this proves Theorem 2 with

$$\kappa = \frac{2N-\frac{1}{4}}{2\sqrt{\{N(N+1)\}}} < 1.$$

Since N may be arbitrarily large, this value of κ may be as near to 1 as we please. The possible values of ν for these forms arise from the non-negative values of

$$(-u^2+N(N+1)v^2)/|\Delta|$$

with $(u, v) = 1$. One such value is

$$\nu = \frac{N}{2\sqrt{\{N(N+1)\}}} = \tfrac{1}{2}\sqrt{\left(\frac{N}{N+1}\right)}$$

arising from $u = N$, $v = 1$. This shows that the result of Theorem 3 cannot be substantially improved when ν is a little less than $\frac{1}{2}$.

For the second proof, based on Lemma 2, we take

$$\alpha u+\beta v+\lambda = u-\theta v-\frac{\theta+2}{N+4}, \quad \gamma u+\delta v+\mu = -\left(u-\theta' v-\frac{\theta'+2}{N+4}\right).$$

Then the inequality

$$0 \leqslant (\alpha u+\beta v+\lambda)(\gamma u+\delta v+\mu) < N-\frac{1}{N+4}$$

has no solution. Since $|\Delta| = |\theta-\theta'| = \sqrt{\{N(N+4)\}}$, this proves Theorem 2 with

$$\kappa = \frac{N-\{1/(N+4)\}}{\sqrt{\{N(N+4)\}}} < 1,$$

and again this value of κ is as near to 1 as we please. One value of ν is

$$\nu = \frac{N}{\sqrt{\{N(N+4)\}}},$$

arising from $u = 0$, $v = 1$. This shows that the result of Theorem 3 cannot be substantially improved when ν is a little less than 1.

7. Lemma 3. *Let N be a positive integer, and $\gamma = 1-1/N$, and suppose $0 < \delta < 1/N$. Then for all integers u, v which satisfy* (A),

$$(u+1-\delta)(\gamma u+v-\delta) > 1-(N+1)\delta.$$

Proof. (A) implies $\gamma u+v-\delta \geqslant N^{-1}-\delta$, hence, if $u \geqslant N$,

$$(u+1-\delta)(\gamma u+v-\delta) \geqslant (N+1-\delta)(N^{-1}-\delta) > 1-(N+1)\delta.$$

Further, $\gamma u+v-\delta = u+v-u/N-\delta$, and (A) implies, for $0 \leqslant u < N$, that $u+v \geqslant 1$. Hence, for $0 \leqslant u \leqslant N-1$,

$$\begin{aligned}(u+1-\delta)(\gamma u+v-\delta) &\geqslant (u+1-\delta)(1-u/N-\delta)\\ &\geqslant \min\{(1-\delta)^2,\ (N-\delta)(1/N-\delta)\}\\ &> 1-(N+1)\delta.\end{aligned}$$

This proves Lemma 3.

A possible value of ν for the particular forms in Lemma 3 is $\nu = 0$, arising from $u=0$, $v=1$. Since δ is arbitrarily small, the constant $1-(N+1)\delta$ is arbitrarily near to 1, and this shows that the conclusion of Theorem 3 would not hold if (B) were replaced by (A).

[*Note added* 21 *August*, 1947. Some of the results of this paper have now been extended by Mr. J. H. H. Chalk. His work is in course of publication in the *Quarterly J. of Math.*]

University College, London.
The University, Bristol.

25.

On the Distribution of the Sequence $n^2\theta \pmod 1$

The Quarterly Journal of Mathematics, Oxford Series, 19 (1948), 249–256
[Commentary on p. 581]

1. THE object of this paper is to prove the

THEOREM. *For every integer $N \geqslant 1$ and every real θ, integers n and g can be found such that*

$$1 \leqslant n \leqslant N, \qquad |n^2\theta - g| \leqslant c(\eta)N^{-\frac{1}{2}+\eta}, \tag{1.1}$$

where η is an arbitrarily small positive number and where $c(\eta)$ depends on η only.

The essential fact is that the inequality (1.1) holds uniformly in θ.

A corresponding result was proved by Vinogradov† with the upper bound $N^{-\frac{2}{5}+\eta}$ instead of $N^{-\frac{1}{2}+\eta}$. It is not impossible that the theorem remains valid if the upper bound is replaced even by $N^{-1+\eta}$, but a considerable improvement of (1.1) appears to be unattainable at present.

My proof of the theorem is not very different from Vinogradov's proof, but a slight change in the fundamental technique enables me to make a more delicate analysis of the error terms which arise.

I wish to thank Mr. H. Todd who pointed out to me the earlier history‡ of this problem and thereby started me on this investigation.

2. N, θ, and η have always the meanings given above. Without loss of generality we assume that θ is irrational. The symbol O implies a constant which depends on η only. We write O_η as an abbreviation for $O(N^\eta)$. As η is arbitrary, the relation

$$O_\eta O_\eta = O_\eta$$

may be applied in a bounded number of places.

Throughout the paper small and capital italics other than e, i, and O denote integers. With the further exception of g, b, a, and A all these integers will be positive.

† *Bull. Acad. Sci. U.R.S.S.* (6) 27 (1927), 567–78.

‡ For further references see J. F. Koksma, *Diophantische Approximationen*, *Ergebnisse der Mathematik und ihrer Grenzgebiete*, 4 (1936), Kapitel X, § 1, 2.

We define $\epsilon = \epsilon(N, \theta)$ by the relation

$$\epsilon = \min_{\substack{1 \leqslant n \leqslant N \\ g}} |n^2\theta - g|,$$

so that $0 < \epsilon < \frac{1}{2}$. We assume at once that $\epsilon > N^{-\frac{1}{2}}$.

The function $\phi_\epsilon(\xi)$ has period one in ξ and is defined for $-\frac{1}{2} < \xi \leqslant \frac{1}{2}$ by

$$\phi_\epsilon(\xi) = \max(0, 1 - \epsilon^{-1}|\xi|).$$

It is easily seen that $\phi_\epsilon(\xi)$ has the Fourier expansion

$$\phi_\epsilon(\xi) = \epsilon + 2\pi^{-2}\epsilon^{-1} \sum_{m=1}^{\infty} m^{-2} \sin^2(\pi m \epsilon) \cos(2\pi m \xi). \tag{2.1}$$

Lemma 1. *For every m we can find four integers a_m, q_m, A_m, Q_m such that*

$$q_m \leqslant N < Q_m, \qquad (a_m, q_m) = 1, \qquad (A_m, Q_m) = 1, \tag{2.2}$$

$$|m\theta - a_m/q_m| < q_m^{-1} Q_m^{-1}, \tag{2.3}$$

$$|m\theta - A_m/Q_m| < Q_m^{-2}. \tag{2.4}$$

Proof. Let a_m/q_m be the last convergent of the continued fraction for $m\theta$ whose denominator does not exceed N, and let A_m/Q_m be the first convergent whose denominator does exceed N. The lemma then follows at once from the theory of continued fractions.

Lemma 2.

$$\sum_{n=1}^{N} e^{2\pi i n^2 m\theta} = O_\eta(N q_m^{-\frac{1}{2}}),$$

$$\sum_{n=1}^{N} e^{2\pi i n^2 m\theta} = O_\eta(Q_m^{\frac{1}{2}}).$$

Proof. It has been well known for a long time† that, if

$$(a, q) = 1, \qquad \alpha = a/q + \beta, \qquad |\beta| \leqslant N^{-1} q^{-1}, \tag{2.5}$$

then

$$\sum_{n=1}^{N} e^{2\pi i n^2 \alpha} = O(N q^{-\frac{1}{2}+\eta} + q^{\frac{1}{2}+\eta}). \tag{2.6}$$

If we put $\alpha = m\theta$, $a = a_m$, $q = q_m$, then, by (2.3) and (2.2),

$$|\beta| < q_m^{-1} Q_m^{-1} < q_m^{-1} N^{-1}.$$

Thus (2.5) holds and (2.6) gives the first part of the lemma since, by (2.2),

$$N q_m^{-\frac{1}{2}+\eta} + q_m^{\frac{1}{2}+\eta} \leqslant 2N^{1+\eta} q_m^{-\frac{1}{2}}.$$

† See Davenport and Heilbronn, *J. of London Math. Soc.* 21 (1946), 185–93, Lemma 6.

If we put $\alpha = m\theta$, $a = A_m$, $q = Q_m$, then, by (2.4) and (2.2),

$$|\beta| < Q_m^{-2} < Q_m^{-1}N^{-1}.$$

Again, (2.5) holds and (2.6) gives the second part of the lemma since, for $Q_m \leqslant N^2$,

$$NQ_m^{-\frac{1}{2}+\eta}+Q_m^{\frac{1}{2}+\eta} \leqslant 2N^{2\eta}Q_m^{\frac{1}{2}},$$

whereas for $Q_m > N^2$ the second part of the lemma is trivial.

LEMMA 3.

$$1 = O_\eta\Big\{\sum_{m\leqslant\epsilon^{-1}} \min(q_m^{-\frac{1}{2}}, N^{-1}Q_m^{\frac{1}{2}})+\epsilon^{-2}\sum_{m>\epsilon^{-1}} m^{-2}\min(q_m^{-\frac{1}{2}}, N^{-1}Q_m^{\frac{1}{2}})\Big\}.$$

Proof. It follows from the definition of ϵ and from (2.1) that

$$0 = \sum_{n=1}^{N}\phi_\epsilon(n^2\theta) = \sum_{n=1}^{N}\Big\{\epsilon+2\pi^{-2}\epsilon^{-1}\sum_{m=1}^{\infty} m^{-2}\sin^2(\pi m\epsilon)\cos(2\pi mn^2\theta)\Big\}$$

$$= N\epsilon+2\pi^{-2}\epsilon^{-1}\sum_{m=1}^{\infty} m^{-2}\sin^2(\pi m\epsilon)\sum_{n=1}^{N}\cos(2\pi n^2 m\theta).$$

Hence, dividing by $N\epsilon$,

$$1 \leqslant N^{-1}\epsilon^{-2}\sum_{m=1}^{\infty} m^{-2}\sin^2(\pi m\epsilon)\Big|\sum_{n=1}^{N} e^{2\pi i n^2 m\theta}\Big|,$$

and the result follows from Lemma 2 if we use the inequalities

$$\sin^2(\pi m\epsilon) \leqslant \pi^2 m^2\epsilon^2 \quad \text{for} \quad m \leqslant \epsilon^{-1},$$

$$\sin^2(\pi m\epsilon) \leqslant 1 \qquad \text{for} \quad m > \epsilon^{-1}.$$

LEMMA 4. If $|\theta-b/r| = \delta$, $1 \leqslant r \leqslant N$, then

$$\epsilon \leqslant r^2\delta.$$

Proof. $$|r^2\theta - rb| = r^2\delta.$$

LEMMA 5. *For* $\omega \geqslant 1$, $\tau > 0$ *let* $\Lambda(\omega,\tau)$ *denote the number of solutions of the inequality*

$$|\theta-b/r| \leqslant \tau$$

subject to $$r \leqslant \omega, \qquad (b,r) = 1.$$

Then $$\Lambda(\omega,\tau) \leqslant 1+2\omega^2\tau.$$

Proof. Any two different pairs b/r, b'/r' have at least the distance

$$1/(rr') \geqslant \omega^{-2}$$

from each other. So there can be at most $1+2\omega^2\tau$ such fractions in an interval of length 2τ.

So far the work has been routine, and has differed from Vinogradov's proof only by the introduction of the function $\phi_\epsilon(\xi)$ whereas Vinogradov employs the function

$$\psi_\epsilon(\xi) = \begin{cases} 1 & (|\xi| < \epsilon), \\ 0 & (\epsilon < |\xi| \leqslant \frac{1}{2}), \end{cases}$$

which has a more slowly convergent Fourier series.

3. I now introduce four more integers l, b, r, d which will depend on m (and N and θ as well). I put

$$l = (a_m, mq_m), \tag{3.1}$$

$$a_m = lb, \tag{3.2}$$

$$mq_m = lr, \tag{3.3}$$

so that
$$(b, r) = 1.$$

Since $l \mid a_m$ and $(a_m, q_m) = 1$, we have $(l, q_m) = 1$; this and (3.1) give $l \mid m$ and we can put

$$m = ld, \tag{3.4}$$

so that by (3.3)
$$r = q_m d. \tag{3.5}$$

Finally we put
$$|\theta - b/r| = \delta. \tag{3.6}$$

I show first that b is uniquely determined by r. Translated into the new notation (2.3) gives

$$|ld\theta - lbdr^{-1}| < d^2 r^{-2} q_m Q_m^{-1},$$

$$|r\theta - b| < dr^{-1} l^{-1} q_m Q_m^{-1} \leqslant (d/r)(q_m/Q_m). \tag{3.7}$$

Since $d \mid r$, we have either $d \leqslant \frac{1}{2}r$,

$$|r\theta - b| < \tfrac{1}{2}, \tag{3.8}$$

or $d = r$, $q_m = 1$, $Q_m \geqslant 2$, and (3.7) again implies (3.8). In either case (3.8) is true, and b depends on r (and θ) only.

The object is to transform Lemma 3 so that instead of summing over m we sum over the three variables r, d, l. In this summation we shall treat r, d, l as independent variables subject to certain restrictions which I am now going to develop.

We note first that by (3.4), (3.2), and (3.5)

$$|m\theta - a_m/q_m| = ld|\theta - b/r| = dl\delta. \tag{3.9}$$

The condition $q_m \leqslant N$ gives

$$rd^{-1} \leqslant N. \tag{3.10}$$

The condition $Q_m > N$ and (2.3) together with (3.5) and (3.9) give

$$N < Q_m < q_m^{-1}|m\theta - a_m/q_m|^{-1} = r^{-1}l^{-1}\delta^{-1} \tag{3.11}$$

and

$$l < N^{-1}r^{-1}\delta^{-1}. \tag{3.12}$$

We restrict m further by the additional condition

$$mq_m < N\epsilon^{-1}. \tag{3.13}$$

The introduction of (3.13) will be justified later. From (3.13) and (3.3) we have

$$l < Nr^{-1}\epsilon^{-1}. \tag{3.14}$$

Since $l \geqslant 1$, we deduce from (3.12) and (3.14)

$$\delta < N^{-1}r^{-1}, \tag{3.15}$$

$$r < N\epsilon^{-1}. \tag{3.16}$$

To these conditions we add by virtue of Lemma 4

$$\delta \geqslant r^{-2}\epsilon \quad \text{for} \quad r \leqslant N. \tag{3.17}$$

4. The first step is the transformation of Lemma 3.

LEMMA 6.

$$1 = O_\eta(N^{-\frac{1}{2}}\epsilon^{-1}) +$$

$$+ O_\eta\left[\sum_{r<N\epsilon^{-1}} \sum_{\substack{d|r \\ d \geqslant N^{-1}r}} \sum_{l} \min\{1, (dl\epsilon)^{-2}\} \min(r^{-\frac{1}{2}}d^{\frac{1}{2}}, N^{-1}r^{-\frac{1}{2}}l^{-\frac{1}{2}}\delta^{-\frac{1}{2}})\right],$$

where the summation is restricted by the inequalities (3.12), (3.14).

Proof. Our first step is the justification of the conditions (3.13). The sum in Lemma 3 extended over those m for which (3.13) is untrue does not exceed

$$O_\eta\left(\sum_{m \leqslant \epsilon^{-1}} N^{-\frac{1}{2}}\epsilon^{\frac{1}{2}}m^{\frac{1}{2}} + \epsilon^{-2}\sum_{m > \epsilon^{-1}} m^{-2}N^{-\frac{1}{2}}\epsilon^{\frac{1}{2}}m^{\frac{1}{2}}\right)$$

$$= O_\eta\{N^{-\frac{1}{2}}\epsilon^{\frac{1}{2}}(\epsilon^{-1})^{\frac{3}{2}} + N^{-\frac{1}{2}}\epsilon^{-\frac{3}{2}}(\epsilon^{-1})^{-\frac{1}{2}}\}$$

$$= O_\eta(N^{-\frac{1}{2}}\epsilon^{-1}).$$

This calculation justifies the restriction of the summation by (3.13).

The restrictions in the summation explicitly mentioned in the formula of Lemma 6 are satisfied by virtue of (3.16), (3.5), and (3.10). So all we have to do is to show that each term in the sum of Lemma 3 is majorized by the corresponding term in the sum of Lemma 6.

By (3.4)

$$1 \leqslant \min\{1, (\epsilon m)^{-2}\} = \min\{1, (dl\epsilon)^{-2}\} \quad \text{for} \quad m \leqslant \epsilon^{-1},$$

$$\epsilon^{-2}m^{-2} \leqslant \min\{1, (\epsilon m)^{-2}\} = \min\{1, (dl\epsilon)^{-2}\} \quad \text{for} \quad m > \epsilon^{-1}.$$

By (3.5) $$q_m^{-\frac{1}{2}} = r^{-\frac{1}{2}}d^{\frac{1}{2}}.$$

By (3.11) $$N^{-1}Q_m^{\frac{1}{2}} < N^{-1}r^{-\frac{1}{2}}l^{-\frac{1}{2}}\delta^{-\frac{1}{2}}.$$

This concludes the proof of Lemma 6.

We now split the sum over r in Lemma 6 into three sums

$$\sum_r = \sum_\alpha + \sum_\beta + \sum_\gamma$$

according to the conditions

(α) $$r \leqslant N;$$

(β) $$r > N, \qquad \delta \geqslant N^{-2}\epsilon;$$

(γ) $$r > N, \qquad \delta < N^{-2}\epsilon.$$

LEMMA 7. $$\sum_\alpha = O_\eta(N^{-\frac{1}{2}}\epsilon^{-1}).$$

Proof. By (3.12),

$$\sum_\alpha \leqslant \sum_{r\leqslant N} \sum_{d|r} \sum_{l<N^{-1}r^{-1}\delta^{-1}} N^{-1}r^{-\frac{1}{2}}l^{-\frac{1}{2}}\delta^{-\frac{1}{2}}$$
$$= N^{-1} \sum_{r\leqslant N} r^{-\frac{1}{2}}\delta^{-\frac{1}{2}} \sum_{d|r} \sum_{l<N^{-1}r^{-1}\delta^{-1}} l^{-\frac{1}{2}}$$
$$= O_\eta\left(N^{-\frac{3}{2}} \sum_{r\leqslant N} r^{-1}\delta^{-1}\right). \tag{4.1}$$

Then (4.1) and (3.17) give

$$\sum_\alpha = O_\eta\left(N^{-\frac{3}{2}} \sum_{r\leqslant N} r\epsilon^{-1}\right)$$
$$= O_\eta\left(N^{-\frac{1}{2}}\epsilon^{-1} \sum_{r\leqslant N} 1\right).$$

But this sum contains at most two terms owing to the condition (3.15). For otherwise we should have three fractions

$$b_j/r_j \quad (j = 1, 2, 3)$$

such that

$$b_1/r_1 < b_2/r_2 < b_3/r_3,$$
$$|\theta - b_j/r_j| < N^{-1}r_j^{-1} \quad \text{for} \quad j = 1, 2, 3.$$

Hence either

$$\theta - b_2/r_2 > 0 \quad \text{or} \quad \theta - b_2/r_2 < 0,$$
$$0 < b_2/r_2 - b_1/r_1 < \theta - b_1/r_1 \quad \text{or} \quad 0 < b_3/r_3 - b_2/r_2 < b_3/r_3 - \theta,$$
$$0 < b_2/r_2 - b_1/r_1 < N^{-1}r_1^{-1} \quad \text{or} \quad 0 < b_3/r_3 - b_2/r_2 < N^{-1}r_3^{-1},$$
$$1 \leqslant b_2r_1 - b_1r_2 < N^{-1}r_2 \quad \text{or} \quad 1 \leqslant b_3r_2 - b_2r_3 < N^{-1}r_2,$$
$$r_2 > N$$

which is a contradiction.

LEMMA 8. $$\sum_{\beta} = O_\eta(N^{-\frac{1}{2}}\epsilon^{-1}).$$

Proof. Repeating *mutatis mutandis* the calculation in the proof of Lemma 7 leading to (4.1) we obtain

$$\sum_{\beta} = O_\eta\Big(N^{-\frac{3}{2}} \sum_{N<r<N\epsilon^{-1}} r^{-1}\delta^{-1}\Big) \tag{4.2}$$

where, by (3.15) and (β),

$$N^{-2}\epsilon \leqslant \delta < N^{-2}.$$

We can cover the rectangle

$$N \leqslant r \leqslant N\epsilon^{-1}, \qquad N^{-2}\epsilon \leqslant \delta \leqslant N^{-2}$$

by $O(\log^2\epsilon^{-1})$ rectangles

$$\Omega \leqslant r \leqslant 2\Omega, \qquad \Delta \leqslant \delta \leqslant 2\Delta$$

allowing for overlapping such that

$$N \leqslant \Omega < N\epsilon^{-1}, \qquad N^{-2}\epsilon \leqslant \Delta < N^{-2}.$$

Hence, by (4.2) and Lemma 5,

$$\begin{aligned}
\sum_{\beta} &= O_\eta\Big[N^{-\frac{3}{2}}\log^2\epsilon^{-1}\max_{\Omega,\Delta}\{\Omega^{-1}\Delta^{-1}\Lambda(2\Omega, 2\Delta)\}\Big] \\
&= O_\eta\Big[N^{-\frac{3}{2}}\max_{\Omega,\Delta}\{\Omega^{-1}\Delta^{-1}(1+16\Omega^2\Delta)\}\Big] \\
&= O_\eta\Big\{N^{-\frac{3}{2}}(\max_{\Omega,\Delta}\Omega^{-1}\Delta^{-1}+16\max_{\Omega}\Omega)\Big\} \\
&= O_\eta\{N^{-\frac{3}{2}}(N\epsilon^{-1}+16N\epsilon^{-1})\} \\
&= O_\eta(N^{-\frac{1}{2}}\epsilon^{-1}).
\end{aligned}$$

LEMMA 9. $$\sum_{\gamma} = O_\eta(N^{-\frac{1}{2}}\epsilon^{-1}).$$

Proof.

$$\begin{aligned}
\sum_{\gamma} &\leqslant \sum_{\substack{N<r<N\epsilon^{-1}\\ \delta<N^{-2}\epsilon}} \sum_{\substack{d|r\\ d>N^{-1}r}} \sum_{l=1}^{\infty} \min\{1, (dl\epsilon)^{-2}\} r^{-\frac{1}{2}}d^{\frac{1}{2}} \\
&= O\Big\{\sum_{\substack{N<r<N\epsilon^{-1}\\ \delta<N^{-2}\epsilon}} r^{-\frac{1}{2}} \sum_{\substack{d|r\\ d>N^{-1}r}} d^{\frac{1}{2}}(d\epsilon)^{-1}\Big\} \\
&= O\Big(\epsilon^{-1} \sum_{\substack{N<r<N\epsilon^{-1}\\ \delta<N^{-2}\epsilon}} r^{-\frac{1}{2}} \sum_{\substack{d|r\\ d\geqslant N^{-1}r}} d^{-\frac{1}{2}}\Big) \\
&= O_\eta\Big(N^{\frac{1}{2}}\epsilon^{-1} \sum_{\substack{N<r<N\epsilon^{-1}\\ \delta<N^{-2}\epsilon}} r^{-1}\Big) \\
&= O_\eta\Big\{N^{\frac{1}{2}}\epsilon^{-1} \int_{N}^{N\epsilon^{-1}} \omega^{-1}\,d\Lambda(\omega, N^{-2}\epsilon)\Big\}.
\end{aligned} \tag{4.3}$$

By Lemma 5

$$\int_{N}^{N\epsilon^{-1}} \omega^{-1} d\Lambda(\omega, N^{-2}\epsilon) \leqslant (N\epsilon^{-1})^{-1}\Lambda(N\epsilon^{-1}, N^{-2}\epsilon) + \int_{N}^{N\epsilon^{-1}} \omega^{-2}\Lambda(\omega, N^{-2}\epsilon)\, d\omega$$

$$\leqslant N^{-1}\epsilon(1+2\epsilon^{-1}) + \int_{N}^{N\epsilon^{-1}} \omega^{-2}(1+2\omega^{2}N^{-2}\epsilon)\, d\omega$$

$$\leqslant N^{-1}\epsilon + 2N^{-1} + N^{-1} + 2(N\epsilon^{-1})(N^{-2}\epsilon) < 6N^{-1}.$$

This inequality and (4.3) prove Lemma 9.

Proof of the theorem. Lemmas 6, 7, 8, 9 give

$$1 = O_{\eta}(N^{-\frac{1}{2}}\epsilon^{-1}),$$

$$\epsilon = O_{\eta}(N^{-\frac{1}{2}}),$$

which is the theorem.

[Received 30 April, 1948]

26.

On Discrete Harmonic Functions

Proceedings of the Cambridge Philosophical Society 45 (1949), 194–206
[Commentary on p. 568]

1. Introduction

A function $f(x_1, x_2)$ of two real variables x_1, x_2 which are restricted to rational integers will be called discrete harmonic (d.h.) if it satisfies the difference equation

$$4f(x_1, x_2) = f(x_1+1, x_2) + f(x_1-1, x_2) + f(x_1, x_2+1) + f(x_1, x_2-1). \qquad (1\cdot1)$$

This equation can be considered as the direct analogue either of the differential equation

$$\frac{\partial^2 f}{\partial x_1^2} + \frac{\partial^2 f}{\partial x_2^2} = 0,$$

or of the integral equation

$$f(x_1, x_2) = \int_0^1 f\{x_1 + r\cos(2\pi\theta),\ x_2 + r\sin(2\pi\theta)\}\, d\theta$$

in the notation normally employed to harmonic functions.

The object of this paper is to develop the elementary theory of d.h. functions and to investigate how far it corresponds to the theory of harmonic functions. We shall find that many but not all the classical theorems remain valid for d.h. functions.

In many cases the theorems will remain true if we generalize our definition to n dimensions. A function of n (rational integer) variables $x_1, \ldots, x_n$ is called d.h. if

$$2nf(x_1, \ldots, x_n) = f(x_1+1, \ldots, x_n) + \ldots + f(x_1, \ldots, x_n - 1). \qquad (1\cdot2)$$

Several authors have considered similar problems, but none of the results seems to be relevant for our purpose.

2. Definitions and notation

The integer $n \geqslant 2$ always denotes the number of dimensions. We shall write $f(x)$ for $f(x_1, \ldots, x_n)$, denoting by x the point with the coordinates $x_1, \ldots, x_n$. Only points with rational integer coordinates will be considered. Two points x and y will be called neighbouring points if

$$\sum_{\nu=1}^{n} (x_\nu - y_\nu)^2 = 1.$$

A set of points is called connected if any two points of the set can be connected by a chain of neighbouring points which belong to the set. The minimum number of links in the chain will be called the distance of the two points in the set.

A domain consists of two types of points. The interior points which may be any connected set, and the boundary points. These are the points which do not belong to the connected set themselves, but which possess at least one neighbour belonging to the connected set.

It should be noted that a domain is not uniquely determined as a set of points, unless some rule is given to distinguish the interior points of the domain.

A function is d.h. in a domain D if it is defined for all points of D and if (1·2) holds for all interior points x of D. Only real functions will be considered.

A domain will be called finite if it contains only a finite number of points, otherwise it will be called infinite.

We introduce the following abbreviations:

$$o = (0, \ldots, 0),$$

$$u_1 = (1, 0, \ldots, 0), \ldots, \quad u_n = (0, 0, \ldots, 1),$$

$$d_\nu f(x) = f(x+u_\nu) - f(x) \quad \text{for} \quad 1 \leqslant \nu \leqslant n,$$

$$\Delta f(x) = \sum_{\nu-1}^{n} (f(x+u_\nu) + f(x-u_\nu) - 2f(x)),$$

$$\sum_D |f'(x)|^2 = \sum_{x,y} (f(x) - f(y))^2,$$

where x and y run through all pairs of neighbours of a finite domain D.

T_R is the domain whose interior points are

$$|x_1| < R, \ldots, |x_R| < R.$$

U_R is the domain whose interior points are

$$|x_1| + \ldots + |x_n| < R.$$

V_R is the domain whose interior points are

$$-R < x_1 \leqslant R, \quad |x_\nu| < R \quad \text{for} \quad 2 \leqslant \nu \leqslant n.$$

For $R > 2$ the domains T_R^* and U_R^* are similarly defined except that the origin is a boundary point.

The constants implied by the symbol O depend on n only.

3. The maximum principle and Dirichlet's principle

Theorem 1. *If $f(x)$ is d.h. on a finite domain D, then $f(x)$ is either a constant or it attains its maximum on D on the boundary only.*

Corollary. *If M is the upper bound of a function $f(x)$ which is d.h., bounded and not constant on an infinite domain D, then $f(x) < M$ for all interior points x of D.*

Proof. Let M be the maximum of $f(x)$ on D, and let x_0 be an interior point of D where $f(x_0) = M$. Then it follows from (1·2) that $f(x) = M$ for all neighbours x of x_0, and by induction that $f(x) = M$ for all points x of D.

Theorem 2. *Let $g(x)$ be a given real function defined on the boundary of a finite domain D. Then there exists one and only one d.h. function $f(x)$ which takes the values $g(x)$ on the boundary of D.*

If $h(x)$ is a real function defined on D which also takes the values $g(x)$ on the boundary of D, then

$$\sum_D |f'|^2 \leqslant \sum_D |h'|^2,$$

and the sign of equality holds only if $f(x) = h(x)$ for all x of D.

Proof. Let $f(x)$ be the function for which $\sum_D |f'|^2$ is a minimum, subject to our boundary condition. Then for each interior point x of D

$$0 = \frac{\partial}{\partial f(x)} \sum_D |f'|^2 = 2\sum_{x'} (f(x)-f(x')),$$

where the sum is extended over all neighbours x' of x; this proves that $f(x)$ is d.h.

If $F(x)$ were another d.h. function satisfying the boundary condition, then $F(x)-f(x) = 0$ by Theorem 1, which shows the uniqueness of $f(x)$ and establishes the theorem.

4. Discrete harmonic polynomials

Theorem 3. *For every integer $k \geqslant 1$ there are exactly*

$$\binom{k-2+n}{n-1}\frac{2k+n-1}{k}$$

linearly independent d.h. polynomials of degree not exceeding k.

Proof. For $k = 1$ the theorem is trivial since the $n+1$ polynomials $1, x_1, \ldots, x_n$ are all d.h. Hence we may assume that $k \geqslant 2$.

An easy count shows that there are

$$\binom{n+k}{n}$$

linearly independent polynomials of degree not exceeding k. Since every polynomial of degree not exceeding $k-2$ can be represented in the form

$$f(x) = \Delta g(x), \quad \deg g(x) \leqslant k,$$

the operator Δ maps the additive group of polynomials of degree not exceeding k on the subgroup of polynomials of degree not exceeding $k-2$. Since

$$\Delta(f(x)+g(x)) = \Delta f(x) + \Delta g(x),$$

this mapping is a homomorphism.

Hence the quotient group is isomorphic to the group of all d.h. polynomials of degree not exceeding k; and the latter contains

$$\binom{k+n}{n} - \binom{k-2+n}{n} = \binom{k-2+n}{n-1}\frac{2k+n-1}{k}$$

linearly independent elements.

Examples. For $n = 2$ the d.h. polynomials are

$$1;\quad x_1, x_2;\quad x_1^2-x_2^2,\ 2x_1x_2;\quad x_1^3-3x_1x_2^2,\ 3x_1^2x_2-x_2^3;$$

$$x_1^4-6x_1^2x_2^2+x_2^4-(x_1^2+x_2^2),\ 4x_1^3x_2-4x_1x_2^3;$$

$$x_1^5-10x_1^3x_2^2+5x_1x_2^4-\tfrac{10}{3}x_1^3,\ \ldots.$$

We notice that in this sequence each polynomial of degree k is of the form

$$\Re(x_1+x_2 i)^k + \text{terms of degree not exceeding } k-2,$$

$$\Re(i^{-1}(x_1+x_2 i)^k) + \text{terms of degree not exceeding } k-2,$$

a fact which is easily verified by direct calculation. It must be pointed out that without further rules the above sequence is not uniquely defined, as there is a considerable degree of freedom in the choice of the lower terms. In this respect the situation differs fundamentally from the continuous case, where the choice of our polynomials is easily made unique by the orthogonality rule.

We also give some examples of the case $n = 3$. By Theorem 3 we have

$$\binom{k+1}{2}\frac{2k+2}{k} = (k+1)^2$$

linearly independent d.h. polynomials of degree not exceeding k, i.e. we have $2k+1$ linearly independent d.h. polynomials of degree k. The first examples are

$$1;\quad x_1, x_2, x_3;\quad x_1x_2, x_2x_3, x_3x_1, x_1^2-x_2^2, x_2^2-x_3^2;$$
$$x_1x_2x_3, x_1^3-3x_1x_2^2, \ldots; x_1x_2(x_1^2+x_2^2-6x_3^2), \ldots,$$
$$x_1^4-6x_1^2x_2^2+x_2^4-(x_1^2+x_2^2), \ldots, x_1x_2(x_1^2-x_2^2), \ldots.$$

The question naturally arises: Is it true that an integer function (i.e. a function d.h. everywhere) can be written as a sum of a unique series of polynomials, at least if $n = 2$? It is difficult to see how the classical theorem on the one-one relation between integer functions and convergent power series can be formulated for d.h. functions, since our sequence of linearly independent polynomials is not uniquely defined. It is trivial that every convergent series of d.h. polynomials converges towards an integer d.h. function. In the opposite direction we start our discussion by proving

THEOREM 4. *If $n = 2$, L is a positive integer, and if $f(x)$ is d.h. on T_L, then we can find a d.h. polynomial $P(x)$ such that $f(x) = P(x)$ on T_L.*

Proof. We make the assertion of the theorem more precise by a further specification of the polynomial $P(x)$. We demand that $P(x)$ shall be a linear combination of polynomials of degree less than $4L-2$ and of the polynomial of degree $4L-2$ of the above sequence which is of the form

$$x_1^{4L-2}-\binom{4L-2}{2}x_1^{4L-4}x_2^2+\ldots.$$

This rule puts $8L-4$ polynomials at our disposal, and the domain T_L has $8L-4$ boundary points. Hence there are two possibilities: Either we can find a linear combination of our $8L-4$ polynomials which assumes the same boundary values as $f(x)$. In this case our theorem is proved. Or the $8L-4$ boundary values of our $8L-4$ polynomials are not linearly independent. In this case there exists a linear combination $Q(x)$ of these polynomials which vanishes on the boundary of T_L, $Q(x)$ not being identically zero. Hence $Q(x)$ vanishes on all points of T_L, and, being d.h., vanishes on all points of U_{2L-1}.

Hence $Q(x)$ has $4L-1$ zeros on the line $x_1 = 0$, and since its degree is less than $4L-1$, $Q(x)$ is either identically zero or divisible by x_1. Similarly, since $Q(x)$ is not identically zero, $Q(x)$ is divisible by x_2. Putting

$$Q(x) = x_1x_2Q_1(x),$$

we have identically

$$Q(x) = x_1x_2Q_1(x) = \alpha\left\{x_1^{4L-2}+\sum_{\nu=1}^{2L-1}(-1)^\nu\binom{4L-2}{2\nu}x_1^{4L-2-2\nu}x_2^{2\nu}\right\}+R(x),$$

where α is a constant and $R(x)$ a polynomial of degree less than $4L-2$. This is only possible if $\alpha = 0$, and if $Q(x)$ is of degree less than $4L-2$; hence $Q_1(x)$ is of degree less than $4L-4$. On each of the four lines $x_1 = \pm 1$, $x_2 = \pm 1$ the polynomial $Q_1(x)$ has at least $4L-4$ zeros, hence it must vanish on these lines identically and

$$Q_1(x_1, x_2) = (x_1^2 - 1)(x_2^2 - 1)\, Q_2(x_1, x_2),$$

where Q_2 is a polynomial of degree less than $4L-8$. Continuing this process we obtain polynomials $Q_l(x_1, x_2)$ of degree less than $4(L-l)$ which satisfy

$$Q_l(x_1, x_2) = (x_1^2 - l^2)(x_2^2 - l^2)\, Q_{l+1}(x_1, x_2)$$

for $1 \leqslant l < L$. Q_{L-1} is of degree less than 4 and has 4 zeros on each of the lines

$$x_1 = \pm(L-1), \quad x_2 = \pm(L-1).$$

Hence Q_{L-1} is identically zero and $Q(x)$ is identically zero, which gives the desired contradiction.

Theorem 4 settles the equation of analytic continuation for a function defined on a square parallel to the axes. It is easily seen that a d.h. function defined on a domain, whose interior points are the lattice points of a convex set in the Euclidean plane, can be continued to a square and therefore is equal to a polynomial.

On the other hand, the function represented by the diagram below (the enclosed points are the interior points of the domain) cannot be continued as a d.h. function to the point marked *; hence it does not equal a d.h. polynomial.

	0				
−1	0	0			
4	1	0	*		
−1	0	0	0	0	
0	0	0	0	0	0
	0	0	0	0	

To sum up: All finite domains can be divided into two classes. Functions d.h. on a domain of the first class can be continued to a square parallel to the axis and can therefore be represented by a d.h. polynomial. To this class belong all convex domains.

Functions d.h. on a domain of the second class cannot always be represented (or even approximated to) by d.h. polynomials. To this class belong the domain in the above diagram and all domains not 'simply connected'.

For infinite domains there is a similar division into two classes; but functions d.h. on an infinite domain of the first class can only be continued to integer functions which can be approximated to by a sequence of polynomials. Naturally this approximation is not uniform, and no analogue to the absolute convergence in the classical theory seems to exist.

5. Liouville's theorem

We next proceed to prove the analogue of Liouville's theorem on bounded integer functions.

Theorem 5. *If $f(x)$ is d.h. everywhere and satisfies the inequality*

$$|f(x)| \leqslant M \tag{5·1}$$

for all x, where M is a constant, then $f(x)$ is constant.

Proof (indirect). We may assume without loss of generality that

$$d_1 f(x) = g(x)$$

is not zero everywhere. Since $|g(x)| \leqslant 2M$, there exists a positive upper bound m such that

$$|g(x)| \leqslant m \leqslant 2M \tag{5·2}$$

everywhere, and without loss of generality

$$g(x) > m - \epsilon$$

somewhere for every $\epsilon > 0$.

We choose an integer l such that $\quad lm > 2M,$

and a positive δ such that $\quad lm(1-(2n)^l \delta) > 2M.$

Then we can find a point $x^{(0)}$ such that

$$g(x^{(0)}) > m - \delta. \tag{5·3}$$

We put $\quad x^{(\lambda)} = x^{(0)} + \lambda u_1 \quad \text{for} \quad 0 < \lambda \leqslant l.$

Then, since $g(x)$ is d.h., $\quad 2ng(x^{(0)}) \leqslant (2n-1)m + g(x^{(1)}),$

and by (5·3) $\quad g(x^{(1)}) \geqslant m - 2n\delta.$

Applying the same argument again, we obtain by induction for $0 \leqslant \lambda \leqslant l$

$$g(x^{(\lambda)}) \geqslant m - (2n)^\lambda \delta.$$

Hence

$$2M \geqslant f(x^{(l)}) - f(x^{(0)}) = \sum_{\lambda=0}^{l-1} g(x^{(\lambda)}) \geqslant \sum_{\lambda=0}^{l-1} (m - (2n)^\lambda \delta)$$

$$> l(m - (2n)^l \delta) > 2M,$$

which is the desired contradiction.

A natural extension of Theorem 5 is

THEOREM 6. *If $f(x)$ is d.h. everywhere and satisfies the inequality*

$$f(x) = O(1 + (|x_1| + \ldots + |x_n|)^k)$$

everywhere, where k is an integer, then $f(x)$ is a polynomial of degree not exceeding k.

A proof of this theorem will be given in the last paragraph.

6. Some special boundary problems

The simplest problem in two dimensions refers to the domain whose interior points are all points except the origin. We prove the following generalization of Theorem 5:

THEOREM 7. *If $n = 2$, and if $f(x)$ is bounded for all x and d.h. for all $x \neq o$, then $f(x)$ is a constant.*

Proof. We may assume without loss of generality that

$$f(o) = 1, \quad 0 \leqslant f(x) \leqslant 2. \tag{6·1}$$

Let $R > 1$ be an integer and let $f_R(x)$ be the function d.h. on U_R^* which satisfies

$$f(o) = 1, \quad f(x) = 0 \quad \text{for} \quad |x_1| + |x_2| = R.$$

Clearly $\quad 0 \leqslant f_R(x) \leqslant f_{R+1}(x) \leqslant 1, \quad f_R(x) \leqslant f(x)$

for all points of U_R^*, hence the sequence $f_1(x), f_2(x), f_3(x), \ldots$ converges towards a limit $f_\infty(x)$ which satisfies $\quad 0 \leqslant f_R(x) \leqslant f_\infty(x) \leqslant 1, \quad f_\infty(x) \leqslant f(x),$

provided that x belongs to U_R^*. If we can show that $f_\infty(x) = 1$ everywhere, it will follow that $f(x) \geq 1$ everywhere, and since (6·1) is symmetric in $f(x)$ and $2-f(x)$, that $2-f(x) \geq 1$, or $f(x) \leq 1$.

We define

$$g_R(x) = 1 - \frac{\log(1+|x_1|+|x_2|)}{\log(1+R)}.$$

Then

$$\sum_{U_R^*} |g_R'(x)|^2 = O\sum_{\nu=1}^{R} \nu\left(\frac{\log(1+\nu)}{\log(1+R)} - \frac{\log \nu}{\log(1+R)}\right)^2$$

$$= O\left(\sum_{\nu=1}^{R} \nu^{-1}\log^{-2} R\right) = O(\log^{-1} R).$$

Since $f_R(x)$ is d.h. on U_R^*, and since $f_R(x)$ has the same boundary values as $g_R(x)$, we have, by Theorem 2,

$$\sum_{U_R^*} |f_R'(x)|^2 \leq \sum_{U_R^*} |g_R'(x)|^2 = O(\log^{-1} R).$$

Hence for each x, as $R \to \infty$

$$d_1 f_R(x) = o(1), \quad d_2 f_R(x) = o(1),$$

$$d_1 f_\infty(x) = d_2 f_\infty(x) = 0, \quad f_\infty(x) = 1.$$

In three or more dimensions the situation is different. We shall prove only

THEOREM 8. *For $n = 3$ there exists a function which is bounded everywhere, d.h. everywhere except at the origin and not a constant.*

Before we proceed to prove this theorem we shall establish a lemma which is well known in the classical calculus of variations.

LEMMA 1. *Let $u(\xi_1, \xi_2, \xi_3)$ be continuous for $1 \leq \xi_1^2+\xi_2^2+\xi_3^2 \leq (2R)^2$ and let u have continuous bounded partial first derivatives almost everywhere in this domain. Let*

$$u(\xi_1,\xi_2,\xi_3) \geq \tfrac{1}{2} \quad \textit{for} \quad \xi_1^2+\xi_2^2+\xi_3^2 = 1$$

and

$$u(\xi_1,\xi_2,\xi_3) = 0 \quad \textit{for} \quad \xi_1^2+\xi_2^2+\xi_3^2 = (2R)^2.$$

Then

$$\iiint \left\{\left(\frac{\partial u}{\partial \xi_1}\right)^2 + \left(\frac{\partial u}{\partial \xi_2}\right)^2 + \left(\frac{\partial u}{\partial \xi_3}\right)^2\right\} d\xi_1 d\xi_2 d\xi_3 \geq \pi \frac{2R}{2R-1}, \quad 1 \leq \xi_1^2+\xi_2^2+\xi_3^2 \leq (2R)^2.$$

Proof. On introducing polar coordinates ρ, θ, λ the integral is transformed into

$$\int_{-\frac{1}{2}\pi}^{\frac{1}{2}\pi} \cos\theta\, d\theta \int_{-\pi}^{\pi} d\lambda \int_1^{2R} \rho^2 \left\{\left(\frac{\partial u}{\partial \rho}\right)^2 + \rho^2\left(\frac{\partial u}{\partial \theta}\right)^2 + \rho^2\cos^2\theta\left(\frac{\partial u}{\partial \lambda}\right)^2\right\} d\rho$$

$$\geq \int_{-\frac{1}{2}\pi}^{\frac{1}{2}\pi} \cos\theta\, d\theta \int_{-\pi}^{\pi} d\lambda \int_1^{2R} \rho^2 \left(\frac{\partial u}{\partial \rho}\right)^2 d\rho.$$

Since, by the Cauchy-Schwarz inequality

$$\frac{2R-1}{2R}\int_1^{2R} \rho^2\left(\frac{\partial u}{\partial \rho}\right)^2 d\rho = \int_1^{2R} \rho^{-2} d\rho \int_1^{2R} \rho^2 \left(\frac{\partial u}{\partial \rho}\right)^2 d\rho \geq \left(\int_1^{2R} \frac{\partial u}{\partial \rho} d\rho\right)^2 \geq \tfrac{1}{4},$$

the result follows.

Proof of Theorem 8. As in the proof of Theorem 7 we define $f_R(x)$ as the function d.h. on U_R^* which satisfies $f_R(o) = 1$ and vanishes at all other points of the boundary of U_R^*. $f_R(x)$ increases with R and tends to a limit $f_\infty(x)$. We want to show that $f_\infty(x) = 1$ is not true everywhere, hence we may assume at once that for sufficiently large R

$$f_R(\pm u_\nu) \geq \tfrac{1}{2} \quad \text{for} \quad 1 \leq \nu \leq 3.$$

We now proceed to construct a continuous function $u_R(\xi_1, \xi_2, \xi_3)$ which will coincide with $f_R(x)$ if $\xi_1 = x_1, \xi_2 = x_2, \xi_3 = x_3$. For every cube parallel to the axes, of unit volume, whose vertices have integer coordinates, we can find 8 constants $\alpha_{000}, \alpha_{100}, \ldots, \alpha_{111}$ such that the trilinear form

$$u_R(\xi_1, \xi_2, \xi_3) = \alpha_{000} + \alpha_{100}\xi_1 + \alpha_{010}\xi_2 + \alpha_{001}\xi_3 + \alpha_{110}\xi_1\xi_2 + \alpha_{101}\xi_1\xi_3 + \alpha_{011}\xi_2\xi_3 + \alpha_{111}\xi_1\xi_2\xi_3$$

takes the same values at the vertices of the cube as the function $f_R(x)$ ($f_R(x)$ being zero outside U_R^*).

This function $u_R(\xi_1, \xi_2, \xi_3)$ is uniquely defined everywhere and satisfies the assumptions of Lemma 1. Further for each of our cubes C

$$\iiint_C \left\{\left(\frac{\partial u_R}{\partial \xi_1}\right)^2 + \left(\frac{\partial u_R}{\partial \xi_2}\right)^2 + \left(\frac{\partial u_R}{\partial \xi_3}\right)^2\right\} d\xi_1 d\xi_2 d\xi_3 \leqslant \sum_C |f_R'(x)|^2.$$

Hence, by Lemma 1, since every edge belongs to four different cubes

$$4\sum_{U_R^*} |f_R'(x)|^2 \geqslant \pi \frac{2R}{2R-1},$$

and, putting $A_R = \sum_{U_R^*} |f_R'(x)|^2$, for $R > 1$, we obtain

$$A_R \geqslant \tfrac{1}{4}\pi.$$

Let σ_R be defined by $\quad \sigma_R = f_R(\pm u_\nu) \quad (1 \leqslant \nu \leqslant 3).$

We now define a function $g_R(x, \rho)$ by

$$g_R(o, \rho) = 1, \quad g_R(x, \rho) = \rho f_R(x) \quad \text{for} \quad x \neq o,\ x \text{ in } U_R^*,$$

where ρ is a positive variable. Then

$$\sum_{U_R^*} |g_R'(x, \rho)|^2 = 6(1-\rho\sigma_R)^2 + \rho^2\{A_R - 6(1-\sigma_R)^2\}.$$

As $g_R(x, 1) = f(x)$ and as $f_R(x)$ is d.h., this expression must be a minimum if $\rho = 1$. Differentiation with respect to ρ gives

$$\frac{d}{d\rho}\sum_{U_R^*} |g_R'(x, \rho)|^2 = -12\sigma_R(1-\rho\sigma_R) + 2\rho\{A_R - 6(1-\sigma_R)^2\}.$$

If we put $\rho = 1$ we obtain

$$0 = 12\sigma_R + 2A_R - 12, \quad \sigma_R = 1 - \tfrac{1}{6}A_R \leqslant 1 - \frac{\pi}{24},$$

whence

$$f_\infty(\pm u_\nu) \leqslant 1 - \frac{\pi}{24} \quad (1 \leqslant \nu \leqslant 3).$$

We conclude this section with the construction of a function which has many useful properties.

THEOREM 9. *For $n \geqslant 2$ there exists a function $h(x)$ d.h. everywhere with the following properties:*

$$h(o) = 1, \tag{6·2}$$

$$h(x) = 0 \quad \textit{for} \quad x_n = 0,\ x \neq o, \tag{6·3}$$

$$h(x) > 0 \quad \textit{for} \quad x_n > 0, \tag{6·4}$$

$$h(x) = O(x_n^{1-n}) \quad \textit{for} \quad x_n > 0. \tag{6·5}$$

Proof. Let $\zeta_1, \ldots, \zeta_{n-1}$ be $n-1$ real continuous variables. We define $\phi(\zeta_1, \ldots, \zeta_{n-1})$ as the smaller root of the quadratic equation

$$\phi + \phi^{-1} + 2\sum_{\nu=1}^{n-1} \cos \zeta_\nu = 2n. \tag{6·6}$$

An elementary calculation shows easily that

$$(4n-2)^{-1} < \phi(\zeta_1, \ldots, \zeta_{n-1}) \leqslant 1.$$

We put

$$h(x) = (2\pi)^{1-n} \int_{-\pi}^{\pi} \ldots \int_{-\pi}^{\pi} \cos(x_1\zeta_1) \ldots \cos(x_{n-1}\zeta_{n-1})\, \phi^{x_n}(\zeta_1, \ldots, \zeta_{n-1})\, d\zeta_1 \ldots d\zeta_{n-1}.$$

Since ϕ satisfies (6·6) we see at once that $h(x)$ is d.h. everywhere. (6·2) and (6·3) are now trivial.

To prove (6·5) we observe that

$$\phi(\zeta_1, \ldots, \zeta_n) = 1 - (\zeta_1^2 + \ldots + \zeta_{n-1}^2)^{\frac{1}{2}} + O(\zeta_1^2 + \ldots + \zeta_n^2).$$

Hence

$$\int_{\zeta_1^2 + \ldots + \zeta_{n-1}^2 \leqslant x_n^{-1}} \cdots\cdots \int \phi^{x_n}(\zeta_1, \ldots, \zeta_{n-1})\, d\zeta_1 \ldots d\zeta_{n-1}$$

$$= O \int_{\zeta_1^2 + \ldots + \zeta_{n-1}^2 \leqslant x_n^{-1}} \cdots\cdots \int \exp[-x_n(\zeta_1^2 + \ldots + \zeta_{n-1}^2)^{\frac{1}{2}}]\, d\zeta_1 \ldots d\zeta_{n-1}$$

$$= O \int_{-\infty}^{\infty} \ldots \int_{-\infty}^{\infty} \exp[-x_n(\zeta_1^2 + \ldots + \zeta_{n-1}^2)^{\frac{1}{2}}]\, d\zeta_1 \ldots d\zeta_{n-1}$$

$$= O(x_n^{1-n}),$$

whereas

$$\underset{\zeta_1^2 + \ldots + \zeta_{n-1}^2 \geqslant x_n^{-1}}{\int_{-\pi}^{\pi} \ldots \int_{-\pi}^{\pi}} \phi^{x_n}(\zeta_1, \ldots, \zeta_{n-1})\, d\zeta_1 \ldots d\zeta_{n-1}$$

$$= O \int_{-\pi}^{\pi} \ldots \int_{-\pi}^{\pi} \{\exp[-x_n^{-\frac{1}{2}} + O(x_n^{-1})]\}^{x_n}\, d\zeta_1 \ldots d\zeta_{n-1}$$

$$= O[\exp(-x_n^{\frac{1}{2}})] = O(x_n^{1-n}).$$

This proves (6·5), since $\qquad h(x) \leqslant h(o + x_n u_n).$

Let $M(u)$ denote the minimum or the greatest lower bound of $h(x)$ for $x_n = u$, $u \geqslant 0$. Clearly

$$2M(u) \geqslant M(u+1) + M(u-1) \quad \text{for} \quad u > 0.$$

Since $M(0) = 0$ and $\lim_{u \to \infty} M(u) = 0$, it follows that $M(u) \geqslant 0$. This proves $h(x) \geqslant 0$ for $x_n \geqslant 0$, and (6·4) follows by Theorem 1.

7. General boundary problems

Theorem 10. *Let S be an infinite domain with at least one boundary point. If $F(x)$ is a bounded function defined for all boundary points of S, then the definition of $F(x)$ can be extended to all points of S such that $F(x)$ is bounded and d.h. in S.*

If $n = 2$ this process is unique; if $n > 2$ it is not in general unique.

Proof. Without loss of generality we may assume that

$$0 \leqslant F(x) \leqslant 1$$

on the boundary of S. Let x_0 be an interior point of S, and let S_1 be the domain which contains only x_0 as interior point. The domains $S_2, S_3, \ldots$ are defined by induction by the following rule: The interior points of S_{l+1} are *all* points of S_l which are interior points of S. Clearly each S_l contains all preceding domains and every x which is an interior point of S belongs to the interior of S_l, if l is sufficiently large. For each $l > 0$ we find the function $f_l(x)$ d.h. on S_l which vanishes on the boundary points of S_l which are interior to S, and which satisfies $f_l(x) = F(x)$ on the boundary points of S_l which are boundary points of S. Clearly $f_l(x)$ is an increasing sequence for fixed x, hence it converges to a function $F(x)$ for all x in S, which has the required properties.

That $F(x)$ need not be unique for $n > 2$ has been demonstrated by Theorem 8.

For $n = 2$ we assume that $|F(x)| \leqslant M$ for all points of S and $F(x) = 0$ on the boundary of S, and we have to show that $F(x)$ vanishes on all points of S. We may assume that o is a boundary point of S. Let W_R for integers $R > 1$ be the largest domains with o as boundary point whose interior points are interior points of U_R and of S. It is clear that W_R exists and is unique for sufficiently large R, and that every boundary point of W_R is either a boundary point of U_R or of S. The function

$$F(x) + M\{1 - f_R(x)\}$$

is d.h. on all interior points x of W_R, $f_R(x)$ having the same meaning as in the proof of Theorem 7. If x lies on the boundary of S, $F(x) = 0$, whereas $f_R(x) = 0$ if x lies on the boundary of U_R. Hence

$$F(x) + M\{1 - f_R(x)\} \geqslant 0$$

for all x of W_R. Since each fixed interior point x of S belongs to W_R if R is sufficiently large, and since $\lim_{R \to \infty} f_R(x) = 1$, it follows that $F(x) \geqslant 0$ for all x of S. In a similar way one proves $F(x) \leqslant 0$ which completes the proof of the theorem.

THEOREM 11. *Let S be an infinite domain with at least one boundary point. If $F(x)$ is a function defined for all boundary points of S, then the definition of $F(x)$ can be extended to all points of S such that $F(x)$ is d.h. in S.*

Proof. For each positive integer m we define $F_m(x)$ on the boundary points of S by

$$F_m(x) = \begin{cases} m & \text{if} \quad F(x) > m, \\ F(x) & \text{if} \quad |F(x)| \leqslant m, \\ -m & \text{if} \quad F(x) < -m. \end{cases}$$

Then the sequence $F_m(x)$ converges towards $F(x)$ on the boundary of S and we can define $F_m(x)$ as a d.h. function in S by virtue of Theorem 10. But it does not follow that the sequence converges for all points of S.

Let $x_1, x_2, x_3, \ldots$ run through all interior points of S in any order. We shall construct for each integer $l \geqslant 0$ an infinite sequence $f_{l,m}(x)$ of functions d.h. on S which converges towards $F(x)$ as $m \to \infty$ and for which the limit

$$\lim_{m \to \infty} f_{l,m}(x_\nu) = F(x_\nu) \quad (1 \leqslant \nu \leqslant l)$$

exists and is independent of l.

For $l = 0$, the sequence $f_{0,m}(x) = F_m(x)$ has the required property. We apply induction and assume that we have constructed sequences $f_{0,m}(x), \ldots, f_{l,m}(x)$ according to our rule. Then the sequence $f_{l,m}(x_{l+1})$ has either a convergent subsequence or it has a subsequence such that the quotient of two consecutive terms tends to infinity. In the first case our proposition can obviously be established for $l+1$. In the second case let $g_{l,m}(x)$ be the subsequence of $f_{l,m}(x)$ with

$$g_{l,m+1}(x_{l+1})/g_{l,m}(x_{l+1}) \to \infty \quad \text{as} \quad m \to \infty.$$

Put
$$\rho_m = \frac{g_{l,m}(x_{l+1})}{g_{l,m}(x_{l+1}) - g_{l,m+1}(x_{l+1})}.$$

Then
$$\rho_m \to 0,$$

and the sequence
$$f_{l+1,m}(x) = (1-\rho_m)\, g_{l,m}(x) + \rho_m g_{l,m+1}(x)$$

has the required property, since

$$\lim_{m\to\infty} f_{l+1,m}(x) = \lim_{m\to\infty} g_{l,m}(x)$$

for all x for which the limit on the right exists, and since

$$f_{l+1,m}(x_{l+1}) = 0 \quad \text{for} \quad m = 1, 2, 3, \ldots.$$

Thus our function $F(x)$ is defined and d.h. everywhere in S.

It is easily seen that in Theorem 11 the function $F(x)$ is never unique. The answer to the question how many linearly independent functions $F(x)$ will satisfy Theorem 11 depends in a rather complicated way on the structure of the domain S. Even the question whether the number of linearly independent solutions is finite or infinite is not easily answered.

A further problem arises if we subject $F(x)$ to some conditions restricting its magnitude. We have seen that if $n = 2$ the boundedness of the solution implies uniqueness, and for certain types of domain a weaker restriction will still preserve uniqueness. This leads to theorems of the Phragmén-Lindelöf type. Many interesting questions arise, but it seems hopeless to formulate a theorem of reasonable generality.

8. Some elementary inequalities

It is easily seen that an analogue of Poisson's formula can be established for d.h. functions. We limit ourselves to a rectangular domain D, whose interior points are given by the inequalities $\alpha_\nu < x_\nu < \beta_\nu \quad (\nu = 1, \ldots, n).$

If $f(x)$ is d.h. on D, and if y is a point of D we have

$$f(y) = \sum_x K(x,y)\, f(x),$$

where x runs through all boundary points of D. Here $K(x,y)$ is the d.h. function of y which vanishes on all boundary points of D except at the point $y = x$ where it assumes the value $K(x,x) = 1$. It follows at once that for every interior point y of D

$$0 < K(x,y) < 1.$$

The exact calculation of $K(x,y)$ is very tedious even for relatively small domains D. We shall restrict ourselves to prove

THEOREM 12. $$K(x,y) = O(|x-y|^{1-n}),$$
where $|x-y|$ *is the Euclidean distance from* x *to* y.

Proof. Without loss of generality we may assume that $x = o$, $y_n \geqslant n^{-1}|y-o|$. Then for all y on the boundary of D $\quad 0 \leqslant K(x,y) \leqslant h(y)$.

Since this inequality is also true for all points y interior to D, the theorem follows from (6·5).

As an easy consequence of Theorem 12 we prove

THEOREM 13. *If* $f(x)$ *is d.h. on a rectangular domain* D, *and if* y *is an interior point of* D *which has a Euclidean distance not less than* $R > 0$ *from every boundary point of* D, *then*

$$f^2(y) = O(R^{1-n}) \sum_x f^2(x),$$

where x *runs through all boundary points of* D.

Proof. Since $$1 = \sum_x K(x,y),$$

Theorem 12 gives $$f^2(y) = \left\{\sum_x K(x,y) f(x)\right\}^2 \leqslant \sum_x K^2(x,y) \sum_x f^2(x)$$
$$\leqslant \max_x K(x,y) \sum_x f^2(x) = O(R^{1-n}) \sum_x f^2(x).$$

For the proof of our last theorem we require

LEMMA 2. *If* $f(x)$ *is d.h. on* V_R *and if*

$$f(x) \geqslant 0 \quad \textit{for} \quad x_1 > 0, \qquad f(x) \leqslant 0 \quad \textit{for} \quad x_1 \leqslant 0 \tag{8·1}$$

on the boundary of V_R, *then* $d_1 f(o) \geqslant 0$.

If $f(x)$ *also satisfies* $$f(x) + f(u_1 - x) = 0 \tag{8·2}$$
on all points of V_R, *then* (8·1) *holds on all points of* V_R.

Proof. Put $$g(x) = \begin{cases} f(x) & \text{if} \quad (x_1 - \frac{1}{2}) f(x) \geqslant 0, \\ 0 & \text{if} \quad (x_1 - \frac{1}{2}) f(x) < 0. \end{cases}$$

Then $g(x) = f(x)$ on the boundary of V_R and, if (8·2) holds,

$$\sum_{V_R} |g'(x)|^2 \leqslant \sum_{V_R} |f'(x)|^2.$$

Since $f(x)$ in d.h. on V_R, it follows that $g(x) = f(x)$ for all points of V_R and (8·1) holds for all points of V_R. This proves the second part of the lemma.

To prove the first part, we put

$$F(x) = f(x) - f(u_1 - x).$$

Then $F(x)$ satisfies (8·2) and (8·1) for all x of V_R; hence

$$d_1 f(o) = F(u_1) \geqslant 0.$$

THEOREM 14. *If* $f(x)$ *is d.h. on* V_R *and if on* V_R

$$|f(x)| \leqslant M,$$

then $$d_1 f(o) = O(MR^{-1}).$$

Proof. Without loss of generality $M = 1$. Let $f_R(x)$ be the function d.h. on V_R which has the boundary values

$$f_R(x) = \begin{cases} 1 & \text{if} \quad x_1 > 0, \\ -1 & \text{if} \quad x_1 \leqslant 0. \end{cases}$$

Clearly $f_R(x)$ satisfies (8·2) and (8·1) on all points of V_R. The function

$$g_R(x) = f_R(x) - \frac{x_1 - \frac{1}{2}}{R + \frac{1}{2}}$$

is also d.h. on V_R and satisfies (8·2) and (8·1) there. In particular,

$$\begin{aligned} g_R(x) &\geqslant 0 \quad \text{for} \quad x_1 = R, \\ g_R(x) &\leqslant 0 \quad \text{for} \quad x_1 = -R+1, \\ g_R(x) &= 0 \quad \text{for} \quad x_1 = R+1 \text{ and } x_1 = -R. \end{aligned}$$

Therefore we have on the boundary of T_R

$$d_1 g_R(x) \begin{cases} = 2 - \dfrac{1}{R+\frac{1}{2}} & \text{for} \quad x_1 = 0, \\ \leqslant 0 & \text{for} \quad |x_1| = R, \\ = -\dfrac{1}{R+\frac{1}{2}} & \text{for} \quad 0 < |x_1| < R. \end{cases}$$

Hence, by Theorem 12, $\qquad d_1 g_R(o) \leqslant O(R^{-1}),$

and, by Lemma 2, $\quad d_1 f(o) \leqslant d_1 f_R(o) = \dfrac{1}{R+\frac{1}{2}} + d_1 g_R(o) \leqslant O(R^{-1}).$

In a similar way it is proved that

$$d_1 f(o) \geqslant O(R^{-1})$$

and the theorem is established.

With the help of Theorem 14, Theorem 6 follows at once in the usual way.

The Royal Fort
Bristol 8

[Received 19 April, 1948]

27.

On Euclid's Algorithm in Cubic Self-Conjugate Fields

Proceedings of the Cambridge Philosophical Society 46 (1950), 377–382
[Commentary on p. 586]

In a paper published in these *Proceedings*† I proved that there are only a finite number of quadratic fields in which Euclid's Algorithm (E.A.) holds. Recently Davenport‡ has found a new proof of this theorem based on the theory of the minima of the product of linear inhomogeneous forms.

He has also been able to extend his result to higher fields, namely, to non-totally real cubic fields and to totally complex quartic fields§. In other words, he has shown that E.A. holds only in a finite number of fields which possess one fundamental unit only.

His results made me take up the problem again, and I shall prove in this paper

THEOREM 1. *E.A. holds only in a finite number of cyclic cubic fields.*

The question of E.A. in totally real non-cyclic cubic fields is thus left open. Though I cannot prove any result in the opposite direction I should be surprised to learn that the analogue of Theorem 1 is true in this case.

In another paper I intend to investigate the case of cyclic fields of degree greater than 3.

If the discriminant of a cyclic cubic field is divisible by t different primes, then the class-number of the field is divisible by 3^{t-1}. Hence it suffices to prove Theorem 1 for fields where discriminant is a power of a prime d. Actually, neglecting the single field of discriminant 81, we may assume that the discriminant equals d^2, where $d \equiv 1 \pmod 6$ is a prime.

A more detailed account of the most important properties of cyclic cubic fields is given in the Appendix.

We introduce the following conventions. Small italics, except e, i and o, denote positive rational integers; d, p, q always denote primes. The symbols O and o involve absolute constants only; unless otherwise stated the o-process refers to the limit as $d \to \infty$.

We assume $d \equiv 1 \pmod 6$, and denote by $\chi(n)$ a cubic character mod d. A, B, C denote the classes of integers for which $\chi(n) = 1$, $\chi(n) = \frac{1}{2}(-1+i\sqrt{3})$, $\chi(n) = \frac{1}{2}(-1-i\sqrt{3})$ respectively. A^* is the subclass of integers a in A which are of the form

$$a = bc, \quad (b,c) = 1, \quad b \text{ in } B, \quad c \text{ in } C.$$

$q_1 < q_2$ are the two smallest primes not in A; r is the smallest number in A^* prime to q_1.

† Vol. 34 (1938), 521–6.

‡ *Proc. London Math. Soc.* (in course of publication).

§ A preliminary account of the cubic case is given in *C.R. Acad. Sci., Paris*, 228 (1949), 883–5. Detailed proofs will appear in *Acta Mathematica*. I am indebted to Prof. Davenport for a private communication of his results.

We introduce a number ϵ which may be arbitrarily small, and will be fixed later. We put

$$x = [d^{\frac{1}{2}+\epsilon}],$$

and define K as the set of numbers n which satisfy

$$d^{\frac{1}{2}-\epsilon} < n \leqslant x.$$

Symbols like KA, $B+C$, etc., have the meaning usually assigned to them in Boolean algebra.

We recall two well-known facts:

$$\sum_{p\leqslant\eta} p^{-1} = \log\log\eta + D + o(1) \quad \text{as} \quad \eta\to\infty, \tag{1}$$

where D is a constant.

If $\psi(n)$ is a non-principal character mod k, then

$$\sum_{n=1}^{y} \psi(n) = O(k^{\frac{1}{2}}\log k) \tag{2}$$

for all y.

Lemma 1. *$q_1 r \leqslant d^{1-\epsilon}$ if d is sufficiently large.*

Proof. We distinguish three cases.

First case. $q_2 \leqslant d^{\frac{1}{2}-\epsilon}$. We assume without loss of generality that q_2 lies in B. Let c be the smallest number in C which is prime to q_1q_2. Since q_2c is prime to q_1 and belongs to A^*, it follows that

$$r \leqslant q_2 c,$$

$$q_1 r \leqslant q_1 q_2 c \leqslant d^{\frac{1}{2}-2\epsilon} c;$$

and the lemma will be proved if we can show that

$$c \leqslant x. \tag{3}$$

If $q_1 \geqslant 7$ it follows from (2) with $k = d$ that there are

$$\tfrac{1}{3}x(1+o(1))$$

numbers not exceeding x which lie in C. Since there are at most

$$(q_1^{-1}+q_2^{-1})\,x < \tfrac{1}{3}x(1+o(1))$$

numbers not exceeding x which are not prime to q_1q_2, (3) follows.

If $q_1 < 7 \leqslant q_2$ it follows from (2) with $k = q_1 d$ that there are

$$\tfrac{1}{3}(1-q_1^{-1})\,x(1+o(1)) \geqslant \tfrac{1}{6}x(1+o(1))$$

numbers not exceeding x which lie in C and are prime to q_1. Since there are at most

$$q_2^{-1}x \leqslant \tfrac{1}{7}x < \tfrac{1}{6}x(1+o(1))$$

numbers not exceeding x which are not prime to q_2, (3) follows again.

If $q_1 < q_2 < 7$ it follows from (2) with $k = q_1q_2d$ that there are

$$\tfrac{1}{3}(1-q_1^{-1})\,(1-q_2^{-1})x\,(1+o(1)) > 0$$

numbers not exceeding x which lie in C and are prime to q_1q_2. This implies (3) again, and the lemma is established in the first case.

Second case. $q_1 > d^{\frac{1}{2}-\epsilon}$. Then all primes $p \leqslant d^{\frac{1}{2}-\epsilon}$ belong to A and we have

$$\sum_{n=1}^{x} \chi(n) = \sum_{n=1}^{x} 1 + \sum_{p \text{ in } K} \sum_{\substack{n=1 \\ p/n}}^{x} (\chi(n)-1) + \sum_{p \text{ in } K} \sum_{p' \text{ in } K} \sum_{\substack{n=1 \\ pp'/n}}^{x} O(1).$$

Unless in the second sum on the right-hand side n/p is divisible by a prime p' in K. we have

$$\chi(n)-1 = \chi(p)-1;$$

hence

$$\begin{aligned} \sum_{n=1}^{x} \chi(n) &= x + \sum_{p \text{ in } K} (\chi(p)-1) \sum_{\substack{n=1 \\ p/n}}^{x} 1 + \sum_{p \text{ in } K} \sum_{p' \text{ in } K} \sum_{\substack{n=1 \\ pp'/n}}^{x} O(1) \\ &= x + \sum_{p \text{ in } K} (\chi(p)-1) \left[\frac{x}{p}\right] + \sum_{p \text{ in } K} \sum_{\substack{p' \text{ in } K \\ pp' \leqslant K}} O\left(\frac{x}{pp'}\right) \\ &= x + \tfrac{1}{2}(-3+i\sqrt{3})x \sum_{p \text{ in } KB} p^{-1} + \tfrac{1}{2}(-3-i\sqrt{3})x \sum_{p \text{ in } KC} p^{-1} \\ &\quad + O(\pi(x)) + O(x(\sum_{d^{\frac{1}{2}-\epsilon} < p \leqslant d^{\frac{1}{2}+2\epsilon}} p^{-1})^2). \end{aligned}$$

Dividing by x, and using (2) and (1), we get

$$o(1) = 1 + \tfrac{1}{2}(-3+i\sqrt{3}) \sum_{p \text{ in } KB} p^{-1} + \tfrac{1}{2}(-3-i\sqrt{3}) \sum_{p \text{ in } KC} p^{-1} + O(\epsilon^2),$$

whence

$$\sum_{p \text{ in } KB} p^{-1} = \tfrac{1}{3} + O(\epsilon^2), \quad \sum_{p \text{ in } KC} p^{-1} = \tfrac{1}{3} + O(\epsilon^2);$$

$$\sum_{p \text{ in } K(B+C)} p^{-1} = \tfrac{2}{3} + O(\epsilon_2).$$

Since by (1) for $0 < \alpha < \frac{1}{2}$

$$\sum_{d^\alpha < p \leqslant d^{\frac{1}{2}+\epsilon}} p^{-1} = \log\left((\tfrac{1}{2}+\epsilon)\alpha^{-1}\right) + o(1),$$

it follows for sufficiently large d that $K(B+C)$ contains at least two primes

$$q_1 < q_2 < d^{(\frac{1}{2}+\epsilon)\exp[-\frac{2}{3}+O(\epsilon)]}, \tag{4}$$

whilst KB and KC contain each at least two primes not exceeding

$$d^{(\frac{1}{2}+\epsilon)\exp[-\frac{1}{3}+O(\epsilon)]}$$

Hence

$$q_1 < d^{(\frac{1}{2}+2\epsilon)\exp[-\frac{2}{3}]}, \quad r < d^{(\frac{1}{2}+2\epsilon)(\exp[-\frac{2}{3}] + \exp[-\frac{1}{3}])},$$

and the lemma follows for sufficiently small ϵ, since

$$e^{-\frac{2}{3}} + \tfrac{1}{2}e^{-\frac{1}{3}} < 1.$$

Third case. $q_1 \leqslant d^{\frac{1}{2}-\epsilon} < q_2$. We repeat the calculation as in the second case, except that n runs only through values prime to q_1. Then by (2)

$$\sum_{\substack{n=1 \\ (n,q_1)=1}}^{x} \chi(n) = \sum_{n=1}^{x} \chi(n) - \chi(q_1) \sum_{m \leqslant xq_1^{-1}} \chi(m) = O(d^{\frac{1}{2}} \log d).$$

In the evaluation of the right-hand side of the first formula each principal term will acquire a factor $1 - q_1^{-1}$, which will cancel in the subsequent calculation. (4) is true again, and the remainder of the argument is the same as before.

LEMMA 2. *d is representable as the sum of two numbers in A^*, if d is sufficiently large.*

Proof. We observe first that if a in A^*, p in A, then ap in A^*.

Since by Lemma 1, $q_1 r < d$, we can write

$$d = sr + tq_1, \quad s < q_1.$$

Since all prime divisors of s lie in A, sr lies in A^*. Since $\chi(tq_1) = \chi(-1)\chi(sr) = 1$, tq_1 will lie in A^* if $(t, q_1) = 1$. Since this condition is not automatically satisfied, we have to overcome the difficulty by consideration of three separate cases.

First case, $q_1 = 2$. We consider the representation

$$d = r + 2t, \quad d = 27r + 2(t - 13r).$$

r and $27r$ lie in A^*, and $2t$ or $2(t-13r)$ lie in A^* if t or $t-13r$ are odd. If they were both even, their difference $13r$ would be even, which is impressible since $(q_1, r) = 1$.

Second case. $q_1 = 3$. We consider the representations

$$d = sr + 3t, \quad d = 4sr + 3(t - sr),$$

where $s \leqslant 2$. Again sr and $4sr$ lie in A^*, and $3t$ or $3(t-sr)$ lie in A^* if t or $t-sr$ are not divisible by 3. If they were both divisible by 3, their difference sr would be divisible by 3, which is impossible since $(q_1, r) = 1$.

Third case. $q_1 \geqslant 5$. We start with the representation

$$d = sr + tq_1,$$

where $s < q_1$. Then we can find a prime p_0 such that†

$$p_0 < q_1, \quad (p_0, s) = 1, \quad p_0 = O(\log d).$$

Hence there exists an $l < p_0 < q_1$ such that

$$s + lq_1 \equiv 0 \pmod{p_0},$$

and we have

$$d = (s + lq_1)r + (t - lr)q_1.$$

Since $s + lq_1 < p_0 q_1$, every prime divisor of $s + lq_1$ lies in A, $(s + lq_1)r$ lies in A^* and tq_1 or $(t - lr)q_1$ lie in A^*, if t or $t - lr$ are not divisible by q_1. If they were both divisible by q_1, their difference lr would be divisible by q_1, which is impossible since $(q_1, r) = 1$, $l < q_1$.

Proof of Theorem 1 (indirect). We assume we have a cyclic cubic field of discriminant d^2 in which E.A. holds, and that d is so large that Lemma 2 applies. This implies that

$$d = a + a',$$

where

$$\chi(a) = 1,$$

and none of the four numbers $\pm a$, $\pm a'$ is norm of an integer in the field.

The ideal d splits up into a cube $(\delta)^3$ of a principal self-conjugate ideal (δ). Since E.A. holds we can find for each n an integer α in the field such that

$$n \equiv \alpha \pmod{\delta}, \quad |N(\alpha)| < |N(\delta)| = d.$$

† See Lemma 4, loc. cit. in (1).

As the ideal (δ) is self-conjugate we have for each conjugate α' of α

$$n \equiv \alpha' \pmod{\delta},$$

and therefore $\qquad n^3 \equiv N(\alpha) \pmod{\delta}, \quad n^3 \equiv N(\alpha) \pmod{d}.$

We now fix n such that

$$n^3 \equiv a \pmod{d}, \quad N(\alpha) \equiv a \pmod{d}.$$

Since $\qquad |N(\alpha)| < d,$

this implies $\qquad N(\alpha) = a \quad \text{or} \quad N(\alpha) = -a',$

which is a contradiction.

APPENDIX

The principal theorem of class-field theory applied to cubic cyclic fields states†.

There is a one-one correspondence between cyclic cubic fields K and pairs of conjugate primitive cubic characters $\chi(n)$ and $\bar{\chi}(n) \pmod{d}$ such that:

(1) d^2 is the discriminant of the field K.

(2) Every rational prime divisor of d is the cube of a self-conjugate prime ideal in the field K.

(3) A prime p splits up into three different prime ideals if and only if $\chi(p) = \bar{\chi}(p) = 1$.

(4) All other rational primes remain primes in K.

It follows from (2), (3) and (4) that a rational integer

$$n = p_1^{l_1} \dots p_s^{l_s}$$

is norm of an ideal in K if and only if

$$\chi(p_j^{l_j}) \geqslant 0 \quad \text{for} \quad 1 \leqslant j \leqslant s.$$

Each primitive cubic character mod d is either a cubic character mod 9 or a cubic character mod p, where $p \equiv 1 \pmod 6$, or the product of a finite number of distinct characters of this type.

Hence by (1) the discriminant of a cyclic cubic field K is either 81 or p^2, where $p \equiv 1 \pmod 6$, or the product of a finite number of distinct factors of this type.

We now assume that

$$d = 9^u p_1 \dots p_{t-u}, \quad u = 0 \text{ or } 1, \quad t > 1,$$

so that the discriminant of K is divisible by t different primes. Let

$$3 < p = \mathfrak{p}^3$$

be such a divisor and α a number in K, not necessarily an integer (‡), which is prime to $\mathfrak{p}$. Then, since $\mathfrak{p}$ is a prime ideal of the first degree

$$\alpha \equiv a \pmod{\mathfrak{p}}$$

† We preserve the convention that small italics denote positive rational integers, but d need no longer be a prime. p continues to denote rational primes.

‡ We define $\alpha_1 \equiv \alpha_2 \pmod{\mathfrak{m}}$, if two integers β_1, β_2 exist such that $\beta_1 \equiv \beta_2 \equiv 1 \pmod{\mathfrak{m}}$ in the usual sense and $\alpha_1\beta_1 = \alpha_2\beta_2$. We call a number prime to d, if it is the quotient of two integers which are both prime to d.

for some rational integer a, and since $\mathfrak{p}$ is self-conjugate we have for each conjugate α' of α:

$$\alpha' \equiv a \pmod{\mathfrak{p}}, \quad N(\alpha) \equiv a^3 \pmod{\mathfrak{p}}, \quad N(\alpha) \equiv a^3 \pmod{p}.$$

Now let α be a number in K which is prime to d; then it follows that $N(\alpha)$ is a cubic residue mod d. We have already verified this for all prime divisors of d, and if $9/d$, it follows from (3) that $\chi(N(\alpha)) = 1$, and therefore $N(\alpha) \equiv \pm 1 \pmod 9$.

In other words there are $3^{-t}\phi(d)$ residue classes mod d which contain the norms of all numbers in K which are prime to d. But by (3) and Dirichlet's theorem about the existence of primes in arithmetic progression it follows that there are exactly $\frac{1}{3}\phi(d)$ residue classes mod d, which contain the norm of an ideal in K prime to d. Hence, if we limit ourselves to ideals prime to d, the class group of K contains a subgroup of index 3^{t-1}, namely the group of all ideals whose norm is a cubic residue mod d.

Since each ideal class in K contains an ideal prime to d, it follows that the class group in the above-mentioned sense is isomorphic to the class group in the usual sense. This implies that 3^{t-1} divides the class number of K.

The Royal Fort
Bristol 8

[Received 11 October, 1949]

28.

On Euclid's Algorithm in Cyclic Fields

Canadian Journal of Mathematics 3 (1951), 257–268
[Commentary on p. 586]

1. Introduction. In two papers I have proved that there are only a finite number of quadratic fields [6] and of cyclic cubic fields [7] in which Euclid's algorithm (E.A.) holds. Davenport has shown by a different method that there are only a finite number of quadratic fields [1, 2], of non-totally real cubic fields [3, 4] and of totally complex quartic fields in which E.A. holds.

The object of this paper is to extend these results to cyclic fields of higher degree. I shall prove

THEOREM 1. *For every $k \geqslant 4$ there are only a finite number of cyclic fields K of degree k whose discriminant Δ is the power of a prime, in which* E.A. *holds.*

The methods employed in this paper could actually furnish a proof of a theorem dealing with a more general type of cyclic field. But the classical theory of abelian fields allows us to name a large number of cyclic fields in which the class-number is greater than 1, and in which therefore E.A. cannot hold. Since these results are difficult to find in the existing literature, they will be quoted and proved in some detail in this paper.

To begin with we recall the two different definitions of the class-number of an algebraic field. H is the number of classes of ideals in an algebraic field if two ideals are considered equivalent provided their quotient is a principal ideal generated by a totally positive number; h is the number of classes of ideals in an algebraic field if two ideals are considered equivalent provided their quotient is any principal ideal. It is clear that $H = h$ for complex abelian fields.

We denote by $w(N)$ the number of distinct rational primes dividing a rational integer $N \neq 0$.

We call a cyclic field K a field of type T_1 if it is the composite field of cyclic fields K_j of degrees k_j and discriminants Δ_j where any two k_j are relatively prime, where any two Δ_j are relatively prime, and where $w(\Delta_j) = 1$.

We call a cyclic field K of degree k a field of type T_2 if it is the composition field of a field K_1 of type T_1 of odd degree, and of a cyclic field K_2 of discriminant Δ_2 of degree 2^l of the following type: $w(\Delta_2) \leqslant k$ and the discriminant of the unique subfield of K_2 of degree 2^{l-1} is a power of a prime, if $l > 1$. (For the purpose of this definition K_1 or K_2 may be the field of rational numbers.) We can now formulate

THEOREM 2. $(k, H) > 1$ *for a cyclic field K not of type T_1.*

THEOREM 3. $h > 1$ *for a cyclic field K not of type T_2.*

Received January 5, 1950.

All results of this paper and of my two previous papers can be summarized as

THEOREM 4. *For each* $k \geqslant 2$ E.A. *holds only in a finite number of cyclic fields* K *of degree* k *and discriminant* Δ, *if only fields of the following types are considered.*

(1) k *a prime.*
(2) $w(k) = 1$, k *odd.*
(3) $w(k) = 1$, K *complex.*
(4) $w(\Delta) = 1$.
(5) $w(\Delta) > w(k)$, k *odd.*
(6) $w(\Delta) > w(k)$, K *complex.*
(7) $w(\Delta) \geqslant k + w(k)$.
(8) $w(\Delta^*) > 1$ *for the discriminant* Δ^* *of every non-rational subfield* K^* *of* K.
(9) k *odd,* K *not of type* T_1.
(10) K *complex and not of type* T_1.
(11) K *not of type* T_2.

Finally I should like to mention two types of cyclic fields for which E.A. may possibly hold in an infinity of cases.

(a) The real quartic field

$$\rho(\sqrt{\tfrac{1}{2}(5 + 5^{\frac{1}{2}})p})$$

of discriminant $125p^2$, where $p \equiv 3 (\text{mod } 20)$ is a prime.

(b) The complex sextic field

$$\rho((e^{2\pi i/9} + e^{-2\pi i/9}),\ (-p)^{\frac{1}{2}})$$

of discriminant -3^8p^3, where $p \equiv 3 (\text{mod } 4)$ is a prime.

We establish the following conventions. Small italics except e, i, and o denote positive rational integers, d, p and q denote positive rational primes.

K, K', K_j etc. denote abelian fields of degrees k, k', k_j etc. and discriminants Δ, Δ', Δ_j etc.

Only absolutely abelian fields will be considered in this paper.

2. Dirichlet characters and Abelian fields. Two Dirichlet characters $\chi(n)$ (mod m) and $\chi'(n)$ (mod m') are said to belong to the same train if and only if $\chi(n) = \chi'(n)$ for all n with $(n, mm') = 1$. Then each train contains exactly one primitive character $\chi_0(n)$ (mod f); f is called the conductor of the train, and also the conductor of all characters in the train. The product of two trains is defined in the obvious way, and it is clear that the trains form an infinite abelian group with respect to multiplication.

If $\chi(n)$ is a character mod m and if

$$m = m_1m_2,\ (m_1, m_2) = 1,$$

then $\chi(n)$ can be written in the form

$$\chi(n) = \chi_1(n)\chi_2(n)$$

where $\chi_1(n)$ and $\chi_2(n)$ are uniquely determined characters mod m_1 and m_2 respectively. In particular, if $\chi(n)$ is primitive, then $\chi_1(n)$ and $\chi_2(n)$ are primitive.

The principal results of class-field theory can easily be expressed in the following way.

Between all finite groups $\mathfrak{G}$ of trains and all abelian fields K there is a one-one relation [5, Theorem 1] which satisfies the following conditions:

I. The group $\mathfrak{G}$ is isomorphic to the Galois group of the field K. [5, Theorem 2.]

II. A field K' contains a field K if and only if the corresponding group $\mathfrak{G}'$ contains the corresponding group $\mathfrak{G}$. [5, Theorem 10.]

III. $|\Delta|$ equals the product of the conductors of the trains in $\mathfrak{G}$. [5, Theorem 16.]

IV. $$\zeta_K(s) = \prod_{\chi} L(s, \chi),$$

where $\zeta_K(s)$ denotes the Dedekind ζ-function of K, and where χ runs through the primitive characters of the trains in $\mathfrak{G}$. [5, Theorem 14.]

V. If $\Delta = \pm p^l$, p becomes in K the kth power of a self-conjugate prime ideal of the first order.

VI. If $(\Delta, p) = 1$, p^l is the norm of an integral ideal in K if and only if $\chi(p^l) = 1$ for all primitive characters of the trains in $\mathfrak{G}$.

VII. If $(\Delta, n) = 1$, n is the norm of an integral ideal in K if and only if in the canonical representation

$$n = p_1^{l_1} \ldots p_s^{l_s}$$

each factor $p_j^{l_j}$ is the norm of an integral ideal in K.

VIII. If n is the norm of an integral ideal in K, then $\chi(n) \geqslant 0$ for all characters of the trains in $\mathfrak{G}$.

IX. If K' is an abelian extension of K of relative discriminant 1, then the class-number H of K is divisible by k'/k. More precisely, the class-group of K contains a subgroup whose quotient group is isomorphic to the Galois group of K' over K. [5, Theorems 2 and 16.]

In addition we require two lemmas about discriminants.

LEMMA 1. *If the fields K_1 and K_2 have discriminants Δ_1 and Δ_2 and if $(\Delta_1, \Delta_2) = 1$, then the composition field $K_0 = K_1, K_2$ has discriminant*

$$\Delta_0 = \Delta_1^{k_2}\Delta_2^{k_1}$$

and degree $k_0 = k_1k_2$. [8, Theorem 88.]

Lemma 2. *If K' is an abelian extension field over K, then*

$$|\Delta'| = |\Delta|^{k'/k}$$

if and only if K' has relative discriminant 1 *over K.* [8, Theorem 39.]

3. Proof of Theorem 2. Let $\chi(n)$ be the primitive character in one of the trains which generate the group $\mathfrak{G}$ corresponding to K, so that k is the order of $\chi(n)$. Then we can write

$$\chi(n) = \chi_1(n) \ldots \chi_s(n),$$

where $\chi_1(n), \ldots, \chi_s(n)$ are primitive characters mod $p_1^{l_1}, \ldots, p_s^{l_s}$ respectively, all the p_j being distinct. Let $k_1, \ldots, k_s$ denote the order of $\chi_1(n), \ldots, \chi_s(n)$ respectively; then the smallest common multiple

$$[k_1, \ldots, k_s] = k.$$

Let P_j denote the product of the conductors of the characters

$$\chi_j(n), \chi_j^2(n), \ldots, \chi_j^{k_j-1}(n) \qquad (1 \leqslant j \leqslant s).$$

Then the product of the conductors of the characters

$$\chi(n), \chi^2(n), \ldots, \chi^k(n)$$

equals

$$P_1^{k/k_1} \ldots P_s^{k/k_s} = |\Delta|$$

by III.

Let us now consider the group $\mathfrak{G}'$ of all trains generated by

$$\chi_1(n), \ldots, \chi_s(n).$$

$\mathfrak{G}'$ contains the train $\chi(n)$, and the order k' of $\mathfrak{G}'$ equals

$$k' = k_1 \ldots k_s.$$

The product of the conductors of all trains in $\mathfrak{G}'$ equals

$$(P_1^{k/k_s} \ldots P_s^{k/k_s})^{k_1 \cdots k_s} = |\Delta^{k'/k}| = |\Delta'|$$

where Δ' is the discriminant of the field K' corresponding to $\mathfrak{G}'$.

It follows by I, II, and Lemma 2 that K' is an extension field of relative discriminant 1 over K. Hence by IX

$$H \equiv 0 \pmod{k'/k}.$$

Hence, if $k' > k$, then $(k, H) > 1$. If $k' = k$, then any two of the numbers $k_1, \ldots, k_s$ are relatively prime, and the field K is of type T_1. This proves Theorem 2.

4. Proof of Theorem 3. We prove first:

Lemma 3. *If K is complex, then $H = h$. If K is real, then $h > 1$ unless* *the class group of K (in the narrow sense) is the direct product of not more than $k-1$ abelian groups of order* 2.

Proof. The first part of the lemma is trivial.

If K is real, then -1 is a non-totally positive unit in K. Therefore the group of all numbers in K, which are products of a unit in K, and of a totally positive number in K, is a subgroup of the group of all numbers ($\neq 0$) in K of index $\leqslant 2^{k-1}$. More precisely, the quotient group is the direct product of at most $k-1$ groups of order 2. Since this quotient group is isomorphic to the quotient group of the two class groups in K, the lemma follows.

Assuming the notation used in the proof of Theorem 2, it suffices by virtue of IX and Lemma 3, to prove that, if the Galois group of K' over K is the direct product of at most $k-1$ groups of order 2, then K is of type T_2.

Let K_0 be the field of largest odd degree k_0 which is contained in K, and let K_e be the field of largest degree $k_e = 2^l$ which is contained in K. Then K_0 and K_e are uniquely determined and we have

$$K = K_0, K_e, \quad k = k_0 k_e.$$

Let K_ϵ be the unique subfield of K_e of degree $k = \frac{1}{2}k_e$. Then we have to prove

(i) $\quad K_0$ is of type T_1.

(ii) $\quad w(\Delta_e) \leqslant k$.

(iii) $\quad w(\Delta_\epsilon) = 1$ if $k_\epsilon > 1$.

We construct the extension field K'_0 over K_0 by the same process which gave us the extension field K' over K. If K_0 were not of type T_1, then $k'_0/k_0 > 1$ and odd. Since K'_0 is a subfield of K', we should have

$$(k'/k'_0)\ (k'_0/k_0) = (k'/k)\ (k/k_0),$$

which is a contradiction, because each factor on the right is a power of 2. This proves (i).

Next we construct the extension field K'_e by the same process. Again K'_e is a subfield of K' and we have

$$(k'/k'_e)\ (k'_e/k_e) = (k'/k)\ (k/k_e).$$

Here

$$k/k_e \equiv 1 \pmod 2, \quad 2^{k-1} \equiv 0 \pmod{k'/k}.$$

If $w(\Delta_e) > k$, then

$$2^k \equiv 0 \pmod{k'_e/k_e}$$

which gives a contradiction. This proves (ii).

Finally if $k_e \geqslant 4$, $w(\Delta_\epsilon) \geqslant 2$, then the absolute Galois group of K'_e would have a subgroup of type (4, 4) by virtue of I. *A fortiori* the absolute Galois group of K' would have a subgroup of type (4, 4). Since the absolute Galois group of K is cyclic, the Galois group of K' over K would contain an element of order 4, which contradicts our hypothesis. This proves (iii).

5. Conventions and notations. We start by proving

LEMMA 4. *If K is cyclic, $w(\Delta) = 1$, $|\Delta| \geqslant k^{3(k-1)}$, then*

$$|\Delta| = d^{k-1}, d \equiv 1 \pmod{k}.$$

Proof. Let $\chi(n)$ be the primitive character in one of the trains which generate the group $\mathfrak{G}$ corresponding to K. By I and III $\chi(n)$ is a primitive character mod d^a (say) of order k. Hence

$$k | \varphi(d^a) = d^{a-1}(d-1),$$

which means that either $d|k$ or $k/d - 1$. In the latter case, if $a > 1$, we should have a number n such that

$$\chi(n) \neq 1, n \equiv 1 (\text{mod } d^{a-1}).$$

For this value of n

$$n^d \equiv 1 (\text{mod } d^a),$$
$$1 = \chi(n^d) = \chi^d(n) = \chi(n) \neq 1,$$

which is a contradiction. Hence $\chi(n)$ is a character mod d; $\chi^j(n)$ is *a fortiori* a character mod d for $1 < j < k$, and it follows from III that

$$|\Delta| = d^{k-1}.$$

If d/k, we proceed as follows. We assume that

$$k = d^b m, \quad (m, d) = 1.$$

Then, for $d > 2$, the group of all characters mod d^a is cyclic of order $\varphi(d^a)$. Hence the number of characters mod d^a of order k equals $\varphi(k)$ if $k/\varphi(d^a)$ and 0 otherwise. Hence there exists a primitive character mod d^a of order k if and only if

$$\varphi(d^a) \equiv 0 \pmod{k}, \quad \varphi(d^{a-1}) \not\equiv 0 \pmod{k}.$$

This implies

$$m|(d-1), \quad a = b + 1,$$
$$d^a \leqslant dk \leqslant k^2, |\Delta| \leqslant k^{2(k-1)}.$$

If $d = 2$, $d|k$, the argument is similar. We may assume at once that $a > 3$. Then the group of characters mod d^a is abelian of type $(2, 2^{a-2})$, and the number of characters of order $k = 2^b$ equals 3 if $b = 1$, 2^b if $2 \leqslant b \leqslant a - 2$, 0 if $b > a - 2$.

Hence there exists a primitive character mod 2^a of order k if and only if

$$b = a - 2.$$

This implies

$$d^a = d^2 k \leqslant k^3, \quad |\Delta| \leqslant k^{3(k-1)}.$$

For the rest of the paper excluding the last paragraph we assume $k \geqslant 4$, K cyclic; hence by virtue of Lemma 4

$$d \equiv 1 \pmod{k},\ |\Delta| = d^{k-1}.$$

$\chi(n)$ is again the primitive character mod d of order k in a train which generates the group $\mathfrak{G}$ corresponding to K. $\chi(n)$ will now be fixed.

Let A_j denote the class of integers n for which

$$\chi(n) = e^{2\pi ij/k} \qquad (0 \leqslant j \leqslant k-1).$$

Let B denote the subclass of integers b in A_0 for which

$$b = b_1 b_2,\ (b_1, b_2) = 1$$

implies b_1 in A_0. Let C denote the sub-class of integers c in A_0 which can be decomposed in the form

$$c = c_1 c_2,\ \ (c_1, c_2) = 1,\ \ \chi(c_1) \neq 1.$$

Clearly every number in A_0 lies either in B or in C. It follows from VI and VII that a number n is norm of an integral ideal in K prime to d if and only if n lies in B.

Also $q_1 < q_2$ are the two smallest primes not in A_0 which do not equal d; and r is the smallest number in C which is prime to q_1 and which satisfies

$$r \equiv -d \pmod{4} \qquad \text{if } q_1 = 2, \tag{1}$$

$$r \equiv -d^2-1 \pmod{9} \quad \text{if } q_1 = 3. \tag{2}$$

For $q_1 \geqslant 5$ no additional condition is imposed upon r.

Let ϵ be a positive number which will be fixed later; it may be arbitrarily small. The constants involved in the symbols O and o will depend on k only. Unless the contrary is stated the symbol o will refer to the limit as $d \to \infty$. We put

$$x = [d^{\frac{1}{2}+\epsilon}],\ \ y = [d^{\frac{1}{4}+\epsilon}].$$

6. Further lemmas.

LEMMA 5. $\sum_{p \leqslant z} p^{-1} = \log\log z + \gamma + o(1)$ *as* $z \to \infty$, *where* γ *is an absolute constant.* [9, Theorem 7].

LEMMA 6. *For each non-principal character* $\chi(n)$ mod m

$$\sum_{n=1}^{z} \chi(n) = O(m^{\frac{1}{2}} \log m) \qquad [10].$$

LEMMA 7. $q_2 \leqslant y$ *if* d *is sufficiently large.*

Proof. We assume $q_2 > y$. Then all primes $\leqslant y$, with the possible exception of q_1, belong to A_0. Hence, if

$$n \leqslant x,\ (n, q_1) = 1,\ p|n,\ y < p,$$

then $\chi(n) = \chi(p)$ unless n is divisible by the product pp' of two primes in the interval $y < p \leqslant x$, $y < p' \leqslant x$.

Therefore we have for $1 \leqslant j \leqslant k-1$

$$\sum_{\substack{n=1\\(n,q_1)=1}}^{x} \chi^j(n) = \sum_{\substack{n=1\\(n,q_1)=1}}^{x} 1 + \sum_{\substack{n=1\\(n,q_1)=1}}^{x} (\chi^j(n)-1)$$

$$= (1-q_1^{-1})x + O(1) + \sum_{\substack{y<p\leqslant x\\p\neq q_1}} (\chi^j(p)-1) \sum_{\substack{m\leqslant x/p\\(m,q_1)=1}} 1 + \sum_{p>y} \sum_{\substack{p'>y\\pp'\leqslant x}} O(x/pp')$$

$$= (1-q_1^{-1})x + O(1) + \sum_{y<p\leqslant x} \{(\chi^j(p)-1)(1-q_1^{-1})xp^{-1} + O(1)\}$$

$$+ O(xy^{-1}) + O(x(\sum_{y<p<d^{\frac{1}{2}+2\epsilon}} p^{-1})^2)$$

$$= (1-q_1^{-1})x\{1 + \sum_{y<p\leqslant x} (\chi^j(p)-1)p^{-1}\} + O(\pi(x))$$

$$+ O\left\{x\left(\log\frac{\frac{1}{4}+2\epsilon}{\frac{1}{4}-2\epsilon} + o(1)\right)^2\right\} \qquad \text{(Lemma 5)}$$

$$= (1-q_1^{-1})x\{1 + \sum_{y<p\leqslant x} (\chi^j(p)-1)p^{-1}\} + O(\epsilon^2 x) + o(x).$$

Applying Lemma 6, this gives, after division by $(1-q_1^{-1})x$,

$$0 = O(\epsilon^2) + o(1) + 1 + \sum_{y<p\leqslant x} (\chi^j(p)-1)p^{-1}.$$

Summing this over $j = 1, \ldots, k-1$ we obtain

$$0 \geqslant O(\epsilon^2) + o(1) + k - 1 - k \sum_{y<p\leqslant x} p^{-1}.$$

Hence

$$\sum_{y<p\leqslant x} p^{-1} \geqslant 1 - k^{-1} + O(\epsilon^2) + o(1).$$

But by Lemma 6

$$\sum_{y\leqslant p\leqslant x} p^{-1} = \log\log x - \log\log y + o(1)$$

$$= \log\frac{\frac{1}{2}+\epsilon}{\frac{1}{4}-\epsilon} + O(1) = \log 2 + O(\epsilon) + o(1).$$

Hence

$$\log 2 \geqslant 1 - k^{-1} + O(\epsilon),$$

which is not true if ϵ is sufficiently small. This proves the lemma.

From now on ϵ is fixed as a function of k.

LEMMA 8. *$q_1 r < d^{1-\epsilon}$, if d is sufficiently large.*

Proof. We assume that d is so large that Lemma 7 applies. If q_2 lies in A_j $(1 \leqslant j \leqslant k-1)$, we choose for u the smallest number in A_{k-j} which satisfies

$$(u, q_1q_2) = 1,$$
$$uq_2 \equiv -d \quad (\text{mod } 4) \text{ if } q_1 = 2,$$
$$uq_2 \equiv -d^2 - 1 \quad (\text{mod } 9) \text{ if } q_1 = 3.$$

If d is sufficiently large, it is easily deduced from Lemma 6 that

$$u < x.$$

(The detailed argument is explicitly developed in [7].) Since uq_2 lies in C it follows from the definition of r that

$$r \leqslant uq_2 < xq_2,$$

and by Lemma 7 that

$$q_1r < q_1(xq_2) < xq_2^2 \leqslant xy^2 \leqslant d^{1-\epsilon}.$$

LEMMA 9. *If $q_1 \geqslant 5$, $s < q_1$, we can find a prime p_0 such that*

$$(p_0, s) = 1, \; p_0 < q_1, \; p_0 \leqslant \log d$$

provided d is sufficiently large [7, Lemma 4].

LEMMA 10. *For sufficiently large d we can write*

$$d = sr + tq_1,$$

where s in B, $(t, q_1) = 1$.

Proof. We distinguish three cases.

First case. $q_1 = 2$. We have

$$d = r + 2t,$$

and it follows from (1) that t is odd.

Second case. $q_1 = 3$. Then we have with $s = 1$ or $s = 2$

$$d = sr + 3t.$$

Clearly s lies in B, since $q_1 = 3$ is the smallest positive integer not in A_0. If t were divisible by 3, we should have by (2)

$$sr \equiv -s(d^2+1) \equiv d \quad (\text{mod } 9),$$
$$(\pm 2s - 1)d \equiv s(d \pm 1)^2 \quad (\text{mod } 9),$$
$$(-4s^2 + 1)d^2 \equiv s^2(d^2 - 1)^2 \quad (\text{mod } 9),$$
$$-4s^2 + 1 \equiv 0 \quad (\text{mod } 9),$$

which is not true for $s = 1$ or $s = 2$.

Third case. $q_1 \geqslant 5$. Again, by Lemma 8, we can find s and t such that

$$d = sr + tq_1, \quad s < q_1.$$

Clearly, s lies in B, as it is not divisible by a prime $\geqslant q_1$.

But q_1 may possibly divide t. If $q_1|t$, we use the prime p_0 of Lemma 9 and denote by n the smallest positive solution of the congruence

$$s + nq_1 \equiv 0 \pmod{p_0}.$$

Then

$$s + nq_1 < q_1 + (p_0 - 1)q_1 = p_0 q_1. \tag{3}$$

We consider the representation

$$d = (s + nq_1)r + (t - nr)q_1.$$

Since $n < q_1$, $t - nr$ is prime to q_1. Since by (3), Lemma 9 and Lemma 8, for sufficiently large d

$$(s + nq_1)r < p_0 q_1 r \leqslant (\log d)d^{1-\epsilon} < d,$$

it follows that

$$t - nr > 0.$$

Finally it follows from (3) and Lemma 9 that no prime $\geqslant q_1$ divides $s + nq_1$. Hence $s + nq_1$ lies in B, and our lemma is proved in all cases.

LEMMA 11. *If d is sufficiently large,*

$$d = c + g,$$

where c lies in C, and g does not lie in B.

Proof. We assume that d is so large that Lemma 10 applies, and put

$$c = sr, \quad g = tq_1.$$

Clearly g does not lie in B, since

$$g = tq_1, \ (t, q_1) = 1, \qquad q_1 \text{ not in } A_0.$$

Since r lies in C, we have a decomposition

$$r = r_1 r_2, \ (r_1, r_2) = 1,$$

where r_1 does not lie in A_0. It follows from the fundamental theorem of arithmetic that we have a decomposition of s such that

$$s = s_1 s_2, \ (s_1, s_2) = 1, \ (r_1, s_2) = (r_2, s_1) = 1.$$

Since s lies in B, s_1 lies in A_0. We have a decomposition

$$c = sr = (s_1 r_1)(s_2 r_2), \ (s_1 r_1, s_2 r_2) = 1,$$

where $s_1 r_1$ does not lie in A_0. Hence c, lying in A_0, lies in C.

7. Proof of Theorem 1. We assume that E.A. holds in K. Then, by condition V, there exists in K a self-conjugate principal prime ideal (δ) of norm d.

We assume that d is so large that Lemma 11 applies. Since c lies in A_0, the congruence

$$n^k \equiv c \pmod{d}$$

has a solution. Since E.A. holds in K, we can find an integer γ in K such that

$$n \equiv \gamma \pmod{\delta}, \quad |N(\gamma)| < |N(\delta)| = d.$$

Since (δ) is self-conjugate, the congruence

$$n \equiv \gamma' \pmod{\delta}$$

holds for each conjugate γ' of γ. Multiplying these k congruences we obtain

$$c \equiv n^k \equiv N(\gamma) \pmod{\delta},$$
$$c \equiv N(\gamma) \pmod{d}.$$

Hence

$$\text{either } N(\gamma) = c \text{ or } N(\gamma) = c - d = -g.$$

This means that the norm of the ideal (γ) equals c or g, which is impossible by Lemma 11 and condition VII.

8. Proof of Theorem 4. We take each individual assertion in Theorem 4, starting from the end.

(11) follows from Theorem 3.

(10) follows from Theorem 2, since $H = h$ if K is complex.

(9) follows from Theorem 3, since for odd k a field of type T_2 is a field of type T_1.

(8) If k is divisible by an odd prime, K_0 is not of type T_1, and therefore K is not of type T_2. If $k = 2^l$, $l \geqslant 2$, the field K_ϵ has discriminant Δ_ϵ with $w(\Delta_\epsilon) > 1$, hence K is not of type T_2. If $k = 2$, the result follows from my first paper [6].

(7) If K were of Type T_2, then

$$w(\Delta_0) \leqslant w(k_0)$$

and

$$w(\Delta_e) \leqslant k \text{ for even } k.$$

Hence

$$w(\Delta) \leqslant w(\Delta_0) + w(\Delta_e) \leqslant \begin{cases} w(k_0) < k + w(k) & \text{for odd } k. \\ w(k_0) + k = k - 1 + w(k) & \text{for even } k. \end{cases}$$

(6) K is not of type T_1, and $h = H > 1$.

(5) Since for odd k a field of type T_2 is a field of Type T_1, K is not of type T_2.

(4) follows from Theorem 1 for $k \geqslant 4$, and from my older results if $k = 2$ or $k = 3$.

(3) follows from (4) and (6).

(2) follows from (4) and (5).

(1) follows from (2) if k is odd, and from my older results if $k = 2$.

References

[1] H. Davenport, *Indefinite binary quadratic forms, and Euclid's Algorithm in real quadratic fields*, Proc. London Math. Soc., in course of publication.

[2] ——— *Indefinite binary quadratic forms*, Quart. J. Math., Oxford Ser. (2), vol. 1 (1950), 54-62.

[3] ——— *Euclid's Algorithm in cubic fields of negative discriminant*, Acta Math., vol. 84 (1950), 159-179.

[4] ——— *Euclid's Algorithm in certain quartic fields*, Trans. Amer. Math. Soc., vol. 68 (1950), 508-532.

[5] H. Hasse, *Bericht über neuere Untersuchungen aus der Theorie der algebraischen Zahlkörper*, Jber. Deutsch. Math. Verein., vol. 35 (1926), 1-55.

[6] H. Heilbronn, *On Euclid's Algorithm in real quadratic fields*, Proc. Cambridge Phil. Soc., vol. 34 (1938), 521-526.

[7] ——— *On Euclid's Algorithm in self-conjugate cubic fields*, Proc. Cambridge Phil. Soc., vol. 46 (1950), 377-382.

[8] D. Hilbert, *Bericht über die Theorie der algebraischen Zahlkörper*, Jber. Deutsch. Math. Verein., vol. 4 (1897), 175-546.

[9] A. E. Ingham, *The distribution of prime numbers* (Cambridge, 1932).

[10] G. Pólya, *Über die Verteilung der quadratischen Reste und Nichtreste*, Nachr. Akad. Wiss. Göttingen, Math. Phys. Kl. 1918, 21-29.

The Royal Fort, Bristol 8

29.

On the Averages of Some Arithmetical Functions of Two Variables

Mathematika 5 (1958), 1–7
[Commentary on p. 599]

1. *Introduction.*

The object of this paper is to prove theorems of the following type.

THEOREM 1. *If p and q are restricted to odd primes, and $\left(\frac{p}{q}\right)$ denotes the Legendre symbol, then for $x > 2$*

$$\sum_{p \leqslant x} \sum_{q \leqslant x} \left(\frac{p}{q}\right) = O\left(x^{7/4}(\log x)^{-5/4}\right).$$

A straightforward generalization is given by

THEOREM 2. *Let $\chi_1, \ldots, \chi_N$ be N distinct Dirichlet characters with moduli not exceeding m. Let g be a natural integer and $x > 1$. Then*

$$\sum_{n=1}^{N} \sum_{p \leqslant x} \chi_n(p) = O(Nx)\{N^{-1/(4g)} + x^{-1/2}(m \log m)^{1/(2g)}\},$$

where p is restricted to primes and the constant implied by the symbol O depends on g only.

It is not difficult to generalize Theorem 2 to algebraic number fields and to prove

THEOREM 3. *Let K be an algebraic number field of degree k and discriminant D. Let $\chi_1, \ldots, \chi_N$ be N distinct abelian characters in K with moduli whose norms do not exceed m. Let g be a natural integer, $x > 1$, $\epsilon > 0$. Then*

$$\sum_{n=1}^{N} \sum_{N(\mathfrak{p}) \leqslant x} \chi_n(\mathfrak{p}) = O(Nx)\{N^{-1/(4g)} x^{\epsilon} + (x^{-g} |D|^{1/2} m)^{1/(g(k+2))+\epsilon}\},$$

where $\mathfrak{p}$ is restricted to prime ideals in K, and the constants implied by the symbol O depend on k, g and ϵ only.

[MATHEMATIKA 5 (1958), 1–7]

The proofs of these three theorems are essentially elementary, being based on a careful application of the Hölder *and* the Cauchy-Schwarz inequality. Readers who are familiar with Vinogradov's work on exponential sums in the 1930's will undoubtedly detect the influence of his ideas in the present paper, but no direct reference is made to his publications.

The interest of Theorem 3 lies in the fact that it leads to a proof of the following generalization of Theorem 1.

THEOREM 4. *Let k and l be two natural integers and let*

$$v = \begin{cases} 1 & \text{if } (2, k, l) = 1, \\ 3/2 & \text{if } k \equiv l \equiv 2 \pmod 4, \\ 2 & \text{if } k \equiv l \equiv 0 \pmod 2),\ kl \equiv 0 \pmod 8. \end{cases}$$

Then the number of pairs of primes p and q not exceeding x which satisfy $p \equiv 1 \pmod k$, $q \equiv 1 \pmod l$ and for which the congruences

$$y^k \equiv q \pmod p, \quad z^l \equiv p \pmod q$$

are soluble in rational integers y and z, equals

$$\left(kl\phi(k)\phi(l)\right)^{-1} v \left\{\int_e^x (\log u)^{-1} du\right\}^2 + O(x^2) \exp\left(-\alpha(\log x)^{1/2}\right),$$

where α and the constants implied by the symbol O depend on k and l only.

The proof of this Theorem, given Theorem 3, is based on routine only, but requires a knowledge of the prime ideal theorem for arithmetic progressions, and familiarity with the basic concepts of class-field theory.

I wish to acknowledge my obligation to Dr. A. Fröhlich, who addressed to me some enquiries, which led to the investigation of the problems solved in this paper.

2. *Preparations.*

We require two lemmas about partial sums of characters in the rational and in the general algebraic fields respectively.

LEMMA 1 (Pólya†). *Let $\chi(n)$ be a non-principal Dirichlet character* mod m. *Then, for $x > 1$,*

$$\sum_{n=1}^{x} \chi(n) = O(m^{1/2} \log m).$$

LEMMA 2. *Let K be an algebraic number field of discriminant D and degree k. Let $\chi(\mathfrak{a})$ be an abelian non-principal character defined in K to a modulus $\mathfrak{m}$; let $m = N(\mathfrak{m})$. Then, for $x > 1$, $\epsilon > 0$,*

$$\sum_{N(\mathfrak{a}) \leq x} \chi(\mathfrak{a}) = O(x^k |D| m)^{1/(k+2)+\epsilon},$$

where the constant implied by the symbol O depends on k and ϵ only.

† G. Pólya, *Göttinger Nachr.* (1918), 21–29. I. Schur, *ibid.*, 30–36.

Proof. We may assume that x is not an integer. We may clearly assume that χ is a primitive character mod $\mathfrak{m}$, *i.e.* that $\mathfrak{m}$ is the conductor of χ.

It was proved by Hecke† that the function $L(s, \chi)$, defined for $\sigma > 1$ by

$$L(s, \chi) = \sum_{\mathfrak{a}} \chi(\mathfrak{a}) N(\mathfrak{a})^{-s},$$

is an integral function and satisfies a functional equation of the type

$$a^s(|D|m)^{\frac{1}{2}s}\Gamma(\tfrac{1}{2}s)^b\,\Gamma\left(\tfrac{1}{2}(s+1)\right)^{k-b}L(s, \chi)$$

$$= Wa^{1-s}(|D|m)^{\frac{1}{2}(1-s)}\Gamma\left(\tfrac{1}{2}(1-s)\right)^b\,\Gamma(1-\tfrac{1}{2}s)^{k-b}L(1-s, \bar{\chi}),$$

where $|W| = 1$, $a > 0$, $\log a = O(k)$, $0 \leqslant b \leqslant k$ and b is a rational integer.

It follows at once from Stirling's formula that for fixed $\epsilon > 0$, $\sigma = -\epsilon$

$$L(s, \chi) = O\left(|D|m(1+|t|)^k\right)^{\frac{1}{2}+\epsilon},$$

and by the Phragmén-Lindelöf principle that this inequality is also true for $\sigma \geqslant -\epsilon$.

Put

$$T = (x^2|D|^{-1}m^{-1})^{1/(k+2)}.$$

If $T < 2$, the Lemma is trivial. Hence we may assume $T \geqslant 2$. We now apply the classical Perron integral formula to $L(s, \chi)$. This gives

$$\sum_{N\mathfrak{a}<x} \chi(\mathfrak{a}) = L(0, \chi) + (2\pi i)^{-1}\int x^s L(s, \chi)\, s^{-1}\,ds,$$

where the path of integration runs along five straight lines from $1+\epsilon-\infty i$ via $1+\epsilon-Ti$, $-\epsilon-Ti$, $-\epsilon+Ti$, $1+\epsilon+Ti$ to $1+\epsilon+\infty i$. By virtue of our inequality for $L(s, \chi)$ the integral from $1+\epsilon-Ti$ to $1+\epsilon+Ti$ contributes at most

$$O(|D|mT^k)^{\frac{1}{2}+\epsilon}\log T.$$

This expression is also an upper bound for $L(0, \chi)$. The integral along the line $\sigma = 1+\epsilon$ contributes at most‡

$$O(x^{1+2\epsilon}T^{-1}).$$

The Lemma follows from our choice of T.

3. *Proof of Theorem* 1.

Let x be a natural integer, $\pi(x)$ the number of primes not exceeding x. Let p, p_1, p_2, q, q_1, q_2 denote odd primes not exceeding x.

---

† E. Hecke, *Göttinger Nachr.* (1917), 299–318, and *Math. Zeitschrift*, 6 (1920), 11–51. Another reference, easier to verify, is E. Landau, *Math. Zeitschrift*, 2 (1918), 52–154, Satz LXI.

‡ For a detailed proof see E. Landau, *Acta Math.*, 35 (1912), 271–294, Hilfssatz 3.

It follows from the Cauchy-Schwarz inequality that

$$\left\{\sum_p \sum_q \left(\frac{p}{q}\right)\right\}^2 \leqslant \sum_q 1 \, . \sum_q \left\{\sum_p \left(\frac{p}{q}\right)\right\}^2 \leqslant \pi(x) \sum_{p_1} \sum_{p_2} \sum_q \left(\frac{p_1 p_2}{q}\right).$$

Again by Cauchy-Schwarz with respect to the summation over p_1, p_2

$$\left\{\sum_p \sum_q \left(\frac{p}{q}\right)\right\}^4 \leqslant \pi^2(x) \sum_{p_1} \sum_{p_2} 1 \, . \sum_{n=1}^{x^2} 2 \left\{\sum_q \left(\frac{n}{q}\right)\right\}^2 \leqslant 2\pi^4(x) \sum_{q_1} \sum_{q_2} \sum_{n=1}^{x^2} \left(\frac{n}{q_1 q_2}\right).$$

The sum over n is at most x^2 if $q_1 = q_2$, and by Pólya's Lemma it is

$$O\left(q_1^{1/2} q_2^{1/2} \log(q_1 q_2)\right) = O(x \log x) \text{ if } q_1 \neq q_2.$$

Hence

$$\left\{\sum_p \sum_q \left(\frac{p}{q}\right)\right\}^4 \leqslant 2\pi^5(x)\, x^2 + O\left(\pi^6(x)\, x \log x\right) = O(x^7 \log x^{-5}).$$

4. *Proofs of Theorems* 2 *and* 3.

These proofs will be given jointly, as they differ only in the last line. It should, however, be understood that $\epsilon > 0$ in the proof of Theorem 3, and that $\epsilon = 0$ in the proof of Theorem 2. x is a natural integer, $\mathfrak{p}$ runs through all prime ideals with $N\mathfrak{p} \leqslant x$, $\mathfrak{a}$ and $\mathfrak{b}$ run through all integer ideals with $0 < N(\mathfrak{a}) \leqslant x^g$, $0 < N(\mathfrak{b}) \leqslant x^g$.

It follows from Hölder's inequality that

$$\left|\sum_{n=1}^{N} \sum_{\mathfrak{p}} \chi_n(\mathfrak{p})\right| \leqslant N^{1-1/(2g)} \left\{\sum_{n=1}^{N} \left|\sum_{\mathfrak{p}} \chi_n(\mathfrak{p})\right|^{2g}\right\}^{1/(2g)}$$

$$= N^{1-1/(2g)} \left\{\sum_{\mathfrak{a}} c(\mathfrak{a}) \sum_{\mathfrak{b}} c(\mathfrak{b}) \sum_{n=1}^{N} \chi_n(\mathfrak{a}) \overline{\chi}_n(\mathfrak{b})\right\}^{1/(2g)},$$

where $0 \leqslant c(\mathfrak{a}) \leqslant g!$.

Again by Cauchy-Schwarz with respect to the summation over $\mathfrak{a}$ and $\mathfrak{b}$

$$\left|\sum_{n=1}^{N} \sum_{\mathfrak{p}} \chi_n(\mathfrak{p})\right| \leqslant N^{1-1/(2g)} \left\{\sum_{\mathfrak{a}} c^2(\mathfrak{a}) \sum_{\mathfrak{b}} c^2(\mathfrak{b})\right\}^{1/(4g)} \left\{\sum_{\mathfrak{a}} \sum_{\mathfrak{b}} \left|\sum_{n=1}^{N} \chi_n(\mathfrak{a}) \overline{\chi}_n(\mathfrak{b})\right|^2\right\}^{1/(4g)}$$

$$\leqslant N^{1-1/(2g)} O(x^{\frac{1}{2}+\epsilon}) \left\{\sum_{n_1=1}^{N} \sum_{n_2=1}^{N} \sum_{\mathfrak{a}} \sum_{\mathfrak{b}} \chi_{n_1}(\mathfrak{a}) \overline{\chi}_{n_1}(\mathfrak{b}) \overline{\chi}_{n_2}(\mathfrak{a}) \chi_{n_2}(\mathfrak{b})\right\}^{1/(4g)}$$

$$= N^{1-1/(2g)} O(x^{\frac{1}{2}+\epsilon}) \left\{\sum_{n_1=1}^{N} \sum_{n_2=1}^{N} \left|\sum_{\mathfrak{a}} \chi_{n_1}(\mathfrak{a}) \overline{\chi}_{n_2}(\mathfrak{a})\right|^2\right\}^{1/(4g)}.$$

Here the sums with respect to $\mathfrak{a}$ are $O(x^{g+\frac{1}{2}\epsilon})$. Using this result if $n_1 = n_2$, we obtain

$$\left|\sum_{n=1}^{N} \sum_{\mathfrak{p}} \chi_n(\mathfrak{p})\right| \leqslant N^{1-1/(4g)} O(x^{1+\epsilon})$$

$$+ N^{1-1/(2g)} O(x^{\frac{1}{2}+\epsilon}) \left\{\sum_{n_1=1}^{N} \sum_{\substack{n_2=1 \\ n_1 \neq n_2}}^{N} \left|\sum_{\mathfrak{a}} \chi_{n_1} \overline{\chi}_{n_2}(\mathfrak{a})\right|^2\right\}^{1/(4g)}.$$

Here the sum over $\mathfrak{a}$ is in the rational case by Lemma 1 at most

$$O(m \log m),$$

and in the general algebraic case by Lemma 2 at most

$$O(x^{gk}|D|m^2)^{1/(k+2)+\varepsilon},$$

and this concludes the proof of Theorems 2 and 3.

5. *Proof of Theorem* 4.

We recall that k and l are natural integers > 1, p a rational prime, $p \equiv 1 \pmod{k}$. We can assume that $p \nmid l$.

We introduce the following number fields. P the rational field, $P_l = P(\sqrt[l]{1})$, $P(p) = P(\sqrt[l]{p})$, P^k the subfield of absolute degree k of $P(\sqrt[p]{1})$. We denote by P_l^k, $P_l(p)$, $P_l^k(p)$ the unions of these fields in the obvious way. Let n be the degree of $P_l^k(p)$ over P_l. We first prove

LEMMA 3. (1) $n = kl$ *if* $(2, k, l) = 1$.

(2) $n = kl$ *if* $k \equiv l \equiv 2 \pmod{4}$, $p \equiv 3 \pmod{4}$.

(3) $n = \frac{1}{2}kl$ *if* $k \equiv l \equiv 0 \pmod{2}$, $p \equiv 1 \pmod{4}$.

(4) $n = \frac{1}{2}kl$ *if* $k \equiv 0 \pmod{2}$, $l \equiv 0 \pmod{4}$.

Proof. It is clear that the degrees of P_l^k and $P_l(p)$ over P_l are k and l respectively. Hence $n = kl$ if every common subfield of P_l^k and $P_l(p)$ is a subfield of P_l. Otherwise, we have a genuine extension field of P_l contained in P_l^k and $P_l(p)$. As a subfield of $P_l(p)$ this field must have the form $P_l(\sqrt[m]{p})$, $m \geqslant 2$, $m \mid l$. As a subfield of the cyclotomic field P_l^k, this extension field is absolutely abelian, and moreover it has the absolutely abelian subfield $P(\sqrt[m]{p})$. This field is absolutely abelian if and only if $m = 2$. Since $P(\sqrt{p})$ is never a subfield of P_l, we are left with only two alternatives. Either, $n = kl$ and $P(\sqrt{p})$ is not a subfield of both P_l^k and $P_l(p)$. Or, $n = \frac{1}{2}kl$ and $P(\sqrt{p})$ lies in P_l^k and $P_l(p)$.

If k or l are odd, P_l^k or $P_l(p)$ are of odd degree over P_l, hence they cannot both contain a quadratic extension of P_l. This proves (1).

If $p \equiv 3 \pmod{4}$, $k \equiv 2 \pmod{4}$, P^k contains $P(\sqrt{-p})$. If P_l^k contained $P(\sqrt{p})$, it would also contain $P(\sqrt{-1})$, which is not true if $4 \nmid l$. This proves (2).

If $k \equiv 0 \pmod{2}$ and $p \equiv 1 \pmod{4}$, P_l^k contains $P(\sqrt{p})$. And if $l \equiv 0 \pmod{2}$, $P_l(p)$ contains $P(\sqrt{p})$. This proves (3).

Finally, if $k \equiv 0 \pmod{2}$ and $l \equiv 0 \pmod{4}$, P_l contains $P(\sqrt{-1})$ and P^k contains either $P(\sqrt{p})$ or $P(\sqrt{-p})$. Hence P_l^k contains $P(\sqrt{p})$ and $P_l(p)$ contains $P(\sqrt{p})$. This proves (4) and completes the proof of Lemma 3.

We now continue the proof of Theorem 4, and assume that $kp \not\equiv 0 \pmod q$. Then q does not divide the discriminant of $P_l^k(p)$, and neglecting all divisors of discriminants, we can state: $q \equiv 1 \pmod l$ if and only if q splits up totally in P_l. In addition $y^l - p \equiv 0 \pmod q$ is soluble if and only if q splits up totally in $P_l(p)$. Finally $z^k - q \equiv 0 \pmod p$ is soluble if and only if q splits up totally in P^k. In short, for each p we have to count all primes $q \leqslant x$ which split up totally in $P_l^k(p)$; or, with a factor $1/\phi(l)$, all prime ideals of absolute degree 1 in P_l which split up totally in $P_l^k(p)$. And the restriction to ideals of degree 1 is clearly not essential, as the error committed for each p is of order $O(x^{\frac{1}{2}+\epsilon})$, $\epsilon > 0$.

As $P_l^k(p)$ is abelian over P_l of degree n, there exist in P_l n abelian characters $\chi_1, \ldots, \chi_n$ such that a prime ideal $\mathfrak{q}$ in P_l splits up totally in $P_l^k(p)$ if and only if

$$\chi_1(\mathfrak{q}) = \ldots = \chi_n(\mathfrak{q}) = 1.$$

Here χ_1 is the principal character, and for fixed p all the χ are distinct, primitive with conductors† dividing $l^2 p$. Hence the norms of the conductors are $O(x^l)$.

The sum we have to evaluate is

$$S = \big(\phi(l)\big)^{-1} \sum_{p \leqslant x} \sum_{N\mathfrak{q} \leqslant x} n^{-1} \sum_{\nu=1}^{n} \chi_\nu(\mathfrak{q}),$$

where p is restricted to primes $\equiv 1 \pmod k$, and $\mathfrak{q}$ to prime ideals in P_l. Some of the characters χ_ν, $\nu > 1$, may actually have conductors prime to p. As their conductors divide l^2, their number t is bounded in terms of l. Call them $\psi_1, \ldots, \psi_t$.

We can now split up the sum as

$$S = S_1 + S_2 + S_3 + S_4$$

so that S_1 contains the contribution of χ_1, S_2 contains the contribution of $\psi_1, \ldots, \psi_t$, S_3 contains the contribution of all the remaining characters for which $n = kl$ and S_4 contains the contribution of all remaining characters for which $n = \frac{1}{2}kl$. Let us consider them in their natural order.

$$S_1 = (kl)^{-1} \sum_{p \leqslant x} \sum_{q \leqslant x} kl/n,$$

where p and q are rational primes $\equiv 1 \pmod k$ or $\equiv 1 \pmod l$ respectively, and where $kl/n = 1$ or 2 according to Lemma 3. Using the prime number theorem for arithmetic progressions, we get

$$S_1 = \big(kl\phi(k)\phi(l)\big)^{-1} v \left\{ \int_e^x (\log u)^{-1} du \right\}^2 + O(x^2) \exp\left(-\alpha (\log x)^{1/2}\right).$$

---

† For a discussion of the conductors of Kummer fields see H. Hasse, *Bericht über neuere Untersuchungen und Probleme aus der Theorie der algebraischen Zahlen*, Teil II (1930), §10, XIII.

S_2 can be written in the form

$$S_2 = \sum_{s=1}^{l} u_s \sum_{N\mathfrak{q} \leqslant x} \psi_s(\mathfrak{q}),$$

where $u_s = O(x)$. Hence by the prime ideal theorem for arithmetic progressions

$$S_2 = O(x^2)\exp\left(-\alpha(\log x)^{1/2}\right).$$

S_3 can be written in the form

$$S_3 = \left(kl\phi(l)\right)^{-1} \sum_{p\leqslant x}{}^{*} \sum_{\nu=1}^{n}{}^{**} \sum_{N(\mathfrak{q})\leqslant x} \chi_\nu(\mathfrak{q}),$$

where the asterisk in the first sum denotes that we restrict ourselves to those $p \equiv 1 \pmod k$, for which $n = kl$, and where the asterisks in the second sum restrict the summation to characters which genuinely depend on p, and are therefore all distinct. Hence we have a sum over $N = O(x)$ distinct characters whose norms do not exceed $m = l^{2l} x^l$. Hence by Theorem 3, with $g = 2l$,

$$\begin{aligned} S_3 &= O(Nx)\{N^{-1/(4g)} + (x^{-1} m^{1/g})^{1/(\phi(l)+2)}\} x^\epsilon \\ &= O(x^2)\left\{x^{-1/(8l)} + x^{-1/(2(\phi(l)+2))}\right\} x^\epsilon \\ &= O(x^{2-1/(9l)}). \end{aligned}$$

The same argument will of course operate for S_4, and this completes the proof of Theorem 4.

Department of Mathematics,
University of Bristol,
Bristol 8.

(*Received 19th November*, 1957.)

30.

On the Representation of Homotopic Classes by Regular Functions

Bulletin de l'Académie Polonaise des Sciences 6 (1958), 181–184
[Commentary on p. 569]

Presented by K. KURATOWSKI on January 27, 1958

Introduction. In a lecture to the London Mathematical Society in 1956, Professor Kuratowski mentioned a conjecture equivalent to the following ([1], p. 319 footnote and p. 344, theorem 5):

THEOREM. *Let A be an open set in the complex plane, $f(x)$ a complex valued continuous function in A which does not take the values 0 or ∞. Then there exists a function $F(x)$, regular in A, and a function $g(x)$, continuous in A such that*

$$f(x) = F(x)e^{g(x)}.$$

After some time I succeeded in finding a proof of this theorem which I communicated to Professor Kuratowski, who returned to me a version of my original proof greatly simplified by Professor Sikorski. This simplified proof is produced in this paper, and I am greatly obliged to Professors Kuratowski and Sikorski for their permission to publish it.

The above theorem is obviously equivalent to the following statement (see [2]):

Let A be an open set in the complex plane, let B be the set of all finite complex $x \neq 0$. Consider the continuous functions which map A into B. Then each homotopy class of these functions contains a function regular in A.

This statement suggests the following

PROBLEM. *Let A and B be any two open sets on the complex sphere. Consider all homotopy classes of continuous functions which map A into B. Is it true that each homotopy class contains a function meromorphic in A?*

The answer is as follows.

1. *If A and B are identical with the whole complex sphere, the answer is negative.* For the only functions meromorphic on A are the rational functions. Their mapping of A into B has non-negative (Brouwer) degree and it is easy to construct continuous functions which map A into B with negative degree.

2. In all other cases, where either A or B is simply connected, there exists only one homotopy class, so the answer is trivially in the affirmative. This result can be proved in a variety of ways. As the proofs are purely topological they are omitted from this paper.

3. If A is at least doubly connected and B consists of the whole plane with two points removed, the answer is in the affirmative.

4. If A is at least doubly connected and if B is either at least trebly connected or doubly connected in such a way that $B+b$ is also at least doubly connected for each individual point b, then the answer is negative.

There remains the case when B can be mapped conformally onto the set B_0: $|x|<1$, $x \neq 0$. In this case the answer is negative if A is not simply connected and if there exists an open set on the sphere which intersects the complement of A in a non-void set of linear measure 0; in particular, if the complement of A contains isolated points. On the other hand, the answer is affirmative if A is of finite connectivity and if the complement of A does not contain isolated points.

But if A is of infinite connectivity, the problem has not been completely solved. Using the above theorem, we can state in the following form this unsolved

PROBLEM. *Let A be an open set in the complex plane of infinite connectivity. Is it always possible to find a non-vanishing function $f(x)$, regular in A, such that $\log|f(x)|+h(x)$ is unbounded above in A for each real function $h(x)$ harmonic in A?*

Preparations. We require the following extension of Riemann's mapping theorem.

Every doubly connected open set can be mapped conformally on one and only one of the following three sets:

a) $x \neq 0, \infty$,

b) $|x|<1$, $x \neq 0$,

c) $1<|x|<R$, $R>1$.

RUNGE'S THEOREM. *Let A be an open set and let $f(x)$ be regular on A. Then there exists a sequence of rational functions which converge uniformly to $f(x)$ on every closed subset of A. Moreover, the poles of these rational functions may be confined to a set T, if T has at least one point in common with every topological component of the complement of A* (see [3], p. 176).

We shall apply the theorem only for sets A of finite connectivity.

MONTEL'S THEOREM. *If the functions $f_n(x)$ $(n=1,2,3,\ldots)$ are regular, $\neq 0$, $\neq 1$ on an open set A, then there exists a subsequence which converges uniformly in every closed subset of A towards a regular function $F(x)$. Moreover, $F(x)$ is either constant (possibly infinite) or $F(x) \neq 0$, $F(x) \neq 1$* (see [3], p. 350).

Finally we require a

LEMMA. *Let A be an open set in the complex plane. Then there exists a sequence of open subsets $A_1, A_2, A_3 \dots$ such that*

(i) *A_{n+1} contains the closure of A_n.*

(ii) *The union of $A_1, A_2, A_3 \dots$ equals A.*

(iii) *A_n is of finite connectivity.*

(iv) *For each $n \geqslant 1$ each topological component of the complement of A_n contains a point not in A.*

This lemma can be easily proved by covering the plane with square gratings of decreasing mesh. The details are left to the reader.

Proof of the Theorem. We construct a sequence of open sets $A_1, A_2 \dots$ as outlined in the Lemma. It is almost obvious that our theorem is true on a set A_n of finite connectivity. More precisely, we can say that for all x on A_n

$$f(x) = r_n(x)\, e^{g_n(x)}, \tag{1}$$

where the $r_n(x)$ are rational functions with poles outside A. If we compare (1) for n and $n+1$, we obtain

$$e^{g_{n+1}(x) - g_n(x)} = r_n(x)\, r_{n+1}^{-1}(x)$$

on A_n. Hence, $g_{n+1}(x) - g_n(x)$ is regular on A_n, and by virtue of Runge's Theorem we can find a rational function $p_n(x)$ such that

$$|g_{n+1}(x) - g_n(x) - p_n(x)| < 2^{-n}$$

on A_n, and such that the poles of $p_n(x)$ do not lie in A. Hence, we have on A_n for $m > n$

$$f(x) = r_n(x)\, e^{g_n(x)} = r_m(x)\, e^{g_n(x) + \sum_{\nu=n}^{m-1} [g_{\nu+1}(x) - g_\nu(x) - p_\nu(x)] + \sum_{\nu=n}^{m-1} p_\nu(x)}$$

This formula can be transformed into

$$f(x) = \left[r_m(x)\, e^{\sum_{\nu=1}^{m-1} p_\nu(x)} \right] \cdot \left[e^{g_n(x) - \sum_{\nu=1}^{n-1} p_\nu(x)} \right] e^{\sum_{\nu=n}^{m-1} [g_{\nu+1}(x) - g_\nu(x) - p_\nu(x)]}$$

The third factor tends to a limit as $m \to \infty$ uniformly on A_n. As the second factor is independent of m, the first factor tends to a limit as $m \to \infty$ uniformly on A_n and therefore this limit is a function $F(x)$, regular in A. Hence, the product of the second and the third factor tends to a limit $e^{g(x)}$ on A_n and this proves our theorem as this last limit is independent of n.

Further proofs. We assume that A is at least doubly connected, and B either at least trebly connected or that B can be mapped conformally on the region

$$1 < |x| < R,$$

where $R > 1$ is a constant. In the first case, we can assume that B does not contain 0, 1, or ∞ and that there exists in B a closed continuous curve $z(\theta)[0 \leqslant \theta \leqslant 2\pi]$ which separates any 2 of these 3 points and which satisfies the relations

$$\int_0^{2\pi} d\arg z = 2\pi, \qquad \int_0^{2\pi} d\arg(z-1) = -2\pi.$$

Clearly, such a curve cannot be continuously transformed in B into an arbitrary small neighbourhood of a point.

In the second case, we can find in B a closed simple Jordan curve $z(\theta)[0 \leqslant \theta \leqslant 2\pi]$ which separates 0 from ∞ and which satisfies the relation

$$\int_0^{2\pi} d\arg z = 2\pi.$$

Again, this curve cannot be continuously transformed in B into an arbitrarily small neighbourhood of a point.

Similarly, as A is at least doubly connected, we can assume that A does not contain 0 or ∞ and that there lies in A a closed rectifiable Jordan curve C which separates 0 from ∞. Introducing in A polar co-ordinates

$$x = re^{i\theta},$$

we define for integers $n \geqslant 1$

$$f_n(x) = z(n\theta),$$

which are continuous functions in A. If these functions were homotopic to functions $F_n(x)$ regular in A, we would have

$$\frac{1}{2\pi i}\int_C \frac{F_n'(x)}{F_n(x)}dx = \frac{1}{2\pi}\int_0^{2\pi} d\arg z(n\theta) = n.$$

Clearly, no subsequence of the functions $F_n(x)$ can converge to a constant, as the curve C is mapped by them into a curve in the B-plane which cannot be contracted into a point. Therefore, by Montel's Theorem, a subsequence must converge towards a limit function $F(x)$ regular in A, so that $F(x)$ maps A into B. But

$$\frac{1}{2\pi i}\int_C \frac{F'}{F}(x)\,dx = \lim_{n\to\infty}\int_C \frac{F_n'}{F_n}(x)\,dx = \infty$$

which is absurd. This proves the statement 4 of the Introduction.

REFERENCES

[1] K. Kuratowski, *Théorèmes sur l'homotopie des fonctions continues de variable complexe et leur rapport à la théorie des fonctions analytiques*, Fund. Math. **33** (1945), 316-367.

[2] S. Eilenberg, *Transformations continues en circonférence et la topologie du plan*, Fund. Math. **26** (1936), 61-112.

[3] S. Saks and A. Zygmund, *Analytic functions*, Monogr. Mat. **28**, Warszawa-Wroclaw, 1952.

[Bristol]

31.

Review of "Mathematische Werke" by Erick Hecke

The Mathematical Gazette 45 (1961), 77–78

Mathematische Werke. By ERICK HECKE. Ed. B. SCHOENEBERG. Pp. 955. 1959. (Vandenhoek and Ruprecht.)

This is a photographic reprint of all the published work of Hecke (1887–1947), excepting only his book on the Theory of Algebraic Numbers. It is prefaced by an obituary speech by J. Nielsen and a short introduction by C. L. Siegel. With a few exceptions, all the papers are concerned with the theory of modular functions and their application to the theory of numbers. In many cases the work is buttressed by his expert knowledge and liberal use of θ-series. Some outstanding results are:—

(i) In his dissertation (1912) Hecke introduced modular functions of two variables in order to construct absolute class fields of real quadratic number fields with the aid of complex functions.

(ii) The proof of the functional equation of the Dedekind Zeta Function (1917).

(iii) The introduction of Grössen Charaktere and the study of their L-series (1918–1920).

(iv) The explicit law of quadratic reciprocity in totally real algebraic number fields (1919).

(v) His systematic study of modular forms connected with Dirichlet series which possess an Euler product and satisfy a functional equation (1935–44).

The list is impressive. In assessing Hecke's work one is struck not only by the depth of his results but also by their vitality and importance for present research work in arithmetic theory. His proofs may strike a mathematician of the younger generation as laborious verification by calculation. But it is the common fate of the great pioneers that their results will be presented in later times more simply in a more abstract language which would have been unintelligble to the original workers in the field, and could not have been invented unless the spade-work had been done in the hard way originally.

H. HEILBRONN

Article was originally printed on two pages and has been presented here on one.

32.

Old Theorems and New Methods in Class-Field Theory *

Journal of the London Mathematical Society 37 (1962), 6–9

I should like to follow the example of my predecessor and say a few words about the affairs of the Society during the last two years. A drive for corporate membership was started which produced an encouraging response from industrial firms, though the reply from universities was disappointing. In April this year an instructional conference on "Functional Analysis and some of its Applications" was held by the Society which was well attended by members of the Society and by scholars from other countries. It is the intention of the Council to repeat such conferences at biennial intervals.

In general the affairs of the Society are in good order and the financial situation is satisfactory, as you have heard from our treasurer. In particular I am glad to report that our former president, L. J. Mordell, has made a gift of £500 to the funds of the Society. But our membership has not increased in the last few years in proportion with the rising number of professional mathematicians.

The number of papers submitted to the Society continues to increase, and the Council has reluctantly decided to expand the size of the *Proceedings* during the present year. But there is clearly a limit beyond which the publications of the Society can not be increased. (I am not thinking of costs only.) Nor is it in the true interests of mathematical learning that our society should have a monopoly of publishing mathematical papers. There are other journals in this country publishing mathematics, and some of them are able and willing to publish more.

Finally, one personal remark. I have been closely concerned with the affairs of the Society for a quarter of a century. But only during my term as president have I learned what an enormous debt the society owes to its honorary secretaries, and I should like to acknowledge this debt in public.

One of the obvious topics of number theory is the provision of an answer to the question: "How does a polynomial $f(x)$ with rational coefficients, irreducible in the rational field, split up mod p, p prime; in particular for which primes p has the congruence

$$f(x) \equiv 0 \pmod{p}$$

solutions?" By Dedekind's theorem, this is essentially equivalent to the problem: Let p be a rational prime, K a finite algebraic number field [generated by a zero of $f(x)$]; how does p split up into prime ideals in the

---

* Presidential Address delivered to the Society on 16 November, 1961.

field K? More generally, let k be a finite algebraic number field, K a finite extension, $\mathfrak{p}$ a prime ideal in k; how does $\mathfrak{p}$ split up in the field K where it is not necessarily prime? A full answer to this problem is available if K has an abelian Galois group over k. The theory which answers this question is called class-field theory, for a reason which will soon appear. It is perhaps the most beautiful part of number theory, at any rate as far as the results are concerned.

A function $\chi(\mathfrak{a})$ defined for all non-zero ideals of the field k is called a character if

$$\text{(i) } \chi(\mathfrak{a}\mathfrak{b}) = \chi(\mathfrak{a}) \times \chi(\mathfrak{b}), \quad \text{(ii) } x\big((\alpha)\big) = 1,$$

for all principal ideals (α) where α is a totally positive integer $\equiv 1 \pmod{\mathfrak{m}}$ for some $\mathfrak{m}$ depending on χ. Two characters are called equivalent if their values differ only for a finite number of prime ideals $\mathfrak{p}$. Equivalent characters can be said to form a train, each train containing one character which vanishes for the minimum number of primes and is called the primitive character of the train. It is easily seen that the trains form an infinite abelian group with multiplication as group operation.

The theory then states that there is a 1–1 relation between all finite abelian extensions K and all finite subgroups of the infinite group of trains such that the subgroup is isomorphic to the Galois group of the extension. Moreover, the Boolean algebras formed by the finite abelian extensions and the subgroups are isomorphic. If we restrict ourselves to those ideals for which the primitive characters of a given subgroup of trains are all $\neq 0$ (in fact all ideals prime to the relative discriminant) the primitive characters in the subgroup are group characters of a quotient group of the group of all ideals in k relatively prime to the relative discriminant with respect to the group of ideals in k for which all primitive characters of the subgroup are equal to 1. The elements of the quotient group, called class group, are the classes of class-field theory. *A prime ideal $\mathfrak{p}$ prime to the relative discriminant splits up into distinct prime ideals of relative degree g where g is the order of the class containing $\mathfrak{p}$ in the class group.* Actually we can say more than this. Artin's reciprocity law states: Let $n(\mathfrak{p})$ be the absolute norm of $\mathfrak{p}$ in k, then there exists a substitution σ of the Galois group such that

$$A^{n(\mathfrak{p})} \equiv \sigma A \pmod{\mathfrak{p}}$$

for all integers A of the extension and this substitution represents an explicit isomorphism between the class group and the Galois group.

The relative discriminant of K/k equals the product of the conductors of the primitive characters. Analytically it follows that

$$\zeta_K(s) = \prod_{\chi} \sum_{\mathfrak{a}} \chi(\mathfrak{a})\, n(\mathfrak{a})^{-s}$$

where $\zeta_K(s)$ is the Dedekind ζ-function of K, χ runs through the primitive characters and $\mathfrak{a}$ through all non-zero integral ideals in k. If k is the rational field, this relation actually gives the decomposition of all primes, *including* the divisors of the discriminant, in terms of the base field. These are the principal results of class-field theory, as they were mostly conjectured by Hilbert and proved by Takagi, Artin and others more than 30 years ago.

A word about the nature of the original proofs. By Abel's classical theorem binomial equations lead to abelian extensions and for binomial extensions the decomposition laws are pretty obvious. By laborious processes of counting, extending and verifying the results were proved, but obviously the proofs were laborious and not as enlightening as one might hope.

In the last 30 years intense efforts were made to remodel the proofs so as to give a clearer picture of the situation. The first important progress was made by Chevalley in 1940 by the use of the idèle, defined as follows.

Let k be a field, $k_{\mathfrak{p}}$ its $\mathfrak{p}$-adic closure. Consider the direct product of all $k_{\mathfrak{p}}$, including the infinite primes. An element a of this field is called an idèle if it is unit for all but a finite number of primes $\mathfrak{p}$ and non-zero for all primes. Clearly the idèles form a group with respect to multiplication, containing the non-zero elements of k as a subgroup of infinite index. Chevalley was able to prove that the characters of the infinite quotient group form a group isomorphic to the group of characters of the Galois group of the maximal (infinite) abelian extension of k. From this fundamental theorem he obtained fairly easily the classical results.

The second reform, dating back to the early 30's and stimulated by Chevalley's success, is the introduction of valuation theory not only to class-field theory but to algebraic number theory in general. As a result it is possible to study class-fields first in the small, *i.e.* $\mathfrak{p}$-adically, and then absolutely.

The third simplification arose from the introduction of cohomology theory, which elucidates many of the old calculations of group indices. Indeed a modern algebraist reading the old proofs will find many traces of cohomology theory contained in them.

Finally there has been an attempt to extend the theory to non-abelian extensions. This attempt has been successful up to a point; many of the structural theorems carry over to the non-abelian case. The relation mentioned above between L-series holds for non-abelian extensions; but no conjecture has been made about a possible formulation of the decomposition law in the non-abelian case.

What is the outlook for the future? Two questions arise. Have today's proofs reached anything like finality? My personal guess is "no"; but I have no suggestions to make about an alternative line of attack. Finally, does there exist a decomposition law in the non-abelian case? As good a witness as Artin has stated that if we only knew what to prove we could prove it today. Far from disagreeing with this, I should like to go further and conjecture that no simple law of decomposition exists except for abelian extensions.

Literature.

H. Hasse, " Bericht über neuere Untersuchungen und Probleme aus der Theorie der algebraischen Zahlkörper ", Pt. I—*Jahresbericht der D.M.V.*, Vol. 35 (1926), Pt. Ia—*Jahresbericht der D.M.V.*, Vol. 36 (1927), Pt. II—*Jahresbericht der D.M.V.*, Vol. 39 (1930).

C. Chevalley, " La théorie du corps de classes ", *Annals of Math.*, 41 (1940), 394–418.

———, *Class-field theory* (Nagoya University, Nagoya, 1954).

J. P. Serre, *Homologies des groupes. Applications arithmétiques* (Collège de France, 1958–59).

———, *Groupes algébriques et corps de classes* (Act. Scient. et Ind., 1959).

E. Artin and J. T. Tate, *Mimeographed lecture notes* (Vol. I (1951), Vol. II (1960), Princeton).

33.

Prof. Jacques Hadamard, For. Mem. R. S.

(With L. Howarth)

Nature 200 (1963), 937–938

JACQUES SALOMON HADAMARD was born in Versailles on December 8, 1865. His mathematical abilities very soon became evident when he won the entrance examination to the École Polytechnique, perhaps the stiffest mathematical entrance examination in the world, with a record of 1.875 marks out of a total of 2,000. After graduation he obtained a teaching post at the University of Bordeaux, but soon returned to Paris as a professor at the Sorbonne. In 1897 he went to the Collège de France, from which he retired in 1935. The Second World War forced him to go to the United States, where he found a temporary home in Princeton. In 1945 he visited Great Britain and gave a charming lecture on personal recollections as a mathematician, which is still remembered with pleasure by many of his British colleagues. He soon returned to Paris, leading a very active life in spite of his old age, until he died on October 17 in his ninety-eighth year.

Most of his work in pure mathematics was done in the 1890's on analysis. Hadamard, desiring to investigate the Riemann ζ-function, developed a theory of integral functions of a complex variable which has been the foundation of all the subsequent work on the subject. In 1896 he proved the prime number theorem which had been conjectured by Gauss and Legendre a hundred years earlier: let $\pi(x)$ be the number of primes not exceeding x, then $\pi(x) \sim x/\log x$. The same result was proved simultaneously by de la Vallée-Poussin. Though a lot of work has since been done on the prime number theorem, Hadamard's proof is still the simplest and most natural proof of the theorem. It must be a unique event in the history of science that a major break-through is achieved simultaneously by two scholars who each survived their achievement by two-thirds of a century.

A few other results can be briefly mentioned:

(1) Hadamard's inequality for real determinants:

$$\|a_{ik}\|^2 \leqslant \prod_{i=1}^{n} \sum_{k=1}^{n} a_{ik}^2$$

(2) Hadamard's "3 circle theorem": let $m(r)$ be the maximum of the absolute value of a regular complex function in a circle with fixed centre and radius r. Then $\log m(r)$ is a convex function of $\log r$.

Article was originally printed as double-column text and has been rearranged here for readability.

(3) Hadamard's gap theorem for a Taylor series of the form

$$\sum_{n=0}^{\infty} a_n x^{b_n}$$

Every point on the boundary of the circle of convergence is a singular point of the function if $b_{n+1} - b_n \to \infty$.

His work in applied mathematics was mainly concerned with wave propagation and is of classical importance. No standard text on the fluid mechanics of high-speed flows, for example, would be complete without basic reference to it, nor would any account of Huygens's principle. Its influence permeates present-day thinking on solutions of the wave equations and allied problems. Starting from the work of Riemann, Kirchhoff and Volterra, he developed in 1903 (ref. 3) and in his later work in 1923 and 1932 (refs. 4 and 5) the theory of hyperbolic linear differential equations for the general case of n independent variables, distinguishing between the cases when n is odd and even. Here in these books are set out the tools in the form of the theory of characteristics and bicharacteristics, the principle of descent and the concept of the finite part of an infinite improper integral, without any of which an understanding of high-speed flows would be impossible. Hadamard was not only interested in existence and uniqueness properties, but also, as Friedrichs remarks, in the property of the continuous dependence of the solution on the data, a property which is required if the problem is to be called correctly posed. With the advent of modern computational techniques the application of the method of characteristics has become a standard method of solution not only for linear but also for quasi-linear hyperbolic differential equations, and three-dimensional aerofoil theory leans heavily on the concept of the finite part of an infinite integral.

Hadamard's name must also be linked with those of Stokes, Earnshaw, Riemann, Rankine, Hugoniot and Rayleigh in developing the ideas of non-linear wave motion which led to the present theory of shock waves and their formation.

Hadamard's merits as a mathematician were fully recognized at home and abroad. He held many honorary doctorates and was an honorary member of many learned societies, in particular a foreign member of the Royal Society and an honorary member of the London Mathematical Society.

Hadamard published the following books: (1) *La série de Taylor et son prolongment analytique* (Paris, 1901); (2) *Notice sur les travaux scientifiques de M. Jacques Hadamard* (Paris, 1912); (3) *Leçons sur la Propagation des Ondes* (Paris, 1903); (4) *Lectures on Cauchy's Problem in Linear Partial Differential Equations* (Yale, 1923; also Dover Publications, 1952); (5) *Le problème de Cauchy et les équations aux derivées partielles linéaires hyperbolique* (Paris, 1932); (6) *An Essay on the Psychology of Invention in the Mathematical Field* (Princeton, 1945).

H. HEILBRONN
L. HOWARTH

34.

On the Representation of a Rational as a Sum of Four Squares by Means of Regular Functions

Journal of the London Mathematical Society 39 (1964) 72–76
[Commentary on p. 582]

1. *Introduction.*

It has been known since the days of Lagrange, if not since the days of Diophantus, that every positive rational x can be written in the form

$$x = y_1^2+y_2^2+y_3^2+y_4^2,$$

where y_1, y_2, y_3, y_4 are rational numbers. I was asked recently whether this theorem remained true under the additional condition that the numbers y_i are continuous functions of x. The affirmative answer is given by

THEOREM 1. *There exist four integral functions $f_j(x)$ which are positive for $x > 0$, satisfy the relation*

$$\sum_{j=1}^{4} f_j^2(x) = 1, \tag{1}$$

and have the property that $x^{\frac{1}{2}} f_j(x)$ is rational for positive rational x.

For higher powers I can only prove a slightly weaker statement. Let $k \geqslant 2$ be a rational integer, and let $s = s(k)$ be the smallest integer such that for each rational $x > 0$ the rational points are everywhere dense on the surface

$$\sum_{\sigma=1}^{s} y_\sigma^k = x, \quad y_\sigma > 0. \tag{2}$$

Let L denote the complex plane exclusive of the negative real axis $x \leqslant 0$. Then I can prove

THEOREM 2. *There exist s functions $g_\sigma(x)$ $(1 \leqslant \sigma \leqslant s)$ regular in L such that*

1. *$g_\sigma(x)$ is rational for rational $x > 0$.*

2. *$\sum_{\sigma=1}^{s} g_\sigma^k(x) = x$ in L.*

For $k = 2, 3, 4$ the numbers $s = 4, 3, 15$ have the required properties as will be indicated in the appendix. And it is well known that 7 is not a sum of 3 rational squares, that 3 is not a sum of 2 rational cubes and that 15 is not a sum of 14 rational fourth powers.

For $k \geqslant 5$ it follows at once from the classical analytical treatment of

Received 9 November, 1962. [JOURNAL LONDON MATH. SOC., 39 (1964), 72–76]

Waring's problem that $s = s(k)$ exists and satisfies $s = O(k \log k)$. On the other hand, I cannot prove that $s > 3$ for a single odd k.

I wish to thank Professor R. Rado, with whom I had some valuable discussions on similar topics before my attention was drawn to the questions raised in this paper.

2. *Proof of Theorem* 1.

For each rational $x > 0$, the surface of the 4-dimensional sphere

$$y_1^2 + y_2^2 + y_3^2 + y_4^2 = x$$

contains a rational point. Any rational line which is not a tangent through this point meets the sphere again in a rational point. Hence the rational points are everywhere dense on the sphere.†

Let $r_1, r_2, r_3, \ldots$ run through all positive rationals subject to the restriction $r_n \leqslant n$. We define 3 integral functions

$$g_i(x) = \arcsin\left(\tfrac{3}{5}\right) + e^{-x-5} \sum_{n=1}^{\infty} a_{i,n} \prod_{\nu=1}^{n} (x - r_\nu),$$

where the $a_{i,n}$ satisfy the inequality

$$0 < a_{i,n} < (n!)^{-2} \text{ for } 1 \leqslant i \leqslant 3,\ n \geqslant 1.$$

Their exact values will be determined later. For all complex x

$$\left| a_{i,n} \prod_{\nu=1}^{n} (x - r_\nu) \right| < (n!)^{-2} \prod_{\nu=1}^{n} (|x| + \nu) \leqslant (n!)^{-1} (|x| + 1)^n$$

which shows that the $g_i(x)$ are integral functions, and that for $x > 0$

$$0 < g_i(x) < \tfrac{1}{2}\pi.$$

We now put

$$f_1(x) = \cos\big(g_1(x)\big) \cos\big(g_2(x)\big) \cos\big(g_3(x)\big),$$
$$f_2(x) = \cos\big(g_1(x)\big) \cos\big(g_2(x)\big) \sin\big(g_3(x)\big),$$
$$f_3(x) = \cos\big(g_1(x)\big) \sin\big(g_2(x)\big),$$
$$f_4(x) = \sin\big(g_1(x)\big),$$

so that (1) is automatically satisfied.

It remains to determine the $a_{i,n}$ by an algorithm with respect to n so that $r_m^{\frac{1}{2}} f_j(r_m)$ is rational for $m \geqslant 1$, $1 \leqslant j \leqslant 4$. Our choice of the constant term in the development of $g_i(x)$ assures that $f_j(r_1)$ is rational irrespective of the choice of the $a_{i,n}$. We now assume that $m > 1$, that our condition is satisfied for $n < m$ and that the $a_{i,n}$ are determined for $n < m-1$; then the 3 values of $a_{i,m-1}$ are at our disposal, subject to their restriction on size. As

$$g_i(r_m) = \arcsin\left(\tfrac{3}{5}\right) + e^{-r_m-5} \sum_{n=1}^{m-1} a_{i,n} \prod_{\nu=1}^{n} (r_m - r_\nu),$$

† Dr Estermann kindly told me this simple proof.

a variation of the $a_{i,m-1}$ over the interior of the cube

$$0 < a_{i,m-1} < \left((m-1)!\right)^{-2}$$

corresponds to a variation of the $y_i(r_m)$ over an open cube, and a variation of the 4 values $r_m^{\frac{1}{2}} f_j(r_m)$ over an open set on the surface of our 4-dimensional sphere. The rational points being everywhere dense on this surface we can choose the $a_{i,m-1}$ so as to fulfil our condition.

3. *Proof of Theorem* 2.

We know that for rational $x > 0$ the rational points are everywhere dense on the surface (2). Let $P = P(x)$ be the projection of the rational points on the plane $y_s = 0$. Then P is everywhere dense on the $n-1$ dimensional set

$$\sum_{\sigma=1}^{s-1} y_\sigma^k < x, \; y_\sigma > 0 \text{ for } 1 \leqslant \sigma \leqslant s-1. \tag{3}$$

Let $r_1 = 1, r_2, r_3, \ldots$ run through all rational numbers in the interval $1 \leqslant x < 2^k$. We put for $1 \leqslant \sigma \leqslant s-1$

$$f_\sigma(x) = x^{1/k} \left\{ a_{\sigma,0} + \sum_{n=1}^{\infty} a_{\sigma,n} \prod_{\nu=1}^{n} \sin^2\left(\frac{\pi}{k \log 2} \log \frac{r_\nu}{x}\right)\right\}$$

where the $a_{\sigma,n}$ are subject to the limitation

$$0 < a_{\sigma,n} < (s 2^{n+1} e^{2\pi^2 n})^{-1}$$

for $n \geqslant 0$, $1 \leqslant \sigma \leqslant s-1$; their exact values will be determined later. In L each factor in the product is absolutely at most

$$e^{2\pi^2/(k \log 2)} < e^{2\pi^2}.$$

Hence the infinite series are uniformly convergent in every bounded subset of L, and the $f_\sigma(x)$ are regular in L. Further, one sees at a glance that

$$2 f_\sigma(x) = f_\sigma(2^k x).$$

Finally, we have the inequality

$$|f_\sigma(x)| < s^{-1} \sum_{n=0}^{\infty} 2^{-n-1} |x|^{1/k} = s^{-1} |x|^{1/k}.$$

Hence, if we define $f_s(x)$ by the relation

$$f_s^k(x) = x - \sum_{\sigma=1}^{s-1} f_\sigma^k(x),$$

$$f_s(x) > 0 \text{ for } x > 0,$$

the function $f_s(x)$ is regular in L.

It remains to prove that we can choose the $a_{\sigma,n}$ so that the $f_\sigma(x)$ are rational for rational $x > 0$. For this it suffices to prove that the $f_\sigma(r_m)$ lie in $P(r_m)$ for $m \geqslant 1$, $1 \leqslant \sigma \leqslant s-1$.

As $f_\sigma(1) = a_{\sigma,0}$ we can clearly choose the $a_{\sigma,0}$ in $P(1)$. Next we assume as before that $a_{\sigma,n}$ have been found for $1 \leqslant \sigma \leqslant s-1$, $0 \leqslant n \leqslant m-2$ such that the $f_\sigma(r_n)$ are in $P(r_n)$ for $1 \leqslant \sigma \leqslant s-1$, $0 \leqslant n \leqslant m-1$. The value of $f_\sigma(r_m)$ is then a function of $a_{\sigma,m-1}$ only. The possible values of $f_\sigma(r_m)$ form an open $s-1$ dimensional cube, which is contained in the cube

$$0 < y^\sigma < s^{-1} r_m^{1/k} \quad (1 \leqslant \sigma \leqslant s-1)$$

which in turn is contained in the set (3) for $x = r_m$. As $P(r_m)$ is everywhere dense in (3), it is everywhere dense in the original cube. Hence we can choose the $a_{\sigma,m-1}$ for $1 \leqslant \sigma \leqslant s-1$ in such a way that $f_\sigma(r_m)$ lies in $P(r_m)$ for $1 \leqslant \sigma \leqslant s-1$. This ensures that $f_\sigma(r_m)$ is rational for $1 \leqslant \sigma \leqslant s$.

4. *Appendix.*

It was first proved by Richmond† that every positive rational is a sum of 3 cubes of positive rationals. Richmond also makes the statement that the rational points are everywhere dense on the surface (2) for $k = s = 3$ and rational $x > 0$. I am unable to follow his proof of this statement. It is, however, easy to prove it from other facts given in Richmond's paper.

For fourth powers the corresponding statement is not so easily obtained. It was proved by Davenport‡ in 1939 that every large integer which is not $\equiv 0$ or 15 (mod 16) can be expressed as the sum of 14 fourth powers of rational integers. From this it follows at once that every positive rational is the sum of 15 fourth powers of rationals. In order to prove that the rational points are everywhere dense on the surface

$$\sum_{i=1}^{15} y_i^4 = x,$$

it suffices to prove Davenport's theorem in the following form.

Let $\delta > 0$, $\alpha_1 > 0, \ldots, \alpha_{14} > 0$,

$$\sum_{i=1}^{14} \alpha_i^4 = 1.$$

Then every sufficiently large integer N which is not $\equiv 0$ *or* 15 (mod 16) *is representable in the form* $N = y_1^4 + \ldots + y_{14}^4$ *where the integers* y_i *satisfy the inequalities*

$$|y_i - \alpha_i N^{\frac{1}{4}}| < \delta N^{\frac{1}{4}} \textit{ for } 1 \leqslant i \leqslant 14. \tag{4}$$

† *Proc. London Math. Soc.* (2), 21 (1923), 401–409.
‡ *Annals of Math.*, 40 (1939), 731–747.

Actually Davenport proves that every large integer $N \not\equiv 0, 15 \pmod{16}$ can be written in the form $N = y_1^4 + \ldots + y_{14}^4$ where the y_i satisfy the inequalities

$$N^{\frac{1}{4}} < y_i < 2N^{\frac{1}{4}} \quad \text{for} \quad 1 \leqslant i \leqslant 6,$$

$$N^{\rho} < y_i < 2N^{\rho} \quad \text{for} \quad 7 \leqslant i \leqslant 8,$$

$$N^{\sigma} < y_i < 2N^{\sigma} \quad \text{for} \quad 9 \leqslant i \leqslant 10,$$

$$N^{\tau} < y_i < 2N^{\tau} \quad \text{for} \quad 11 \leqslant i \leqslant 14.$$

Here ρ, σ, τ are explicitly defined constants such that $\frac{1}{4} > \rho > \sigma > \tau > 0$.

If one chooses 3 distinct primes $p \equiv q \equiv r \equiv -1 \pmod 4$ such that

$$p \sim N^{\frac{1}{4}-\rho}, \quad q \sim N^{\rho-\sigma}, \quad r \sim N^{\sigma-\tau}$$

it is possible to modify Davenport's proof such that (4) is satisfied; this is achieved by imposing on the y_i the divisibility conditions

$$p \mid y_i \text{ for } i \geqslant 7, \quad q \mid y_i \text{ for } i \geqslant 9, \quad r \mid y_i \text{ for } i \geqslant 11.$$

The detailed proof should present no serious difficulty to the expert, but requires modification not only on the minor arcs, but also on the major arcs. I feel therefore justified in omitting it from this paper.

Department of Mathematics,
University of Bristol.

35.

On the Addition of Residue Classes mod p

(With P. Erdös)

Acta Arithmetica 9 (1964), 149–159
[Commentary on p. 600]

In this paper we investigate the following question. Let p be a prime, $a_1, \ldots, a_k$ distinct non-zero residue classes mod p, N a residue class mod p. Let

$$F(N) = F(N; p; a_1, \ldots, a_k)$$

denote the number of solutions of the congruence

$$e_1 a_1 + \ldots + e_k a_k \equiv N \pmod{p}$$

where the $e_1, \ldots, e_k$ are restricted to the values 0 and 1. What can be said about the function $F(N)$?

We prove two theorems.

THEOREM I. $F(N) > 0$ *if* $k \geqslant 3(6p)^{1/2}$.

THEOREM II. $F(N) = 2^k p^{-1}(1+o(1))$ *if* $k^3 p^{-2} \to \infty$ *as* $p \to \infty$.

Theorem I is almost best possible. Put

$$a_1 = 1,\ a_2 = -1,\ a_3 = 2,\ a_4 = -2,\ \ldots,\ a_k = (-1)^{k-1}[\tfrac{1}{2}(k+1)].$$

Then it follows from an easy calculation that $F(\frac{1}{2}(p-1)) = 0$ if $k < 2(p^{1/2}-1)$. Theorem II is best possible. Define $a_1, \ldots, a_k$ as above and assume that $p^{2/3} < k = O(p^{2/3})$. Then it follows from our analysis that

$$\lim_{p \to \infty} p 2^{-k} F(0) > 1.$$

In the method of proof the two theorems differ considerably. The proof of Theorem I is elementary, depending entirely on the manipulation of residue classes mod p, whereas the proof of Theorem II is based on the application of finite Fourier series and simple considerations on diophantine approximations.

In an appendix we state various further conjectures which we are not able to prove.

Proof of Theorem I. We start with a definition. Let $b_1, \ldots, b_l$ be l distinct residue classes $\bmod p$. Then $B(x)$ denotes the number of solutions of the congruence

$$x \equiv b_i - b_j \pmod{p}, \quad 1 \leqslant i \leqslant l,\ 1 \leqslant j \leqslant l.$$

We recall the inequality

$$B(x+y) \geqslant -l + B(x) + B(y), \tag{I.1}$$

which is easily proved as follows. Assume that

$$x \equiv b_i - b_j \pmod{p}, \quad y \equiv b_g - b_h \pmod{p}.$$

If $j = g$, this implies that

$$x + y \equiv b_i - b_h \pmod{p}.$$

As there are only l possible values for b_j, (I.1) follows. It can also be written in the form

$$\big(l - B(x+y)\big) \leqslant \big(l - B(x)\big) + \big(l - B(y)\big). \tag{I.2}$$

LEMMA I.1. *Let* $1 < m \leqslant l < \frac{1}{2}p$; $a_1, \ldots, a_m$ *are distinct non-zero residue classes* $\bmod p$. *Then there exists an* i *in* $1 \leqslant i \leqslant k$ *such that*

$$B(a_i) < l - \tfrac{1}{6}m.$$

Proof. Put $r = 1 + [2l/m]$. By Davenport's theorem [1] about the addition of residue classes $\bmod p$, applied to the residue classes $0, a_1, \ldots, a_m$ we obtain $t \geqslant \mathrm{Min}(p-1, rm)$ distinct non-zero residue classes $c_1, \ldots, c_t$ which can be expressed as the sum of at most r residue classes a_j $(1 \leqslant j \leqslant m)$, which need not have distinct indices j.

As

$$\sum_{s=1}^{t} B(c_s) \leqslant \sum_{z=1}^{p-1} B(z) = l(l-1),$$

it follows that there exists an s such that

$$B(c_s) \leqslant l(l-1)t^{-1} \leqslant l(l-1)\,\mathrm{Max}\big((p-1)^{-1}, (rm)^{-1}\big) < \tfrac{1}{2}l,$$

or

$$l - B(c_s) > \tfrac{1}{2}l.$$

Hence, by (I.2), there exists an a_i such that

$$l - B(a_i) > \tfrac{1}{2}lr^{-1} \geqslant \tfrac{1}{2}\,lm(m+2l)^{-1} \geqslant \tfrac{1}{6}\,m,$$

which completes the proof of the lemma.

Proof of Theorem I. We begin with a definition. If

$$1 \leqslant u \leqslant \tfrac{1}{2}k$$

we consider all possible subsets S_u of u elements of the classes

$$a_1, a_2, \ldots, a_{2u-1}, a_{2u}.$$

For each subset S_u we consider the number $L(S_u)$ of distinct residue classes which can be written in the form

$$e_1 a_1 + \ldots + e_{2u} a_{2u},$$

where

$$e_i = \begin{cases} 0 \text{ or } 1 & \text{if} \quad a_i \text{ lies in } S_u, \\ 0 & \text{if} \quad a_i \text{ does not lie in } S_u. \end{cases}$$

Next we put

$$L(u) = \max L(S_u),$$

where S_u ranges over all subsets of u elements. It is easily verified that $L(1) = 2$, $L(2) = 4$, and that $L(u) \geqslant u+2$ for $u \geqslant 2$. It is also clear that $L(u+1) \geqslant L(u)$.

Our next step is to prove the inequality

$$\text{(I.3)} \qquad L(u+1) \geqslant L(u) + \tfrac{1}{6}(u+2) \quad \text{for} \quad 2 \leqslant u \leqslant \tfrac{1}{2}k - 1$$

provided that $L(u) < \frac{1}{2}p$.

We assume that S_u is the set for which $L(S_u) = L(u)$. Then we have $L(u)$ residue classes $b_1, \ldots, b_{L(u)}$ which are representable as linear combinations of the a_j in S_u with coefficients 0 or 1. We also have at our disposal $m = u+2$ residue classes a_i not in S_u with $1 \leqslant i \leqslant 2u+2$. Lemma I.1 is applicable as

$$m = u+2 \leqslant L(u) < \tfrac{1}{2}p.$$

So we obtain an i in $1 \leqslant i \leqslant 2u+2$ such that a_i not in S_u,

$$B(a_i) < \tfrac{1}{6} m.$$

We now define S_{u+1} as the union of S_u and a_i. Then, by Lemma I.1,

$$L(u+1) \geqslant L(S_{u+1}) = L(u) + \big(l - B(a_i)\big) > L(u) + \tfrac{1}{6} m$$

which proves (I.3).

By addition, it follows immediately from (I.3) that either $L(u) \geqslant \frac{1}{2}p$ or that

$$L(u) \geqslant 4 + \sum_{n=2}^{u-1} \tfrac{1}{6}(n+2) > \tfrac{1}{12}(u+1)(u+2)$$

for all $u \leqslant \frac{1}{2}k$. Hence, putting $t = [(6p)^{1/2}]$, we have in any case

$$L(t) \geqslant \tfrac{1}{2}p.$$

Further we may assume that S_t contains $a_1, \ldots, a_t$.

We now apply the same argument to the $2t$ residue classes $a_{t+1}, \ldots, a_{3t}$. Again a linear combination of at most t of them will represent at least half the residue classes mod p.

Thus we have 2 (not necessarily disjoint) sets each containing at least half the residue classes mod p. From this it follows at once that every N is representable as a sum of an element of the first set and of an element of the second set. This completes the proof of Theorem I.

Proof of Theorem II. We start by introducing some notations. Small latin letters denote rational integers, and therefore by implication residue classes mod p. Small greek letters denote real numbers.

$$\Lambda = \log p, \qquad p^{2/3} < k < p,$$

but until we reach Lemma II.5 it will be assumed that $k < p^{2/3}\Lambda$. The letter m with or without suffices will denote an integer in the interval $\frac{1}{4}k \leqslant m \leqslant k$. S_k is a given sequence of k non-zero distinct residue classes mod p, denoted by $a_1, \ldots, a_k$. For some permissible values of m we shall introduce subsequences S_m which we denote, without fear of misunderstanding, by $a_1, \ldots, a_m$.

For $r \not\equiv 0 \pmod p$ we put

$$\sigma(r) = \sigma(r, S_m) = \sum_{n=1}^{m} \sin^2(\pi r a_n/p),$$

$$\gamma(r) = \gamma(r, S_m) = \sigma(r, S_m)(m^3 p^{-2})^{-1}.$$

We note that $\gamma(r) \geqslant \gamma_0 > 0$, where γ_0 is an absolute constant. For given S_m we call r *critical* if $\gamma(r, S_m) < \Lambda$.

The symbol O implies absolute constants only. The symbol o refers to $p \to \infty$ uniformly in all other variables, unless stated otherwise.

If for S_k no value of r is critical, we take no further steps until we reach Lemma II.5. Otherwise we define

$$\mu = \operatorname{Min} \gamma(r, S_m)(\Lambda^6 + k - m),$$

where we admit all residue classes $r \not\equiv 0 \pmod p$, all m in $\frac{1}{2}k \leqslant m \leqslant k$ and all subsequences S_m of S_k containing m terms. For the remainder of the paper let s, m, S_m be the residue class s, the number m and the subsequence S_m for which the minimum is attained.

As some r is critical for S_k, it follows that

$$\mu \leqslant \operatorname*{Min}_{r \not\equiv 0} \gamma(r, S_k)\Lambda^6 < \Lambda^7.$$

As

$$\mu \geqslant \gamma_0(\Lambda^6 + k - m),$$

we have

$$\gamma_0(k - m + \Lambda^6) < \Lambda^7, \qquad m > k - \gamma_0^{-1}\Lambda^7.$$

Further, for each subsequence $S_{m'}$ of S_m where $m' \geqslant \frac{1}{2}k$ we have

$$\gamma(r, S_{m'}) \geqslant \gamma(s, S_m).$$

LEMMA II.1. *Let $r \not\equiv s \pmod p$ be a critical value of S_m. Then there exist integers u and v such that $vr \equiv us \pmod p$, $(u, v) = 1$, $1 \leqslant v \leqslant \Lambda$, $1 \leqslant u \leqslant \Lambda^2$.*

Further, assuming that the residue classes sa_n $(1 \leqslant n \leqslant m)$ are represented by numbers in the interval $[-\frac{1}{2}p, \frac{1}{2}p]$, these numbers are divisible by v with at most $2\Lambda^5 m^3 p^{-2}$ exceptions.

Proof. Without loss of generality we may assume that $s = 1$ and that $|a_n| < \frac{1}{2}p$ for $1 \leqslant n \leqslant m$.

From Dirichlet's principle it follows by a classical argument that we can solve the congruence $vr \equiv u \pmod p$ subject to

$$1 \leqslant v \leqslant \Lambda, \quad 1 \leqslant |u| \leqslant p\Lambda^{-1}, \quad (u, v) = 1.$$

We write

$$vr = u + qp.$$

Because $s = 1$ is critical, the inequality

$$\sin^2(\pi a_n/p) \geqslant 4\Lambda m^2 p^{-2}$$

has at most $\frac{1}{4}m$ solutions. Similarly, because r is critical, the inequality

$$\sin^2(\pi r a_n/p) \geqslant 4\Lambda m^2 p^{-2}$$

has at most $\frac{1}{4}m$ solutions. Hence, for at least $m^* \geqslant \frac{1}{2}m$ values of a_n (say $a_1, \ldots, a_{m^*}$) we have

$$\sin^2(\pi a_n/p) < 4\Lambda m^2 p^{-2}, \quad \sin^2(\pi r a_n/p) < 4\Lambda m^2 p^{-2};$$

$$|a_n| < \Lambda^{1/2} m, \quad |ra_n - pg_n| < \Lambda^{1/2} m.$$

The last inequality, multiplied with v, gives

$$|ua_n - p(vg_n - qa_n)| < \Lambda^{1/2} m v \leqslant \Lambda^{3/2} m.$$

Putting

$$h_n = vg_n - qa_n,$$

this becomes

$$|ua_n - ph_n| < \Lambda^{3/2} m. \tag{II.1}$$

The sequence a_n contains m^* terms confined to the interval $[-\Lambda^{1/2} m, \Lambda^{1/2} m]$; hence it contains two terms a', a'' such that

$$1 \leqslant a'' - a' \leqslant 2\Lambda^{1/2} m (m^* - 1)^{-1} \leqslant 4\Lambda^{1/2} + o(1).$$

As by (II.1) for some h

$$|u(a''-a')-ph| < 2A^{3/2}m = o(p),$$

it follows that $h = 0$ since

$$|u(a''-a')| \leqslant |u|(4A^{1/2}+o(1)) \leqslant 4pA^{-1/2}+o(p) = o(p).$$

And $h = 0$ implies

$$|u| \leqslant |u|(a''-a') < 2A^{3/2}m.$$

If $|u| \leqslant A^2$, the first part of our lemma is proved. Hence we may assume

$$A^2 < |u| < 2A^{3/2}m. \tag{II.2}$$

We now consider all integers of the form ux where $|x| < A^{1/2}m$. They contain the sequence ua_n, $1 \leqslant n \leqslant m^*$.

We proceed to count how many of these x satisfy

$$|ux-ph_x| < A^{3/2}m \tag{II.3}$$

for some suitable integer h_x. If h_x is fixed, the number of x in the interval (II.3) is obviously

$$\leqslant 1+2|u|^{-1}A^{3/2}m.$$

On the other hand, it follows from $|x| < A^{1/2}m$ and (II.3) that

$$|h_x| \leqslant |u|A^{1/2}mp^{-1}+A^{3/2}mp^{-1}.$$

Hence the number of x in $|x| < A^{1/2}m$ satisfying (II.3) does not exceed

$$\begin{aligned}(1+2|u|^{-1}A^{3/2}m)&(1+2|u|A^{1/2}mp^{-1}+2A^{3/2}mp^{-1})\\ &\leqslant (1+2|u|^{-1}A^{3/2}m)(2+2|u|A^{1/2}mp^{-1})\\ &= 2+4A^2m^2p^{-1}+2|u|A^{1/2}mp^{-1}+4|u|^{-1}A^{3/2}m\\ &\leqslant 2+4A^2m^2p^{-1}+4A^2m^2p^{-1}+4A^{-1/2}m\\ &= o(m) < m^*.\end{aligned}$$

As the set of ux with $|x| < A^{1/2}m$ contains the set ua_n with $1 \leqslant n \leqslant m^*$, (II.1) is not true for all $n \leqslant m^*$. Thus (II.2) is disproved, and the first part of our lemma is established.

Next we note that $h_n = 0$ implies $v \mid a_n$. We now return to our original sequence S_m and remove from it all terms for which either

$$\sin^2(\pi a_n/p) \geqslant A^{-4} \quad \text{or} \quad \sin^2(\pi r a_n/p) \geqslant A^{-4}.$$

Then we have for the remaining terms

$$|a_n| \leqslant \pi^{-1}\Lambda^{-2}p(1+o(1)) \quad \text{and} \quad |ra_n - pg_n| \leqslant \pi^{-1}\Lambda^{-2}p(1+o(1))$$

or, after multiplication with v, using our previous notation,

$$|ua_n - ph_n| \leqslant \pi^{-1}\Lambda^{-1}p(1+o(1)).$$

Hence

$$p\,|h_n| \leqslant |ua_n| + o(p) \leqslant (\pi^{-1}+o(1))p < p,$$

$$h_n = 0, \qquad v \mid a_n.$$

The number of terms we have omitted is

$$\leqslant \Lambda^4(\sigma(1)+\sigma(r)) \leqslant 2\Lambda^5 m^3 p^{-2}.$$

This finishes the proof of the lemma.

LEMMA II.2. $v = 1$ *under the conditions of Lemma* II. 1.

Proof. We have by Lemma II.1 a subsequence S_{m^*} of S_m represented by $a_1, \ldots, a_{m^*}$ say, such that

$$m^* \geqslant m - 2\Lambda^5 m^3 p^{-2},$$

$$-\tfrac{1}{2}p < sa_n < \tfrac{1}{2}p, \qquad v \mid sa_n \quad \text{for} \quad 1 \leqslant n \leqslant m^*.$$

For this subsequence we have

$$\sigma(v^{-1}s) = \sum_{n=1}^{m^*} \sin^2(\pi sa_n/(vp)) \leqslant v^{-2}\sum_{n=1}^{m^*}(\pi sa_n/p)^2$$

$$\leqslant v^{-2}(\tfrac{1}{2}\pi)^2 \sum_{n=1}^{m^*} \sin^2(\pi sa_n/p) \leqslant (\tfrac{1}{2}v^{-1}\pi)^2 \sum_{n=1}^{m} \sin^2(\pi sa_n/p)$$

$$= (\tfrac{1}{2}v^{-1}\pi)^2 \mu m^3 p^{-2} < \mu m^{*3} p^{-2}$$

which for $v \geqslant 2$ contradicts the minimum definition of μ as

$$m^* \geqslant m(1 - 2\Lambda^5 m^2 p^{-2}) = m + o(m) \geqslant \tfrac{1}{2}k.$$

LEMMA II.3. *There exists an* m_0 *in the interval*

$$m - \Lambda^{21} m^3 p^{-2} \leqslant m_0 \leqslant m$$

and a subsequence S_{m_0} *of* S_m, *say* $a_1, \ldots, a_{m_0}$, *such that*

$$\sum_{n=1}^{m_0} \sin^4(\pi sa_n/p) \leqslant \Lambda^{-19} m^3 p^{-2}.$$

Proof. From the series

$$\sigma(s, S_m) = \sum_{n=1}^{m} \sin^2(\pi sa_n/p) \leqslant \Lambda m^3 p^{-2}$$

we remove all terms for which

$$|\sin(\pi s a_n/p)| \geqslant \Lambda^{-10}.$$

The number of terms removed is

$$m - m_0 \leqslant (\Lambda^{10})^2 \sigma(s, S_m) \leqslant \Lambda^{21} m^3 p^{-2}$$

and

$$\sum_{n=1}^{m_0} \sin^4(\pi s a_n/p) \leqslant \Lambda^{-20} \sum_{n=1}^{m_0} \sin^2(\pi s a_n/p)$$

$$\leqslant \Lambda^{-20} \sigma(s, S_m) \leqslant \Lambda^{-19} m^3 p^{-2}.$$

LEMMA II.4.

(II.4) $$\sigma(s, S_m) = \mu m^3 p^{-2};$$

(II.5) $$\sigma(us, S_m) \geqslant u^2 \mu m_0^3 p^{-2} + O(\Lambda^{-11} m^3 p^{-2}) \quad \textit{for} \quad 1 \leqslant |u| \leqslant \Lambda^2,$$

where m_0 *is defined by Lemma* II.3;

(II.6) $$\sigma(r, S_m) \geqslant \Lambda m^3 p^{-2} \quad \textit{for the other } r \not\equiv 0 \pmod p).$$

Proof. (II.4) follows from the minimum definition. (II.6) is a consequence of Lemma II.1 and Lemma II.2.

To prove (II.5) we note that for all $a, t \neq 0$,

(II.7) $$\sin^2(ta) - t^2 \sin^2(a) = O(t^4 \sin^4 a).$$

(II.7) is true because for $0 \leqslant a \leqslant |t|^{-1}$

$$\sin^2(ta) = t^2 a^2 + O(t^4 a^4), \qquad t^2 \sin^2 a = t^2 a^2 + O(t^2 a^4),$$

whereas for $|t|^{-1} < a \leqslant \frac{1}{2}\pi$

$$\sin^2(ta) \leqslant 1 = O(t^2 \sin^2 a) = O(t^4 \sin^4 a).$$

From (II.7) and Lemma II.3 we obtain for $t \neq 0$

$$\sigma(ts, S_{m_0}) - t^2 \sigma(s, S_{m_0}) = O(t^4 \Lambda^{-19} m^3 p^{-2}).$$

This gives (II.5) as

$$\sigma(us, S_{m_0}) \leqslant \sigma(us, S_m), \qquad \sigma(s, S_{m_0}) \geqslant \mu m_0^3 p^{-2}.$$

LEMMA II.5. *If* $\beta_r = \prod_{n=1}^{k} \cos(\pi r a_n/p)$, *then*

$$\sum_{r=1}^{p-1} |\beta_r| = o(1)$$

as $p \to \infty$, $kp^{-2/3} \to \infty$.

Proof. We note first that if $k < \Lambda p^{2/3}$, then m and m_0 are defined and

$$\lim_{p\to\infty} mk^{-1} = 1, \qquad \lim_{p\to\infty} m_0 m^{-1} = 1.$$

For $r \not\equiv 0 (\bmod p)$ we have

$$|\beta_r| \leqslant \prod_{n=1}^{m} |\cos(\pi r a_n/p)| \leqslant \left\{m^{-1} \sum_{n=1}^{m} \cos^2(\pi r a_n/p)\right\}^{m/2}$$

$$= \{1 - m^{-1}\sigma(r, S_m)\}^{m/2} \leqslant e^{-(1/2)\sigma(r,S_m)}.$$

Hence if r is not critical for S_m, (II.6) is applicable and

$$|\beta_r| \leqslant e^{-(1/2)\Lambda m^3 p^{-2}} \leqslant p^{-2},$$

as eventually $m^3 p^{-2} \geqslant 4$.

If (II.4) is applicable, it gives

$$|\beta_s| = |\beta_{-s}| \leqslant e^{-(1/2)\mu m^3/p^2} \leqslant e^{-(1/2)\gamma_0 m^3/p^2} = o(1),$$

whereas (II.5) if applicable gives for $2 \leqslant |u| \leqslant \Lambda^2$

$$|\beta_{us}| \leqslant e^{-(1/2)u^2 \mu m_0^3 p^{-2} + O(\Lambda^{-11} m^3 p^{-2}} \leqslant e^{-(1/2)|u|\gamma_0 m^3 p^{-2}},$$

$$\sum_{2\leqslant|u|\leqslant\Lambda^2} |\beta_{us}| \leqslant 2\sum_{u=2}^{\infty} e^{-(1/2)u\gamma_0 m^3 p^{-2}} = 2e^{-\gamma_0 m^3 p^{-2}}(1 - e^{-(1/2)\gamma_0 m^3 p^{-2}})^{-1} = o(1).$$

This completes the proof of Lemma II.5 if $k < \Lambda p^{2/3}$ and if at least one r is critical for S_k.

Otherwise, we still have for $r \not\equiv 0 (\bmod p)$

$$|\beta_r| \leqslant e^{-(1/2)\sigma(r,S_k)}.$$

If $k < \Lambda p^{2/3}$ and no critical r exists, we have

$$|\beta_r| < e^{-(1/2)\sigma(r,S_k)} \leqslant e^{-(1/2)\Lambda k^3 p^{-2}} < p^{-2}$$

eventually. Finally, if $k \geqslant \Lambda p^{2/3}$,

$$\sigma(r, S_k) \geqslant 2 \sum_{1\leqslant n\leqslant(k+1)/2} \sin^2(\pi n/p) \geqslant 8 \sum_{1\leqslant n\leqslant(k+1)/2} n^2 p^{-2}$$

$$= \tfrac{1}{3}k^3 p^{-2}(1 + o(1)) > \tfrac{1}{4}\Lambda^3$$

and

$$|\beta_r| < e^{-\Lambda^3/8} < e^{-2\Lambda} = p^{-2}$$

eventually. This completes the proof of the lemma.

Proof of Theorem II. Put $A = \sum_{n=1}^{k} a_n$. Then

$$F(N) = p^{-1} \sum_{r=0}^{p-1} e^{-2\pi i r N/p} \prod_{n=1}^{k} (1 + e^{2\pi i r a_n/p})$$

$$= p^{-1} 2^k \sum_{r=0}^{p-1} e^{\pi i r (A-2N)/p} \beta_r,$$

$$|F(N) - p^{-1}2^k| \leqslant p^{-1}2^k \sum_{r=1}^{p-1} |\beta_r| = o(p^{-1}2^k)$$

by Lemma II.5. This proves the theorem.

Finally, if k is even, $p^{2/3} \leqslant k \leqslant O(p^{2/3})$

$$a_1 = 1,\ a_2 = -1,\ a_3 = 2,\ a_4 = -2,\ \ldots,\ a_{k-1} = \tfrac{1}{2}k,\ a_k = -\tfrac{1}{2}k,$$

then $A = 0$, $\beta_r \geqslant 0$. Hence

$$F(0) = p^{-1}2^k \Big(1 + \sum_{r=1}^{p-1} \beta_r\Big) \geqslant p^{-1}2^k (1 + \beta_1).$$

An easy calculation shows that

$$\beta_1 = \prod_{n=1}^{k/2} \cos^2(\pi n/p) \sim e^{-(24)^{-1}\pi^2 k^3 p^{-2}},$$

which does not tend to zero. This shows that Theorem II is best possible.

Unproved Conjectures.

CONJECTURE 1. *It is possible to replace the constant* $3 \cdot 6^{1/2}$ *in Theorem* I *by the constant* 2.

This is fairly plausible. Let S_k^* be the sequence

$$a_1 = 1,\ a_2 = -1,\ a_3 = 2,\ a_4 = -2,\ \ldots,\ a_k = (-1)^{k-1}[\tfrac{1}{2}(k+1)]$$

and let $G(S_k)$ be the number of residue classes N for which

$$F(N; p; S_k) = F(N; p; a_1, \ldots, a_k) > 0.$$

Then we can state

CONJECTURE 2. $G(S_k) \geqslant G(S_k^*)$ *for all* $k \geqslant 1$.

This would of course imply Conjecture 1.

For composite moduli Theorem I and II cease to be true. It is however reasonable to formulate

CONJECTURE 3. $F(0) > 0$ *for* $k > 2p^{1/2}$, *where* p *is not necessarily a prime.*

This conjecture may also be true for finite abelian groups of composite order p, and possibly even, *mutatis mutandis*, for non-abelian groups.

Finally we mention a more complicated, but probably easier problem.

CONJECTURE 4. *Let $n, s, l_1, \ldots, l_s$ be positive integers, such that $l_1+\ldots+l_s = n$. Let $a_\lambda^{(\sigma)}$ $(1 \leqslant \sigma \leqslant s,\ 1 \leqslant \lambda \leqslant l_\sigma)$ be n residue classes* mod n *such that $a_\lambda^{(\sigma)} \not\equiv a_\mu^{(\sigma)} (\operatorname{mod} n)$ for $1 \leqslant \mu < \lambda \leqslant \sigma$. Then there exists a non-void subset T of the integers $1 \leqslant \sigma \leqslant s$, such that for σ in T we can choose a $\lambda(\sigma)$ in $1 \leqslant \lambda \leqslant l_s$ with the effect that*

$$\sum_{\sigma \text{ in } T} a_{\lambda(\sigma)}^{(\sigma)} \equiv 0 (\operatorname{mod} n).$$

As the paper goes to press Dr Flor informs us that Conjecture 4 follows from a recent result by P. Scherk [2]. We also want to draw the attention of the reader to a theorem by P. Erdös, A. Ginzburg and A. Ziv [3] which states that each set of $2n-1$ integers contains a sub-set of n integers, the sum of which is divisible by n.

References

[1] H. Davenport, *On the addition of residue classes*, Journ. London Math. Soc. 10 (1935), pp. 30-32.

[2] P. Scherk, *Distinct elements in a set of sums*, Amer. Math. Monthly 62 (1955), pp. 46-47.

[3] P. Erdös, A. Ginzburg and A. Ziv, *Theorem in the additive number theory*, Bull. Research Council Israel, 10F (1961), pp. 41-43.

Reçu par la Rédaction le 22. 8. 1963

[Bristol]

36a.

Introduction to Papers on Combinatorial Analysis and Sums of Squares

Collected Papers of G. H. Hardy (edited),
Oxford at the Clarendon Press, 1966, volume I, 263–264.

It is no accident that of all Hardy's papers on additive number theory, those on modular forms come first in chronological order. The general idea is to determine (either exactly or approximately) an arithmetical function $r(n)$ by means of the formula

$$r(n) = \frac{1}{2\pi i}\int f(x)x^{-n-1}\,dx,$$

where the 'generating function'

$$f(x) = \sum_{n=0}^{\infty} r(n)x^n$$

is regular in the interior of the unit circle and the integral is taken along a circle $|x| = r < 1$.

The most interesting case arises when the unit circle is the natural boundary of the function $f(x)$. In many important cases the function $f(e^{\pi i\tau})$ is a modular form in τ in the upper half-plane, so that the behaviour of $f(e^{\pi i\tau})$ as τ approaches a rational number h/k is easily determined from the behaviour of $f(e^{\pi i\tau})$ as τ approaches 0.

The first example of this type which Hardy dealt with, in collaboration with S. Ramanujan, was the partition function p_n, the number of partitions of n into positive integers without respect to the order of the terms, but with repetitions allowed.* The generating function in this case is

$$\prod_{m=1}^{\infty}(1-x^m)^{-1} = 1+\sum_{n=1}^{\infty} p_n x^n,$$

which is closely related to the well-known modular 'discriminant'

$$\Delta(\tau) = e^{\pi i\tau}\prod_{m=1}^{\infty}(1-e^{\pi i m\tau})^{24}.$$

In their first paper on the subject (1917, 4) Hardy and Ramanujan had already obtained the result

$$\log p_n \sim \pi(2n/3)^{\frac{1}{2}}.$$

Their method consisted of the application of a Tauberian theorem to $f(x)$ near $x = 1$ only. Though the method can and did lead to fairly general theorems, it could not produce results as deep as those found in their later work.

In the paper (1918, 5), for which the papers (1916, 10) and (1917, 1) are preliminary announcements,† Hardy and Ramanujan obtained a rapidly convergent asymptotic formula giving p_n with an error term $O(n^{-\frac{1}{4}})$. Twenty years later H. Rademacher (*Proc. London Math. Soc.* (2), 43 (1937), 241–54), by a slight but far-reaching modification of the proof, replaced the asymptotic expansion by a convergent series for p_n.

* J. V. Uspensky also investigated the asymptotic behaviour of p_n by means of the contour integral; see *Bull. de l'Acad. des Sciences de Russie*, (6) 14 (1920), 199–218.

† An abstract of the paper itself was published in *Proc. London Math. Soc.* (2), 16 (1917), xxii.

‡ There was also an announcement in *Proc. London Math. Soc.* (2), 17 (1918) xxii–xxiv.

The next paper (1918, 2) deals with functions $f(x)$ based on modular functions which have poles, but no essential singularities, in the upper half-plane.

The paper (1920, 10) deals with the number $r_s(n)$ of representations of n as a sum of s squares. [(1918, 10) is a preliminary announcement.‡] The relevant function is

$$\theta(\tau) = \sum_{n=-\infty}^{\infty} e^{\pi i n^2 \tau},$$

which satisfies the functional equation

$$\left(\frac{\tau}{i}\right)^{\frac{1}{2}} \theta(\tau) = \theta\left(-\frac{1}{\tau}\right).$$

Hardy and Littlewood had already studied the asymptotic behaviour of $\theta(\tau)$, particularly in (1914, 3).

By the use of the formula

$$\theta(\tau)^s = \sum_{n=0}^{\infty} r_s(n) e^{\pi i n \tau}$$

and Cauchy's integral formula, Hardy obtained an exact expression for $r_s(n)$ if $5 \leqslant s \leqslant 8$ and an asymptotic expression if $s \geqslant 9$. Hardy was, of course, aware of the fact that his results for $5 \leqslant s \leqslant 8$ were not new, though he stresses quite rightly that his method deals simultaneously with even and odd values of s and presents a unified approach to the problem for all values of s, though the classical cases $s = 4$ and especially $s = 3$ present some special difficulties. On the case $s = 3$, see T. Estermann, *Proc. London Math. Soc.* (3), 9 (1959), 575–94.

The whole matter was put into a more general context by C. L. Siegel [*Annals of Math.* (2) 36 (1935), 527–606]. He proved not only that Hardy's formula holds, *mutatis mutandis*, asymptotically for the representation of a number by any definite quadratic form with rational integral coefficients, but that it holds exactly if a system of forms is considered, which has one representative from each class in a fixed genus. As the number of classes of positive definite quadratic forms of discriminant 1 is one if and only if $s \leqslant 8$, it is now apparent why Hardy's formula holds exactly for $s \leqslant 8$ only.

H. H.

NOTE

The four papers [1917, 1; 1917, 4; 1918, 2; 1918, 5] written in collaboration with Ramanujan are reprinted from *Collected Papers of S. Ramanujan* (Cambridge, 1927).

The reader is referred to Hardy's own comments given there, and to Chapters 3, 8, 9 of his *Ramanujan* (Cambridge, 1940).

36b.

Introduction to Papers on Waring's Problem

Collected Papers of G. H. Hardy (edited),
Oxford at the Clarendon Press, 1966, volume I, 377–381

Hardy's papers on Waring's problem were all written in collaboration with J. E. Littlewood. In a sense, they followed naturally from his earlier papers on the representation of a number as a sum of squares, but the new difficulties which had to be overcome were severe.

The starting-point of the method is the same. For each integer $k \geqslant 3$ the generating function is defined as

$$f(x) = 1+2 \sum_{n=1}^{\infty} x^{n^k},$$

so that

$$f(x)^s = \sum_{n=0}^{\infty} r_{k,s}(n) x^n,$$

where $r_{k,s}(n)$ is the number of representations of n as the sum of s absolute values of integral kth powers. For even k this is, of course, the same as the number of representations by kth powers. Then by Cauchy's Theorem the number $r(n)$ is given by

$$r(n) = r_{k,s}(n) = \frac{1}{2\pi i} \int (f(x))^s x^{-n-1}\, dx,$$

where the integral is taken along the circle $|x| = e^{-1/n}$. This circle, $x = e^{-n^{-1}+2\pi i\theta}$, is divided into 'Farey arcs' by first marking all rational points $\theta = p/q$ with

$$1 \leqslant q \leqslant n^{1-k^{-1}}, \qquad (p, q) = 1, \qquad 0 \leqslant p < q.$$

Two neighbouring rationals $\theta = p/q$, $\theta = p'/q'$ are separated by their mediant $\theta = (p+p')/(q+q')$, so that to each rational p/q there belongs a zone of influence, bounded by its two neighbouring mediants. These zones are the Farey arcs $M_{p,q}$. In the treatment of Waring's Problem they fall into two classes, *major arcs* if $q \leqslant n^{1/k}$, and *minor arcs* if $n^{1/k} < q \leqslant n^{1-1/k}$.

On each major arc Hardy and Littlewood introduced an approximating function

$$F_{p,q}(x) = C\Gamma(s/k)(q^{-1}S_{p,q})^s\{\log(e^{2\pi i p/q}x^{-1})\}^{-s/k},$$

where

$$C = \{2\Gamma(1+1/k)\}^s\{\Gamma(s/k)\}^{-1},$$

$$S_{p,q} = \sum_{h=0}^{q-1} e^{2\pi i h^k p/q}.$$

They were then able to show that on each major arc $M_{p,q}$ the function $f^s(x)$ can be approximated by $F_{p,q}(x)$ with a sufficiently good error term not to upset the final result. The difficulties they had to overcome were formidable and their first paper on the subject (1920, 2) makes impressive reading even today. In one respect they were perhaps fortunate. H. Weyl's celebrated paper (*Math. Annalen*, 77 (1916), 313–52) on exponential sums had just appeared and the methods developed in it made the work more manageable.

The second difficulty arose on the minor arcs, which do not occur at all in the case of squares. Whereas Weyl's inequality was a convenience on the major arcs, it was and to a large extent still is a necessity on the minor arcs, because the error in the approximation to $f^s(x)$ by $F_{k,s}(x)$ becomes too large; indeed it is larger than the principal term.

The third difficulty was to deal with the *singular series*. Having replaced $f^s(x)$ by $F_{p,q}(x)$ on all major arcs, Hardy and Littlewood obtained for $s > k2^{k-1}$ the asymptotic formula

$$r(n) \sim Cn^{-1+s/k}S,$$

where

$$S = \sum_{q=1}^{\infty} q^{-s} \sum_{\substack{p=1 \\ (p,q)=1}}^{q} (S_{p,q})^s e^{-2\pi inp/q}$$

is the singular series.

In their first paper (1920, 2) they did no more than prove that the series is absolutely convergent and greater than a positive constant for $k = 4$, $s = 33$ and all n, though they suggested that there would be no serious difficulties in extending this result to general $k \geqslant 3$ if $s > k2^{k-1}$. This programme was carried through in (1920, 5),* the first of their famous papers on 'partitio numerorum'.

In P.N. II (1921, 1), the asymptotic formula is proved for $k = 4$, $s = 21$ and it is shown that the singular series is uniformly positive. Apart from a detailed study of the singular series the paper introduces an improvement in minor arc technique, namely the use of the inequality ($\mathfrak{m}$ denoting the union of all the minor arcs)

$$\int_{\mathfrak{m}} |f(re^{2\pi i\theta})|^{2s}\,d\theta \leqslant \max_{\theta \text{ in } \mathfrak{m}} |f(re^{2\pi i\theta})|^{2(s-2)} \int_0^1 |f(re^{2\pi i\theta})|^4\,d\theta.$$

For the first factor on the right Weyl's estimate is available, and the second factor can easily be estimated by elementary means.

In P.N. IV (1922, 4) the authors prove their asymptotic formula for $r(n)$ for all $s > (k-2)2^{k-1}+5$. But the real interest of the paper lies in the study of the singular series. It is now commonplace to realize that the inequality $S > 0$ implies, subject to absolute convergence, that the congruence

$$x_1^k+\ldots+x_s^k \equiv n \pmod m$$

is soluble for every modulus m; conversely if the congruence is soluble for all m, then $S > 0$. In fact we can be more precise. If in the summation formula for S we restrict q to the powers of a fixed prime ϖ, the series turns out to be finite for $n \neq 0$ and represents $\varpi^{-t(s-1)}$ times the number of solutions of the above congruence mod ϖ^t, for sufficiently large t. Moreover, S equals the product of the restricted sums extended over all primes ϖ.

This idea, which is fundamental not only for Waring's Problem, but for all applications of the Hardy–Littlewood method, appears first in P.N. IV. It is historically interesting that the important relation between the singular series and the congruence was formulated by the authors as a lemma only (Lemma 2).

* There was a preliminary announcement in *Proc. London Math. Soc.* (2), 18 (1920), vii–viii.

In P.N. VI (1925, 1)† the authors break new ground. They prove first that almost all positive integers are the sum of fifteen fourth powers, and more generally the sum of

$$(\tfrac{1}{2}k-1)2^{k-1}+3$$

non-negative kth powers for $k=3$ and $k \geqslant 5$. Secondly they prove that all large integers are the sum of

$$(\tfrac{1}{2}k-1)2^{k-1}+k+5+\left[\frac{(k-2)\log 2-\log k+\log(k-2)}{\log k-\log(k-1)}\right]$$

non-negative kth powers. The idea of the proof is to introduce a sequence of integers which have a relatively large density and are representable as the sum of few kth powers. It was the first step in the combination of analytical with elementary methods, serving as a signpost to the future, and in particular to the work of Vinogradov.

Finally they introduced in P.N. VI the Hypothesis K, which asserts that the number of solutions of

$$n = x_1^k+...+x_k^k$$

is $O(n^\epsilon)$ for each $\epsilon > 0$. Under the assumption of Hypothesis K they proved that their asymptotic formula for $r(n)$ holds for $s \geqslant 2k+1$, in particular that each large integer is a sum of $2k+1$ non-negative kth powers, if k is not a power of 2. Unfortunately, K. Mahler proved (*Journal London Math. Soc.* 11 (1936), 136–8) that Hypothesis K does not hold for $k=3$ by means of the simple identity

$$(9x^4)^3+(3xy^3-9x^4)^3+(y^4-9x^3y)^3 = y^{12}.$$

Whether the hypothesis is true for $k \geqslant 4$ is still an unanswered question. For the application to Waring's Problem it would suffice if the hypothesis held in a weaker 'mean square' form.

P.N. VIII (1928, 4) differs from the preceding papers in posing a purely arithmetical problem. Let $\Gamma(k)$ be the smallest integer s such that for all n and all primes p, and all $m > 0$, the congruence

$$x_1^k+...+x_s^k \equiv n \pmod{p^m}$$

has a solution in which not all the x_σ are $\equiv 0 \pmod{p}$.

The authors determine five distinct classes of k for which $\Gamma(k) > k$ and calculate $\Gamma(k)$ explicitly. They also prove that $\Gamma(k) = k$ if $2k+1$ is a prime. For all other cases they prove the inequality $\Gamma(k) \leqslant k$.

As a conjecture they mention the innocent-looking relation

$$\varliminf_{k\to\infty} \Gamma(k) \geqslant 4,$$

which has remained unproved. It would be sufficient to find for all large k a prime π such that

$$\pi \equiv 1 \pmod{k}, \qquad \pi < k^{4/3}.$$

Finally, a footnote should be quoted:

'We may add that "P.N. 7", which is still unpublished, contains an application of our methods to the problem of the order of magnitude of the difference between

† An abstract appeared in *Proc. London Math. Soc.* (2), 23 (1925), xx–xxi.

consecutive primes. We prove (subject to our generalized form of the Riemann hypothesis) that

$$\varliminf_{n\to\infty} \frac{p_{n+1}-p_n}{\log n} \leqslant \tfrac{2}{3}.\text{'}$$

The paper was never published, but R. A. Rankin (*Proc. Camb. Phil. Soc.* 36 (1940), 255–66) proved that the lower limit in question does not exceed 3/5 under the generalized Riemann Hypothesis, whereas P. Erdős (*Duke Math. J.* 6 (1940), 438–41) proved without any assumptions that it is less than 1. Later R. A. Rankin (*Journal London Math. Soc.* 22 (1947), 226–30) proved that the lower limit is less than 57/59.

The subsequent major improvements upon the Hardy–Littlewood results in Waring's Problem are due to I. M. Vinogradov. An account of his results is given in his book, *The Method of Trigonometrical Sums in the Theory of Numbers*, English translation by K. F. Roth and A. Davenport, 1954.

He proved first (*Annals of Math.* 36 (1935), 395–405) that

$$G(k) \leqslant 6k\log k+(4+\log 216)k,$$

where $G(k)$ is defined as the smallest s such that $\varliminf_{n\to\infty} r_{k,s}(n) > 0$. A simplified version was given by H. Heilbronn (*Acta Arithmetica*, 1 (1935), 212–21). Subsequently I. M. Vinogradov improved this result progressively in various papers, and in his book he proves

$$G(k) < k(3\log k+11).$$

He also in a series of papers considered the validity of the asymptotic formula for $r_{k,s}(n)$. The best result given in his book is that the formula holds for

$$s > 10k^2\log k-1.$$

L. K. Hua had previously (*Quarterly Journal*, 9 (1938), 199–202) proved the asymptotic formula for $s \geqslant 2^k+1$.

The inequality $G(3) \leqslant 7$ was first proved by U. V. Linnik (*Recueil Math.* 12 (1943), 218–24) and subsequently a simpler proof was given by G. L. Watson (*Journal London Math. Soc.* 26 (1951), 153–6).

H. Davenport proved that almost all numbers are sums of four positive cubes (*Acta Math.* 71 (1939), 123–43); that all large integers not $\equiv 0$ or $15 \pmod{16}$ are sums of fourteen fourth powers (*Annals of Math.* 40 (1939), 731–47); and that $G(5) \leqslant 23$ and $G(6) \leqslant 36$ (*American J. of Math.* 64 (1942), 199–207).

Finally, the problem raised in P.N. VIII, that of improving the upper bound for $\Gamma(k)$, has not received much attention. It was, however, proved by I. Chowla (*Proc. Nat. Acad. Sci. India*, sect. A 13 (1943), 195–220) that $\Gamma(k) < k^c$, for some c with $0 < c < 1$, in all cases in which Hardy and Littlewood had not shown that $\Gamma(k) \geqslant k$.

In recent years there has been renewed interest in the Hardy–Littlewood method, which has been successfully applied to prove the existence of integral solutions of homogeneous algebraic equations with integral coefficients in a sufficiently large number of variables (subject to certain conditions if the degree is even). The proofs are very sophisticated, requiring a good deal of 'old-fashioned' algebraic geometry.

For some of these results, and for references, the reader may consult H. Davenport, *Analytic methods for Diophantine equations and Diophantine inequalities*, Ann Arbor, Michigan, 1963.

H. H.

36c.

Introduction to Papers on Goldbach's Problem

Collected Papers of G. H. Hardy (edited),
Oxford at the Clarendon Press, 1966, volume I, 533–534

The main papers in this section are P.N. III (1922, 3) and P.N. V (1924, 6). The other three papers are historical reviews and a preliminary announcement.

In P.N. III and P.N. V Hardy and Littlewood deal with Goldbach's Problem on the assumption that the following hypothesis is true.

Hypothesis R. There exists a real number $\Theta < 3/4$ such that all zeros of all L-series $L(s, \chi)$ formed with Dirichlet characters lie in the half-plane $\sigma \leqslant \Theta$.

Hypothesis R is weaker than the original Riemann hypothesis, but is asserted for a larger class of functions. It remains unproved today. Throughout P.N. III and P.N. V Hypothesis R is taken for granted.

Let $N_3(n)$ be the number of representations of the odd positive integer n as a sum of three primes. Let

$$C_3 = \prod_{\varpi} (1+(\varpi-1)^{-3}),$$

where ϖ runs through all odd primes. The authors prove that

$$N_3(n) \sim C_3 n^2(\log n)^{-3} \prod_p \frac{(p-1)(p-2)}{p^2-3p+3},$$

where p runs through all odd prime divisors of n. In particular it follows that $N_3(n) > 0$ for large n.

The proof runs on lines similar to those of the preceding sections, though new difficulties cropped up and had to be overcome. The generating function is the third power of

$$f(x) = \sum_{\varpi} (\log \varpi)x^{\varpi},$$

where ϖ runs through all odd primes.

The introduction of the factor $\log \varpi$ is a familiar device to make the use of the prime number theorem for arithmetical progressions easier. The Farey dissection works as before; there are no minor arcs. If

$$x = e^{-Y+2\pi i p/q},$$

it follows from Mellin's formula that, on a Farey arc around p/q,

$$x^{\varpi} = e^{2\pi i p\varpi/q} \frac{1}{2\pi i} \int_{2-i\infty}^{2+i\infty} Y^{-s}\Gamma(s)\varpi^{-s}\, ds,$$

and apart from negligible terms, $f(x)$ can be expressed as a linear combination of integrals

$$\frac{1}{2\pi i} \int_{2-i\infty}^{2+i\infty} Y^{-s}\Gamma(s) \frac{-L'(s,\chi)}{L(s,\chi)}\, ds,$$

where χ runs through all characters (primitive or not) mod q. The integrand is regular for $\sigma > \Theta$, except for a pole at $s = 1$ when χ is the principal character, and is meromorphic for all s.

Hence Cauchy's theorem can be applied so as to shift the path of integration to the line $\sigma = -\frac{1}{4}$. The residue at $s = 1$ gives the principal terms in the approximation for $f(x)$; the contributions resulting from the zeros of $L(s, \chi)$ give simple poles in the strip $0 \leqslant \sigma \leqslant \Theta$. A careful estimate of their contribution to the value of the integral gives an approximation for $f(x)$ on the Farey arc. Then the older techniques can be applied; the singular series, though troublesome, does not present serious difficulties. And the asymptotic formula is proved.

In the rest of the paper, the authors apply their method heuristically to a large number of problems which are exceedingly difficult. Of their large number of conjectures only one has so far been proved, namely by Linnik, who showed in his book *The dispersion method for binary additive problems* (Leningrad, 1961) that every large integer is a sum of two squares and a prime. The same result had been proved earlier by C. Hooley (*Acta Math.* 97 (1957), 189–210) under the assumption of the generalized Riemann hypothesis.

In P.N. V (1924, 6)† the authors assume that Hypothesis R holds with $\Theta = \frac{1}{2}$, and deduce that there are at most $O(n^{\frac{1}{2}+\epsilon})$ even positive integers not exceeding n which are not the sum of 2 odd primes. The paper stands to P.N. III in the same relation as P.N. VI to the preceding papers on Waring's problem.

Since the time of Hardy and Littlewood much progress has been made in Goldbach's problem.

L. Schnirelmann (*Iswestija Donskowo Polytechn. Inst.* (Novotscherkask) 14 (1930), 3–28) proved the existence of a constant γ such that every integer > 1 is the sum of at most γ primes. I. M. Vinogradov (*Rec. Math. Moscou,* (2) 2 (1937), 179–95) proved that every large odd number is the sum of three odd primes, that is, he proved the Hardy–Littlewood theorem without the use of Hypothesis R. His proof is based on a clever introduction of an exponential sum, essentially transferring the sieve method to exponential level. U. V. Linnik (*Rec. Math.* [*Math. Sbornik*], N.S. 19 (61) (1946), 3–8) proved the result again. His proof reverts to the original Hardy–Little wood pattern, but instead of using Hypothesis R he obtains an estimate for the total number of zeros, in a particular region, of all the L-functions to a given modulus.

For expositions of modern work on Goldbach's problem, the reader is referred to:

T. Estermann, *Introduction to modern prime number theory* (Cambridge Tract No. 41), Cambridge, 1952.

L. K. Hua, *Additive Primzahltheorie*, Leipzig, 1959.

K. Prachar, *Primzahlverteilung*, Berlin–Göttingen–Heidelberg, 1957.

I. M. Vinogradov, *The method of trigonometrical sums in the theory of numbers* (trans. K. F. Roth and A. Davenport), London, 1954.

H. H.

† An abstract appeared in *Proc. London Math. Soc.* (2) 22 (1924), xi.

37.

Zeta-Functions and L-Functions

Algebraic Number Theory (J. W. S. Cassels and A. Fröhlich, eds.), Academic Press, London, 1967, 204–230

1. Characters 204
2. Dirichlet L-series and Density Theorems 209
3. L-functions for Non-abelian Extensions 218
References 230

The object of this Chapter is to study the distribution of prime ideals in various algebraic number fields, the principal results being embodied in several so-called "density theorems" such as the Prime Ideal Theorem (Theorem 3) and Čebotarev's Theorem (Section 3). As in the study of the distribution of rational primes, L-series formed with certain group characters play an important part in such investigations.

Much of what we have to say goes back to the early decades of this century and is described in Hasse's "Bericht über neuere Untersuchungen und Probleme aus der Theorie der algebraischen Zahlenkörper" (Hasse 1926, 1927 and 1930), henceforward to be referred to as Hasse's Bericht.

1. Characters

Throughout, let k be a finite extension of **Q**, of degree ν. Let S denote a set consisting of the infinite primes (archimedean valuations) of k together with a finite number of other primes (non-archimedean valuations) of k. S is sometimes referred to as the *exceptional* set.

In slight modification of earlier notation, an idèle x of k will be written as

$$x = (x_{\mathfrak{p}_1}, \ldots, x_{\mathfrak{p}_a}; x_{\mathfrak{p}_{a+1}}, \ldots, x_{\mathfrak{p}_b}; x_{\mathfrak{p}_{b+1}}, \ldots),$$

where $\mathfrak{p}_1, \ldots, \mathfrak{p}_a$ are the infinite primes, $\mathfrak{p}_{a+1}, \ldots, \mathfrak{p}_b$ are the finite primes of S, and the remaining $\mathfrak{p}_i$ are the primes not in S. Let J denote the idèle group.

By a character ψ of the idèle class group we understand a homomorphism from J into the unit circle of the complex plane satisfying

(1) $\psi(x) = 1$ if $x \in k^*$ (i.e. $x_\mathfrak{p} = x$ for all $\mathfrak{p}$);
(2) $\psi(x)$ is continuous on J (in the idèle topology);
(3) $\psi(x) = 1$ if $x_\mathfrak{p} = 1$ for $\mathfrak{p} \in S$ and $|x_\mathfrak{p}|_\mathfrak{p} = 1$ for $\mathfrak{p} \notin S$.

As was pointed out in Chapter VII, "Global Class Field Theory", §4, ψ generates a character of the ideal group I^S (the free abelian group on the set of all $\mathfrak{p} \notin S$) in the following way:

Let

$$\mathfrak{a} = \prod_{i>b} \mathfrak{p}_i^{\alpha_i} \qquad (\alpha_i \in \mathbf{Z})$$

be a general element of I^S (so that almost all the exponents α_i are 0). For each $i > b$, let π_i be an element of k with $|\pi_i|_{\mathfrak{p}_i} < 1$ and maximal. With $\mathfrak{a}$ we associate the idèle $x^{\mathfrak{a}}$ having

$$(x^{\mathfrak{a}})_{\mathfrak{p}_i} = \begin{cases} 1, & i = 1, 2, \ldots, b, \\ \pi_i^{\alpha_i}, & i = b+1, \ldots; \end{cases}$$

then we define

$$\chi(\mathfrak{a}) = \psi(x^{\mathfrak{a}}).$$

We observe that although $x^{\mathfrak{a}}$ is not unique, $\chi(\mathfrak{a})$ is, nevertheless, well-defined in view of property (3) above. Clearly χ is a multiplicative function in the sense that for all ideals $\mathfrak{a}, \mathfrak{b}$ coprime with S,

$$\chi(\mathfrak{a}\mathfrak{b}) = \chi(\mathfrak{a})\chi(\mathfrak{b}).$$

Any ideal character χ defined in this way is called a Grössencharacter, and such characters were first studied by Hecke (1920) and later, in the idèle setting, by Chevalley (1940).

Let S, S' be two exceptional sets and χ, χ' characters of $I^S, I^{S'}$ respectively. We say that χ and χ' are *co-trained* if $\chi(\mathfrak{a}) = \chi'(\mathfrak{a})$ whenever both $\chi(\mathfrak{a}), \chi'(\mathfrak{a})$ are defined. This definition determines an equivalence relation† among characters. In the equivalence class of χ there is a unique χ' corresponding to the least possible exceptional set S' (S' being the intersection of all exceptional sets corresponding to the characters co-trained with χ); we refer to this character χ' as the *primitive* character co-trained with χ.

The principal character χ_0 of I^S is defined by $\chi_0(\mathfrak{a}) = 1$ for all $\mathfrak{a} \in I^S$. The principal characters form a co-trained equivalence class, and the primitive member of this class is 1 for all non-zero finite ideals.

We shall now look more closely into the structure of a character of the ideal group I^S, with a view to describing the important class of Hilbert and Dirichlet characters.

We can write each idèle x in the form

$$x = \prod_{\mathfrak{p}} x(\mathfrak{p})$$

where each factor $x(\mathfrak{p})$ is the idèle defined by

$$x(\mathfrak{p})_{\mathfrak{p}_i} = \begin{cases} x_{\mathfrak{p}_i}, & \mathfrak{p} = \mathfrak{p}_i \\ 1, & \mathfrak{p} \neq \mathfrak{p}_i. \end{cases}$$

† To prove the transitivity of the relation it suffices to show that if $\psi(y) = 1$ for all idèles y for which $y_{\mathfrak{p}_i} = 1$, $i = 1, 2, \ldots, m$, then $\psi(x) = 1$ for all idèles x. However, for any $\varepsilon > 0$ there is an $\alpha \in k^*$ such that

$$|\alpha - x_{\mathfrak{p}_i}|_{\mathfrak{p}_i} < \varepsilon, \qquad i = 1, 2, \ldots, m,$$

by the Chinese Remainder theorem and thus $\psi(x) = \psi(\alpha^{-1}x)$ tends to 1 as $\varepsilon \to 0$.

Then

$$\psi(x) = \prod_{\mathfrak{p}} \psi_{\mathfrak{p}}(x),$$

where $\psi_{\mathfrak{p}}$ is a character of J defined by

$$\psi_{\mathfrak{p}}(x) = \psi(x(\mathfrak{p}))$$

and is referred to as a *local component* of ψ (Chevalley, 1940). We note that $\psi_{\mathfrak{p}}$ is continuous and satisfies (3). We first follow up some consequences of the continuity of ψ.

Let $\mathcal{N}$ be a neighbourhood of 1 containing no subgroup of $\psi(J)$ except 1. By continuity of ψ, there exists a neighbourhood $\mathcal{N}'$ of 1 in J, say

$$|x_{\mathfrak{p}} - 1|_{\mathfrak{p}} < \varepsilon_{\mathfrak{p}}, \qquad \mathfrak{p} \in E, \tag{1.1}$$

$$|x_{\mathfrak{p}}|_{\mathfrak{p}} = 1, \qquad \mathfrak{p} \notin E, \tag{1.2}$$

where E is some finite set of valuations, such that

$$\psi(\mathcal{N}') \subset \mathcal{N}.$$

We choose $\mathcal{N}'$ first to make the set E minimal, and then to make the $\varepsilon_{\mathfrak{p}}$'s maximal; so that for the finite $\mathfrak{p}$ in E we have $\varepsilon_{\mathfrak{p}} \leqslant 1$. For finite $\mathfrak{p}$, (1.1) is now equivalent to saying that $x_{\mathfrak{p}} \in 1 + \mathfrak{p}^{\mu_{\mathfrak{p}}}$ where $\mu_{\mathfrak{p}}$ is a positive integer. But since $1 + \mathfrak{p}^{\mu_{\mathfrak{p}}}$ is a group, so is $\psi_{\mathfrak{p}}(1 + \mathfrak{p}^{\mu_{\mathfrak{p}}})$, whence $\psi_{\mathfrak{p}}(1 + \mathfrak{p}^{\mu_{\mathfrak{p}}}) = 1$, and this holds for no smaller integer than $\mu_{\mathfrak{p}}$. We write

$$\mathfrak{f}_{\chi} = \prod_{\substack{\mathfrak{p}\ \text{finite} \\ \mathfrak{p} \in E}} \mathfrak{p}^{\mu_{\mathfrak{p}}}$$

and we refer to $\mathfrak{f}_{\chi}$ as the *conductor* of the character χ derived from ψ. If $\mathfrak{p}$ is finite and belongs to E, (1.1) implies, by virtue of (3), that $\mathfrak{p} \in S$. Moreover, if $\mathfrak{p}$ is finite and not in E, (1.2) shows that it is not necessary, in view of (3), to include $\mathfrak{p}$ in S. Thus the finite $\mathfrak{p}$ in E are just those non-archimedean primes in the exceptional set of the primitive character co-trained with χ.

Let $\mathfrak{m}$ be a given integral ideal of k and let χ be any character such that $\mathfrak{f}_{\chi} | \mathfrak{m}$. Then if $\mathfrak{a} = (\alpha)$ where $\alpha \equiv 1 \pmod{\mathfrak{m}}$,

$$\chi(\mathfrak{a}) = \psi(1, \ldots, 1; 1, \ldots, 1; \alpha, \alpha, \ldots)$$

$$= \psi(\alpha^{-1}, \ldots, \alpha^{-1}; \alpha^{-1}, \ldots, \alpha^{-1}; 1, 1, \ldots$$

by (1),

$$= \psi(\alpha^{-1}, \ldots, \alpha^{-1}; 1, \ldots, 1; 1, 1, \ldots$$

since $\alpha^{-1} \in 1 + \mathfrak{p}^{\mu_{\mathfrak{p}}}$ for finite $\mathfrak{p} \in S$. Hence

$$\chi(\mathfrak{a}) = \prod_{\mathfrak{p} \in S_0} \psi_{\mathfrak{p}}(\alpha^{-1})$$

where S_0 is the set of archimedean primes.

We now restrict attention to those characters χ for which $\psi_{\mathfrak{p}}(J)$ is a discrete sub-set of the unit circle for all $\mathfrak{p} \in S_0$. Consider a particular $\mathfrak{p} \in S_0$. There exists an integer m such that

$$\psi_{\mathfrak{p}}(x^m) = 1 \qquad \text{for all } x \in J,$$

and, in particular, for all $x \in k^*$. The completion $k_{\mathfrak{p}}$ is $\mathbf{R}$ or $\mathbf{C}$ according as $\mathfrak{p}$ is real or complex, with each x of k^* mapped onto the $\mathfrak{p}$-conjugate $x_{\mathfrak{p}}$ of x. However, a homomorphic mapping of $\mathbf{C}^*$ into the unit circle and having finite order must map $\mathbf{C}^*$ onto 1 and similarly, a homomorphic map of $\mathbf{R}^*$, of finite order, into the unit circle either maps $\mathbf{R}^*$ onto 1 or maps $\mathbf{R}^*$ onto ± 1 by $x \to \operatorname{sgn} x$. Thus, if $\mathfrak{p}$ is complex, $\psi_{\mathfrak{p}}$ is identically 1, and if $\mathfrak{p}$ is real $\psi_{\mathfrak{p}}$ is either identically 1 or is ± 1; in the latter case, for $x \in k^*$ we have

$$\psi_{\mathfrak{p}}(x) = \begin{cases} 1, & x^{(\mathfrak{p})} > 0, \\ -1, & x^{(\mathfrak{p})} < 0. \end{cases}$$

If $x^{(\mathfrak{p})} > 0$ for all real $\mathfrak{p} \in S_0$, x is said to be *totally positive* and we write $x \gg 0$. Thus, if ψ has discrete infinite components, χ is a character determined by the subgroup of totally positive principal ideals $\equiv 1 \pmod{\mathfrak{m}}$. Such a character is called a *Dirichlet character modulo* $\mathfrak{m}$; if $\mathfrak{m} = 1$, χ is called a *Hilbert character.*

Conversely, we shall now show that any character χ of the ideal group I^S which is 1 on the subgroup $H^{(\mathfrak{m})}$ of totally positive principal ideals congruent 1 modulo $\mathfrak{m}$ (where S consists of the archimedean primes together with the primes dividing $\mathfrak{m}$) arises in this way from an idèle class character ψ with exceptional set S. Let χ be such a character, and let χ^*, χ_1 be the restrictions of χ to the group A^S of principal ideals coprime with $\mathfrak{m}$, and to the group $B^{(\mathfrak{m})}$ of principal ideals $\equiv 1 (\text{mod } \mathfrak{m})$ respectively. Then $\chi_1(\alpha)$ is determined by the signs of the real conjugates of α. We extend this definition to all elements α of k^* in the obvious way. Then we write $\chi_2 = \chi^* \chi_1^{-1}$; clearly χ_2 is a character of A^S equal to 1 on $B^{(\mathfrak{m})}$.

We define ψ corresponding to χ by stages. For any idèle x, and each $\mathfrak{p}$ not in S, define

$$\psi_{\mathfrak{p}}(x) = \chi(\mathfrak{p}^{\nu_{\mathfrak{p}}}) \qquad \text{where† } \mathfrak{p}^{\nu_{\mathfrak{p}}} \| x_{\mathfrak{p}}. \tag{1.3}$$

This part of the definition ensures that ψ satisfies (3).

We next consider together all the finite primes of S, that is, all $\mathfrak{p} \in S - S_0$. We choose $y \in k^*$ such that $y \equiv x_{\mathfrak{p}} \pmod{\mathfrak{p}^{\nu_{\mathfrak{p}}}}$ for each $\mathfrak{p} \in S - S_0$, where $\mathfrak{p}^{\nu_{\mathfrak{p}}} \| \mathfrak{m}$ (this is possible by the Chinese Remainder Theorem). We then define

$$\prod_{\mathfrak{p} \in S - S_0} \psi_{\mathfrak{p}}(x) = \chi_2^{-1}(y) \tag{1.4}$$

for those idèles x having $|x_{\mathfrak{p}}|_{\mathfrak{p}} = 1$ for all $\mathfrak{p} \in S - S_0$.

† $\mathfrak{p}^{\nu} \| x_{\mathfrak{p}}$ means that $x_{\mathfrak{p}} \in \mathfrak{p}^{\nu} - \mathfrak{p}^{\nu+1}$.

To complete the definition of ψ on this sub-set of J, we need to define $\psi_\mathfrak{p}(x)$ for $\mathfrak{p} \in S_0$. For real $\mathfrak{p} \in S_0$ let $\chi^{(\mathfrak{p})}$ be the character on k^* defined by $\chi^{(\mathfrak{p})}(\alpha) = \operatorname{sgn} \alpha_\mathfrak{p}$. Then χ_1 is the product of some sub-set of these characters $\chi^{(\mathfrak{p})}$, say

$$\chi_1 = \prod_{\mathfrak{p} \in T} \chi^{(\mathfrak{p})} \qquad (T \subset S_0) \tag{1.5}$$

and we define $\psi_\mathfrak{p}(x)$ for $\mathfrak{p} \in S_0$ by

$$\psi_\mathfrak{p}(x) = \begin{cases} \chi^{(\mathfrak{p})}(x_\mathfrak{p})^{-1} & \text{if } \mathfrak{p} \in T \\ 1 & \text{if } \mathfrak{p} \in S_0 - T, \end{cases} \tag{1.6}$$

provided $x_\mathfrak{p} \in k^*$, and we extend this definition to $k_\mathfrak{p}^*$ by continuity.

We postpone defining ψ for those idèles x for which $|x_\mathfrak{p}|_\mathfrak{p} \neq 1$ for some $\mathfrak{p} \in S - S_0$.

It is clear from our construction that ψ is multiplicative. We proceed to verify that ψ satisfies (1) and (2), (3) having been satisfied by construction. Consider (1) first. Then

$$\prod_{\mathfrak{p} \in S_0} \psi_\mathfrak{p}(\alpha) = \chi_1^{-1}(\alpha)$$

by (1.5) and (1.6),

$$\prod_{\mathfrak{p} \in S - S_0} \psi_\mathfrak{p}(\alpha) = \chi_2^{-1}(\alpha)$$

by (1.4), and

$$\prod_{\mathfrak{p} \notin S} \psi_\mathfrak{p}(\alpha) = \chi(\alpha) = \chi^*(\alpha)$$

by (1.3). Since $\chi^* = \chi_1 \chi_2$, ψ is seen to be 1 on k^*.

It remains to consider (2). But $\psi(x) = 1$ if

$$\begin{aligned} |x_\mathfrak{p} - 1|_\mathfrak{p} &< 1, & \mathfrak{p} &\in S_0, \\ x_\mathfrak{p} &\in 1 + \mathfrak{p}^{u_\mathfrak{p}}, & \mathfrak{p} &\in S - S_0, \\ |x_\mathfrak{p}|_\mathfrak{p} &= 1, & \mathfrak{p} &\notin S, \end{aligned}$$

and this is an open set in the idèle topology.

Finally, we need to extend the definition of ψ to idèles x for which, for some $\mathfrak{p} \in S - S_0$, $|x_\mathfrak{p}|_\mathfrak{p} \neq 1$. Let x be such an idèle and suppose that $\mathfrak{p} \in S - S_0$. If $\mathfrak{p}^{u_\mathfrak{p}} \| x_\mathfrak{p}$ (where $u_\mathfrak{p}$ is, of course, not necessarily positive), choose $\alpha \in k^*$ such that

$$\mathfrak{p}^{-u_\mathfrak{p}} \| \alpha \qquad \text{for each } \mathfrak{p} \in S - S_0.$$

We now define

$$\psi(x) = \psi(\alpha x),$$

noting that the right-hand side has already been defined. Hence ψ is well defined; for, if β is another element of k^* such that $\mathfrak{p}^{-u_\mathfrak{p}} \| \beta$ for each $\mathfrak{p} \in S - S_0$,

$$\psi(\beta x) = \psi(\beta/\alpha)\psi(\alpha x) = \psi(\alpha x),$$

since $\beta/\alpha \in k^*$ and ψ is multiplicative. This completes the definition of ψ as an idèle class character.

We recall that I^S is the group of all ideals of k relatively prime to S, A^S denotes the subgroup of principal ideals in I^S, $B^{(\mathfrak{m})}$ the subgroup of A^S consisting of those principal ideals (α) where $\alpha \equiv 1 \pmod{\mathfrak{m}}$, and $H^{(\mathfrak{m})}$ the subgroup of $B^{(\mathfrak{m})}$ consisting of those principal ideals (α) which satisfy

$$\alpha \equiv 1 \pmod{\mathfrak{m}}, \qquad \alpha \gg 0.$$

We note that $H^{(\mathfrak{m})}$ is the intersection of the kernels of all the distinct Dirichlet characters modulo $\mathfrak{m}$, and the number $h_{\mathfrak{m}}$ of such characters is thus the index of $H^{(\mathfrak{m})}$ in I^S.

The index of A^S in I^S is equal to the so-called *absolute class number h*. The index of $B^{(\mathfrak{m})}$ in A^S is the number of units in the residue class ring mod $\mathfrak{m}$, denoted by $\phi(\mathfrak{m})$; and the index of $H^{(\mathfrak{m})}$ in $B^{(\mathfrak{m})}$ is the number of totally positive units $\equiv 1 \pmod{\mathfrak{m}}$, equal to $H/(h\phi_1(\mathfrak{m}))$ where $\phi_1(\mathfrak{m})$ is the number of residue classes mod $\mathfrak{m}$ containing totally positive units and H is the number of ideal classes in the "narrow" sense, i.e. relative to the subgroup of I^S consisting of all totally positive principal ideals of k. Thus the number $h_{\mathfrak{m}}$ of distinct Dirichlet characters mod $\mathfrak{m}$ is given by

$$h_{\mathfrak{m}} = \frac{H}{\phi_1(\mathfrak{m})}\phi(\mathfrak{m}).$$

2. Dirichlet L-series and Density Theorems

In this section we shall take "character" to mean "Dirichlet character". If χ is a character, we define χ on S by $\chi(\mathfrak{p}) = 0$ if $\mathfrak{p} \in S$.

Let $\mathfrak{a}$ denote a general integral ideal, with the prime decomposition of $\mathfrak{a}$ in k given by

$$\mathfrak{a} = \prod_{\mathfrak{p}} \mathfrak{p}^{v_{\mathfrak{p}}}, \qquad v_{\mathfrak{p}} \geqslant 0,\ v_{\mathfrak{p}} = 0 \text{ for almost all } \mathfrak{p}.$$

By $N(\mathfrak{a})$ we shall understand the absolute norm of $\mathfrak{a}$, i.e. $N_{k/\mathbf{Q}}(\mathfrak{a})$.

We define the Dedekind zeta-function $\zeta_k(s)$ by

$$\zeta_k(s) = \sum_{\mathfrak{a} \neq 0} \frac{1}{N(\mathfrak{a})^s} = \prod_{\mathfrak{p}} \left(1 - \frac{1}{N(\mathfrak{p})^s}\right)^{-1} \qquad (s = \sigma + it),$$

and with each character χ we associate a so-called L-series

$$L(s, \chi) = \sum_{\mathfrak{a} \neq 0} \frac{\chi(\mathfrak{a})}{N(\mathfrak{a})^s} = \prod_{\mathfrak{p}} \left(1 - \frac{\chi(\mathfrak{p})}{N(\mathfrak{p})^s}\right)^{-1}.$$

Since each rational prime p is the product of at most $[k:\mathbf{Q}]$ primes $\mathfrak{p}$, the convergence of both sum and product in each case, and the equality between them, for $\sigma > 1$ all follow from the absolute convergence of sum and product for $\sigma > 1$.

We observe that $L(s, \chi_0)$ and $\zeta_k(s)$ differ only by the factors corresponding to the ramified primes, and that $\zeta_k(s)$ is the L-series corresponding to the primitive character co-trained with χ_0.

As in classical rational number theory, L-functions are introduced as a means of proving various density theorems (such as Theorem 3 below). An important property of these functions is that their range of definition can be extended analytically to the left of the line $\sigma = 1$. The continuation into the region $\sigma > 1 - \frac{1}{\nu}$ can be effected in an elementary way (using only Abel summation and the fact that $\zeta(s)$ is regular apart from a simple pole at $s = 1$) by virtue of the following

THEOREM 1

$$\sum_{N(\mathfrak{a}) \leq x} \chi(\mathfrak{a}) = \begin{cases} O(x^{1-1/\nu}), & \chi \neq \chi_0, \\ \kappa x + O(x^{1-1/\nu}), & \chi = \chi_0, \end{cases}$$

where κ depends on the degree, class number, discriminant and units of k.

We indicate briefly the proof of this result. Let C denote a typical coset of $H^{(\mathfrak{m})}$ in I^S. Then the sum considered in the theorem is equal to

$$\sum_C \chi(C) \sum_{\substack{N(\mathfrak{a}) \leq x \\ \mathfrak{a} \in C}} 1,$$

and since

$$\sum_C \chi(C) = \begin{cases} h_\mathfrak{m}, & \chi = \chi_0, \\ 0 & \text{otherwise,} \end{cases}$$

it suffices to estimate the inner sum. Let $\mathfrak{b}$ be a fixed ideal in C^{-1}. Then $\mathfrak{a}\mathfrak{b} = (a)$ where a is a totally positive element of k such that $a \equiv 1 \pmod{\mathfrak{m}}$. As $\mathfrak{a}$ varies, a is a variable integer in the ideal $\mathfrak{b}$. Also,

$$|N(a)| = |N((a))| = N(\mathfrak{a})N(\mathfrak{b}),$$

so that the inner sum may be written in the form

$$\sum_{\substack{|N(a)| \leq xN(\mathfrak{b}) \\ a \in \mathfrak{b}}}^{*} 1$$

where the asterisk indicates that $a \equiv 1 \pmod{\mathfrak{m}}$, $a \gg 0$, and that in each set of associates only one element is to be counted. The integers a of the ideal $\mathfrak{b}$ may be represented by the points of a certain n-dimensional lattice, and the problem is essentially the classical one of estimating the number of these points in a certain simplex (Dedekind, 1871–94; Weber, 1896 and Hecke, 1954). The estimate takes the form†

$$\frac{\kappa}{h_\mathfrak{m}} x + O(x^{1-1/\nu}).$$

† A better error term can be derived as in Landau (1927), Satz 210.

A result of this type may also be derived for Grössencharacters, but the proof is complicated (Hecke, 1920).

As in classical theory, deeper methods lead to analytic continuation of L-functions into the whole complex plane, and thence to a functional equation for such functions. In the case of the ordinary ζ-function, one obtains†

$$\pi^{-s/2}\Gamma\left(\frac{s}{2}\right)\zeta(s) = \frac{1}{s-1} - \frac{1}{s} + \int_1^\infty (x^{-\frac{1}{2}s-\frac{1}{2}} + x^{\frac{1}{2}s-1})\tfrac{1}{2}(\theta(x)-1)\,dx = \Phi(s),$$

where

$$\theta(x) = \sum_{m=-\infty}^{\infty} e^{-\pi m^2 x},$$

and one derives at once the functional equation

$$\Phi(s) = \Phi(1-s).$$

Hecke (1920) extended this classical result in a far-reaching way to L-series formed with Grössencharacters; and, more recently, Tate (Chapter XV) generalized Hecke's result to general spaces. For Dirichlet characters, Hecke's result is summarized below (Hasse's Bericht, 1926, 1927 and 1930).

Let

$$\Phi(s,\chi) = \prod_{q=1}^{r_1} \Gamma\left(\frac{s+a_q}{2}\right) . \Gamma(s)^{r_2} \left\{\frac{|d|N(\mathfrak{f}_\chi)}{4^{r_2}\pi^{\nu}}\right\}^{s/2} L(s,\chi);$$

then $\Phi(s,\chi)$ is meromorphic and satisfies the functional equation

$$\Phi(s,\chi) = W(\chi)\Phi(1-s,\bar{\chi})$$

where W is a constant of absolute value 1, and

- d denotes the discriminant;
- r_1, r_2 are the numbers of real and complex valuations respectively;
- $a_q = 0$ or 1 according as the value of χ in the domain of all principal ideals (α) with $\alpha \equiv 1 \pmod{\mathfrak{f}_\chi}$ does or does not depend on the sign of the qth real conjugate of α.

An explicit expression for W is given, e.g. in Hasse's Bericht (Section 9, Satz 15); its presence in the functional equation leads to interesting information of an algebraic nature, for instance, about generalized Gaussian sums (Bericht, Section 9.2 *et seq*). It follows from the proof of the functional equation that $L(s,\chi)$ is an integral function, except for a simple pole at $s = 1$ when $\chi = \chi_0$.

We shall see that information about the zeros of L-functions plays an important part in applications to density results. The distribution of zeros in the "critical" strip $0 < \sigma \leqslant 1$ is particularly important. According to the

† Titchmarsh (1951).

generalized Riemann hypothesis, $L(s, \chi) \neq 0$ if $\sigma > \frac{1}{2}$; but this conjecture is far from being settled.

Nevertheless, significant information can be extracted from the much weaker

THEOREM 2

$$L(s, \chi) \neq 0 \qquad \text{if} \quad \sigma \geqslant 1.$$

Proof. In view of the product representation of L for $\sigma > 1$ we may restrict ourselves to the case $\sigma = 1$. The proof falls into two parts, the first due to Hadamard, the second to Landau. We suppose on the contrary that $1 + it$ is a zero of $L(s, \chi)$.

(i) (Hadamard) Assume that $\chi^2 \neq \chi_0$ if $t = 0$. If $\sigma > 1$, we have, from the product representation, that

$$L(s, \chi) = \exp\left\{\sum_{\mathfrak{p}} \sum_{m=1}^{\infty} \frac{1}{m} N(\mathfrak{p})^{-m\sigma} e^{-itm \log N(\mathfrak{p})} \chi(\mathfrak{p}^m)\right\}.$$

When $\mathfrak{p}$ is prime to $\mathfrak{f}_\chi$, we write $\chi(\mathfrak{p}) = e^{ic_\mathfrak{p}}$, $\beta_\mathfrak{p} = t \log N(\mathfrak{p}) + c_\mathfrak{p}$ and consider the function

$$\begin{aligned}|L^3(\sigma, \chi_0)L^4(\sigma+it, \chi)L(\sigma+2it, \chi^2)| &= \exp\left\{\sum_{\mathfrak{p} \text{ prime to } \mathfrak{f}_\chi} \sum_{m=1}^{\infty} \frac{1}{m} N(\mathfrak{p})^{-m\sigma}(3 + 4\cos m\beta_\mathfrak{p} + \cos 2m\beta_\mathfrak{p})\right\} \\ &\geqslant 1\end{aligned}$$

since $3 + 4\cos\omega + \cos 2\omega \geqslant 0$ for all real ω. Keeping t fixed, we now let $\sigma \to 1+0$. Of the terms on the left, the first is $O((\sigma-1)^{-3})$, the second is $O((\sigma-1)^4)$ by hypothesis and the third is $O(1)$ because if $t = 0$, $\chi^2 \neq \chi_0$. Hence the expression on the left tends to 0 as $\sigma \to 1+0$, and we arrive at a contradiction.

(ii) (Landau) It remains to consider the case when $\chi^2 = \chi_0$ (so that χ is real) and, if possible, $L(1, \chi) = 0$. We consider the product

$$\zeta_k(s)L(s, \chi) = \sum_{\mathfrak{a}} N(\mathfrak{a})^{-s}\lambda(\mathfrak{a}) = \sum_{u=1}^{\infty} a_u . u^{-s}$$

where

$$a_u = \sum_{N(\mathfrak{a})=u} \lambda(\mathfrak{a})$$

and

$$\begin{aligned}\lambda(\mathfrak{a}) = \sum_{\mathfrak{b}|\mathfrak{a}} \chi(\mathfrak{b}) &= \prod_{\mathfrak{p}^m \| \mathfrak{a}} \{1 + \chi(\mathfrak{p}) + \ldots \chi(\mathfrak{p}^m)\} \\ &\geqslant 0,\end{aligned}$$

where $\mathfrak{p}^m \| \mathfrak{a}$ signifies that $m = m_\mathfrak{p}$ is the highest exponent to which $\mathfrak{p}$ divides $\mathfrak{a}$. Moreover, if $m_\mathfrak{p}$ is even for all $\mathfrak{p}$ dividing $\mathfrak{a}$,

$$\lambda(\mathfrak{a}) \geqslant 1.$$

Hence, if σ is real and both series converge,

$$\sum_{\mathfrak{a}} N(\mathfrak{a})^{-\sigma}\lambda(\mathfrak{a}) \geqslant \sum_{\mathfrak{a}} N(\mathfrak{a})^{-2\sigma}. \tag{2.1}$$

We now utilize the following simple result from the theory of Dirichlet series:

A Dirichlet series of type

$$f(s) = \sum_{u=1}^{\infty} a_u . u^{-s}, \qquad a_u \geqslant 0 \quad (u = 1, 2, \ldots),$$

has a half-plane Re $s > \sigma_0$ *as its domain of convergence, and if* σ_0 *is finite, then* $f(s)$ *is non-regular at* $s = \sigma_0$. (Titchmarsh, 1939, see the theorem of Section 9.2.)

By hypothesis, $f(s) = \zeta_k(s)L(s, \chi)$ is everywhere regular (the hypothetical zero of $L(s, \chi)$ at $s = 1$ counteracting the pole of $\zeta_k(s)$); hence the series on the left of (2.1) converges for all real σ. However, the series on the right of (2.1) is equal to $\zeta_k(2\sigma)$ which tends to $+\infty$ as $\sigma \to \frac{1}{2}+0$. Hence we arrive at a contradiction.

The Landau proof is purely existential, whereas Hadamard's argument can be used to yield quantitative results about zero-free regions and orders of magnitude of L-functions.†

We are now in a position to prove

THEOREM 3 (*Prime Ideal Theorem*)

$$\sum_{N(\mathfrak{p}) \leqslant x} \chi(\mathfrak{p}) = \begin{cases} \dfrac{x}{\log x}\{1 + o(1)\}, & \chi = \chi_0, \\ o\left(\dfrac{x}{\log x}\right), & \chi \neq \chi_0. \end{cases}$$

Proof. For Re $s > 1$, form the logarithmic derivative of $L(s, \chi)$, namely

$$-\frac{L'(s,\chi)}{L(s,\chi)} = \sum_{\mathfrak{p}} \sum_{m=1}^{\infty} N(\mathfrak{p})^{-ms} \log N(\mathfrak{p}) . \chi(\mathfrak{p}^m)$$

$$= \sum_{\mathfrak{p}} \chi(\mathfrak{p}) \frac{\log N(\mathfrak{p})}{N(\mathfrak{p})^s} + g(s, \chi)$$

where g is a function regular for Re $s > \frac{1}{2}$. Our first, and major, step will be to estimate the coefficient sum

$$\sum_{N(\mathfrak{p}) \leqslant x} \chi(\mathfrak{p}) \log N(\mathfrak{p}).$$

One way of doing this is to express the sum as a contour integral of

$$\frac{x^s}{s} \frac{L'(s,\chi)}{L(s,\chi)}$$

† See Estermann (1952) and Landau (1907).

along the line $(c-i\infty, c+i\infty)$ with some $c > 1$, and then to shift the line of integration to the parallel line $\operatorname{Re} s = 1$, with an indent round $s = 1$. This we now know to be the only point at which a pole can occur (and does occur precisely when $\chi = \chi_0$, the residue in this case then leading to the dominant term). For an account of this treatment, see Landau (1927). It is clear from even this sketch that the non-vanishing of $L(s, \chi)$ on $\operatorname{Re} s = 1$ is vital.

An alternative approach is via the Wiener–Ikehara tauberian theorem: *Suppose that*

$$f(s) = \sum_{m=1}^{\infty} \frac{a_m}{m^s} \quad (a_m \geqslant 0), \qquad g(s) = \sum_{m=1}^{\infty} \frac{b_m}{m^s}$$

are Dirichlet series, convergent for $\operatorname{Re} s > 1$, *regular on* $\operatorname{Re} s = 1$ *with simple poles at* $s = 1$, *of residue* 1 *in the case of* f, *and* η *in the case of* g, *where* η *may be* 0. *Assume that there exists a constant* c *such that* $|b_m| \leqslant c a_m$. *Then*

$$\sum_{m \leqslant x} b_m \sim \eta x, \quad \text{as } x \to \infty.$$

We apply this result with

$$f(s) = -\frac{\zeta'(s)}{\zeta(s)}, \qquad g(s) = -\frac{L'(s, \chi)}{L(s, \chi)}, \qquad c = \lceil k:$$

Clearly $\eta = 1$ when $\chi = \chi_0$, and otherwise $\eta = 0$.

Finally, it is an easy matter to show that

$$\frac{1}{\log x} \sum_{N(\mathfrak{p}) \leqslant x} \chi(\mathfrak{p}) \log N(\mathfrak{p}) \sim \sum_{N(\mathfrak{p}) \leqslant x} \chi(\mathfrak{p}),$$

e.g. by partial summation; and this proves the theorem. (The tauberian approach is described in Lang (1964).)

It is worth remarking that the analytic method can be made to give sharper estimates of the error terms (Hecke, 1920). Both methods extend to Grössencharacters.

In close analogy to Dirichlet's famous proof of the infinitude of rational primes in arithmetic progressions, one can show that every Dirichlet ideal class C contains infinitely many primes $\mathfrak{p}$ of k. Indeed, by means of the prime ideal theorem we can show that the primes $\mathfrak{p}$ are equally distributed among the classes C; we have, in particular, that

THEOREM 4

$$\sum_{\substack{N(\mathfrak{p}) \leqslant x \\ \mathfrak{p} \in C}} 1 \sim \frac{1}{h_{\mathfrak{m}}} \frac{x}{\log x}, \qquad x \to \infty.$$

Proof. We have only to remark that

$$\sum_{\chi} \bar{\chi}(C) \chi(\mathfrak{p}) = \begin{cases} h_{\mathfrak{m}}, & \mathfrak{p} \in C, \\ 0 & \text{otherwise.} \end{cases}$$

By virtue of this orthogonality formula, we have

$$h_{\mathfrak{m}} \sum_{\substack{N(\mathfrak{p}) \leqslant x \\ \mathfrak{p} \in C}} 1 = \sum_{\chi} \bar{\chi}(C) \sum_{N(\mathfrak{p}) \leqslant x} \chi(\mathfrak{p});$$

on the other hand, the sum on the right is asymptotic to $x/\log x$ as $x \to \infty$ by the prime ideal theorem.

If B is any set of ideals in k, and

$$\lim_{x \to \infty} \frac{\log x}{x} \operatorname{card} \{\mathfrak{p} \in B \mid N(\mathfrak{p}) \leqslant x\} =$$

exists, then l is called the *density* of primes in B. Thus the density of the set of all primes of k is 1.

To investigate the density of primes in a given set, it suffices to consider only primes $\mathfrak{p}$ of absolute first degree, that is, those $\mathfrak{p}$ for which $N_{k/\mathbf{Q}}(\mathfrak{p})$ is a rational prime; for all primes $\mathfrak{p}$ whose norms $N(\mathfrak{p})$ are powers of rational primes greater than the first, and satisfy $N(\mathfrak{p}) \leqslant x$, are $O(x^{\frac{1}{2}})$ in number. For the same reason we may clearly disregard the finite number of ramified primes.

We shall use this remark in proving the next result, known in classical parlance as the *first fundamental inequality of class field theory* (see Chapter VII, "Global Class Field Theory", §9, and Chapter XI, "The History of Class Field Theory", §1).

THEOREM 5. *Let K be a finite normal extension of k, and denote $[K:k]$ by n. Let S be the exceptional set in k. K determines the subgroup H_S of I^S, of finite index h_S in I^S, composed of those cosets of $H^{\mathfrak{m}}$ which contain norms (relative to k) of ideals of K coprime with S. Then*

$$h_s \leqslant n.$$

Proof. We identify $1/n$ and $1/h_S$ as the densities of two sets of primes in k.

First of all, if C is the set of all primes $\mathfrak{p}$ (of k) in H_S, then, by the preceeding theorem, the density of C is $1/h_S$.

Next, since K/k is normal, the prime decomposition of any $\mathfrak{p}$ in K is of the form

$$\mathfrak{p} = (\mathfrak{P}_1 \ldots \mathfrak{P}_r)^e$$

where all the primes $\mathfrak{P}_i$ have the *same* degree f relative to $\mathfrak{p}$, and

$$efr = n.$$

Since the number of ramified $\mathfrak{p}$ is finite, we need consider only primes $\mathfrak{p}$ for which $e = 1$. Now consider the set B of all primes $\mathfrak{p}$ whose prime decomposition in K is characterized by $e = f = 1$, that is, by $r = n$. On the one hand, by the prime ideal theorem for K (noting that the primes $\mathfrak{P}_i$ are all the primes of K of absolute first degree) the density of B is clearly $1/n$; on the other hand, if $\mathfrak{p} \in B$ and $\mathfrak{P}|\mathfrak{p}$, then $N_{K/k}(\mathfrak{P}) = \mathfrak{p}$, so that $B \subset H_S$. The result follows at once.

Note. It is worth remarking that Theorem 5 can be proved without appeal to the relatively deep Theorems 3 and 4 and, in particular, to the non-vanishing of $L(s, \chi)$ on $\operatorname{Re} s = 1$. A more elementary argument, using only real variable theory, runs as follows:

Let s be real and > 1. By Theorem 1 and partial summation we have

(a) $$\lim_{s \to 1} (s-1)L(s, \chi_0) = \kappa$$

and

(b) $$\lim_{s \to 1} L(s, \chi) \text{ exists and is finite if } \chi \neq \chi_0.$$

From the product representation of $L(s, \chi)$ it follows that

$$\log L(s, \chi) = \sum \frac{\chi(\mathfrak{p})}{N(\mathfrak{p})^s} + g(s, \chi)$$

where $g(s, \chi)$ is a Dirichlet series absolutely convergent for $s > \frac{1}{2}$. Using the orthogonality properties of the characters χ, we conclude that

(c) $$\sum_{\mathfrak{p} \in H_s} \frac{1}{N(\mathfrak{p})^s} = \frac{1}{h_s} \log \frac{1}{s-1} + f(s)$$

where

$$h_s f(s) = \sum_{\chi \neq \chi_0} \{\log L(s, \chi) - g(s, \chi)\} + \log (s-1)L(s, \chi_0) - g(s, \chi_0).$$

By (a) and (b), $\limsup_{s \to 1} f(s)$ is not $+\infty$ and, indeed, is finite unless one of the $L(s, \chi)\,(\chi \neq \chi_0)$ vanishes at $s = 1$.

Now let K be as described in the statement of Theorem 5. We have, analogously to (a), that $\lim_{s \to 1} \zeta_K(s).(s-1)$ exists and is finite. Taking logarithms of the product representation of $\zeta_K(s)$, gives

(d) $$\sum_{\mathfrak{p} \in B} \frac{1}{N(\mathfrak{p})^s} = \frac{1}{n} \log \frac{1}{s-1} + G(s)$$

where $\lim_{s \to 1} G(s)$ exists and is finite. Subtracting (d) from (c), and using $B \subset H_S$, we obtain

$$\left(\frac{1}{h_s} - \frac{1}{n}\right) \log \frac{1}{s-1} + f(s) - G(s) \geqslant 0$$

for all $s > 1$. Letting $s \to 1+0$, the result follows.

For the next theorem we require some results from class-field theory.

THEOREM 6. *If K is a finite abelian extension of k, then*

$$\zeta_K(s) = \prod_\chi L(s, \chi, k) \tag{2.2}$$

where, in the notation of the preceding theorem, the product on the right extends over the primitive characters co-trained with the characters of the class group I^S/H_S.

Proof. The proof is carried out in terms of local factors, and we consider separately non-ramified and ramified primes $\mathfrak{p}$ of k.

(i) Let $\mathfrak{p}$ be a non-ramified prime of k, so that

$$\mathfrak{p} = \mathfrak{P}_1 \ldots \mathfrak{P}_l$$

where $\mathfrak{P}_1, \ldots, \mathfrak{P}_l$ are distinct primes of K. From class field theory,

$$N_{K/\mathbf{Q}}(\mathfrak{P}_i) = N_{k/\mathbf{Q}}(\mathfrak{p})^f$$

where $lf = [K:k] = n$.

Thus the corresponding local factor on the left is

$$(1-N(\mathfrak{p})^{-fs})^{-n/f},$$

whilst the corresponding local factor on the right is

$$\prod_\chi (1-\chi(\mathfrak{p})N(\mathfrak{p})^{-s})^{-1}.$$

Since f is the least positive integer such that $\chi(\mathfrak{p}^f) = 1$ for all χ, we have the easily verifiable identity (take logs of both sides and use $h_S = n$)

$$(1-y^f)^{-n/f} = \prod_\chi (1-\chi(\mathfrak{p}).y)^{-1};$$

from this, with $y = N(\mathfrak{p})^{-s}$, equality of the local factors follows at once.

(ii) The proof for ramified primes is more difficult, and depends on the functional equations satisfied by the various L-functions. We begin by writing

$$\zeta_K(s) = g(s) \prod_\chi L(s, \chi, k),$$

and prove that $g(s)$ is identically 1. From above, $g(s)$ is equal to the finite product over ramified $\mathfrak{p}$ of the expressions

$$\frac{\prod_\chi \{1-\chi(\mathfrak{p})N(\mathfrak{p})^{-s}\}}{\prod_{\mathfrak{P}|\mathfrak{p}} \{1-N(\mathfrak{P})^{-s}\}}.$$

If this product is not constant, then it has a pole or zero at a pure imaginary point it_0, $t_0 \neq 0$. In view of the functional equations, $g(1-s)/g(s)$ is a quotient of gamma functions and so can have only real poles and zeros. Thus $1-it_0$ is also a pole or zero of g. But we know that $1-it_0$ is not a singularity or zero of any of the L-series or of $\zeta_K(s)$. Hence g is constant, and so equal to 1.

Example 1.† Let $k = \mathbf{Q}$, $K = \mathbf{Q}(\omega)$ where ω is a primitive mth root of unity. Then

$$\zeta_K(s) = \zeta(s) \prod_{\chi} L(s, \chi)$$

where the product on the right extends over the primitive characters co-trained with all non-principal rational characters mod m.

Proof. See Chapter III, §1.

Example 2. Let $k = \mathbf{Q}$, $K = \mathbf{Q}(\sqrt{d})$ where d is the discriminant of K. Then

$$\zeta_K(s) = \zeta(s) \sum_{m=1}^{\infty} \left(\frac{d}{m}\right) m^{-s},$$

where $\left(\frac{d}{m}\right)$ denotes the Kronecker symbol (Hecke, 1954, Chapter VII).

We touch upon another consequence of this theorem. We recall that the functional equation of an L-function contains a factor

$$\{|d| N(\mathfrak{f}_\chi)\}^{s/2}.$$

Applying the functional equation to $\zeta_K(s)$ on the left of (2.2) and to each of the L-series on the right, one can derive the relation

$$\prod_{\chi} |d| N(\mathfrak{f}_\chi) = |D|,$$

where D is the discriminant of K (Hasse's Bericht, Section 9.3). If we assume for the moment that $k = \mathbf{Q}$, we obtain

$$\prod_{\chi} \mathfrak{f}_\chi = \text{Disc}(K/k). \tag{2.3}$$

This inference is not as easy to justify in the general case when $k \neq \mathbf{Q}$, because two distinct ideals in k can have equal norms. However, it can be proved (Hasse's Bericht, Section 9.3, formula (12) and Note 44) that (2.3) is valid in general.

3. L-functions for Non-abelian Extensions

Suppose now that K is a finite, normal but not necessarily abelian extension of k of degree n. The problem is to develop in this case an analogue of the above theory of L-series formed with abelian group characters.

As usual, let G be the Galois group of K over k. Let $\{M(\mu)\}_{\mu \in G}$ be a representation of G into matrices over the complex field. Thus to each element μ of G corresponds a matrix $M(\mu)$. The character $\chi(\mu)$ of μ is defined to be

† By (2.2), since $\zeta_K(s)$ and $L(s, \chi_0, k) = \zeta_k(s)$ have simple poles at $s = 1$, and the remaining $L(s, \chi, k)$ on the right are regular there, it follows that these $L(s, \chi, k)$ do not vanish at $s = 1$. In particular, in the situation of Example 1, it follows that all the L-functions formed with non-principal Dirichlet characters mod m do not vanish at $s = 1$. This is an alternative proof of the key step in the proof of Dirichlet's theorem on the infinitude of primes in arithmetic progressions mod m.

the trace of $M(\mu)$ and depends only on the conjugacy class $\langle\mu\rangle$ in which μ lies.

Two representations† $\{M(\mu)\}_{\mu\in G}$, $\{N(\mu)\}_{\mu\in G}$ are said to be *equivalent* if there exists a non-singular matrix P such that

$$PM(\mu)P^{-1} = N(\mu) \qquad \text{for all } \mu \in G.$$

A representation $\{M(\mu)\}$ is said to be *reducible* if and only if it is equivalent to a representation $\{N(\mu)\}$ such that

$$N(\mu) = \begin{pmatrix} N^{(1)}(\mu) & 0 \\ 0 & N^{(2)}(\mu) \end{pmatrix} \qquad \text{for all } \mu \in G,$$

where $\{N^{(1)}(\mu)\}$, $\{N^{(2)}(\mu)\}$ are themselves representations of G.

The character of an *irreducible* representation is said to be *simple.* From the general theory of group representations (for an account of this theory see e.g. Hall, 1959) the number g of simple characters of G is equal to the number of conjugacy classes of G, and the following orthogonality relations are known to hold:

$$\sum_{\mu\in G} \chi(\mu)\bar{\chi}'(\mu) = \begin{cases} n, & \chi = \chi', \\ 0, & \chi \neq \chi'; \end{cases} \tag{3.1}$$

in particular

$$\sum_{\mu\in G} \chi(\mu) = \begin{cases} n, & \chi \text{ principal}, \\ 0, & \text{otherwise}, \end{cases}$$

where the *principal* character is the character of the representation $M(\mu) = 1$ for all $\mu \in G$, to be denoted henceforward by χ_0. Also, if $\psi_1, \psi_2, \ldots, \psi_g$ are the simple characters of G, then

$$\sum_{i=1}^{g} \psi_i(\mu)\bar{\psi}_i(\mu') = \begin{cases} n/l_\mu, & \mu' \in \langle\mu\rangle, \\ 0, & \mu' \notin \langle\mu\rangle, \end{cases} \tag{3.2}$$

where l_μ is the number of elements in the conjugacy class $\langle\mu\rangle$ of μ. In particular, taking $\mu' = \mu = 1$,

$$\sum_{i=1}^{g} n_i^2 = n, \tag{3.3}$$

where n_i is the degree of ψ_i (that is, ψ_i is the character of a representation by $n_i \times n_i$ matrices).

If G is abelian, then each $l_\mu = 1$, so that $g = n$ and each $n_i = 1$. Thus each character ψ_i is a homomorphism of G into the unit circle, and so an abelian character.

Let $\mathfrak{P}$ be a non-ramified prime of K, and let $\mathfrak{p}$ be the prime of k lying under $\mathfrak{P}$. We shall denote the Frobenius automorphism of K/k relative to $\mathfrak{P}$ (for

† It should be clear from the context that N is not used here to denote a norm!

definition see Chapter VII §2) by $\left[\frac{K/k}{\mathfrak{P}}\right]$. If

$$|I-M(\mu)x|$$

is the characteristic polynomial of $M(\mu)$—so that I denotes the unit matrix—we take as local factor of our L-function the expression

$$\left|I-M\left(\left[\frac{K/k}{\mathfrak{P}}\right]\right)N_{k/\mathbf{Q}}(\mathfrak{p})^{-s}\right|^{-1}.$$

We note that this definition cannot be extended to ramified $\mathfrak{P}$ since there is no corresponding Frobenius automorphism.

It is important to note that this local factor depends only on the character χ of the representation, and not on the explicit representation matrix. For any similarity transformation of $M(\mu)$ leaves its characteristic polynomial invariant and, since $M(\mu)$ has finite order, the Jordan canonical form of $M(\mu)$ must therefore be a diagonal matrix all of whose diagonal elements are roots of unity. Thus, without loss of generality,

$$M\left(\left[\frac{K/k}{\mathfrak{P}}\right]\right)$$

may be taken to be

$$M\left(\left[\frac{K/k}{\mathfrak{P}}\right]\right)=\begin{pmatrix}\varepsilon_1 & & 0\\ & \ddots & \\ 0 & & \varepsilon_b\end{pmatrix}.$$

The local factor is then equal to

$$\prod_{i=1}^{b}(1-\varepsilon_i N(\mathfrak{p})^{-s})^{-1}=\exp\left\{\sum_{i=1}^{b}\sum_{m=1}^{\infty}\frac{1}{m}\varepsilon_i^m N(\mathfrak{p})^{-ms}\right\}$$

$$=\exp\left\{\sum_{m=1}^{\infty}\frac{1}{m}\chi\left(\left[\frac{K/k}{\mathfrak{P}}\right]^m\right)N(\mathfrak{p})^{-ms}\right\} \qquad (3.4)$$

since $\sum_{i=1}^{b}\varepsilon_i^m$ is equal to the trace of $\left\{M\left[\frac{K/k}{\mathfrak{P}}\right]\right\}^m$ which, by definition, is equal to $\chi\left(\left[\frac{K/k}{\mathfrak{P}}\right]^m\right)$.

We collect up the local factors corresponding to non-ramified $\mathfrak{P}$ and define the so-called Artin L-function† essentially by

$$L(s,\chi)=L(s,\chi,K/k)=\prod_{\substack{\text{non-ram.}\\ \mathfrak{p}}}\left|I-M\left(\left[\frac{K/k}{\mathfrak{P}}\right]\right)N(\mathfrak{p})^{-s}\right|^{-1};$$

we shall introduce later a factor derived from the ramified primes.

† See also Artin (1930).

We now list a number of observations about $L(s, \chi)$:

(I) $L(s, \chi)$ is regular for $\sigma > 1$, since the product is absolutely and uniformly convergent in every closed sub-set of the half-plane $\mathrm{Re}\, s > 1$.

(II) If K/k is abelian, and if χ is simple then, apart from the factor corresponding to the ramified primes, this definition coincides with that given in Section 2 above.

(III) Suppose that Ω is a field intermediate between K and k, normal over k. Let $H = \mathrm{Gal}\,(K/\Omega)$, so that H is a normal subgroup of G and

$$G/H = \mathrm{Gal}\,(\Omega/k).$$

Then if χ is a character of G/H, it can be regarded in an obvious way as a character of G, and

$$L(s, \chi, K/k) = L(s, \chi, \Omega/k).$$

Proof. Take the character over G defined by the representation

$$M'(\mu) = M(\mu H).$$

We have

$$X^{\left[\frac{K/k}{\mathfrak{P}}\right]} \equiv X^{N(\mathfrak{p})} \pmod{\mathfrak{P}} \tag{3.5}$$

for all integers $X \in K$, and in particular for all integers $X \in \Omega$. Since Ω is a normal extension of k,

$$X \in \Omega \Rightarrow X^{\mu} \in \Omega \qquad \text{for all } \mu \in G.$$

Hence (3.5) is a congruence in Ω, and since $\mathfrak{P}$ is unramified, if $\mathfrak{P}$ lies over $\mathfrak{q}$ in Ω we have

$$X^{\left[\frac{K/k}{\mathfrak{P}}\right]} \equiv X^{N(\mathfrak{p})} \pmod{\mathfrak{q}}.$$

Hence also

$$X^{\left[\frac{K/k}{\mathfrak{P}}\right]H} \equiv X^{N(\mathfrak{p})} \pmod{\mathfrak{q}},$$

i.e. $\left[\frac{K/k}{\mathfrak{P}}\right] H$ is the Frobenius automorphism in Ω.

Thus

$$\left| I - M'\left(\left[\frac{K/k}{\mathfrak{P}}\right]\right) N(\mathfrak{p})^{-s} \right| = \left| I - M\left(\left[\frac{K/k}{\mathfrak{P}}\right] H\right) N(\mathfrak{p})^{-s} \right|$$
$$= \left| I - M\left(\left[\frac{\Omega/k}{\mathfrak{P}}\right]\right) N(\mathfrak{p})^{-s} \right|.$$

(IV) Suppose that χ is a non-simple character of G, say $\chi = \chi_1 + \chi_2$. Then

$$L(s, \chi) = L(s, \chi_1) L(s, \chi_2),$$

since $\log L$ is linear in χ by (3.4).

(V) Suppose again that Ω is a field intermediate between K and k, this time not necessarily normal over k. Let $H = \mathrm{Gal}(K/\Omega)$, and suppose that

$$G = \sum_i H\alpha_i$$

is the partition of G into right cosets of H. To each character χ of H there corresponds an *induced* character χ^* of G, given by

$$\chi^*(\mu) = \sum_{\substack{i \\ \alpha_i\mu\alpha_i^{-1} \in H}} \chi(\alpha_i\mu\alpha_i^{-1}), \qquad \mu \in G; \tag{3.6}$$

then

$$L(s, \chi^*, K/k) = L(s, \chi, K/\Omega).$$

Proof.† Let $\mathfrak{P}$ be a non-ramified prime of K, and $\mathfrak{p}$ the prime of k under $\mathfrak{P}$. Suppose that, in Ω, $\mathfrak{p}$ has the prime decomposition

$$\mathfrak{p} = \prod_{i=1}^{r} \mathfrak{q}_i, \qquad N_{\Omega/\mathbf{Q}}(\mathfrak{q}_i) = (N_{k/\mathbf{Q}}\mathfrak{p})^{f_i}; \tag{3.7}$$

and suppose that τ_i is an element of G such that $\mathfrak{q}_i$ lies under $\tau_i\mathfrak{P}$. Then (Hasse's Bericht, Section 23, I) G can be decomposed in the form

$$G = \sum_{i=1}^{r} \sum_{x_i=0}^{f_i-1} H\tau_i\mu_0^{x_i},$$

where

$$\mu_0 = \left[\frac{K/k}{\mathfrak{P}}\right].$$

From the definition of induced character given above, we have

$$\begin{aligned}
\chi^*(\mu^m) &= \sum_{i=1}^{r} \sum_{\substack{x_i=0 \\ \tau_i\mu_0^{x_i}(\mu_0^m)\mu_0^{-x_i}\tau_i^{-1} \in H}}^{f_i-1} \chi(\tau_i\mu_0^{x_i}\mu_0^m\mu_0^{-x_i}\tau_i^{-1}) \\
&= \sum_{\substack{i=1 \\ \tau_i\mu_0^m\tau_i^{-1} \in H}}^{r} f_i\chi(\tau_i\mu_0^m\tau_i^{-1}) \\
&= \sum_{\substack{i=1 \\ f_i|m}}^{r} f_i\chi(\tau_i\mu_0^m\tau_i^{-1})
\end{aligned}$$

since $(\tau_i\mu_0\tau_i^{-1})^m \in H$ if and only if $f_i|m$.

The logarithm of the $\mathfrak{p}$-component ($\mathfrak{p}$ non-ramified) of $L(s, \chi^*, K/k)$ is

$$\begin{aligned}
\sum_{m=1}^{\infty} \frac{1}{m}\chi^*(\mu_0^m)N(\mathfrak{p})^{-ms} &= \sum_{m=1}^{\infty} \frac{1}{m}N(\mathfrak{p})^{-ms} \sum_{\substack{i=1 \\ f_i|m}}^{r} f_i\chi(\tau_i\mu_0^m\tau_i^{-1}) \\
&= \sum_{i=1}^{r} \sum_{\substack{m=1 \\ f_i|m}}^{\infty} \frac{f_i}{m}\chi(\tau_i\mu_0^m\tau_i^{-1})N(\mathfrak{p})^{-ms} \\
&= \sum_{i=1}^{r} \sum_{t=1}^{\infty} t^{-1}\{\chi(\tau_i\mu_0^{f_i}\tau_i^{-1})\}^t N(\mathfrak{q}_i)^{-ts}
\end{aligned}$$

† For an alternative proof see Hasse's Bericht, Section 27, VII.

by (3.7), on writing $m = f_i t$ in the inner summation; and this is nothing but the sum of the logarithms of those $\mathfrak{q}$-components of $L(s, \chi, K/\Omega)$ which, by (3.7), correspond to $\mathfrak{p}$.

We proceed to consider some special cases of (V).

V(i) $\Omega = K$. Then $H = \mathrm{Gal}(K/\Omega) = (1)$ and we have only one character, the principal character χ_0. In this case, then, $L(s, \chi_0, K/\Omega)$ reduces to $\zeta_K(s)$. By (3.6) the corresponding induced character, $\chi_0{}^*$, of G is given by

$$\chi_0^*(\mu) = \begin{cases} n \text{ if } \mu \text{ is the unit element of } G, \\ 0 \text{ otherwise.} \end{cases} \tag{3.8}$$

Let $\psi_1, \psi_2, \ldots, \psi_g$ be all the simple characters of G. By (3.2), with $\mu' = 1$, we have $l_1 = 1$ and arrive, by (3.8), at

$$\sum_{i=1}^{g} \psi_i(\mu)\psi_i(1) = \chi_0^*(\mu). \tag{3.9}$$

We observe that, of course, $\psi_i(1)$ must be a positive integer. By repeated application of (IV), it follows that

$$\zeta_K(s) = \prod_{i=1}^{g} L(s, \psi_i, K/k)^{\psi_i(1)}. \tag{3.10}$$

V(ii) $\Omega = k$. By (III) (with $\Omega = k$ and therefore $G = H$),

$$\begin{aligned} L(s, \chi_0, K/k) &= L(s, \chi_0, k/k) \\ &= \zeta_k(s). \end{aligned}$$

We shall now deduce from the preceding theorems the remarkable result that a general Artin L-function $L(s, \chi, K/k)$ can be expressed as a product of rational powers of abelian L-functions $L(s, \psi, K/\Omega)$, where the Ω's are fields intermediate between k and K with K/Ω abelian.

With the notation used in (i), each character χ of G can be written in the form

$$\chi = \sum_{i=1}^{g} r_i \psi_i, \tag{3.11}$$

where the r_i's are non-negative rational integers. Hence, by repeated application of (IV), it suffices to consider the *simple* Artin L-functions

$$L(s, \psi_i, K/k).$$

Let H be any subgroup of G, and let ξ_j run through the simple characters of H. Each ξ_j induces a character ξ_j^* of G, given, let us say, by

$$\xi_j^*(\mu) = \sum_i r_{ji}\psi_i(\mu) \qquad \text{for all } \mu \in G, \tag{3.12}$$

in accordance with (3.11).

The restriction of ψ_i to H is, itself, a character of H and therefore has an expression of type (3.11) in terms of the ξ_j's. Moreover, by the theory of induced characters, this representation actually takes the form

$$\psi_i(\tau) = \sum_j r_{ji}\xi_j(\tau), \qquad \text{all } \tau \in H, \tag{3.13}$$

where the coefficients r_{ji} are the *same* as in (3.12).

Now take H to be a *cyclic* subgroup of G. Then if Ω is the subfield of K consisting of elements invariant under H, K/Ω is abelian and so, by (V),

$$L(s, \xi_j^*, K/k) = L(s, \xi_j, K/\Omega),$$

which is an abelian L-function. To prove our result, it therefore suffices to express each ψ_i as a linear combination of characters of type ξ_j^*.

Each element γ of G generates a cyclic subgroup H_γ of G and so, by (3.12), denoting the simple characters of H_γ by $\xi_{\gamma;j}$, we have

$$\xi_{\gamma;j}^*(\mu) = \sum_{i=1}^{g} r_{\gamma;ji}\psi_i(\mu), \qquad \text{all } \mu \in G. \tag{3.14}$$

The system of equations (3.14) is described by a matrix with each fixed pair (γ, j) determining a row and each i a column and we shall prove that its rank is equal to g.

Suppose on the contrary that the rank is less than g. Then the columns are linearly dependent, i.e. there exist integers $c_1, c_2, \ldots, c_g$ not all zero, such that

$$\sum_{i=1}^{g} c_i r_{\gamma;ji} = 0 \qquad \text{for all } \gamma \in G \text{ and all } j.$$

Thus, by (3.13), interpreted in relation to a particular H_γ,

$$\begin{aligned}\sum_i c_i\psi_i(\tau) &= \sum_i c_i \sum_j r_{\gamma;ji}\xi_{\gamma;j}(\tau), &&\text{all } \tau \in H_\gamma,\\ &= 0, &&\text{all } \tau \in H_\gamma.\end{aligned}$$

In particular, taking $\tau = \gamma$, we arrive at

$$\sum_i c_i\psi_i(\gamma) = 0 \qquad \text{for all } \gamma \in G, \tag{3.15}$$

contradicting the linear independence of the simple characters $\psi_1, \psi_2, \ldots, \psi_g$. Following on from (3.15), multiply this relation by $\bar{\psi}_k(\gamma)$ and sum over all the elements γ of G. From (3.1) it follows that

$$nc_k = 0 \qquad (k = 1, 2, \ldots, g) \tag{3.16}$$

whence $c_1 = c_2 = \ldots = c_g = 0$ (so that, incidentally, we have here another way of arriving at a contradiction). Either way, we have now that the system of equations (3.14) can be solved for ψ_i, giving

$$\psi_i = \sum_{\gamma \in G} \sum_j u_{\gamma;ji}\xi_{\gamma;j}^*$$

where the coefficients $u_{\gamma;ji} \in \mathbf{Q}$.

The argument centering on (3.16) can be carried out modulo any prime p which does not divide n, whence it follows that the coefficients $u_{\gamma;\,jl}$ have denominators composed of prime factors of n.

We have thus proved the following

THEOREM 7. *For any character χ of G, the Artin L-function $L(s, \chi, K/k)$ is given by*

$$L(s,\chi,K/k)=\prod_{l}\prod_{j}L(s,\xi_{lj},K/\Omega_l)^{n_{lj}}$$

where each $\operatorname{Gal}(K/\Omega_l)$ *is cyclic, the ξ_{lj}'s are (abelian) characters of* $\operatorname{Gal}(K/\Omega_l)$ *and each n_{lj} is rational with denominator composed of prime factors of n.*

It has been proved by Brauer (1947) that the exponents in the above theorem can be taken to be rational integers, and it follows, in particular, that the Artin L-functions are meromorphic. Furthermore, if χ is non-principal, then the ξ's in the above product representation of $L(s,\chi,K/k)$ can also be chosen non-principal. Hence Artin L-functions formed with non-principal characters are, in addition, regular and non-zero for $\sigma \geqslant 1$.

If, on the other hand, $\chi=\chi_0$ is the principal character, then $L(s,\chi_0,K/k)$ has a simple pole at $s=1$. Artin has conjectured that, apart from this simple pole, in the case $\chi=\chi_0$, the $L(s,\chi,K/k)$ are integral functions.

This conjecture would imply that $\zeta_K(s)/\zeta_k(s)$ is an integral function whenever $k \subset K$. If K/k is normal, this is true by (3.10), (V(ii)) and a fundamental theorem on group characters due to Brauer (see Lang, 1964 p. 139).

Artin's conjecture has been verified when G is one of the following special groups: S_3 and, more generally, any group of squarefree order; any group of prime power order; any group whose commutator subgroup is abelian (Speiser, 1927); S_4 (Artin, 1924). The conjecture has not been confirmed for $G=A_5$.

By Theorem 7, each Artin's L-function $L(s,\chi)$ is seen to satisfy a functional equation by virtue of the fact that each abelian L-function occuring on the right satisfies its own functional equation. This equation will be of the form

$$\Phi(s,\chi)=W(\chi)\Phi(1-s,\bar{\chi})$$

where W is a constant of absolute value 1, and Φ is of the form

$$\Phi(s,\chi)=A(\chi)^s\,\Gamma\left(\frac{s}{2}\right)^{a(\chi)}\Gamma\left(\frac{s+1}{2}\right)^{b(\chi)}L(s,\chi)$$

with a,b in $\mathbf{Q}$ and A a positive constant.

At this point we recall that, in the definition of Artin's L-function, we omitted to include a factor corresponding to the ramified primes. Now we introduce on the right-hand side of (3.17) the ramified local factors corresponding to the abelian L-functions (see beginning of Section 2), and take the new product as the definition of L. Then the above functional equation is satisfied automatically by the redefined L-function.

What is lacking in the theory of Artin L-functions is a non-local approach (provided, in the abelian case, by a series definition of L-functions). This lack causes the difficulties in extending these L-functions to the complex plane.

Example: The case $G = S_3$.

The elements of S_3 fall into three conjugacy classes

$$C_1\colon (1);\quad C_2\colon (1,2,3),(3,2,1);\quad C_3\colon (1,2),(2,3),(3,1).$$

Hence there are three simple characters. Let ψ_1 be the principal character and ψ_2 the other character determined by the subgroup $C_1 \cup C_2$. These are both of first degree and so, by (3.3), ψ_3 is of degree 2. Thus, by (3.2) with $\mu' = 1$, we have that

$$\psi_1(\mu)+\psi_2(\mu)+2\psi_3(\mu) = \begin{cases} 6, & \mu = 1, \\ 0, & \text{otherwise.} \end{cases}$$

Hence the table of simple characters of S_3 can be set out as follows:

	ψ_1	ψ_2	ψ_3
C_1	1	1	2
C_2	1	1	−1
C_3	1	−1	0.

Working from (3.6), we obtain the induced characters χ^* corresponding to the characters χ of

(i) $H = A_3$:

	χ_1^*	χ_2^*	χ_3^*
C_1	2	2	2
C_2	2	−1	−1
C_3	0	0	0;

(ii) $H = \{1,(1,2)\}$:

	χ_4^*	χ_5^*
C_1	3	3
C_2	0	0
C_3	1	−1;

(iii) $H = \{1\}$:

	χ_6^*
C_1	6
C_2	0
C_3	0.

From these tables we see that

$$\chi_1^* = \psi_1+\psi_2,\qquad \chi_2^* = \chi_3^* = \psi_3,$$
$$\chi_4^* = \psi_1+\psi_3,\qquad \chi_5^* = \psi_2+\psi_3,$$
$$\chi_6^* = \psi_1+\psi_2+2\psi_3.$$

S_3 is the Galois group of any non-abelian normal extension K of k of degree 6. Let Ω_1, Ω_2 be the two intermediate fields fixed under the subgroups $A_3, \{(1), (1, 2)\}$ respectively. Thus Ω_1 is a quadratic extension and Ω_2 a cubic extension of k, and K is therefore an abelian extension of each of them. Also, Ω_1 is an abelian extension of k. Thus, by (V(ii)), and (IV),

$$\begin{aligned} \zeta_K(s) &= L(s, \chi_0, K/K) = L(s, \psi_1+\psi_2+2\psi_3, K/k) \\ &= L_{\psi_1} L_{\psi_2} L^2_{\psi_3}, \\ \zeta_{\Omega_1}(s) &= L(s, \chi_0, K/\Omega_1) = L(s, \psi_1+\psi_2, K/k) \\ &= L_{\psi_1} L_{\psi_2}, \\ \zeta_{\Omega_2}(s) &= L(s, \chi_0, K/\Omega_2) = L(s, \psi_1+\psi_3, K/k) \\ &= L_{\psi_1} L_{\psi_3}, \end{aligned}$$

and

$$\zeta_k(s) = L(s, \chi_0, K/k) = L_{\psi_1}.$$

We remark that

$$L_{\psi_2} = L(s, \psi_2, K/k) = L(s, \chi_5, \Omega_1/k)$$

by (III) and that

$$L_{\psi_3} = L(s, \psi_3, K/k) = L(s, \chi_2, K/\Omega_1)$$

by (V). Hence L_{ψ_2} and L_{ψ_3} are integral functions.

Incidentally, this example verifies Artin's conjecture for $\mathrm{Gal}\,(K/k) = S_3$.

We conclude this brief survey of the non-abelian case by quoting *Čebotarev's density theorem:*

Let K/k be normal and let $G = \mathrm{Gal}\,(K/k)$. Let C denote a given conjugacy class in G. Then the class of all non-ramified primes $\mathfrak{p}$, each having the property that $\left[\dfrac{K/k}{\mathfrak{P}}\right] \in C$ for some† prime $\mathfrak{P}$ of K lying over $\mathfrak{p}$, has density equal to

$$\frac{\mathrm{card}\, C}{\mathrm{card}\, G}.$$

Using arguments of the kind described above (see Section 2), this result follows from the fact than an Artin L-function has no zeros on the line $\mathrm{Re}\, s = 1$.

Example. As an illustration of the theorem, let us consider the case of a non-normal cubic extension K_3/k. A (non-ramified) prime $\mathfrak{p}$ in k factorizes in K_3 in one of the following three ways:

(1) $\mathfrak{p} = \mathfrak{P}_1 \mathfrak{P}_2 \mathfrak{P}_3$

(2) $\mathfrak{p} = \mathfrak{P}_1 \mathfrak{P}_2$ where the $\deg \mathfrak{P}_1/\mathfrak{p} = 1$, $\deg \mathfrak{P}_2/\mathfrak{p} = 2$

(3) $\mathfrak{p} = \mathfrak{P}$.

† In fact, C consists of the Frobenius automorphisms $\left[\dfrac{K/k}{\mathfrak{P}}\right]$ corresponding to those $\mathfrak{P}_i$ lying over $\mathfrak{p}$.

We wish to determine the density of primes $\mathfrak{p}$ falling into each of these categories.

Let K_6 denote the minimal extension of K_3 that is *normal* over k. We study the factorization of $\mathfrak{p}$ in K_6, bearing in mind that in any decomposition of $\mathfrak{p}$ into primes of K_6, these all have equal order. Thus in (1) $\mathfrak{p}$ decomposes either in six primes (in K_6) of degree 1, or $\mathfrak{P}_1, \mathfrak{P}_2, \mathfrak{P}_3$ are themselves primes of K_6, each of degree 2. The latter possibility is ruled out because K_3/k is not normal.

[Suppose, on the contrary that, each $\mathfrak{P}_i$ remains a prime in K_6, and hence is of degree 2 in K_6. Write

$$\sigma_i = \left[\frac{K/k}{\mathfrak{P}_i}\right].$$

Then each σ_i is of order 2, and hence is a 2-cycle (note that Gal $(K_6/k) = S_3$). Further, the σ_i's span a conjugacy class, and hence are distinct transpositions.

Now K_3 is the set of elements of K_6 invariant under some 2-cycle, say σ_1. Thus

$$\sigma_1(\mathfrak{P}_i \cap K_3) = \mathfrak{P}_i \cap K_3.$$

But since there are three distinct prime ideals in K_3, and since each has a unique extension to K_6 (for each remains a prime in K_6),

$$\sigma_1 \mathfrak{P}_i = \mathfrak{P}_i \qquad (i = 1, 2, 3).$$

Let the subfields invariant under σ_2 and σ_3 be K_3' and K_3''. Then, in a similar way we have (since K_3' and K_3'' are the conjugates of K_3)

$$\sigma_j \mathfrak{P}_i = \mathfrak{P}_i \qquad (i, j, = 1, 2, 3).$$

But the transpositions generate S_3. Hence

$$S_3 \mathfrak{P}_i = \mathfrak{P}_i.$$

Thus there is no automorphism τ such that

$$\tau \mathfrak{P}_1 = \mathfrak{P}_2,$$

and this is impossible.]

In (2), we must have $\mathfrak{p} = \mathfrak{q}_1 \mathfrak{q}_2 \mathfrak{P}_1$ as the prime factorization over K_6, each factor being of degree 2, with $\mathfrak{q}_1 \mathfrak{q}_2 = \mathfrak{P}_2$. In (3), either $\mathfrak{p}$ remains prime, or $\mathfrak{p}$ decomposes into two primes of degree 3. The former would imply that $\left[\frac{K_6/k}{\mathfrak{p}}\right]$ generates S_3, which is impossible. Hence the latter decomposition must hold.

We now apply the fact (Chapter VII, §2) that the order of $\left[\frac{K_6/k}{\mathfrak{q}}\right]$ equals the degree of $\mathfrak{q}$. From this it follows for each prime $\mathfrak{q}$ of K_6 lying over $\mathfrak{p}$ that $\left[\frac{K_6/k}{\mathfrak{q}}\right]$ is an i-cycle in case (i) ($i = 1, 2, 3$). Hence, by Čebotarev's theorem, the required densities in cases (1), (2) and (3) are $\frac{1}{6}$, $\frac{1}{2}$ and $\frac{1}{3}$ respectively.

The following result is a direct consequence of Kummer's theorem (see Chapter III, Appendix, or Weiss (1963), 4.9 and in particular 4.9.2).

"Suppose $f \in k[x]$, is irreducible, and θ is a zero of f. Let $\mathfrak{p}$ denote a prime of k. Then, for almost all primes $\mathfrak{p}$ of k, $\mathfrak{p}$ splits in $k(\theta)$ as $f(x)$ splits over the residue class field of k modulo $\mathfrak{p}$."

Now let $n(\mathfrak{p},f)$ denote the number of solutions of the congruence

$$f(x) \equiv 0 \pmod{\mathfrak{p}};$$

then, for almost all $\mathfrak{p}$, $n(\mathfrak{p},f)$ may be regarded, alternatively, as the number of prime factors from $k(\theta)$ of $\mathfrak{p}$ having degree 1 relative to k. Applying the prime ideal theorem (Theorem 3) to primes of $k(\theta)$, we arrive therefore at the following

THEOREM 8. *Using the notation defined above,*

$$\sum_{N(\mathfrak{p}) \leqslant x} n(\mathfrak{p},f) \sim \frac{x}{\log x} \qquad \text{as } x \to \infty.$$

Corollary 1. More generally, if f is the product of l distinct irreducible factors, then

$$\sum_{N(\mathfrak{p}) \leqslant x} n(\mathfrak{p},f) \sim l\frac{x}{\log x} \qquad \text{as } x \to \infty.$$

Corollary 2. If, for almost all $\mathfrak{p}$, f splits completely mod $\mathfrak{p}$, then f factorizes completely over $k[x]$.

Proof. If n' denotes the degree of f, then we are given that $n(\mathfrak{p},f) = n'$ for almost all $\mathfrak{p}$. Hence, by the prime ideal theorem for k, $l = n'$.

It would seem plausible that if f has at least one linear factor mod $\mathfrak{p}$ for almost all $\mathfrak{p}$, then f has at least one linear factor in $k[x]$. However, the following counter-examples show this to be false.

(i) $$f(x) = (x^2-a)(x^2-b)(x^2-c)$$

where abc is a perfect square in $\mathbf{Z}$, with none of a, b, c a perfect square.

For, if a, b are quadratic non-residues mod p, then c is a quadratic residue mod p and hence f has two linear factors mod p.

(ii) $$f(x) = (x^2+3)(x^3+2).$$

For if $p \equiv 1 \pmod 3$, x^2+3 factorizes mod p, and if $p \equiv 2 \pmod 3$, x^3+2 factorizes mod p.

However, we do have the following,

THEOREM 9. *If f is a non-linear polynomial over k which has at least one linear factor modulo $\mathfrak{p}$ for almost all $\mathfrak{p}$, then f is reducible in $k[x]$.*

Proof. Assume that f is irreducible. Then it follows from Theorem 8 and the prime ideal theorem for k (Theorem 3) that

$$\sum_{N(\mathfrak{p})\leqslant x} (n(\mathfrak{p},f)-1) = o\left(\frac{x}{\log x}\right).$$

Hence, since $n(\mathfrak{p},f) \geqslant 1$ almost everywhere, $n(\mathfrak{p},f) = 1$ almost everywhere.

Let K be the root field of f over k, and $\mathfrak{q}$ a prime ideal in k which splits totally in K. Then $\mathfrak{q}$ splits totally in $k(\theta)$ and $n(\mathfrak{q},f)$ is equal to the degree of f, with a finite number of exceptions. These $\mathfrak{q}$ have positive density by the prime ideal theorem (Theorem 3) in K. Hence f is linear.

REFERENCES

Artin, E. (1923). Über eine neue Art von *L*-Reihen. *Abh. math. Semin. Univ. Hamburg*, **3**, 89–108. (Collected Papers (1965), 105–124. Addison-Wesley.)

Artin, E. (1930). Zur Theorie der *L*-Reihen mit allgemeinen Gruppencharakteren. *Abh. math. Semin. Univ. Hamburg*, **8**, 292–306. (Collected Papers (1965), 165–179. Addison-Wesley.)

Brauer, R. (1947). On Artin's *L*-series with general group characters. *Ann. Math.* (2) **48**, 502–514.

Chevalley, C. (1940). La théorie du corps de classes. *Ann. Math.* (2) **41**, 394–418; and other papers referred to in the above.

Dedekind, R. Vorlesungen über Zahlentheorie von P. G. Lejeune Dirichlet. Braunschweig, 1871–1894. Supplement 11.

Estermann, T. (1952). Introduction to modern prime number theory. *Camb. Tracts in Math.* **41.**

Hall, M., Jr. (1959). "The Theory of Groups". Macmillan, New York.

Hasse, H. Bericht über neuere Untersuchungen und Probleme aus der Theorie der algebraischen Zahlkörper. *Jber. dt. Mat Verein.*, **35** (1926), **36** (1927) and **39** (1930).

Hecke, E. (1920). Eine neue Art von Zetafunktionen und ihre Beziehungen zur Verteilung der Primzahlen (Zweite Mitteilung). *Math. Z.* **6**, 11–51 (Mathematische Werke (1959), 249–289. Vandenhoeck und Ruprecht).

Hecke, E. (1954). Vorlesungen über die Theorie der algebraischen Zahlen, 2. Unveränderte Aufl., Leipzig.

Landau, E. (1907). Über die Verteilung der Primideale in den Idealklassen eines algebraischen Zahlkörpers. *Math. Annln*, **63**, 145–204.

Landau, E. (1927). Einführung in die elementare und analytische Theorie der algebraischen Zahlen und der Ideale, 2. Aufl. Leipzig.

Lang, S. (1964). "Algebraic Numbers". Addison-Wesley.

Speiser, A. (1927). "Die Theorie der Gruppen von endlicher Ordnung", 2. Aufl. Berlin.

Titchmarsh, E. C. (1939). "The Theory of Functions", 2nd edition. Oxford University Press, London.

Titchmarsh, E. C. (1951). "The Theory of the Riemann Zeta-function". Oxford University Press, London.

Weber, H. (1896). Über einen in der Zahlentheorie angewandten Satz der Integralrechnung. *Göttinger Nachrichten*, 275–281.

Weiss, E. (1963). "Algebraic Number Theory". McGraw-Hill, New York.

38.

Burnside Metabelian Groups

(With S. Bachmuth and H. Y. Mochizuki)

Proceedings of the Royal Society, London, A 307 (1968), 235–250
[Commentary on p. 600]

If B is a group of prime-power exponent p^e and solubility class 2, then B has nilpotency class at most $e(p^e - p^{e-1}) + 1$ provided the number of generators of B are at most $p+1$. Representations of B are constructed which in the case of two generators and prime exponent is a faithful representation of the free group of the variety under study and for prime-power exponent show the existence of a group with nilpotency class $e(p^e - p^{e-1})$. In the general situation where B as above has exponent n, and n is not a prime-power, the place where the lower central series of G becomes stationary is determined by a knowledge of the nilpotency class of the groups of prime-power exponent for all prime divisors of n. The bound $e(p^e - p^{e-1}) + 1$ on the nilpotency class is a consequence of the following: Let G be a direct product of at most $p-1$ cyclic groups of order p^e and R the group ring of G over the integers modulo p^e. Then the $e(p^e - p^{e-1})$th power of the augmentation ideal of R is contained in the ideal of R generated by all 'cyclotomic' polynomials $\sum_{i=0}^{p^e-1} g^i$ for all g in G. If G is a direct product of more than $p+1$ cyclic groups, then this result is no longer true unless $e = 1$.

1. Introduction

Let F be a free group, F' its commutator subgroup, F'' the second commutator subgroup, and F^n the subgroup of F generated by all nth powers of elements of F. In this paper we study the Burnside metabelian group $F/F''F^n$. We prove conjecture 1 of Bachmuth & Mochizuki (1966) and also related material not mentioned in that paper. For convenience we now formulate most of these results.

If H is a group, then H_m will denote the mth term of the lower central series of H, beginning with $H_1 = H$. H is of nilpotency class m if $H_{m+1} = 1$ and $H_m \neq 1$. We also say that H has solubility class m if $H^{(m)} = 1$ and $H^{(m-1)} \neq 1$ in the derived series of H.

Theorem A. *Let p be a prime and let F be free of rank ρ. (1) $F/F''F^p$ has nilpotency class $(p-1)$ if $\rho = 2$ and nilpotency class p if $2 < \rho$. (2) $F/F''F^{p^e}$ has nilpotency class $e(p^e - p^{e-1})$ or $e(p^e - p^{e-1}) + 1$ for any integers $e \geq 1$ and $2 \leq \rho \leq p+1$.*

Part (1) of theorem A is due to Meier-Wunderlei (1951) and we set it apart from (2) not merely because it gives the exact result, but also because this extra piece of knowledge enables us to exhibit a faithful matrix representation of $F/F''F^p$ (see § 5). This representation is the analogue of Magnus's representation of F/F'' and $F/F''(F')^n$ (see Magnus 1939; Bachmuth & Hughes 1966) and is a derivative of these. The existence of such a representation leads to the possibility of some interesting applications.

Theorem A, as will be made clear, is an immediate consequence of a ring theoretic result which is of interest in its own right. In what follows, let n be an arbitrary but fixed integer. We put $G = F/F'F^n$ and denote the integral group ring of G by $Z(G)$.

The augmentation ideal Σ of $Z(G)$ (also commonly referred to as the fundamental or Magnus ideal in the literature) is the ideal generated by all elements of the form $1-g$ where g runs through all elements of the group G. The cyclotomic ideal $\mathfrak{J}(n)$ is the ideal generated by the elements $C(u,n) = 1+u+u^2+\ldots+u^{n-1}$, where u runs through all elements of G (including $u=1$, and hence the ideal (n) of Z is contained in $\mathfrak{J}(n)$).

THEOREM B. *$\Sigma^{e(p^e-p^{e-1})} \subseteq \mathfrak{J}(p^e)$, but $\Sigma^{e(p^e-p^{e-1})-1} \nsubseteq \mathfrak{J}(p^e)$ in $Z_{p^e}(F/F'F^{p^e})$ for any prime power p^e.*

Remark. If $e=1$, then we obviously have equality $\Sigma^{p-1} = \mathfrak{J}(p)$, since containment in the other direction is clear.

In the case where n is not a prime power, a result such as theorem B cannot hold since (to give a group theoretical reason) $F/F''F^n$ is not nilpotent. In this situation theorem A becomes the following

THEOREM C. *Suppose $n = p_1^{e_1}p_2^{e_2}\ldots p_k^{e_k}$ is the factorization of n into prime powers. Then the lower central series of $H = F/F''F^n$ is stationary at*

$$d = 2+\max\left[e_i(p_i^{e_i}-p_i^{e_i-1})\right],$$

i.e. $H_d = H_{d+1}$. (It is possible that $H_{d-1} = H_d$, but $H_{d-2} \neq H_{d-1}$.)

In order to deduce results of the nature of theorem A for groups more general than metabelian groups, one requires theorems like theorem B for group rings of non-abelian groups. However, there is one special case covered by theorem A, namely the investigation of Engel identities, for which one is always in the abelian situation. Therefore theorem A may be iterated for arbitrary soluble groups. We recall that Engel identities are commutator identities involving only two elements of the group. If H is a group and a, b are any elements of H, we define

$$[a,b;1] = [a,b] = aba^{-1}b^{-1},$$

and inductively

$$[a,b;n+1] = [[a,b;n],b]$$

for $n = 1,2,\ldots$. We say H has Engel length at most n if $[a,b;n] = 1$ for all a,b in H. Because of the general interest in Engel identities, we single out the following theorem, although it is a special case of theorems A and B. The following is essentially theorem 3·3 of Glauberman, Krause & Struik (1966).

THEOREM D. (1) *Let G be a cyclic group of order p^e. Then $(1-x)^{e(p^e-p^{e-1})}$ is in $\mathfrak{J}(p^e)$ for any x in G. Hence we have* (2) *Let A be an abelian normal subgroup of a group H of prime power exponent p^e. Then $[a,x;e(p^e-p^{e-1})] = 1$. In fact*

$$[a,x^{p^{e-k}};k(p^k-p^{k-1})]^{p^{e-k}} = 1 \quad \textit{and also} \quad [a,x;(e-k)(p^e-p^{e-1})]^{p^k} = 1,$$

for any a in A and x in H and $k = 0,1,2,\ldots,e$. (3) *A group of solubility class k and exponent p^e has Engel length at most $(k-1)e(p^e-p^{e-1})+1$.*

Conspicuous for its absence in theorem D are lower bounds for Engel lengths in groups of prime power exponent. Of course the proof of theorem A will establish the existence of a group H of exponent p^e and an abelian normal subgroup A of H such that

$$[a,x^{p^{e-k}};k(p^k-p^{k-1})]^{p^{e-k-2}} \neq 1 \quad \text{and also} \quad [a,x;(e-k)(p^e-p^{e-1})]^{p^k-2} \neq 1,$$

but the bound $(k-1)e(p^e-p^{e-1})+1$ for soluble groups of class k appears to be rather large. The only general result known to us is a theorem of Kostrikin (1960) which states that there exists a group $G(p)$ of prime exponent p with Engel length at least $\frac{1}{2}(3p-5)$. We know that $G(p)$ can be taken to be soluble of class at most $p-1$ and still retain the bound $\frac{1}{2}(3p-5)$. But there remains a large gap between $\frac{1}{2}(3p-5)$ and $(p-2)(p-1)$.

We give an independent proof of theorem D in § 3. In § 4 we will prove theorem B, in § 5 we will prove theorem A, and theorem C is proved in § 6.

2. Motivation and preliminary remarks

For a ring R and a group G we denote the group ring of G over R by $R(G)$. Z refers to the integers and Z_n to the integers modulo n.

Suppose H is group of exponent n and A is an abelian normal subgroup of H. Then, in a natural manner, we may view A as an $Z(H/A)$-module by putting $a^x = xax^{-1}$, $a^{-x} = xa^{-1}x^{-1}$ and $a^{R+S} = a^R a^S$ for a in A, x in H/A and R, S in $Z(H/A)$. We note that $a^{(1-x)} = [a, x]$ and in general $a^{(1-x_1)\dots(1-x_m)} = [a, x_1, \dots, x_m]$. Since A has exponent n, we may also consider A as an $Z_n(H/A)$-module. (Beginning with § 4, we shall specialize by taking $H = F/F''F^n$ and $A = H'$. This will mean that H/A will be abelian.) At the present state of our knowledge, this module is very little understood, but a useful observation one can make is that for any a in A and x in H/A, we have $a^{1+x+x^2+\dots+x^{n-1}} = (ax)^n$. Thus one may as well factor $Z_n(H/A)$ by the cyclotomic ideal $\mathfrak{J}(n)$ and instead of considering the action of $Z_n(H/A)$ on A, we consider $Z_n(H/A)/\mathfrak{J}(n)$ as acting on A. In §§ 4 and 5, we will see that this viewpoint is very 'close' to the true picture at least in the case where H/A is abelian.

As a final remark in this section, we note that n is in the ideal $\mathfrak{J}(n)$ of $Z(G)$ (where $G = H/A$ is of exponent n) and that the image of the ideal $\mathfrak{J}(n)$ modulo n is the ideal $\mathfrak{J}(n)$ of $Z_n(G)$. We shall state and prove theorems in the context of $Z(G)$ or $Z_n(G)$, whichever is more convenient.

3. Engel identities in groups of prime power exponent

In this section G is a cyclic group of order p^e and $\mathfrak{J}(p^e)$ is the ideal of $Z_{p^e}(G)$ or $Z(G)$ generated by all polynomials

$$C(g, p^e) = 1 + g + \dots + g^{p^e-1}$$

for g in G.

Lemma. *$(1-x)^{e(p^e-p^{e-1})}$ is in $\mathfrak{J}(p^e)$ for any x in G.*

Proof. We compute directly and use the following evident facts:

$$(1) \quad (1-x)^{p^{e-1}} = (1-x^{p^{e-1}}) + pQ,$$

where Q is a polynomial in x.

$$(2) \quad (1-y)^{p-1} = 1 + y + y^2 + \dots + y^{p-1} + pQ',$$

15-2

where Q' is a polynomial in y and y is any element of G. We have therefore

$$\begin{aligned}(1-x)^{e(p^e-p^{e-1})} &= [(1-x^{p^{e-1}})+pQ]^{e(p-1)} \quad \text{by (1)}\\ &= [(1-x^{p^{e-1}})^{p-1}+pR]^e, \quad \text{where } R \text{ is a polynomial in } x\\ &= [(1+x^{p^{e-1}}+x^{2p^{e-1}}+\dots+x^{(p-1)p^{e-1}})+pS]^e,\\ &\qquad \text{by (2), where } S \text{ is a polynomial in } x.\end{aligned}$$

This latter expression is in $\mathfrak{J}(p^e)$ since each term in the expansion is of the form

$$T = (1+x^{p^{e-1}}+\dots+x^{(p-1)p^{e-1}})^i p^j S^j \quad \text{where} \quad i+j=e.$$

But by an easy induction argument, using the fact that

$$(x^{p^{e-1}})^k(1+x^{p^{e-1}}+\dots+x^{(p-1)p^{e-1}}) = 1+x^{p^{e-1}}+\dots+x^{(p-1)p^{e-1}}$$

we obtain

$$(1+x^{p^{e-1}}+\dots+x^{(p-1)p^{e-1}})^i = p^{i-1}(1+x^{p^{e-1}}+\dots+x^{(p-1)p^{e-1}}).$$

Hence
$$\begin{aligned}T &= S^j p^{e-1}(1+x^{p^{e-1}}+\dots+x^{(p-1)p^{e-1}})\\ &= S^j\sum_{k=0}^{p^e-1}(x^{p^{e-1}})^k, \quad \text{which is in } \mathfrak{J}(p^e).\end{aligned}$$

COROLLARY. (1) $p^k(1-x^{p^k})^{(e-k)(p^{e-k}-p^{e-k-1})}$ *is in* $\mathfrak{J}(p^e)$ *for* $k=0,1,\dots,e$;

(2) $p^k(1-x)^{(e-k)(p^e-p^{e-1})}$ *is in* $\mathfrak{J}(p^e)$ *for* $k=0,1,\dots,e$.

Proof. (1) is equivalent to the lemma for groups of exponent p^{e-k} and (2) follows from the same argument as the lemma. That is

$$p^k(1-x)^{(e-k)(p^e-p^{e-1})} = p^k[1+x^{p^{e-1}}+x^{2p^{e-1}}+\dots+x^{(p-1)p^{e-1}}+pS]^{e-k},$$

where S is a polynomial in x, and now we finish up as in the lemma.

As a corollary to the lemma and corollary, we therefore have

THEOREM D. (1) *Let A be an abelian normal subgroup of a group H of prime power exponent p^e. If $a \in A$, then*

(i) $[a,x;(e-k)(p^e-p^{e-1})]^{p^k} = 1$ *for* $k=0,1,2,\dots,e$;

(ii) $[a,x^{p^{e-k}};k(p^k-p^{k-1})]^{p^{e-k}} = 1$ *for* $k=0,1,2,\dots,e$.

(2) *A group of solubility class k and exponent p^e has Engel length at most*

$$(k-1)e(p^e-p^{e-1})+1.$$

Proof. $[a,x;(e-k)(p^e-p^{e-1})]^{p^k} = a^{p^k(1-x)^{(e-k)(p^e-p^{e-1})}}$, and $p^k(1-x)^{(e-k)(p^e-p^{e-1})}$ is in $\mathfrak{J}(p^e)$ by the previous corollary. But for any a in A,

$$a^P = 1 \quad \text{for} \quad P = (1+g+\dots+g^{p^e-1})P',$$

where $g \in H/A$ and $P' \in Z(H/A)$, and thus (1) (i) follows. (1) (ii) has a similar proof.

(2) follows from (1) by going down the derived series of the soluble group G, i.e. if a is in $G^{(i)}$, then $[a,x;e(p^e-p^{e-1})]$ is in $G^{(i+1)}$ for any x in G.

4. Cyclotomic ideals in abelian group rings and Burnside metabelian groups

In this section we prove theorem B and thereby achieve upper bounds on the nilpotency class of ρ-generator metabelian groups of exponent p^e, $2 \leqslant \rho \leqslant p+1$. Throughout this section, G is a direct product of ρ cyclic groups of order p^e and we put $\phi(p^i) = p^i - p^{i-1}$, $i \geqslant 1$.

Lemma 1. *If a is a positive integer prime to p, then for any x in G, $1 - x^a = (1-x)u$, where u is a unit in $Z_{p^e}(G)$.*

Proof. $1 - x^a = (1-x)u$, where $u = (1+x+x^2+\ldots+x^{a-1})$. Adding and subtracting a to u, we have

$$u = a + (1-1) + (x-1) + (x^2-1) + \ldots + (x^{a-1}-1)$$
$$= a + R,$$

where R is in Σ and hence is a nilpotent element. Since $(a, p^e) = 1$, a is a unit in $Z_{p^e}(G)$, and hence $a+R$, being the sum of a unit plus a nilpotent element, is a unit.

Corollary. *If $(a,p) = 1$, $(1 - x^{p^l a}) = (1 - x^{p^l})u$, where u is a unit.*

Lemma 2. $(1-x)^{\gamma_0}(1-y)^{\phi(p^e)-\gamma_0}(1-x^p)^{\gamma_1}(1-y^p)^{\phi(p^{e-1})-\gamma_1}$

$$\ldots(1-x^{p^{e-1}})^{\gamma_{e-1}}(1-y^{p^{e-1}})^{\phi(p)-\gamma_{e-1}}$$

is in $\mathfrak{J}(p^e)$ for any elements x, y in G and any integers γ_i $(i = 0, 1, \ldots, e-1)$ satisfying $0 \leqslant \gamma_i \leqslant \phi(p^{e-i})$.

Proof. Let a_i be the canonical representations of the residue classes of the integers modulo p^e which are prime to p, i.e.

$$1 = a_1 < a_2 < \ldots < a_{\phi(p^e)} < p^e \quad \text{and} \quad (a_i, p) = 1.$$

We remark that for all integers n satisfying $1 \leqslant n < p^e$, x^n is in the set

$$\bigcup_{j=0}^{e-1} \{x^{p^j a_i}\colon 1 \leqslant i \leqslant \gamma_j\},$$

or y^n is in the set

$$\bigcup_{j=0}^{e-1} \{y^{p^j a_i}\colon \gamma_j < i \leqslant \phi(p^{e-j})\}.$$

For arbitrary $\gamma_0, \gamma_1, \ldots, \gamma_{e-1}$ as specified in the statement of lemma 2, form

$$F_n(x,y) = \prod_{1 \leqslant i \leqslant \gamma_0} (x^n - x^{a_i}) \prod_{\gamma_0 < i \leqslant \phi(p^e)} (y^n - y^{a_i}) \prod_{1 \leqslant i \leqslant \gamma_1} (x^n - x^{pa_i})$$
$$\times \prod_{\gamma_1 < i \leqslant \phi(p^{e-1})} (y^n - y^{pa_i}) \ldots \prod_{1 \leqslant i \leqslant \gamma_{e-1}} (x^n - x^{p^{e-1}a_i}) \prod_{\gamma_{e-1} < i \leqslant \phi(p)} (y^n - y^{p^{e-1}a_i}).$$

Let
$$F(x,y) = \sum_{n=0}^{p^e-1} F_n(x,y).$$

If we multiply out the products in each $F_n(x,y)$ we get a sum of elements of the form $x^{kn+h}y^{k'n+h'}$ and thus $F(x,y)$ is a sum of cyclotomic polynomials

$$x^h y^{h'} \sum_{n=0}^{p^e-1} x^{kn} y^{k'n}$$

and therefore $F(x, y)$ is in $\mathfrak{J}(p^e)$. But for $n \neq 0$, it is clear, because of the above remark, that $F_n(x, y)$ vanishes identically. Hence

$$F_0(x, y) = \prod_{k=0}^{e-1} \{ \prod_{1 \leqslant i \leqslant \gamma_k} (1 - x^{p^k a_i}) \prod_{\gamma_k \leqslant i \leqslant \phi(p^{e-k})} (1 - y^{p^k a_i})\}$$

is in $\mathfrak{J}(p^e)$. Now apply lemma 1, and lemma 2 follows.

COROLLARY. *Let x, y be elements of G. Then*

$$p^j Q = p^j \prod_{i=j}^{e-1} (1 - x^{p^i})^{\gamma_i} (1 - y^{p^i})^{\phi(p^{e-i}) - \gamma_i}$$

is in $\mathfrak{J}(p^e)$ for any integers γ_i satisfying $0 \leqslant \gamma_i \leqslant \phi(p^{e-i})$.

Proof. p^j in front of the expression allows us to work modulo p^{e-j}. The elements $x^{p^i}, y^{p^i}, j \leqslant i \leqslant e-1$, all lie in a subgroup H of exponent p^{e-j}. Hence lemma 2 applied to H tells us that Q is in $\mathfrak{J}(p^{e-j}) \subseteq Z(H)$. But, $p^j \mathfrak{J}(p^{e-j}) \subseteq \mathfrak{J}(p^e)$ since

$$p^j(1 + g + g^2 + \ldots + g^{p^{e-j}}) = 1 + g + g^2 + \ldots + g^{p^{e-1}}$$

for any g in H. This completes the proof.

Lemma 2 and the above corollary remain valid for any finite number of variables by exactly the same arguments. For convenience we now restate lemma 2 in the form it will be used.

LEMMA 2′. *Let $x_1, \ldots, x_k$ be elements in any finitely generated abelian group of exponent p^e. For each i in the range $0 \leqslant i \leqslant e-1$, let*

$$\sum_{\lambda=1}^{k} p^i a_{\lambda, i} = \phi(p^{e-j}).$$

Then
$$p^j \prod_{i=j}^{e-1} \prod_{\lambda=1}^{k} (1 - x_{\lambda, i}^{p^i})^{a_{\lambda, i}}$$

is in $\mathfrak{J}(p^e)$.

We next observe the following corollary of lemma 2′.

COROLLARY. *In the case $e = 1$, $\Sigma^{p-1} = \mathfrak{J}(p)$ in the group ring of any finitely generated abelian group of exponent p over Z_p.*

Proof. Lemma 2 shows that $\Sigma^{p-1} \subseteq \mathfrak{J}(p)$. But clearly $\Sigma^{p-1} \supseteq \mathfrak{J}(p)$ since $(1-x)^{p-1} \equiv 1 + x + x^2 + \ldots + x^{p-1} \pmod{p}$.

This corollary is theorem B for the case $e = 1$. We may therefore restrict our attention to the cases $e \geqslant 2$ in the proof of theorem B. For these cases we must require that our group have $\rho \leqslant p+1$ generators since unlike the case $e = 1$, we shall show that theorem B is false if $e \geqslant 2$ and $\rho > p+1$.

THEOREM B. *$\Sigma^{e\phi(p^e)} \subseteq \mathfrak{J}(p^e)$ in $Z_{p^e}(G)$, G an abelian group of exponent p^e, generated by $\rho \leqslant p+1$ elements.*

We introduce the following notation to be used in the remainder of this section.

G, $e > 1$, p as before, and we fix $\rho \leqslant p+1$. $x_1, \ldots, x_\rho$ are elements of G. We consider matrices of e columns and ρ rows

$$A = \begin{pmatrix} a_{1,0} & pa_{1,1} & \cdots & p^{e-1}a_{1,e-1} \\ \vdots & \vdots & & \vdots \\ a_{\rho,0} & pa_{\rho,1} & \cdots & p^{e-1}a_{\rho,e-1} \end{pmatrix},$$

where the $a_{\lambda,i} \geqslant 0$ in Z, and assign to them the polynomials

$$f_A = \prod_{\lambda=1}^{\rho} \prod_{i=0}^{e-1} (x_\lambda^{p^i} - 1)^{a_{\lambda,i}}.$$

We denote by H_λ the sum

$$H_\lambda(A) = H_\lambda = \sum_{i=0}^{e-1} p^i a_{\lambda,i} \quad \text{for} \quad 1 \leqslant \lambda \leqslant \rho,$$

and finally
$$T = T(A) = \sum_{\lambda=1}^{\rho} H_\lambda.$$

We prove theorem B by establishing the following result (theorem B is the case $j = 0$).

Proposition. Let $e \geqslant 2$ and $0 \leqslant j \leqslant e-1$. If

$$T(A) \geqslant e(p^e - p^{e-1}) - j(p^{e-1} - p^{e-2}), \quad \text{then} \quad p^j f_A \in \mathfrak{J}(p^e).$$

The proof of this proposition is broken up into the following seven lemmas.

LEMMA 3. *Let $H_\lambda(A) \geqslant H_\lambda(B)$ for λ satisfying $1 \leqslant \lambda \leqslant \rho$. Assume that for all D with $T(D) \geqslant T(B) - \phi(p^{e-1})$, $p^{j+1} f_D$ is in $\mathfrak{J}(p^e)$. Then $p^j f_B \in \mathfrak{J}(p^e)$ implies $p^j f_A \in \mathfrak{J}(p^e)$.*

Proof. We shall show that modulo $\mathfrak{J}(p^e)$, $p^j f_A$ is equal to $p^j f_C$, where $0 \leqslant b_{\lambda,i} \leqslant c_{\lambda,i}$, $0 \leqslant i \leqslant e-1$, for all λ satisfying $1 \leqslant \lambda \leqslant \rho$. From this it follows that if $p^j f_B \in \mathfrak{J}(p^e)$, then $p^j f_C \in \mathfrak{J}(p^e)$ since $p^j f_B$ is a factor of $p^j f_C$, whence $p^j f_A$ is in $\mathfrak{J}(p^e)$.

We first prove that for $0 \leqslant i \leqslant (e-2)$, any factor $(1 - x_\mu^{p^i})^p$ can be replaced modulo $\mathfrak{J}(p^e)$ by $(1 - x_\mu^{p^{i+1}})$ in any $p^j f_A$ such that $H_\lambda(A) \geqslant H_\lambda(B)$ for $1 \leqslant \lambda \leqslant \rho$.

$$(1 - x_\mu^{p^i})^p = (1 - x_\mu^{p^{i+1}}) + \sum_{l=1}^{p-1} \binom{p}{l} (-1)^l x_\mu^{p^i l}$$

$$= (1 - x_\mu^{p^{i+1}}) + \sum_{l=1}^{p-1} \binom{p}{l} (-1)^l (1 - x_\mu^{p^i l}).$$

Since $(1 - x_\mu^{p^i l}) = u(1 - x_\mu^{p^i})$, where u is a unit, the error term (after discarding units) has the form $p^{j+1} f_E$ where

$$f_E = (1 - x_\mu^{p^i}) \prod_{l \neq i} (1 - x_\mu^{p^l})^{a_{\mu,l}} \left[\prod_{\lambda \neq \mu} \prod_{l=0}^{e-1} (1 - x_\lambda^{p^l})^{a_{\lambda,l}} \right],$$

$$T(E) = T(A) - (p^{i+1} - p^i) \geqslant T(A) - \phi(p^e) \geqslant T(B) - \phi(p^e).$$

Thus by hypothesis, $p^{j+1} f_E \in \mathfrak{J}(p^e)$. This means the error term is in $\mathfrak{J}(p^e)$. In a similar manner, we can replace $(1 - x_\mu^{p^{i+1}})$ by $(1 - x_\mu^{p^i})^p$ modulo $\mathfrak{J}(p^e)$ in $p^j f_A$. We note that the matrix A' obtained by the above replacement satisfies $H_\lambda(A') \geqslant H_\lambda(B)$ for $1 \leqslant \lambda \leqslant \rho$ and $p^j f_{A'} \equiv p^j f_A \pmod{\mathfrak{J}(p^e)}$. It is therefore clear that by successive replacement, we can obtain $p^j f_C$ such that $0 \leqslant b_{\lambda,i} \leqslant c_{\lambda,i}$, $0 \leqslant i \leqslant (e-1)$, $1 \leqslant \lambda \leqslant \rho$. This completes the proof of Lemma 3.

LEMMA 4. *Let $\psi(\eta) = (e - \eta)(p^e - p^{e-1}) + (\rho - 1)(p^\eta - 1)$ for η a real number, $1 \leqslant \rho \leqslant (p+1)$. For $0 \leqslant j \leqslant (e-1)$, $\psi(\eta) \leqslant \psi(j)$ if $j \leqslant \eta \leqslant (e-1)$ and η is an integer.*

Proof. We may assume $0 \leqslant j < (e-1)$. $d^2\psi/d\eta^2 = (\rho-1)p^\eta(\log p)^2 > 0$. Thus ψ is a convex function. Since

$$\psi(e-2)-\psi(e-1) = (p^e - p^{e-1}) - (\rho-1)(p^{e-1}-p^{e-2})$$
$$= (p-\rho+1)(p^{e-1}-p^{e-2}) \geqslant 0,$$

the function is increasing for decreasing η, $\eta \leqslant (e-2)$, and our result follows.

LEMMA 5. *Let*

$$T = \sum_{\lambda=1}^{\rho} H_\lambda \geqslant (e-j)(p^e-p^{e-1}) + (\rho-1)(p^j-1),$$

where $1 \leqslant \rho \leqslant p+1$. *Then there exists a matrix* B *such that* $H_\lambda \geqslant H_\lambda(B)$ *and* $p^j f_B \in \mathfrak{J}(p^e)$.

Proof. It follows from lemma 4 that for $j \leqslant i \leqslant (e-1)$,

$$T \geqslant (e-i)(p^e-p^{e-1}) + (\rho-1)(p^i-1).$$

If W is a real number, let $[W]$ be the greatest integer $\leqslant W$. Then

$$[H_\lambda p^{-i}] \geqslant H_\lambda p^{-i} - (p^i-1)p^{-i},$$

$$\sum_{\lambda=1}^{k} p^i[H_\lambda p^{-i}] \geqslant (e-i)(p^e-p^{e-1}) + (\rho-1)(p^i-1) - \rho(p^i-1)$$
$$\geqslant (e-i)(p^e-p^{e-1}) - (p^i-1).$$

Thus,
$$\sum_{\lambda=1}^{k} p^i[H_\lambda p^{-i}] - (e-i)(p^e-p^{e-1}) \geqslant -(p^i-1) > -p^i.$$

But the left side is a multiple of p^i, implying that

$$\sum_{\lambda=1}^{k} p^i[H_\lambda p^{-i}] - (e-i)(p^e-p^{e-1}) \geqslant 0. \tag{1}$$

We will now successively define $b_{\lambda,e-1}, b_{\lambda,e-2}, \ldots, b_{\lambda,j}$ for $1 \leqslant \lambda \leqslant \rho$ such that $b_{\lambda,i} \geqslant 0$,

$$\sum_{\lambda=1}^{\rho} p^i b_{\lambda,i} = (p-1)p^{e-1} \quad \text{for} \quad i = (e-1), (e-2), \ldots, j,$$

$$H_\lambda \geqslant \sum_{i=j}^{e-1} p^i b_{\lambda,i}.$$

For $1 \leqslant \lambda \leqslant \rho$, we choose $b_{\lambda,e-1}$ such that $[H_\lambda p^{-e+1}] \geqslant b_{\lambda,e-1}$ and

$$\sum_{\lambda=1}^{k} p^{e-1} b_{\lambda,e-1} = (p-1)p^{e-1}.$$

That this may be done is a consequence of (1) with $i = e-1$. For $1 \leqslant \lambda \leqslant \rho$, we choose $b_{\lambda,e-2}$ such that

$$[H_\lambda p^{-e+2}] - p b_{\lambda,e-1} \geqslant b_{\lambda,e-2}$$

and
$$\sum_{\lambda=1}^{\rho} p^{e-2} b_{\lambda,e-2} = (p-1)p^{e-1}.$$

From (1) and the construction of the $b_{\lambda, e-1}$,

$$\sum_{\lambda=1}^{\rho} p^{e-2}[H_\lambda p^{-e+2}] - \sum_{\lambda=1}^{\rho} p^{e-1} b_{\lambda, e-1}$$
$$\geqslant 2(p^e - p^{e-1}) - (p^e - p^{e-1})$$
$$= (p^e - p^{e-1}).$$

Thus the construction of the $b_{\lambda, e-2}$ is possible.

By an induction argument for $j \leqslant i \leqslant (e-1)$, we choose $b_{\lambda, i}$, $1 \leqslant \lambda \leqslant \rho$, such that

$$[H_\lambda p^{-i}] - p^{e-1-i} b_{\lambda, e-1} - p^{e-2-i} b_{\lambda, e-2} - \dots - p b_{\lambda, i+1} \geqslant b_{\lambda, i}$$

and
$$\sum_{\lambda=1}^{\rho} p^i b_{\lambda, i} = (p-1) p^{e-1}.$$

We have, with the understanding that $b_{\lambda, i} = 0$ if $0 \leqslant i < j$,

$$H_\lambda \geqslant [H_\lambda p^{-j}] p^j$$
$$\geqslant p^j (p^{e-1-j} b_{\lambda, e-1} + p^{e-2-j} b_{\lambda, e-2} + \dots + b_{\lambda, j})$$
$$= \sum_{i=j}^{e-1} p^i b_{\lambda, i}$$
$$= H_\lambda(B).$$

Finally,
$$p^j f_B = p^j \prod_{i=j}^{e-1} \left[\prod_{\lambda=1}^{\rho} (1 - x_\lambda^{p^i})^{b_{\lambda, i}} \right],$$

and the $b_{\lambda, i}$ satisfy the conditions of lemma 2′. Thus $p^j f_B$ is in $\mathfrak{J}(p^e)$ and the proof of lemma 5 is complete.

Remark. By throwing in more factors onto f_B, it is clear that we can choose B such that $H_\lambda = H_\lambda(B)$ for each λ.

LEMMA 6. *Let* $e = 2$. *If* $T_A \geqslant 2p^2 - 3p + 1$, *then* $pf_A \in \mathfrak{J}(p^2)$.

Proof. Set $T_A^* = \sum_{\lambda=1}^{\rho} p[H_\lambda p^{-1}]$.

$$T_A^* \geqslant T_A - \rho(p-1)$$
$$\geqslant 2p^2 - 3p + 1 - (p+1)(p-1)$$
$$\geqslant (p^2 - 3p + 2)$$
$$> p(p-3).$$

Since $T_A^* \equiv O \pmod{p}$, we see that $T_A^* \geqslant p(p-2)$.

If $T_A^* \geqslant p(p-1)$, then T_A^* satisfies condition (1) in the proof of lemma 5, where $i = 1$, $e = 2$. But it is this condition which enables us to construct a matrix B such that $H_\lambda(A) \geqslant H_\lambda(B)$, $1 \leqslant \lambda \leqslant \rho$, and $pf_B \in \mathfrak{J}(p^2)$. We now apply lemma 3 to conclude $pf_A \in \mathfrak{J}(p^2)$.

Hence we may assume $T_A^* = p(p-2)$. For at least one λ, $H_\lambda(A) \equiv (p-1) \pmod{p}$. Otherwise,

$$T_A \leqslant T_A^* + \rho(p-2) \leqslant p(p-2) + (p+1)(p-2) = 2p^2 - 3p - 2 < T_A,$$

a contradiction. We may suppose $H_1(A) \equiv (p-1) \pmod{p}$. Write $H_1(A) = a_{10} + pa_{11}$. Since

$$p(1-x_1)^{a_{10}}(1-x_1^p)^{a_{11}} \equiv p(1-x_1)^{a_{10}}(1-x_1)^{pa_{11}} \pmod{p^2}$$
$$= p(1-x_1)^{a_{10}+pa_{11}} \pmod{p^2},$$

we may regard pf_A as having a factor $p(1-x_1)^{p-1}$. Because

$$(1-x_1)^{p^2+p-1} = (1-x_1)^{p^2}(1-x_1)^{p-1}$$
$$= \left(1-x_1^{p^2}+(-1)^p\binom{p^2}{p}x_1^p\right)(1-x_1)^{p-1}$$
$$= up(1-x_1)^{p-1},$$

where u is a unit, we have

$$pf_A = u^{-1}(1-x_1)^{p^2}f_A = u^{-1}f_C,$$

where $T_C = T_A + p^2 \geqslant 3p^2 - 3p + 1$. We now apply lemma 5 with $j = 0$ to find matrix B such that $H_\lambda(B) = H_\lambda(C)$, $1 \leqslant \lambda \leqslant \rho$ and $f_B \in \mathfrak{J}(p^2)$.

Suppose $T(D) \geqslant T(B) - (p^2 - p)$. Then

$$T(D) \geqslant 2(p^2-p) = (p^2-p) + p(p-1) \geqslant (p^2-p) + (\rho-1)(p-1).$$

Thus, we may apply lemma 5 with $j = 1$ to find matrix E such that $H_\lambda(D) \geqslant H_\lambda(E)$ and $pf_E \in \mathfrak{J}(p^2)$. Applying lemma 3 with $j = 1$, we find that $pf_D \in \mathfrak{J}(p^2)$. Hence C and B satisfy the hypotheses of lemma 3 with $j = 0$ and with C and B the matrices A and B of the lemma respectively. We have $f_C \in \mathfrak{J}(p^2)$ and therefore $pf_A \in \mathfrak{J}(p^2)$. This completes the proof.

LEMMA 7. *For* $e = 2$, $2(1-x_1)(1-x_2)(1-x_3) \in \mathfrak{J}(4)$.
For $e = 3$, $4(1-x_1)^2(1-x_2)^2(1-x_3)^2 \in \mathfrak{J}(8)$.
For $e = 4$, $8(1-x_1)^4(1-x_2)^4(1-x_3)^4 \in \mathfrak{J}(16)$.

Proof. Since $3 = 2p^2 - 3p + 1$ for $p = 2$, the first assertion follows from lemma 6. If $e = 3$, then

$$4(1-x_1)^2(1-x_2)^2(1-x_3)^2 \equiv 2 \cdot 2(1-x_1^2)(1-x_2^2)(1-x_3^2) \pmod{8}.$$

Thus we use the first assertion (by putting $x_i^2 = u_i$) to see that

$$4(1-x_1)^2(1-x_2)^2(1-x_3)^2$$

is equal modulo $\mathfrak{J}(p^3)$ to a linear combination of cyclotomic polynomials of form $1 + y + y^2 + y^3$, where y has exponent 4 in G, with a common coefficient of 2. Since $2(1+y+y^2+y^3) = (1+y+\ldots+y^7) \in \mathfrak{J}(8)$, the second assertion follows. The third assertion follows from the second in the same way. This completes the proof of lemma 7.

LEMMA 8. *For* $e \geqslant 3$, $0 \leqslant j \leqslant (e-1)$ *and* $2 \leqslant \rho \leqslant (p+1)$, *except for* $p = 2$, $e = 3$ *or* 4, $\rho = 3$ *and* $j = e-1$,

$$e(p^e - p^{e-1}) - j(p^{e-1} - p^{e-2}) \geqslant (e-j)(p^e - p^{e-1}) + (\rho-1)(p^j - 1).$$

Proof. The proof of this inequality reduces to showing that

$$j(p-1)^2 p^{e-2} \geqslant (\rho-1)(p^j - 1).$$

We first note that the inequality fails by exactly the margin 2 in the exceptional cases. It is clear that we may assume $\rho = (p+1)$.

Our inequality then reduces to

$$j(1-p^{-1})^2 \geqslant p^{j+1-e}(1-p^{-j}),$$

which is obviously true for $j = 0$. For $1 \leqslant j \leqslant (e-2)$, the inequality follows from

$$\begin{aligned} j(1-p^{-1})^2 &= jp^{-1}(p-2+p^{-1}) \\ &\geqslant p^{-1}(1-p^{-j}). \end{aligned}$$

This latter inequality is obvious for $j > 1$ and also for $j = 1$ and $p > 2$. For the case $j = 1$, $p = 2$, we have equality.

Finally, for $j = (e-1)$, the inequality reduces to

$$(e-1)(1-p^{-1})^2 \geqslant (1-p^{-e+1}).$$

For $p = 2$ and $e \geqslant 5$, this last inequality holds as well as for $p \geqslant 3$ and $e \geqslant 3$. This completes the proof of lemma 8.

LEMMA 9. *If* $T_A \geqslant e(p^e - p^{e-1}) - (e-1)(p^{e-1} - p^{e-2})$, *then* $p^{e-1} f_A \in \mathfrak{J}(p^e)$.

Proof. For $e = 2$, this is exactly lemma 6. For $e \geqslant 3$, except $p = 2$, $e = 3$ or 4, and $\rho = p+1$, the lemma follows from Lemmas 8, 5 and 3. In the two exceptional cases, we have $T_A \geqslant 8$ or $T_A \geqslant 20$. In the case where $T_A \geqslant 8$, we either have one of the $H_\lambda(A)$ having the value 4 or more, or else the $H_\lambda(A)$ have the values 3, 3, 2. In the case where one $H_\mu(A) \geqslant 4$ we observe that $4(1-x_\mu)^4 \equiv 4(1-x_\mu^4) \pmod 8$ and hence is in $\mathfrak{J}(8)$. Thus for $H_\lambda(B) = H_\lambda(A)$ for $\lambda \neq \mu$ and $H_\mu(B) = 4 \cdot 2^0 + 0 \cdot 2^1 + \ldots$, we have $4f_B$ is in $\mathfrak{J}(8)$ and we apply lemma 3 to conclude that $4f_A$ is in $\mathfrak{J}(8)$. In the case where the $H_\lambda(A)$ have the values 3, 3, 2, we use lemmas 7 and 3 to conclude $4f_A$ is in $\mathfrak{J}(8)$. In a similar manner, the case $T_A \geqslant 20$ reduces to the case one $H_\lambda(A) \geqslant 8$ or else the $H_\lambda(A)$ have the values 7, 7, 6. In the latter situation we use lemmas 7 and 3 to conclude $8f_A$ is in $\mathfrak{J}(16)$. In the case where one $H_\lambda(A) \geqslant 8$, we observe that $8(1-x_\lambda)^8 \equiv 8(1-x_\lambda^8) \pmod{16}$ and hence is in $\mathfrak{J}(16)$. We thus use this observation together with lemma 3 to conclude $8f_A$ is in $\mathfrak{J}(16)$. This completes the proof of lemma 9.

Proposition. Let $e \geqslant 2$ and $0 \leqslant j \leqslant (e-1)$. If $T(A) \geqslant e(p^e - p^{e-1}) - j(p^{e-1} - p^{e-2})$, then $p^j f_A \in \mathfrak{J}(p^e)$.

Proof. For $e = 2$, $j = 1$, this is lemma 9. For $e = 2$, $j = 0$, we use lemma 5 to produce a matrix B such that $f_B \in \mathfrak{J}(p^2)$. Using the established result for $j = 1$, we may apply lemma 3 to conclude $f_A \in \mathfrak{J}(p^2)$.

For $e \geqslant 3$, we use the same idea; i.e. we induce on decreasing j with e fixed. The case $j = e-1$ is just lemma 9. For $0 \leqslant j < e-2$, we use lemma 8 to conclude $T(A) \geqslant (e-j)(p^e - p^{e-1}) + (\rho - 1)(p^j - 1)$. Hence by lemma 5, there exists a matrix B such that $H_\lambda(A) \geqslant H_\lambda(B)$ for each λ and $p^j f_B \in \mathfrak{J}(p^e)$. We now use the induction hypothesis in conjunction with lemma 3 to conclude that $p^j f_A \in \mathfrak{J}(p^e)$. This completes the proof.

As a corollary we can now state one part of theorem A.

Proposition. Let F have rank $\leqslant p+1$. Then $F/F''F^{p^e}$ has nilpotency class at most $e\phi(p^e)+1$.

Proof. Suppose x_λ are generators of $F/F''F^{p^e}$ and consider $[c, u_1, u_2, \ldots, u_{e\phi(p^e)}]$ where u_i is one of the x_λ and c is in the commutator subgroup of $F/F''F^{p^e}$. This is the same as $c^{\Pi(1-x_\lambda)^{a_\lambda}}$, where $\Sigma a_\lambda = T = e\phi(p^e)$, and thus by theorem B can be written as the product of elements of the form $c^{(1+u+\cdots+u^{p^e-1})Q}$, where Q is in $Z_{p^e}(G)$; the latter form is just the identity element. Now if the u_i are not generators, but rather products of generators, we can reduce to the above case either by using the standard commutator identities or by observing that $(1-uv) = (1-u)v+(1-v)$. Hence $[c, uv] = c^{(1-u)v}c^{(1-v)}$, and thus in general we can write

$$[c, u_1, \ldots, u_{e\phi(p^e)}] = c^{\Sigma\Pi Q_\lambda(1-x_\lambda)^{a_\lambda}} = 1,$$

where the Q_λ are in $Z_{p^e}(G)$. This completes the proof.

Although, as remarked, theorem B remains true for any finite number of generators for $e = 1$, it is no longer generally true for $e > 1$. The following indicates that the power of Σ which lies in $\mathfrak{J}(p^e)$ depends upon p, e and the number of generators.

Proposition. Let G be the direct product of $p+2$ cyclic groups of order p^2. Then $\Sigma^k \notin \mathfrak{J}(p^2)$, where $k = 2p^2-p-2 \geqslant 2(p^2-p)$.

Proof. Let
$$f = (1-x_1)^{p^2-p-1}\prod_{\lambda=2}^{p+2}(1-x_\lambda)^{p-1}.$$

We note that the degree of this polynomial is

$$(p^2-p-1)+(p+1)(p-1) = 2p^2-p-2 \geqslant 2(p^2-p).$$

Let $\mathfrak{K}$ be the ideal generated by $(1-x_\lambda)^p$, $1 \leqslant \lambda \leqslant p+2$. We note that f is not in $(\mathfrak{K}^{p-1}, p)$, the ideal generated by $\mathfrak{K}^{p-1}$ and p. This follows, for example, from the fact that working modulo p, the polynomial

$$g = (1-x_1)^p\prod_{\lambda=2}^{p+2}(1-x_\lambda)^{p^2-p}$$

annihilates $\mathfrak{K}^{p-1}$, but $g.f$ is not the zero element of $Z_p(G)$. Hence to complete the proof of the Proposition we will show that $\mathfrak{J}(p^2) \subseteq (\mathfrak{K}^{p-1}, p)$.

The generators of $\mathfrak{J}(p^2)$ are $1+g+g^2+\ldots+g^{p^2-1} \equiv (1-g)^{p^2-1} \pmod p$ for $g \in G$. Since $1-g = \Sigma r_\lambda(1-x_\lambda)$ for suitable r_λ in $Z(G)$, we have $(1-g)^p \in (\mathfrak{K}, p)$ and hence $(1-g)^{p^2-p} \in (\mathfrak{K}^{p-1}, p)$. Thus the generators of $\mathfrak{J}(p^2)$ are in $(\mathfrak{K}^{p-1}, p)$ and hence $\mathfrak{J}(p^2) \subseteq (\mathfrak{K}^{p-1}, p)$. This completes the proof.

5. Lower bounds for the class of $F/F''F^{p^e}$ and a representation of $F/F''F^p$

The notation is as in §4. We begin with the following:

Lemma. *Let G be as in §4. Then $(1-x)^{e\phi(p^e)-1}$ is not in the ideal $\mathfrak{J}(p^e)$ of $Z(G)$.*

Proof. Let ω be a primitive p^eth root of unity. In $Z[\omega]$ we have

$$1+\omega^i+(\omega^i)^2+\ldots+(\omega^i)^{p^e-1} = 0 \quad (i = 1, 2, \ldots, p^e-1).$$

The natural map from $Z(G)$ into $Z[\omega]$ defined by $x \to \omega$, $y \to 1$, sends $\mathfrak{J}(p^e)$ into the ideal (p^e) of $Z[\omega]$ generated by p^e. In order to show that $(1-x)^{e\phi(p^e)-1}$ is not in $\mathfrak{J}(p^e)$, we shall show that $(1-\omega)^{e\phi(p^e)-1}$ is not in (p^e).

We have the well-known identity (see, for example, Weiss 1963, §7.4)

$$(p^e) = ((1-\omega)^{e\phi(p^e)}) = ((1-\omega))^{e\phi(p^e)},$$

where $((1-\omega))^{e\phi(p^e)}$ means the $e\phi(p^e)$ power of the ideal $(1-\omega)$. Since $Z[\omega]$ is a Dedekind domain, each ideal in $Z[\omega]$ is the 'unique' product of prime ideals. Thus it is evident that $((1-\omega))^{e\phi(p^e)-1}$ properly contains (p^e), and hence $(1-\omega)^{e\phi(p^e)-1}$ is not in (p^e). This completes the proof.

Before proving theorem A we need a theorem due to Magnus (1939). Let R be a normal subgroup of F, $R \neq 1$. The letter u with subscripts refers to elements of F and also to their images in F/R' under the natural map. The corresponding elements in F/R under the natural map are denoted by s with the corresponding subscripts. Note that R/R' is an abelian normal subgroup of F/R' and hence is a $Z(F/R)$-module.

THEOREM (*Magnus*). *Let u_1, u_2 be free generators of F, let t_1, t_2 be indeterminates which commute with all elements of $Z(F/R)$ and let s_1, s_2 be free generators of F/R. Then the mapping of F/R' into 2×2 matrices over $Z(F/R)[t_1, t_2]$ defined by*

$$u_i \to \begin{pmatrix} s_i & t_i \\ 0 & 1 \end{pmatrix} \quad (i = 1, 2)$$

is an isomorphism.

The following remarks will be helpful in understanding the proofs of the theorems below.

Any element c in the commutator subgroup of F/R' is represented by a matrix of form

$$\begin{pmatrix} 1 & P_1 t_1 + P_2 t_2 \\ 0 & 1 \end{pmatrix},$$

where P_1, P_2 are in the augmentation ideal of $Z(F/R)$. In particular, $[u_1, u_2]$ corresponds to

$$\begin{pmatrix} 1 & (1-s_2)t_1 + (1-s_1)t_2 \\ 0 & 1 \end{pmatrix}.$$

Furthermore, c^Q corresponds to

$$\begin{pmatrix} 1 & QP_1 t_1 + QP_2 t_2 \\ 0 & 1 \end{pmatrix}.$$

THEOREM A. *$F/F''F^{p^e}$ has nilpotency class (and also Engel length) $e(p^e - p^{e-1})$ or $e(p^e - p^{e-1}) + 1$.*

Proof. We have already shown that the class is less than or equal to $e(p^e - p^{e-1}) + 1$ and hence we need only show that it is greater than or equal to $e(p^e - p^{e-1})$. To show this, consider the above Magnus representation of F/F'' where F has rank 2. If one sets $s_1^{p^e} = s_2^{p^e} = 1$ and $ut_1 = ut_2 = 0$ for any u in $\mathfrak{J}(p^e)$, then the matrices become a (not necessarily faithful) representation of $F/F''F^{p^e}$. This follows since an element u of F/F'' has the form

$$u = \begin{pmatrix} S & v_1 t_1 + v_2 t_2 \\ 0 & 1 \end{pmatrix},$$

where S is some element of F/F' and v_1, v_2 are elements of $Z(F/F')$. But

$$u^{p^e} = \begin{pmatrix} S^{p^e} & (1+S+S^2+\ldots+S^{p^e-1})(v_1t_1+v_2t_2) \\ 0 & 1 \end{pmatrix} = \begin{pmatrix} 1 & 0 \\ 0 & 1 \end{pmatrix}$$

and thus the homomorphic property is verified. We now assert that the image of $[u_1, u_2]^{(1-u_1)^{e\phi(p^e)-2}}$ does not equal 1, since the coefficient of t_2 is $(1-s_1)^{e\phi(p^e)-1}$ and by the above lemma it is not in $\mathfrak{J}(p^e)$. Therefore the above commutator does not represent the identity element, and thus $(F/F''F^{p^e})_{e\phi(p^e)} \neq 1$. This completes the proof of theorem A.

We have already mentioned the theorem of Meier-Wunderlei (1951) which gives the exact value for the case $e = 1$.

Theorem (*Meier-Wunderlei*). *$F/F''F^p$ has nilpotency class $p-1$, where F has rank 2.*

If we knew that the Magnus representation of $F/F''F^p$ is faithful we would be able to give another proof of Meier-Wunderlei's theorem. But we can reverse the procedure and use Meier-Wunderlei's theorem to show that the Magnus representation of $F/F''F^p$ is faithful. This is of interest since we may now expect some interesting applications (cf. Bachmuth & Mochizuki 1967 and references listed therein). However, we mention this word of caution without citing any details. The following theorem is false in the case where F has rank larger than 2.

Theorem. *In the Magnus representation of F/F'' where F has rank 2, setting $s_1^p = s_2^p = 1$ and $ut_1 = ut_2 = 0$ for any u in $\mathfrak{J}(p)$ yields a faithful representation of $F/F''F^p$.*

Proof. Setting $s_1^p = s_2^p = 1$ in the Magnus representation of F/F'' gives a faithful representation of $G = F/(F'F^p)'$ (i.e. $R = F'F^p$ in Magnus's theorem). Setting $Pt_1 = Pt_2 = 0$ for any P in $\mathfrak{J}(p)$ is equivalent to factoring G by G_p, since $\Sigma^{p-1} = \mathfrak{J}(p)$ and the subgroup corresponding to G_p consists exactly of those matrices in the image of G of the form

$$\begin{pmatrix} 1 & P_1t_1+P_2t_2 \\ 0 & 1 \end{pmatrix},$$

where P_1 and P_2 are in Σ^{p-1}. But

$$G/G_p \simeq F/F''F_p[F^p, F^p][F', F^p],$$

and by Meier-Wunderlei's theorem, $F_p \subseteq F''F^p$. Hence

$$F/F''F^p[F^p, F^p][F', F^p] = F/F''F^p$$

is a homomorphic image of G/G_p and thus a factor group of the matrix group defined in the statement of the theorem. But in the proof of theorem A, we saw that this matrix group is a factor group of $F/F''F^p$. All groups are finite, whence our result follows.

6. Reduction to the prime power case

We wish to prove

THEOREM C. *Suppose* $n = p_1^{e_1} \ldots p_m^{e_m}$ *and* $H = F/F''F^n$. *Then* $H_d = H_{d+1}$, *where* $d = 2 + \max e_i\phi(p_i^{e_i})$.

Suppose A is the commutator subgroup of $H = F/F''F^n$ and Σ is the augmentation ideal of $Z_n(H/A)$. Then it is clear that the lower central series of H becomes stationary at the smallest integer $(k+2)$, where Σ^k has the same action as Σ^{k+1} on A. Thus theorem C is an immediate corollary of the more general proposition below.

We shall use the following notation: $p_1, p_2, \ldots, p_m$ are distinct prime integers, $e_1, e_2, \ldots, e_m$ are positive integers and $n = p_1^{e_1} \ldots p_m^{e_m}$. $G = G(1) \times \ldots \times G(m)$ where each $G(i)$ is an abelian group of exponent $p_i^{e_i}$. Σ is the augmentation ideal of $Z_n(G)$. $\mathfrak{J} = \mathfrak{J}(n)$ is the cyclotomic (degree $n-1$) ideal of $Z_n(G)$. Σ_i is the augmentation ideal of $Z_{p_i^{e_i}}(G)$. $\mathfrak{J}_i = \mathfrak{J}_i(n)$ is the cyclotomic (degree $n-1$) ideal of $Z_{p_i^{e_i}}(G)$.

$$H(i) = G(1) \times \ldots \times \widehat{G(i)} \times \ldots \times G(m)$$

($\wedge$ means omit). c_i is the least power of the augmentation ideal of $Z_{p_i^{e_i}}(G_i)$ which is contained in the cyclotomic (degree $p_i^{e_i}-1$) ideal of $Z_{p_i^{e_i}}(G_i)$.

Proposition. If $c = \max\{c_1, \ldots, c_m\}$, then $\Sigma^c = \Sigma^{c+1}$ modulo $\mathfrak{J}$.

Proof. $Z_m = \mathfrak{K}_1 \oplus \ldots \oplus \mathfrak{K}_n$ where the $\mathfrak{K}_i$ are ideals generated by orthogonal idempotents ϵ_i. $\mathfrak{K}_i \simeq Z_{p_i^{e_i}}$ as rings, and ϵ_i is the identity of $\mathfrak{K}_i$.

We have the following evident facts:

(1) $Z_n(G) = \mathfrak{K}_1(G) \oplus \ldots \oplus \mathfrak{K}_m(G)$.

(2) $\Sigma^k = \epsilon_1 \Sigma^k \oplus \ldots \oplus \epsilon_m \Sigma^k \quad (k = 1, 2, \ldots)$.

(3) $\mathfrak{J} = \epsilon_1 \mathfrak{J} \oplus \ldots \oplus \epsilon_m \mathfrak{J}$.

(4) $Z_n(G)/\mathfrak{J} \simeq \mathfrak{K}_1(G)/\epsilon_1\mathfrak{J} + \ldots + \mathfrak{K}_m(G)/\epsilon_m\mathfrak{J}$.

(5) $\epsilon_i \Sigma$ and $\epsilon_i \mathfrak{J}$ are the augmentation ideal and cyclotomic ideal (degree $n-1$) of $\mathfrak{K}_i(G)$, considered as a group ring.

(6) $(\Sigma^k + \mathfrak{J})/\mathfrak{J} \simeq (\epsilon_1 \Sigma^k + \epsilon_1 \mathfrak{J})/\epsilon_1 \mathfrak{J} + \ldots + (\epsilon_m \Sigma^k + \epsilon_m \mathfrak{J})/\epsilon_m \mathfrak{J} \quad (k = 1, 2, \ldots)$.

From (6), what we need to show is that $(\epsilon_i \Sigma^c + \epsilon_i \mathfrak{J})/\epsilon_i \mathfrak{J} = (\epsilon_i \Sigma^{c+1} + \epsilon_i \mathfrak{J})/\epsilon_i \mathfrak{J}$, $i = 1, \ldots, m$. Identifying $Z_{p_i^{e_i}}(G)$ and $\mathfrak{K}_i(G)$ and using the notation introduced above, we want to prove, equivalently, that $\Sigma_i^c = \Sigma_i^{c+1} \bmod \mathfrak{J}_i$. This is proved by the following sequence of lemmas.

Let $\mathfrak{T}_{i,1}$ denote the ideal of $Z_{p_i^{e_i}}(G)$ generated by the $(1-g)$, $g \in G(i)$, and $\mathfrak{T}_{i,2}$ denote the ideal of $Z_{p_i^{e_i}}(G)$ generated by the $1-g$, $g \in H(i)$.

LEMMA 1. $\Sigma_i = \mathfrak{T}_{i,1} + \mathfrak{T}_{i,2}$.

Proof. Letting q_i be the order of $H(i)$, we can find integers r and s such that $rp_i^{e_i} + sq_i = 1$. If $g \in G$, then $(1-g) = (1-g^{sq_i}) + g^{sq_i}(1-g^{rp_i^{e_i}})$ where $(1-g^{sq_i}) \in \mathfrak{T}_{i,1}$ and $g^{sq_i}(1-g^{rp_i^{e_i}}) \in \mathfrak{T}_{i,2}$. Here we use the fact that $G(i)$ is the subgroup of all elements of G of order dividing $p_i^{e_i}$ and similarly that $H(i)$ is the subgroup of all elements of G of order dividing q_i. Thus, $\Sigma_i \subseteq \mathfrak{T}_{i,1} + \mathfrak{T}_{i,2}$. Since $\Sigma_i \supseteq \mathfrak{T}_{i,1} + \mathfrak{T}_{i,2}$, the proof is complete.

Let $\epsilon = 1/q_i \Sigma h \ (h \in H(i))$. By a familiar argument, ϵ and $(1-\epsilon)$ are orthogonal idempotents. Recall that q_i is the order of $H(i)$.

LEMMA 2. $\mathfrak{T}_{i,2} = (1-\epsilon) Z_{p_i^{e_i}}(G)$.

Proof. $1/q_i \Sigma (1-h)\,(h \in H(i)) = (1-\epsilon) \in \mathfrak{T}_{i,2}$. Thus, $\mathfrak{T}_{i,2} \supseteq (1-\epsilon) Z_{p_i^{e_i}}(G)$. Suppose $h \in H(i)$. Then, $(1-h)\epsilon = \epsilon - h\epsilon = \epsilon - \epsilon = 0$ since $h\epsilon = \epsilon$. Hence,

$$(1-h) = (1-h)(1-\epsilon) \in (1-\epsilon) Z_{p_i^{e_i}}(G).$$

Thus, the generators of $\mathfrak{T}_{i,2}$ are in $(1-\epsilon) Z_{p_i^{e_i}}(G)$, and we have $\mathfrak{T}_{1,2} \subseteq (1-\epsilon) Z_{p_i^{e_i}}(G)$. This completes the proof.

From Lemmas 1 and 2, it is clear that $\mathfrak{T}_{i,2}^k = \mathfrak{T}_{i,2}$ and $\Sigma_i^k = \mathfrak{T}_{i,1}^k + \mathfrak{T}_{i,2}$, $k = 1, 2, \ldots$. Thus, we need only prove

LEMMA 3. $\mathfrak{T}_{i,1}^c \equiv 0$ (*modulo* $\mathfrak{J}_i$).

Proof. If $g \in G(i)$, then

$$(1+g+\ldots+g^{n-1}) = (1+g+\ldots+g^{p_i^{e_i}-1})$$
$$+(1+g+\ldots+g^{p_i^{e_i}-1})+\ldots+(1+g+\ldots+g^{p_i^{e_i}-1}) = q_i(1+g+\ldots+g^{p_i^{e_i}-1}).$$

Since q_i is a unit in $Z_{p_i^{e_i}}(G(i))$, the ideal generated by cyclotomics of degree $(p_i^{e_i}-1)$ of $Z_{p_i^{e_i}}(G(i))$ is a subset of $\mathfrak{J}_i$. If $g_1, \ldots, g_{c_i} \in G(i)$, then by hypothesis $(1-g_1)\ldots(1-g_{c_i})$ is in the ideal generated by cyclotomics of degree $(p_i^{e_i}-1)$ and hence in $\mathfrak{J}_i$. Thus $\mathfrak{T}_{i,1}^c \equiv 0$ modulo $\mathfrak{J}_i$. This completes the proof of the proposition.

REFERENCES

Bachmuth, S. & Hughes, I. 1966 Centers of certain presentations of finite groups. *Notices Am. Math. Soc.* **13**, no. 5, abstract (636-4).

Bachmuth, S. & Mochizuki, H. Y. 1967 Cyclotomic ideals in group rings. *Bull. Am. Math. Soc.* **72**, 1018–1020.

Bachmuth, S. & Mochizuki, H. Y. 1968*b* Automorphisms of a class of metabelian groups. II. *Trans. Am. Math. Soc.* **127**, 294–301.

Glauberman, G., Krause, E. & Struik, R. 1966 Engel congruences in groups of prime-power exponent. *Can. J. Math.* **18**, 579–588.

Kostrikin, A. I. 1960 On Engel properties of groups with the identical relation $x^{p^\alpha} = 1$. *Dokl. Akad. Nauk SSSR* **135**, pp. 524–526 (Russian); translated as *Soviet Math. Dokl.* **1**, 1282–1284 (1961).

Magnus, W. 1939 On a theorem of Marshall Hall. *Annals Math.* (2) **40**, 764–768.

Meier-Wunderlei, H. 1951 Metabelsche Gruppen. *Comment. Math. Helv.* **25**, 1–10.

Weiss, E. 1963 *Algebraic number theory.* New York: McGraw-Hill Book Co.

[Department of Mathematics, University of California, Santa Barbara, USA]

[Department of Mathematics, University of Toronto, Canada]

[Department of Mathematics, University of California, Santa Barbara, USA]

[Received 2 February, 1968]

39.

On the Average Length of a Class of Finite Continued Fractions

Abhandlungen aus Zahlentheorie und Analysis zur Erinnerung an Edmund Landau (P. Turán, ed.),
VEB Deutscher Verlag der Wissenschaften, Berlin, 1969, 89–96
[Commentary on p. 602]

1. Introduction

Many years ago Dr. J. Gillis asked me the following question: Let N and a be coprime natural integers, $1 \leqq a < N$, so that a/N can be represented by a finite continued fraction

$$a/N = 1/c_1 + 1/c_2 + \cdots + 1/c_{n(a)},$$

where the c_i are natural integers depending on N and a, and where $c_{n(a)} > 1$ (to make the representation unique). What can be said about the sum

$$L(N) = \sum_{\substack{a=1 \\ (a,N)=1}}^{N} n(a)?$$

At the time I was able to make only the trivial statement that

$$L(N) = O(N \log N).$$

Recently I discovered a connection between $L(N)$ and the number $r(N)$ of representation of N by the bilinear form $N = xx' + yy'$, where the natural integers x, x', y, y' are subject to the restrictions $x > y$, $x' > y'$, $(x, y) = 1$, $(x', y') = 1$.

Theorem 1.

$$L(N) = \frac{3}{2}\varphi(N) + 2r(N) \quad \textit{for} \quad N > 2.$$

This raises the question, what can be said about the behaviour of $r(N)$. The answer is given by

Theorem 2.

$$r(N) = 6\pi^{-2} \log 2\ \varphi(N) \log N + O(N\sigma_{-1}^3(N)),$$

where $\sigma_{-1}(N)$ denotes the sum of the reciprocals of the positive divisors of N.

It is clear that the main term dominates the error term by a factor at least of the order $\log N(\log\log N)^{-4}$. It appears very difficult to obtain a substantially better error term, though numerical evidence suggests that the error term is much too large.

Combining the two theorems leads to

Theorem 3.

$$L(N) = 12\pi^{-2}\log 2\varphi(N)\log N + O(N\sigma_{-1}^{3}(N)).$$

A slight extension of the method of the proof of theorem 2 leads to the following result. Let $L_c(N)$ be the number of times that the denominator c occurs in the continued fraction of the rationals a/N, $1 < a < N$, $(a, N) = 1$. Then we have

Theorem 4.

$$L_c(N) = 12\pi^{-2}\log(1 - (c+1)^{-2})^{-1}\ \varphi(N)\log N + O(N\sigma_{-1}^{3}(N)).$$

This theorem suggests that in some sense the frequency of the denominator c equals

$$\log(1 - (c+1)^{-2})^{-1}/\log 2,$$

if we consider the continued fraction of all real numbers. This is indeed the case, as Khintchine has shown [1].

2. Preliminaries

Small roman letters with or without indices are restricted to positive rational integers. The symbols $\varphi(n)$, $\mu(n)$, $\sigma_\tau(n)$, $d(n)$ have the meaning usual in elementary number theory, i.e. $\varphi(n)$ denotes the Euler function, $\mu(n)$ the Moebius function, $\sigma_\tau(n)$ the sum of the τth powers of the positive divisors of n, $d(n) = \sigma_0(n)$. The symbol O holds uniformly in all variables except possibly in $\varepsilon > 0$.

We shall make frequent use of the Moebius inversion formula, and such well known results as

$$d(n) = O(n^{\varepsilon}),$$

$$\sum_{n=1}^{z}\varphi(n)\,n^{-1} = 6\pi^{-2}z + O(\log z),$$

$$\sigma_{-1}(n) = O(\operatorname{loglog} N),$$

$$6\pi^{-2} < \varphi(n)\,\sigma_{-1}(n)\,n^{-1} \leqq 1.$$

We have already defined $r(N)$ as the number of solutions of $N = xx' + yy'$ subject to

$$x > y, \quad x' > y', \quad (x, y) = (x', y') = 1.$$

We further define $R(N)$ as the number of solutions subject to

$$x > y, \quad x' > y';$$

and for each $d \geqq 1$ we define $\varrho(N, d)$ as the number of solutions subject to

$$x > y, \quad x' > y', \quad (x, y) = 1, \quad x' > dx.$$

Then

$$R(N) = \sum_{bb'/N} r(N(bb')^{-1}),$$

and by a repeated application of the Moebius formula

$$r(N) = \sum_{bb'/N} \mu(b)\, \mu(b')\, R(N(bb')^{-1}).$$

Utilizing the symmetry in the definition of $R(N)$ between the primed and the 'unprimed' variables, we obtain

$$R(N) = 2 \sum_{d/N} \varrho(N d^{-1}, d) + O \sum_{x^2 < N} d(N - x^2).$$

$$r(N) = 2 \sum_{dbb'/N} \mu(b)\, \mu(b')\, \varrho(N(dbb')^{-1}, d) + O(N^{1/2+\varepsilon}). \tag{2}$$

3. Continued fractions

We define the function $[c_1, \ldots, c_n]$ in the usual way by

$$[\,] = 1, \quad [c_1] = c_1, \quad [c_1, c_2] = c_2 c_1 + 1,$$

$$[c_1, \ldots, c_n] = c_n[c_1, \ldots, c_{n-1}] + [c_1, \ldots, c_{n-2}] \quad \text{for} \quad n \geqq 3.$$

Our first object is to prove the formula

$$[c_1, \ldots, c_n] = [c_1, \ldots, c_m]\,[c_{m+1}, \ldots, c_n] + [c_1, \ldots, c_{m-1}]\,[c_{m+2}, \ldots, c_n]$$

for $1 \leqq m < n$. (I don't believe this formula is new, but I have not been able to find it in the literature.) The formula is obviously true for $1 \leqq m = n - 1$ and also, as a little calculation shows, for $1 \leqq m = n - 2$. Hence by induction with respect to m, it is true for $m \geqq 1$ and all $n > m$. Our formula makes it evident that

$$[c_1, \ldots, c_n] = [c_n, \ldots, c_1].$$

Now we introduce a pair N, a with $a < \frac{1}{2}N$, $(N, a) = 1$ and develop a/N as a continued fraction

$$a/N = 1/c_1 + 1/c_2 + \cdots + 1/c_n, \quad c_n \geqq 2, \tag{3}$$

and we have automatically

$$c_1 \geqq 2, \quad N = [c_1, \ldots, c_n].$$

If $a \neq 1$, then $n = n(a) > 1$ and we can choose m in the interval $1 \leqq m \leqq n - 1$ in $n - 1$ different ways. Put

$$x = [c_1, \ldots, c_m], \quad y = [c_1, \ldots, c_{m-1}], \tag{4}$$

$$x' = [c_n, \ldots, c_{m+1}], \quad y' = [c_n, \ldots, c_{m+2}]. \tag{5}$$

These integers, by virtue of our identity, satisfy the relation

$$N = xx' + yy',$$

and they also fulfil the conditions

$$x > y, \quad x' > y', \quad (x, y) = (x', y') = 1.$$

(If $m = 1$ or $m = n - 1$, remember $c_1 \geqq 2$ or $c_n \geqq 2$.)

Moreover we have for $m > 2$

$$\frac{y}{x} = \frac{[c_1, \ldots, c_{m-1}]}{[c_1, \ldots, c_m]} = \frac{[c_1, \ldots, c_{m-1}]}{c_m[c_1, \ldots, c_{m-1}] + [c_1, \ldots, c_{m-2}]}$$

$$= \left(c_m + \frac{[c_1, \ldots, c_{m-2}]}{[c_1, \ldots, c_{m-1}]}\right)^{-1}$$

and hence by induction

$$y/x = 1/c_m + \cdots + 1/c_1; \tag{6}$$

similarly

$$y'/x' = 1/c_{m+1} + \cdots + 1/c_n. \tag{7}$$

Conversely, given a representation of N by our bilinear form with our restrictions, we can find a unique sequence $c_1, \ldots, c_n$ not starting or finishing with 1 such that (6), (7), (4), (5) are satisfied and

$$N = [c_1, \ldots, c_n].$$

Putting $a = [c_2, \ldots, c_n]$, it is clear that (3) holds.

To sum up, we have a $(1-1)$ relation between all suitably restricted representations of N by the bilinear form, and all pairs of sequences

$$c_1, \ldots, c_m; \quad c_{m+1}, \ldots, c_n \quad \text{with} \quad c_1 \geqq 2, \quad c_n \geqq 2, \quad 1 \leqq m < n.$$

Thus, for $N > 2$

$$r(N) = \sum_{\substack{1<a<N/2\\(a,N)=1}} (n(a)-1) = \sum_{\substack{a<N/2\\(a,N)=1}} n(a) - \frac{1}{2}\varphi(N).$$

As for $0 < \alpha < \frac{1}{2}$

$$1-\alpha = \cfrac{1}{1+\cfrac{1}{\cfrac{1}{\alpha}-1}},$$

it follows that, for $a < \frac{1}{2}N$, $(a, N) = 1$, (3) implies

$$(N-a)\,N = 1/1 + 1/(c_1-1) + 1/c_2 + \cdots + 1/c_n.$$

Hence

$$\sum_{\substack{N/2<a<N\\(a,N)=1}} n(a) = \frac{1}{2}\varphi(N) + \sum_{\substack{a<N/2\\(a,N)=1}} n(a),$$

and theorem 1 is proved.

4. Proof of theorem 2

We require the following

Lemma. $\varrho(N,d) = 3\pi^{-2}\log 2\; N\log(Nd^{-1}) + O(N)$ *for* $d \leqq N$.

Proof. Fix a pair x, y with

$$y < x < (Nd^{-1})^{1/2}, \quad (x, y) = 1. \tag{8}$$

We have to count the number of positive integers x' which satisfy

$$xx' \equiv N \pmod{y}, \quad x' > x, \quad x' > y' = y^{-1}(N - xx') > 0.$$

The last two inequalities can be written as

$$N(x+y)^{-1} < x' < Nx^{-1}.$$

This means we have to find the number, say $P(N, d, x, y)$ of solutions of the congruence in the interval

$$\operatorname{Max}(xd, N(x+y)^{-1}) < x' < Nx^{-1}.$$

Clearly

$$|P(N, d, x, y) - y^{-1}(Nx^{-1} - \operatorname{Max}(xd, N(x+y)^{-1}))| < 1$$

and

$$\varrho(N, d) = \sum_{x,y} P(N, d, x, y)$$

where the sum is extended over all x, y satisfying (8). We distinguish two cases.

Case 1. $xd \geqq N(x+y)^{-1}$.

Then

$$P(N, d, x, y) < 1 + y^{-1}(Nx^{-1} - xd)$$
$$\leqq 1 + y^{-1}(xd(x+y)x^{-1} - xd) = 1 + d,$$

Hence, in this case, summing over x, y satisfying (8)

$$\sum_{x,y} P(N, d, x, y) \leqq (1+d) \sum_{x<(Nd^{-1})^{1/2}} x = O(N).$$

Case 2. $xd < N(x+y)^{-1}$.

This implies $6d < N$. If $6d \geqq N$, the lemma is trivial. Keeping y fixed, x is now restricted by

$$(x, y) = 1, \quad x > y, \quad x(x+y)\,d < N. \tag{9}$$

As

$$P(N, d, x, y) = O(1) + y^{-1}(Nx^{-1} - N(x+y)^{-1}),$$

we obtain, summing over all relevant x,

$$\sum_x P(N, d, x, y) = O(N^{1/2}d^{-(1/2)}) + y^{-1}N \sum_{\substack{y<x<2y \\ (x,y)=1}} x^{-1} - y^{-1}N \sum_{\substack{x_0 \leqq x < x_0+y \\ (x,y)=1}} x^{-1}, \tag{10}$$

where x_0 is the smallest integer for which $x_0(x_0+y)\,d \geqq N$. As $x_0 > \left(\frac{1}{2}Nd^{-1}\right)^{1/2}$, the last sum including the factor $y^{-1}N$ is at most $O(N^{1/2}\,d^{1/2})$. Further

$$\sum_{\substack{y<x<2y \\ (x,y)=1}} x^{-1} = \sum_{b/y} \mu(b)\,b^{-1} \sum_{r=yb^{-1}+1}^{2yb^{-1}-1} r^{-1} = \varphi(y)\,y^{-1}\log 2 + O(d(y)\,y^{-1}).$$

Hence

$$\sum_x P(N, d, x, y) = N\varphi(y)\,y^{-2}\log 2 + O(N^{1/2}d^{1/2}) + O(Nd(y)\,y^{-2}).$$

This expression has to be summed over all $y < \left(\frac{1}{2} Nd^{-1}\right)^{1/2}$. Thus

$$\sum_{x,y} P(N, d, x, y) = O(N) + N \log 2 \sum_{y<(Nd^{-1}/2)^{1/2}} \varphi(y)\, y^{-2},$$

and a simple summation by parts gives the lemma.

It is now easy to obtain theorem 2. As

$$r(N) = 2 \sum_{dbb'/N} \mu(b)\, \mu(b')\, \varrho(N(dbb')^{-1} d) + O(N^{1/2+\varepsilon})$$

$$= 6\pi^{-2} \log 2\, N \sum_{dbb'/N} \mu(b)\, \mu(b')\, (dbb')^{-1} (\log N - \log (d^2bb')) + O(1).$$

From this theorem 2 follows as the error term has the required form, and as

$$\sum_{dbb'/N} \mu(b)\, \mu(b')\, (dbb')^{-1} = \sum_{b/N} \mu(b)\, b^{-1} \sum_{n/Nb^{-1}} n^{-1} \sum_{b'/n} \mu(b')$$

$$= \sum_{b/N} \mu(b)\, b^{-1} = \varphi(N)\, N^{-1},$$

whilst

$$\sum_{dbb'/N} (dbb')^{-1}\, \mu(b)\, \mu(b') \log (d^2bb')$$

$$= 2 \sum_{dbb'/N} (dbb')^{-1}\, \mu(b)\, \mu(b') \log (db')$$

$$= 2 \sum_{b/N} \mu(b)\, b^{-1} \sum_{n/Nb^{-1}} n^{-1} \log n \sum_{b/N} \varphi(b') = 0.$$

5. Proof of theorem 4

Assume that a fixed number c is given. We denote by $r_c(N)$, $R_c(N)$, $\varrho_c(N, d)$ the number of representations of N by the bilinear form subject to the restriction mentioned before and the additional restriction

$$cy \leqq x < (c + 1)\, y. \tag{11}$$

Reverting to the pattern of the proof of theorem 1, we see that $r_c(N)$ equals the number of times that c is a denominator in the continued fraction of the rationals a/N, $1 < a < \frac{1}{2}N$, $(N, a) = 1$, not counting any possible $c_{n(a)}$, as $m < n$. These exceptional values are at most $\varphi(N)$. If we include the rationals a/N, $\frac{1}{2}N < a < N$, $(N, a) = 1$, every c_m will occur as often as before, except for changes in the first or second place. Hence we have

$$L_c(N) = 2r_c(N) + O(\varphi(N)).$$

Formula (1) works as before if we replace R and r by R_c and r_c respectively.

Further our argument continues to be true if we replace the new restriction (11) by

$$cy' \leqq x' < (c+1)\,y',$$

because with the sequence $c_1, \ldots, c_n$ the sequence $c_n, \ldots, c_1$ also occurs. Hence $R_c(N)$ equals twice the number of solutions subject to the restrictions

$$cy \leqq x < (c+1)\,y, \quad x' > y', \quad x' > x,$$

with an error $O(N^{1/2+\varepsilon})$. This implies that (2) remains true if we replace r and ϱ by r_c and ϱ_c.

The proof of the lemma goes as before but (9) has to be replaced by (11) and $(x, y) = 1$, $x(x+y)\,d < N$. Then (10) has to be replaced by

$$\begin{aligned}
\sum_x P(N, d, x, y) &= O(N^{1/2}d^{-(1/2)}) + y^{-1}N \sum_{\substack{cy<x<(c+1)y \\ (x,y)+1}} (x^{-1} - (x+y)^{-1}) \\
&= O(N^{1/2}d^{-(1/2)}) + y^{-1}N \sum_{b/y} \mu(b)\, b^{-1} \\
&\quad \times \sum_{r=cyb^{-1}+1}^{(c+1)yb^{-1}-1} (r^{-1} - (r+yb^{-1})^{-1}) \\
&= O(N^{1/2}d^{-(1/2)}) + \varphi(y)\, y^{-2} N\, (\log(1+c^{-1})) \\
&\quad - \log(1+(c+1)^{-1})) + O(Nd(y)\, y^{-2}) \\
&= O(N^{1/2}d^{-(1/2)}) + O(Nd(y)y^{-2}) + N\varphi(y)y^{-2}\log(1-(c+1)^{-2})^{-1}.
\end{aligned}$$

The rest of the calculation goes as before, with $\log(1-(c+1)^{-2})^{-1}$ in place of $\log 2$.

Reference

[1] A. Khintchine, Metrische Kettenbruchprobleme, Compositio Mathematica **1** (1935), 361—382.

40.

Sums of Complexes in Torsion-Free Abelian Groups

(With P. Scherk)

Canadian Mathematical Bulletin 12 (1969), 479–480

Let A, B denote two non-void finite complexes (= subsets) of the torsion free abelian group G,

$$A + B = \{a + b \mid a \in A,\ b \in B\} .$$

Let $d(A), \ldots$ denote the maximum number of linearly independent elements of $A, \ldots$ and let $n = n(A, B)$ denote the number of elements of $A + B$ whose representation in the form $a + b$ is unique. In the preceding paper, Tarwater and Entringer [1] proved that $n \geq d(A)$. We wish to show by an entirely different and perhaps simpler method that $n \geq d = d(A \cup B)$.

Given A and B, we may replace G by the subgroup generated by $A \cup B$. Then G may be interpreted as a d-dimensional vector lattice over the ring of the integers. Imbed G into a d-dimensional real vector space and construct its affine d-space R.

Let $\mathfrak{C}$ denote the set of those points of R whose radius vectors belong to $A + B$. Since $d(A + B) = d(A \cup B) = d$, we have $\dim \mathfrak{C} = d - 1$ or d. The convex closure $\mathfrak{H}(\mathfrak{C})$ of $\mathfrak{C}$ is a convex polytope in R.

If ξ is an extremal point of $\mathfrak{H}(\mathfrak{C})$, ξ is not the barycenter of other points of $\mathfrak{H}(\mathfrak{C})$, in particular of $\mathfrak{C}$. Hence $\xi \in \mathfrak{C}$.

Suppose the radius vector x of the point ξ has two distinct representations $x = a + b = a' + b'$ where $\{a, a'\} \subset A$, $\{b, b'\} \subset B$. The points with the radius vector $a + b'$ and $a' + b$ would lie in $\mathfrak{C}$ and ξ would be the centre of the connecting segment. In particular, ξ could not be an extremal point. Thus every extremal point of $\mathfrak{H}(\mathfrak{C})$ has a radius vector in $A + B$ with a unique representation $a + b$.

Since $\mathfrak{H}(\mathfrak{C})$ is a convex polytope of dimension $\geq d - 1$ and since every convex polytope is the convex closure of the set of its extremal points, $\mathfrak{H}(\mathfrak{C})$ has not less than d extremal points. This proves $n \geq d$.

Remark. If B contains at least two elements, $\mathcal{H}(C)$ has not less than $2(d(A) - 1)$ extremal points. We then have $n \geq 2(d(A) - 1)$.

REFERENCE

1. J.D. Tarwater and R.C. Entringer, Sums of complexes in torsion free abelian groups. Canad. Math. Bull. 12 (1969) 475-478.

University of Toronto

[Received 15 February, 1969]

41.

On the Density of Discriminants of Cubic Fields

(With H. Davenport)

Bulletin of the London Mathematical Society 1 (1969), 345–348
[Commentary on p. 592]

1. *Introduction*

Several years ago we investigated the following question. Let $P(x)$ denote the number of cubic number fields over $\mathbf{Q}$ with discriminant d in the range $0 < d < x$; let $N(x)$ be similarly defined for the range $-x < d < 0$. Here and in the sequel triplets of conjugate fields are counted once only. Can it be proved that $\lim_{x\to\infty} x^{-1} P(x)$ and $\lim_{x\to\infty} x^{-1} N(x)$ exist and that they are positive? We were unable to answer this question, but could prove the following results.

THEOREM 1.
$$\overline{\lim_{x\to\infty}}\, x^{-1} P(x) \leqslant (5/4)\, \pi^{-2},$$
$$\overline{\lim_{x\to\infty}}\, x^{-1} N(x) \leqslant (15/4)\, \pi^{-2}.$$

THEOREM 2.
$$\underline{\lim_{x\to\infty}}\, x^{-1} P(x) \geqslant (1/240)\, \pi^{-2},$$
$$\underline{\lim_{x\to\infty}}\, x^{-1} N(x) \geqslant \tfrac{1}{4}\pi^{-2}.$$

The proof of Theorem 1 is based on Davenport's [**1**, **2**] estimate of the number of classes of binary cubic forms of discriminant d. The proof of Theorem 2 uses class-field theory for the construction of cubic fields of given discriminant. The constants in Theorem 2 could be improved substantially, but the computing effort would not be negligible and would not narrow the gap considerably. It is also possible to show that the constants in Theorem 1 can be improved. This is supported by the numerical evidence. Godwin and Samet [**3**] have shown that

$$P(20000) = 830 = 20000(5/12)\, \pi^{-2}\, 1{\cdot}0016.$$

2. *Proof of Theorem* 1. Let k be a given cubic number field of discriminant d and basis $1, \omega, \theta$. We put $K = k(\surd d)$, so that $K = k$ if k is cyclic, and K is a sextic field if k is not cyclic; in this case the Galois group of $K/\mathbf{Q}$ is isomorphic to the S_3. We write $\mathfrak{d}(\alpha)$ for the discriminant of α over $\mathbf{Q}$, where α is in k; by $\alpha, \alpha', \alpha''$ we denote the conjugates of α. We assign to k the binary cubic form

$$F_k(x, y) = d^{-\frac{1}{2}}\, \mathfrak{d}^{\frac{1}{2}}(\omega x + \theta y),$$

or explicitly, after factorization in K,

$$F_k(x, y) = d^{-\frac{1}{2}}((\omega-\omega')\, x + (\theta-\theta')\, y)((\omega'-\omega'')\, x + (\theta'-\theta'')\, y)$$
$$((\omega''-\omega)\, x + (\theta''-\theta)\, y),$$

Received 17 June, 1969. [BULL. LONDON MATH. SOC., 1 (1969), 345–348]

where the sign of $d^{-\frac{1}{2}}$ is arbitrary, but fixed. We assert

1. $F_K(x, y)$ has coefficients in $\mathbf{Z}$.
2. $F_K(x, y)$ is irreducible over $\mathbf{Q}$.
3. $F_K(x, y)$ has discriminant d.
4. $F_k(x, y)$ is primitive, i.e. the coefficients of $F_k(x, y)$ are coprime.
5. If k_1 is a cubic field not conjugate to k, then the forms $F_k(x, y)$ and $F_{k_1}(x, y)$ are not equivalent. (Admitting linear substitutions with coefficients in $\mathbf{Z}$ and determinant ± 1.)
6. $F_k(x, y)$ and $F_k(-x, y)$ are not properly equivalent. (Admitting such substitutions with determinant $+1$.)

We now proceed to verify these assertions.

1. The coefficient of x^3 is $d^{-\frac{1}{2}}\mathfrak{d}^{\frac{1}{2}}(\omega)$ which lies in $\mathbf{Z}$ as ω is an integer. Further we can put

$$\theta = (\omega^2 + a\omega + b)/c; \qquad a \in \mathbf{Z}, \quad b \in \mathbf{Z}; \qquad \mathfrak{d}(\omega) = dc^2.$$

Then the coefficient of $x^2 y$ equals

$$d^{-\frac{1}{2}}\mathfrak{d}^{\frac{1}{2}}(\omega)\{(\theta-\theta')/(\omega-\omega') + (\theta'-\theta'')/(\omega'-\omega'') + (\theta''-\theta)/(\omega''-\omega)\}$$
$$= d^{-\frac{1}{2}}\mathfrak{d}^{\frac{1}{2}}(\omega)\, c^{-1}\, Tr_{k/\mathbf{Q}}(2\omega+a) = Tr_{k/\mathbf{Q}}(2\omega+a)$$

which is an integer. Hence, by symmetry, $F_k(x, y)$ has coefficients in $\mathbf{Z}$.

2. If $F_k(x, y)$ were reducible over $\mathbf{Q}$, we could find a pair of rationals $x_0, y_0 \neq 0, 0$ such that $F_k(x_0, y_0) = 0$. But $\mathfrak{d}(x_0\,\omega + y_0\,\theta) \neq 0$, as $x_0\,\omega + y_0\,\theta$ is irrational.

3. A routine calculation shows that the discriminant of $F_k(x, y)$ equals

$$d^{-2}((\omega-\omega')(\theta'-\theta'') - (\theta-\theta')(\omega'-\omega''))^2 \text{ times two similar terms.}$$

Here

$$((\omega-\omega')(\theta'-\theta'') - (\theta-\theta')(\omega'-\omega''))^2 = \begin{vmatrix} 0 & \omega-\omega' & \theta-\theta' \\ 0 & \omega'-\omega'' & \theta'-\theta'' \\ 1 & \omega'' & \theta'' \end{vmatrix}^2$$

$$= \begin{vmatrix} 1 & \omega & \theta \\ 1 & \omega' & \theta' \\ 1 & \omega'' & \theta'' \end{vmatrix}^2 = d.$$

The "two similar terms" also equal d each, and the result follows.

4. If a rational prime p should divide all coefficients of $F_k(x, y)$, p would be a non-essential discriminant divisor in k [**4**; p. 332]. This can only happen if $p = 2$ and d is odd. But d is a polynomial in the coefficients of the form over $\mathbf{Z}$. Hence, if all the coefficients were even, d would be even.

5. The zeros of the polynomial $F_k(x, 1)$ lie in K, and so do the zeros of $F_{k_1}(x, 1)$ if the two forms are equivalent. But K does not contain k_1.

6. If the forms $F_k(x, y)$ and $F_k(-x, y)$ were properly equivalent, there would exist a bilinear inhomogeneous function $R(x)$ with positive determinant and rational

coefficients such that the unordered triplet ρ_i of the zeros of $F_k(x, 1)$ would equal the triplet $R(-\rho_i)$ for $1 \leqslant i \leqslant 3$. But for each individual i $\rho_i = P(-\rho_i)$ is impossible, as ρ_i would be at most quadratic over $\mathbf{Q}$. Hence the substitution $x \to R(-x)$ would give a cyclic permutation of the three ρ_i; the cube of this substitution would have each ρ_i as a fixed point, and would equal the identity. But this is impossible as the bilinear function $R(-x)$ has negative determinant.

Let $h(d)$ denote the number of classes of strictly equivalent irreducible primitive cubic forms over $\mathbf{Z}$ of discriminant d. Davenport has proved **[1, 2]**.

THEOREM 3.

$$\sum_{d=1}^{x} h(d) = (5/2)\,\pi^{-2}\,x + O[x^{15/16}],$$

$$\sum_{d=-x}^{-1} h(d) = \tfrac{1}{2}\,15\pi^{-2}\,x + O[x^{15/16}].$$

As we have shown that to each cubic field there " belong " two such forms, Theorem 1 follows immediately.

3. *Proof of Theorem* 2. We start with a simple Lemma on the density of discriminants of certain *quadratic* number fields.

LEMMA. *Let $P_2(x)$ denote the number of quadratic number fields of discriminant d_2 satisfying*

$$0 < d_2 < x, \quad d_2 \equiv 5 \pmod 8, \quad d_2 \equiv \pm 2 \pmod 5.$$

Let $N_2(x)$ denote the number of quadratic number fields of discriminant d_2 satisfying

$$0 > d_2 > -x, \quad d_2 \equiv 5 \pmod 8.$$

Then

$$\lim_{x\to\infty} x^{-1} P_2(x) = (5/12)\,\pi^{-2}, \quad \lim_{x\to\infty} x^{-1} N_2(x) = \pi^{-2}.$$

Proof. The generating Dirichlet series for the d_2 counted in $P_2(x)$ is

$$\sum_{\substack{n=1 \\ n\equiv 13 \text{ or } 37(40)}}^{\infty} \mu^2(n)\,n^{-s} = (1/16) \sum_{\chi} (\bar{\chi}(13)+\bar{\chi}(37))\, L(s,\chi)/L(2s,\chi^2),$$

where χ runs through the 16 characters mod 40. If χ is not principal, the quotient of the L-series is regular for $\sigma \geqslant \frac{1}{2}$. The term involving the principal character has the form

$$\tfrac{1}{8}(1+2^{-s})^{-1}(1+5^{-s})^{-1}\zeta(s)\zeta^{-1}(2s).$$

This function has a simple pole with residue $(5/12)\,\pi^{-2}$ at $s = 1$. The first result now follows by the usual methods of analytic number theory; and the second is proved similarly.

Now we apply class field theory. The reader who is not familiar with the generation of cubic fields may find it useful to consult Hasse **[4]**. Let d_2 be square free, $d_2 < -3$, $d_2 \equiv 5 \pmod 8$. If and only if the class-number h of $\mathbf{Q}(\sqrt{d_2})$ is divisible by 3, then there exists at least one cubic field of discriminant $d = d_2$. If $h \not\equiv 0 \pmod 3$, we argue as follows. The rational prime 2 remains prime in $\mathbf{Q}(\sqrt{d_2})$, and there are 3 non-zero residue classes mod 2. Let G be the group of ideals prime to 2 in $\mathbf{Q}(\sqrt{d_2})$, and let H_1 be the subgroup of all principal ideals $\equiv 1 \pmod 2$. Then index $(G/H_1) = 3h$ and we can insert between G and H_1 a subgroup H such that index $(G/H) = 3$. H is unique and contains all rationals in G. Hence the class-field K corresponding to H is a sextic field of discriminant $16 d_2^3$, which is normal and not abelian over $\mathbf{Q}$. Let d be the discriminant of the cubic subfield k of K, which is unique apart from conjugacy. As $d^2 | 16 d_2^3$, $d \neq d_2$ it follows that $d = 4d_2$. Thus we have shown: If $d_2 \equiv 5 \pmod 8$, $d_2 < -3$ is discriminant of a quadratic field, then either d_2 or $4d_2$ is discriminant of a cubic field, and the second assertion of Theorem 2 follows from the Lemma.

For positive discriminants the proof proceeds on similar lines, but the units create a complication. Let $d_2 > 0$ be discriminant of a quadratic field, $d_2 \equiv 5 \pmod 8$, $d_2 \equiv \pm 2 \pmod 5$. If and only if the class-number h of the field $\mathbf{Q}(\sqrt{d_2})$ is divisible by 3, then there exists at least one cubic field of discriminant d_2. Otherwise, we note that the rational primes 2 and 5 remain prime in $\mathbf{Q}(\sqrt{d_2})$. The multiplicative residue class group mod 10 has order 72 and is of type (3, 3, 8). As the positive units form a cyclic group, there exists a subgroup F of 24 residue classes mod 10, which contains all units. F need not be unique. Let G be the group of all ideals in $\mathbf{Q}(\sqrt{d_2})$ which are prime to 10. Let H_1 be the group of all principal ideals in F. Again index $(G/H_1) = 3h$, and we can insert between G and H_1 a group H such that index $(G/H) = 3$. Again all rationals prime to 10 lie in H_1, and *a fortiori* in H. Hence the class-field K corresponding to H is normal, but not abelian over $\mathbf{Q}$. The conductor of the 2 cubic characters of G which are trivial on H is either 2 or 5 or 10. Hence the discriminant of K is either $2^4 d_2^3$ or $5^4 d_2^3$ or $10^4 d_2^3$; and the discriminant d of k, the cubic subfield of K, satisfies $d = 4d_2$ or $d = 25d_2$ or $d = 100d_2$. In any case to each d_2 corresponds at least one cubic field k of discriminant $d = m^2 d_2$, where $m | 10$. This fact and the lemma establish the first assertion of Theorem 2.

References

1. H. Davenport, " On the class-number of binary cubic form (I) ", *J. London Math. Soc.*, 26 (1951), 183–192. *Corrigendum, ibidem*, 27 (1952), 512.
2. ———, " On the class-number of binary cubic forms (II)", *J. London Math. Soc.*, 26 (1951), 192–198.
3. H. J. Godwin and P. A. Samet, " A table of real cubic fields ", *J. London Math. Soc.*, 34 (1959), 108–110.
4. H. Hasse. " Arithmetische Theorie der kubischen Zahlkörper auf klassenkörpertheoretischer Grundlage ", *Math. Z.*, 31 (1930), 565–582.
5. ———, *Zahlentheorie* (Akademie-Verlag, Berlin, 1949).

Trinity College,
Cambridge.

Department of Mathematics,
University of Toronto.

42.

On the Density of Discriminants of Cubic Fields II

(With H. Davenport)

Proceedings of the Royal Society, London A 322 (1971), 405–420
[Commentary on p. 592]

An asymptotic formula is proved for the number of cubic fields of discriminant $\mathfrak{d}$ in $0 < \mathfrak{d} < X$; and in $-X < \mathfrak{d} < 0$.

1. Introduction

Let $N_3(\xi, \eta)$ denote the number of cubic fields K with discriminant $\mathfrak{d}_K$ satisfying $\xi < \mathfrak{d}_K < \eta$, where a triplet of conjugate fields is counted once only. The main purpose of this paper is to prove

THEOREM 1.
$$X^{-1}N_3(0, X) \to (12\zeta(3))^{-1} \quad as \quad X \to \infty,$$
$$X^{-1}N_3(-X, 0) \to (4\zeta(3))^{-1} \quad as \quad X \to \infty.$$

In a previous paper (Davenport & Heilbronn 1969) we proved the weaker result that the upper and lower limits are finite and positive. This proof is a refinement of our previous method. We showed then that there exists a discriminant-preserving 1–1 relation between cubic fields and a subset U of the classes of irreducible primitive cubic binary forms $F(x, y)$ with coefficients in $\mathbf{Z}$. In this paper U will be determined explicitly by congruence conditions on the coefficients of F. Using an easy generalization of Davenport's earlier results on the class-number of binary cubic forms (Davenport 1951 *a*, *b*) we obtain an estimate of the cardinality of U, and thus theorem 1.

As a by-product, two further results will be obtained. Let K_6 be the sextic normal extension of the non-cyclic cubic field K, and let p be a rational prime unramified in K (and hence in K_6). Then the Frobenius–Artin symbol $\{(K_6/Q)/p\}$ is defined as a conjugacy class of the S_3, its values being I or $A_3 - I$ or $S_3 - A_3$, where I is the identity class of S_3. Then it is a consequence of the Frobenius–Chebotarev density theorem that for fixed K and varying p (unramified in K) the values I, $A_3 - I$, $S_3 - A_3$ occur with relative frequency $1:2:3$. We shall prove

THEOREM 2. *Let p be a fixed prime, and let K run through the cubic non-cyclic fields in which p does not ramify, the fields being ordered by the size of the discriminants. Then the Frobenius–Artin symbol $\{(K_6/Q)/p\}$ takes the values I, $A_3 - I$, $S_3 - A_3$ with relative frequency $1:2:3$.*

Actually we shall do a little more. We shall also determine for each p the density of cubic fields K in which p is totally ramified, and the density of fields K in which p is partially ramified.

Another application of the method of this paper deals with the 3-class-number of quadratic fields. Let $h_3^*(\Delta_2)$ be the number of those ideal classes in the quadratic field of discriminant Δ_2 whose cube is the unit class. We shall prove

THEOREM 3.
$$\sum_{0<\Delta_2<X} h_3^*(\Delta_2) \sim \tfrac{4}{3} \sum_{0<\Delta_2<X} 1 \quad as \quad X \to \infty,$$
$$\sum_{-X<\Delta_2<0} h_3^*(\Delta_2) \sim 2 \sum_{-X<\Delta_2<0} 1 \quad as \quad X \to \infty.$$

This theorem suggests the possibility that the relative density of positive and negative discriminants Δ_2 for which the congruence $h_3^*(\Delta_2) \equiv 0 (\mathrm{mod}\, 3^n)$ holds, is 3^{-2n} and 3^{1-2n} respectively for $n > 0$. But at the moment there does not seem to be any hope of proving results of this nature.

2. Notation and definitions

Small roman letters are reserved for rational integers, p is always a positive prime.

Φ is the set of all irreducible primitive binary cubic forms
$$F(x,y) = ax^3 + bx^2y + cxy^2 + dy^3$$
of discriminant
$$D = b^2c^2 + 18abcd - 27a^2d^2 - 4b^3d - 4c^3a.$$

The letters a, b, c, d and D will always be reserved for the coefficients and discriminant of the form F.

Two forms $F(x, y)$ and $F'(x', y')$ are called equivalent, or integrally equivalent, if there exists a unimodular 2 by 2 matrix M of determinant ± 1 such that the substitution $(x', y') = M(x, y)$ transforms F' into F. For quadratic forms we retain the classical definition of equivalence, which requires that $\det(M) = 1$.

Two forms $F(x, y)$ and $F'(x', y')$ in Φ are called rationally equivalent if there exists a non-singular 2 by 2 matrix M over $\boldsymbol{Z}$ such that the substitution $(x', y') = M(x, y)$ transforms F' into δF, where $\delta \neq 0$ is rational. This definition will only be used in §6.

The congruence $F_1(x, y) \equiv F_2(x, y)\,(\mathrm{Mod}\, m)$ will denote that each coefficient of F_1 is congruent $(\mathrm{mod}\, m)$ to the corresponding coefficient of F_2, whereas
$$F_1(x, y) \equiv F_2(x, y)\,(\mathrm{mod}\, m)$$
will imply only that for each pair $x, y \in \boldsymbol{Z}$ the forms assume values congruent to each other $(\mathrm{mod}\, m)$.

Now we define the symbol (F, p) for $F \in \Phi$. We put
$$(F,p) = (111) \quad \text{if} \quad F \equiv \lambda_1(x,y)\,\lambda_2(x,y)\,\lambda_3(x,y) \quad (\mathrm{Mod}\, p),$$
where λ_1, λ_2, λ_3 are linear forms mod p, no two of which have a constant quotient.
$$(F,p) = (12) \quad \text{if} \quad F(x,y) \equiv \lambda(x,y)\,\kappa(x,y) \quad (\mathrm{Mod}\, p),$$

where $\lambda(x,y)$ is a linear form and $\kappa(x,y)$ is a quadratic form which is irreducible Mod p.

$$(F,p) = (3) \quad \text{if} \quad F(x,y) \equiv \kappa(x,y) \quad (\text{Mod}\, p),$$

where $\kappa(x,y)$ is irreducible Mod p.

$$(F,p) = (1^3) \quad \text{if} \quad F(x,y) \equiv \alpha\lambda^3(x,y) \quad (\text{Mod}\, p),$$

where $\lambda(x,y)$ is a linear form, and α a constant mod p.

$$(F,p) = (1^2\,1) \quad \text{if} \quad F(x,y) \equiv \lambda_1^2(x,y)\,\lambda_2(x,y) \quad (\text{Mod}\, p),$$

where $\lambda_1(x,y)$ and $\lambda_2(x,y)$ are linear forms with a non-constant quotient.

If F_1 and F_2 are either equivalent or congruent (Mod p) clearly $(F_1,p) = (F_2,p)$. Note also that $p|D$ if and only if $(F,p) = (1^3)$ or $(F,p) = (1^2\,1)$; further that $(F,p) = (1^3)$ implies $p^2|D$. By $T_p(111)$, $T_p(12)$, etc., we denote the set of $F \in \Phi$ for which $(F,p) = (111)$, $(F,p) = (12)$, etc. (Clearly each set T_p consists of classes of equivalent forms.) We define W_p by the relation

$$F \in W_p \Leftrightarrow D \equiv 0 \quad (\text{mod}\, p^2).$$

Next we define for each p subsets V_p and U_p of Φ. $F \in V_2$ if $D \equiv 1$ (mod 4) or if $D \equiv 8$ or 12 (mod 16). $F \in V_p$ for $p \neq 2$ if $F \notin W_p$. $F \in U_p$ if $F \in V_p$ or if $(F,p) = (1^3)$ and if the congruence $F(x,y) \equiv ep$ (mod p^2) has a solution for some $e \not\equiv 0$ (mod p). Finally we put

$$V = \bigcap_p V_p, \quad U = \bigcap_p U_p.$$

Clearly all the sets V_p, U_p, V and U consist of complete classes of equivalent forms.

By the letter K we denote a cubic number field, by $\mathfrak{d}_K$ the discriminant of K. If $\alpha \in K$, we denote by $\text{Nm}(\alpha)$, $\text{tr}(\alpha)$, $\mathfrak{d}(\alpha)$ the norm, trace and discriminant of α taken in K over $\boldsymbol{Q}$.

Let S be a subset of Φ consisting of complete equivalence classes. Then we denote by $N(\xi,\eta;S)$ the number of classes in S whose forms have a discriminant D with $\xi < D < \eta$.

Let $\Delta_2 \in \boldsymbol{Z}$, $\Delta_2 \equiv 0$ or 1 (mod 4), Δ_2 not a square. Then $h_3^*(\Delta_2)$ denotes the number of those classes of primitive quadratic form of discriminant Δ_2 whose cube is the unit class. If Δ_2 is a field discriminant, this definition agrees with the definition given in the introduction.

$\tau(n)$ denotes the number of positive divisors of n.

Constants implied in the symbol O are independent of all parameters.

3. Local densities

In this section we consider forms $F \in \Phi$ over the residue class ring mod p^r for $r = 1$ and $r = 2$. Naturally, we neglect irreducibility over $\boldsymbol{Q}$. The number of such forms is $p^{4r}(1-p^{-4})$. Let S be a set of forms in Φ. We denote by $A(S;p^r)$ the number of residue classes mod p^r occupied by forms in S, divided by $p^{4r}(1-p^{-4})$.

LEMMA 1. *For* $r = 1$ *and* $r = 2$

$$A(T_p(111); p^r) = \tfrac{1}{6}p(p-1)(p^2+1)^{-1},$$
$$A(T_p(12); p^r) = \tfrac{1}{2}p(p-1)(p^2+1)^{-1},$$
$$A(T_p(3); p^r) = \tfrac{1}{3}p(p-1)(p^2+1)^{-1},$$
$$A(T_p(1^3); p^r) = (p^2+1)^{-1},$$
$$A(T_p(1^21); p^r) = p(p^2+1)^{-1}.$$

Proof. As the definition of (F,p) depends only on the residue-class of $F\,(\mathrm{Mod}\,p)$, it suffices to prove the lemma for $r = 1$. Call a form normalized if the highest non-vanishing coefficient equals 1. It is well known that the number of normalized homogeneous polynomials in x and y irreducible Mod p of degree 1, 2 and 3 equals $p+1$, $\tfrac{1}{2}p(p-1)$ and $\tfrac{1}{3}p(p-1)(p+1)$ respectively. The lemma now follows by an elementary counting process.

DEFINITION (only used in this section). $S_1 = S_{1,p}$ *denotes the set of forms* $F \in \Phi$ *satisfying*

$$a \not\equiv 0\,(\mathrm{mod}\,p), \quad b \equiv c \equiv 0\,(\mathrm{mod}\,p), \quad d \equiv 0\,(\mathrm{mod}\,p^2).$$

$S_2 = S_{2,p}$ *denotes the set of forms* $F \in \Phi$ *satisfying*

$$b \not\equiv 0\,(\mathrm{mod}\,p), \quad a \equiv c \equiv 0\,(\mathrm{mod}\,p), \quad d \equiv 0\,(\mathrm{mod}\,p^2).$$

Σ_1 *and* Σ_2 *denote the set of forms in* Φ *which are equivalent to at least one* F *in* S_1 *and* S_2 *respectively.*

Note that $F \in \Sigma_1 \Rightarrow (F,p) = (1^3)$ and $F \in \Sigma_2 \Rightarrow (F,p) = (1^2\,1)$.

LEMMA 2.
$$A(\Sigma_1; p^2) = p^{-1}(p^2+1)^{-2},$$
$$A(\Sigma_2; p^2) = (p^2+1)^{-2}.$$

Proof. It is clear that

$$A(S_1;p^2) = A(S_2;p^2) = p^{-1}(p+1)^{-1}(p^2+1)^{-1}.$$

Let $\begin{pmatrix} k & l \\ m & n \end{pmatrix}$ be a linear substitution mod p^2 of determinant ± 1. Then if $F \in S_1$,

$$F(kx+ly,\, mx+ny) \equiv a(kx+ly)^3 \quad (\mathrm{Mod}\,p);$$

so this form lies in S_1 only if $l \equiv 0\,(\mathrm{mod}\,p)$. Conversely, if $l \equiv 0\,(\mathrm{mod}\,p)$,

$$F(kx+ly, mx+ny) \equiv a(kx+ly)^3 + ckx(mx+ny)^2 + b(kx)^2(mx+ny) \quad (\mathrm{Mod}\,p^2)$$

and the form lies in S_1. The unimodular substitutions mod p^2 with $l \equiv 0\,(\mathrm{mod}\,p)$ form a subgroup of index $p+1$ of the group of all unimodular substitutions mod p^2. Hence

$$A(\Sigma_1;p^2) = (p+1)\,A(S_1;p^2) = p^{-1}(p^2+1)^{-1}.$$

Similarly, if $F \in S_2$,

$$\begin{aligned} F(kx+ly,\, mx+ny) &\equiv b(kx+ly)^2(mx+ny) \\ &\equiv bk^2mx^3 + bk(2lm+kn)\,x^2y + bl(lm+2kn)\,xy^2 + bl^2ny^3 \\ &\equiv a'x^3 + b'x^2y + c'xy^2 + d'y^3 \quad (\mathrm{Mod}\,p) \quad \text{say.} \end{aligned}$$

Assume this form lies in S_2. Then $p\nmid b'$, hence $p\nmid k$. As $p|a'$, $p\nmid b$, we have $p|m$. As $p\nmid b'$, $p|m$, we have $p\nmid n$. As $p|d'$, $p\nmid bn$, we have $p|l$.

Conversely, if $l \equiv m \equiv 0 \pmod p$,

$$\begin{aligned} F(kx+ly, mx+ny) &\equiv ak^3x^3 + b(kx+ly)^2(mx+ny) + ckn^2xy^2 + dn^3y^3 \\ &\equiv (ak^3+bk^2m)x^3 + b(k^2n+2klm)x^2y \\ &\quad + (b(2kln+l^2m)+ckn^2)xy^2 + (bl^2n+dn^3)y^3 \pmod{p^2}. \end{aligned}$$

Thus this form belongs to S_2. The unimodular matrices with $l \equiv m \equiv 0 \pmod p$ form a subgroup of index $p(p+1)$ in the group of all unimodular matrices mod p^2. Hence

$$A(\Sigma_2; p^2) = p(p+1)A(S_2; p^2) = (p^2+1)^{-2}.$$

LEMMA 3. *$\Phi = V_p \cup T_p(1^3) \cup \Sigma_2$ for all p, and no two sets on the right have an element in common.*

Proof. It is clear that each F with $(F,p) \neq (1^2 1)$ belongs to one and only one of these sets. Hence we need only prove the lemma for $F \in T(1^2 1)$. Such F may be assumed to have coefficients a, b, c, d such that

$$a \equiv c \equiv d \equiv 0 \pmod p, \quad b \not\equiv 0 \pmod p.$$

Then

$$D \equiv -4b^3d \pmod{p^2}.$$

Thus for $p \neq 2$, $D \equiv 0 \pmod{p^2}$ if and only if $d \equiv 0 \pmod{p^2}$. This shows that every form of $T_p(1^2 1)$ lies either in V_p or in Σ_2.

For $p = 2$ we have

$$D \equiv b^2c^2 - 4b^3d \equiv 4((\tfrac{1}{2}c)^2 - bd) \pmod{16}.$$

Thus $d \equiv 0 \pmod 4$ if and only if $D \equiv 0$ or $4 \pmod{16}$. This proves the lemma.

LEMMA 4. $A(V_p; p^2) = (p^2-1)(p^2+1)^{-1}$ *for all p.*

Proof. By lemma 3

$$1 = A(V_p; p^2) + A(T_p(1^3); p^2) + A(\Sigma_2; p^2).$$

By lemmas 1 and 2

$$A(T_p(1^3); p^2) = (p^2+1)^{-1}, \quad A(\Sigma_2; p^2) = (p^2+1)^{-1},$$

and the result follows.

LEMMA 5. $A(U_p; p^2) = (p^3-1)p^{-1}(p^2+1)^{-1}$ *for all p.*

Proof. It follows from the definition of U_p that

$$T_p(1^3) = (T_p(1^3) \cap U_p) \cup \Sigma_1, \quad U_p = V_p \cup (T_p(1^3) \cap U_p).$$

As $\Sigma_1 \cap U_p$ is empty, we have

$$\begin{aligned} U_p \cup \Sigma_1 &= V_p \cup T_p(1^3), \\ A(U_p; p^2) &= A(V_p; p^2) + A(T_p(1^3); p) - A(\Sigma_1; p^2) \\ &= (p^2-1)(p^2+1)^{-1} + (p^2+1)^{-2} - p^{-1}(p^2+1)^{-1} \end{aligned}$$

by lemmas 4, 1 and 2. Hence the assertion follows.

LEMMA 6. *If* $(F, p) = (1^3)$, $p \neq 3$ *then* $F \in U_p$ *if and only if* $D \not\equiv 0 \pmod{p^3}$. *If* $(F, 3) = (1^3)$, $F \in U_3$, *then* $D \not\equiv 0 \pmod{729}$.

Proof. Assume $a \not\equiv 0 \pmod p$, $b \equiv c \equiv d \equiv 0 \pmod p$. Then for $p \neq 3$

$$D \equiv -27a^2d^2 \pmod{p^3}.$$

Hence $D \equiv 0 \pmod{p^3}$ if and only if $d \equiv 0 \pmod{p^2}$.

For $p = 3$, put $b = 3\beta$, $c = 3\gamma$, $d = 3\delta$, so that $3 \nmid \delta$. Then

$$D = 81\beta^2\gamma^2 + 486a\beta\gamma\delta - 243a^2\delta^2 - 324\beta^3\delta - 108\gamma^3 a.$$

If $3 \nmid \gamma$, $$D \equiv -108\gamma^3 a \pmod{81}.$$

If $3 | \gamma$, $$D \equiv -81\delta(3a^2\delta - 4\beta^3) \pmod{729}.$$

Hence in either case $D \not\equiv 0 \pmod{729}$.

4. AN AUXILIARY PROPOSITION

In order to apply a simple sieve method later, we require

PROPOSITION 1. $N(-X, X; W_p) = O(xp^{-2})$ *as* $X \to \infty$.

We first prove

LEMMA 7. $$\sum_{|\Delta_2| < X} h_3^*(\Delta_2) = O(X) \text{ as } X \to \infty,$$

where Δ_2 *runs through the discriminants of quadratic fields.*

Proof. This lemma follows from our old theorem

$$N_3(-X, X) = O(X) \quad \text{as} \quad X \to \infty$$

(Davenport & Heilbronn 1969) as theorem 3 will follow from theorem 1. (See §7.)

We now introduce the Hessian $H(x, y)$ of a given cubic form $F(x, y)$. H is defined by the relation

$$H(x,y) = -\tfrac{1}{4}(F_{xx}F_{yy} - F_{xy}^2),$$

where the lower indices denote partial derivatives. It is well known that $H(x, y)$ is a covariant of $F(x, y)$ with respect to linear substitutions of determinant 1. A simple calculation gives

$$H(x,y) = (bx+cy)^2 - (3ax+by)(cx+3dy)$$
$$= Px^2 + Qxy + Ry^2, \quad \text{say},$$

where $P = b^2 - 3ac$, $Q = bc - 9ad$, $R = c^2 - 3bd$. An easy calculation shows the discriminant Δ of H is given by

$$\Delta = Q^2 - 4PR = -3D.$$

The class of H is uniquely determined by the class of F, but the converse is not necessarily true. The formula for Δ shows H is reducible if and only if $-3D$ is a square. H is primitive if and only if for all primes p $(F, p) \neq (1^3)$. So we put

$$M = (P, Q, R), \quad P = MP_1, \quad Q = MQ_1, \quad R = MR_1,$$

$$H_1(x, y) = P_1 x^2 + Q_1 xy + R_1 y^2,$$

and this quadratic form has discriminant

$$\Delta_1 = Q_1^2 - 4P_1 R_1 = M^{-2}\Delta = -3M^{-2}D.$$

The explicit definition of $H(x, y)$ leads immediately to the identities

$$H_1(b, -3a) = MP_1^2,$$

$$H_1(c, -b) = MP_1 R_1,$$

$$H_1(3d, -c) = MR_1^2.$$

LEMMA 8. *Let $k > 0$, $M > 0$, $M \in \mathbf{Z}$. Let $B = B(k, M)$ denote the number of classes of forms in Φ with Hessian $H(x, y) = M(kx + ly)y$, where $0 \leqslant l < k$, $(l, k) = 1$. Then*

$$B \leqslant 2k\tau(M).$$

Moreover, if p is a prime such that $p|k$, $p^2 \nmid M$, then

$$B \leqslant 6kp^{-1}\tau(M).$$

Proof. Let F be a form in Φ with Hessian

$$H(x, y) = M(kx + ly)y = MH_1(x, y), \quad \text{say.}$$

We may assume that $a > 0$. The equations

$$H_1(b, -3a) = (kb - 3al)(-3a) = MP_1^2 = 0,$$

$$H_1(c, -b) = (kc - bl)(-b) = MP_1 R_1 = 0$$

yield $b = 3k^{-1}la$ and, if $l \neq 0$, $c = 3k^{-2}l^2 a$. If $l = 0$, the third equation

$$H_1(3d, -c) = (3kd - cl)(-c) = MR_1^2 = Ml^2$$

yields $c = 0$ because $d \neq 0$. Hence F has the form

$$F(x, y) = a(x + k^{-1}ly)^3 \pm (9a)^{-1} Mky^3,$$

the last coefficient being determined by the value of

$$D = -\tfrac{1}{3}M^2k^2 = -27a^2((9a)^{-1}Mk)^2.$$

As the coefficients of F are integers, we obtain the congruences

$$3al^2 \equiv 0 \pmod{k^2}, \quad 9a^2l^3 \pm Mk^4 \equiv 0 \pmod{9ak^3}.$$

If $k = 1$, the second congruence shows that $a|M$, so that we have $\tau(M)$ choices for a and one choice for l which proves our result.

If $k > 1$, the first congruence shows that $k^2|3a$, so we can put $3a = sk^2$. The second congruence now reads

$$s^2l^3 \pm M \equiv 0 \pmod{3sk}.$$

This implies that $s|M$ and we can find at most $\tau(M)$ values of a and at most k values of l. This proves our first result for $k > 1$.

Now assume the existence of p with $p|k$, $p^2 \nmid M$. Then $p \nmid s$ and the congruence

$$s^2 l^3 \pm M \equiv 0 \pmod p$$

has at most six solutions mod p. Hence the original congruence has at most $6kp^{-1}$ solutions in $0 < l < k$. This proves the last assertion of the lemma.

LEMMA 9. *If $M > 0$ and $H_1(x,y)$ are given, and if Δ_1 is not a square, then there are at most $18\tau(M)$ classes of irreducible primitive cubic forms with Hessian equivalent to $MH_1(x,y)$.*

Proof. As $H_1(x,y)$ is primitive we may assume that P_1 is a prime. Assume first that $\Delta_1 < 0$. Then

$$H_1(b, -3a) = MP_1^2.$$

Hence by the theory of definite primitive quadratic forms, the number of representations of MP_1^2 is at most $6\tau(MP_1^2) \leqslant 18\tau(M)$.

Thus there are at most $18\tau(M)$ choices for a, b. As a, b, P_1, Q_1 determine c and d uniquely (since $a \neq 0$), the lemma follows for $\Delta < 0$.

For a positive Δ the situation is not so simple, as the form $H_1(x,y)$ has a cyclic infinite group of automorphs.

We write $H(x,y)$ in the form

$$H(x,y) = MH_1(x,y) = MP_1(x+\theta y)(x+\theta' y)$$

where
$$\theta = (2P_1)^{-1}(Q_1+\sqrt{\Delta_1}),\ \theta' = (2P_1)^{-1}(Q_1-\sqrt{\Delta_1}).$$

If $H(x,y)$ is the Hessian of $F(x,y)$, we have

$$3(\theta-\theta')F(x,y) = (b-3a\theta')(x+\theta y)^3-(b-3a\theta)(x+\theta' y)^3.$$

Let $\epsilon > 1$ be the smallest unit in $\boldsymbol{Q}(\sqrt{\Delta_1})$ which can be written in the form

$$\epsilon = \tfrac{1}{2}(e_1+e_2\sqrt{\Delta_1}).$$

The non-trivial automorphs of $H(x,y)$ are then generated by the substitution S

$$x^*+\theta y^* = \epsilon(x+\theta y),$$
$$x^*+\theta' y^* = \epsilon^{-1}(x+\theta' y).$$

Hence
$$b^*-3a^*\theta = \epsilon^3(b-3a\theta),$$
$$b^*-3a^*\theta' = \epsilon^{-3}(b-3a\theta').$$

This shows that if the x, y space is transformed by S, the b, $-3a$ space is transformed by S^3. Thus we need only count solutions of

$$H_1(b,-3a) = MP_1^2$$

subject to equivalence by S^{3n}, as two solutions which differ only by S^{3n} lead to equivalent forms F. The number of solutions not equivalent by S^n are at most $2\tau(MP_1^2)$, hence the number of solutions not equivalent by S^{3n} is at most $6\tau(MP_1^2) \leqslant 18\tau(M)$, as P_1 may be assumed to be a prime.

LEMMA 10. *Let $M > 0$ and $\Delta_1 \equiv 0$ or $1 \pmod 4$ be elements of $\boldsymbol{Z}$, Δ_1 not a square. Then there exist at most $3\tau(M)\,h_3^*(\Delta_1)$ classes of primitive quadratic forms*

$$H_1(x,y) = P_1x^2+Q_1xy+R_1y^2 \quad \textit{with} \quad Q_1^2-4P_1R_1 = \Delta_1,$$

such that MH_1 is the Hessian of a form $F \in \Phi$.

Proof. Let $F(x, y)$ be a form in Φ with Hessian $MH_1(x, y)$. Then we have

$$P_1 b^2 - 3Q_1 ba + 9R_1 a^2 = MP_1^2.$$

Without loss of generality we may assume that P_1 is a prime.

We now consider classes of equivalent primitive quadratic forms of discriminant Δ_1. Let η be the class of H_1 and let $\mu_1, \ldots, \mu_t$ be the classes which represent M. It follows from the theory of composition of quadratic forms that $1 \leqslant t \leqslant \tau(M)$. Hence there exists at least one s in $1 \leqslant s \leqslant t$ such that at least one of the following three relations holds:

$$\eta = \mu_s \quad \text{or} \quad \eta = \mu_s \eta^2 \quad \text{or} \quad \eta = \mu_s \eta^{-2}.$$

The number of such η is at most

$$t(2 + h_3^*(\Delta_1)) \leqslant \tau(M)(2 + h_3^*(\Delta_1)) \leqslant 3\tau(M) h_3^*(\Delta_1).$$

Proof of proposition 1. We first deal with those classes for which $-3D$ is a square. We have to find an upper bound for the sum

$$\sum_{\substack{Mk < (3X)^{\frac{1}{2}} \\ p \mid Mk}} B(k, M) = \sum_{Mk < (3X)^{\frac{1}{2}} p^{-1}} B(k, pM) + \sum_{\substack{Mk < (3X)^{\frac{1}{2}} p^{-1} \\ p \nmid M}} B(pk, M).$$

To the first sum we apply the first estimate in lemma 8, to the second sum the second estimate. Then our bound is

$$\begin{aligned} &\leqslant \sum_{kM < (3X)^{\frac{1}{2}} p^{-1}} (2k\tau(pM) + 6k\tau(M)) \\ &\leqslant 10 \sum_{M < (3X)^{\frac{1}{2}} p^{-1}} \tau(M) \sum_{k < (3X)^{\frac{1}{2}} p^{-1} M^{-1}} k \\ &\leqslant 10(3X) p^{-2} \sum_{M=1}^{\infty} \tau(M) M^{-2} \\ &= O(Xp^{-2}). \end{aligned}$$

Now we have to count those classes for which $-3D$ is not a square and the Hessian is irreducible. That means, by virtue of lemma 9 and lemma 10 we have to find an upper bound for the sum

$$\sum_{\substack{|M^2 \Delta_1| \leqslant 3X \\ p^2 \mid M^2 \Delta_1}} 54\tau^2(M) h_3^*(\Delta_1),$$

where Δ_1 is restricted to discriminants of quadratic forms. Each such Δ_1 can be factorized uniquely in the form $\Delta_1 = L^2 \Delta_2$, where $L > 0$, $L \in \mathbf{Z}$ and Δ_2 is discriminant of a quadratic field. For $p = 2$ the proposition follows from Davenport's theorem, so we may assume $p \neq 2$. Hence $p^2 \nmid \Delta_2$, and $p^2 \mid M^2 \Delta_1$ implies $p \mid ML$.

To express $h_3^*(\Delta_1)$ by $h_3^*(\Delta_2)$ exactly is difficult; it is however well known that

$$h_3^*(\Delta_1) \mid 3^n h_3^*(\Delta_2),$$

where n denotes the number of distinct prime divisors of L. Hence

$$h_3^*(\Delta_1) \leqslant \tau^2(L) h_3^*(\Delta_2).$$

Substituting this in our formula for the upper bound we obtain

$$54 \sum_{\substack{|M^2L^2\Delta_2|<3X \\ p|ML}} \tau^2(M)\tau^2(L)h_3^*(\Delta_2).$$

By virtue of lemma 7 this is majorized by

$$O(X) \sum_{\substack{M=1 \\ p|ML}}^{\infty} \sum_{L=1}^{\infty} \tau^2(M)\tau^2(L)M^{-2}L^{-2} = O(Xp^{-2}).$$

5. Global densities

The starting-point of this section is the

THEOREM (Davenport 1951 *a, b*)

$$N(0,X;\Phi) = \tfrac{5}{4}\pi^{-2}X + O(X^{\frac{15}{16}}),$$
$$N(-X,0;\Phi) = \tfrac{15}{4}\pi^{-2}X + O(X^{\frac{15}{16}}).$$

Actually we require a refinement of this theorem. Let $m \geqslant 1$ and S_m be a set of forms in ϕ which are defined by conditions on the residue classes of $a, b, c, d \pmod m$. Moreover let S_m be a union of equivalence classes of Φ. Then

$$\lim_{X\to\infty} X^{-1}N(0,X;S_m) = \tfrac{5}{4}\pi^{-2}A(S_m;m),$$
$$\lim_{X\to\infty} X^{-1}N(-X,0;S_m) = \tfrac{15}{4}\pi^{-2}A(S_m;m).$$

This extension is proved in exactly the same way as the original theorem. It does not hold uniformly in m.

Let Y be a large integer in $\boldsymbol{Z}$, and let

$$P_Y = \prod_{p<Y} p.$$

Then as $X \to \infty$, for fixed Y,

$$\begin{aligned} X^{-1}N(X,0;\textstyle\bigcap_{p<Y} U_p) &\to \tfrac{5}{4}\pi^{-2}A(\textstyle\bigcap_{p<Y} U_p;P_Y^2) \\ &= \tfrac{5}{4}\pi^{-2}\prod_{p<Y} A(U_p;p^2) \\ &= \tfrac{5}{4}\pi^{-2}\prod_{p<Y}(p^3-1)p^{-1}(p^2+1)^{-1} \end{aligned}$$

by lemma 5. Thus

$$\limsup_{X\to\infty} X^{-1}N(X,0;U) \leqslant \tfrac{5}{4}\pi^{-2}\prod_{p<Y}(p^3-1)p^{-1}(p^2+1)^{-1}.$$

As this is true for all $Y > 0$, we may replace the product by the infinite product over all primes. This gives

$$\begin{aligned} \limsup_{X\to\infty} X^{-1}N(X,0;U) &\leqslant \tfrac{5}{4}\pi^{-2}\prod_p (1-p^{-3})(1+p^{-2})^{-1} \\ &= \tfrac{5}{4}\pi^{-2}\zeta(3)^{-1}\zeta(2)^{-1}\zeta(4) = \tfrac{5}{4}\pi^{-2}\zeta(3)^{-1}(6\pi^{-2})(\pi^4/90) \\ &= (12\zeta(3))^{-1}. \end{aligned}$$

To obtain a lower bound for $N(0, X; U)$ we observe that

$$\bigcap_{p<Y} U_p \subset (U \cup \bigcup_{p \geq Y} W_p).$$

Hence, using proposition 1,

$$\tfrac{5}{4}\pi^{-2} \prod_{p<Y} (p^3-1)p^{-1}(p^2+1)^{-1} \leq \liminf_{X\to\infty} (X^{-1}N(0,X;U) + X^{-1} \sum_{p\geq Y} N(0,X;W_p))$$

$$\leq \liminf_{X\to\infty} (X^{-1}N(0,X;U)) + O \sum_{p\geq Y} p^{-2}.$$

Letting Y tend to infinity, this gives

$$\liminf_{X\to\infty} X^{-1}N(0,X;U) \geq \tfrac{5}{4}\pi^{-2} \prod_{p} (p^3-1)p^{-1}(p^2+1)^{-1} = (12\zeta(3))^{-1}.$$

The same argument works for negative discriminants. We have thus proved

PROPOSITION 2.
$$\lim_{X\to\infty} X^{-1} N(0,X;U) = (12\zeta(3))^{-1},$$
$$\lim_{X\to\infty} X^{-1} N(-X,0;U) = (4\zeta(3))^{-1}.$$

Applying the same argument to V instead of U, we note that the relation

$$\bigcap_{p<Y} V_p \subset (V \cup \bigcup_{p\geq Y} W_p)$$

still holds. Also by lemma 4

$$A(V_p; p^2) = (p^2-1)(p^2+1)^{-1},$$

$$\tfrac{5}{4}\pi^{-2} \prod_{p} (1-p^{-2})(1+p^{-2})^{-1} = \tfrac{5}{4}\pi^{-2}\zeta(4)\zeta(2)^{-2}$$
$$= \tfrac{5}{4}\pi^{-2}(\pi^4/90)(36/\pi^4) = \tfrac{1}{2}\pi^{-2}.$$

This gives

PROPOSITION 3.
$$\lim_{X\to\infty} X^{-1}N(0,X;V) = (2\pi^2)^{-1},$$
$$\lim_{X\to\infty} X^{-1}N(-X,0;V) = 3(2\pi^2)^{-1}.$$

6. The fundamental mapping

Let K be a cubic field over Q. In our previous paper we attached to each K a binary cubic form in the following way. Let 1, ω, ν be an integral basis of K. Put

$$F_K(x,y) = \mathfrak{d}_K^{-\frac{1}{2}}\,\mathfrak{d}^{\frac{1}{2}}(\omega x + \nu y),$$

where $\mathfrak{d}_K$ denotes the absolute discriminant of K. We proved

(1) $F_K \in \Phi$.
(2) F_K is uniquely determined by K apart from equivalence.
(3) If K' is conjugate to K, $F_{K'}$ is equivalent to F_K.
(4) $D(F_K) = \mathfrak{d}_K$.
(5) If K_1 is not conjugate to K, then F_{K_1} is not even rationally equivalent to F_K.

Lemma 11. *The rational prime p factorizes in K according to the following table*:

$$\begin{aligned}
(p) &= \mathfrak{p}_1\mathfrak{p}_2\mathfrak{p}_3 && \text{if } (F_K, p) = (111),\\
(p) &= \mathfrak{p}_1\mathfrak{p}_2 && \text{if } (F_K, p) = (12),\\
(p) &= (p) && \text{if } (F_K, p) = (3),\\
(p) &= \mathfrak{p}^3 && \text{if } (F_K, p) = (1^3),\\
(p) &= \mathfrak{p}_1^2\mathfrak{p}_2 && \text{if } (F_K, p) = (1^2 1).
\end{aligned}$$

Proof. Assume first that a, the coefficient of x^3 in F_K, is not divisible by p. Consider the polynomial

$$f(x) = x^3+bx^2+acx+a^2 d.$$

This polynomial is irreducible over Q, and has a zero in K. Its discriminant equals $a^2\mathfrak{d}_K$. Hence, by the Kummer–Dedekind theorem, $f(x)$ factorizes Mod p in the same way as p factorizes in K. As $f(x)$ factorizes Mod p in the same way as $F_K(x, y)$, our lemma is proved.

It remains to deal with the case that $p|a$ for all forms equivalent to F_K. This happens only if $p^2\mathfrak{d}_K|\mathfrak{d}(\alpha)$ for all integers α in K, i.e. if p is a 'non-essential divisor' of the discriminant of K. It is well known that this case arises only if $p = 2$, $\mathfrak{d}_K \equiv 1 \pmod 2$ and 2 factorizes completely in K. Then $a \equiv d \equiv 0 \pmod 2$, $D \equiv 1 \pmod 2$, hence $b \equiv c \equiv 1 \pmod 2$; $F_K(x, y) \equiv xy(x+y)$ (Mod 2), i.e. $(F, 2) \equiv (111)$. This observation completes the proof of the lemma.

Lemma 12. $F_K \in U$.

Proof. We state a few well-known facts on cubic fields (Hasse 1930). If K is cyclic, the discriminant $\mathfrak{d}_K$ of K has the form $\mathfrak{d}_K = f^2$; if K is not cyclic, $\mathfrak{d}_K$ has the form $\mathfrak{d}_K = \Delta_2 f^2$, where Δ_2 is the discriminant of a quadratic field. In both cases $p^2 \nmid f$ if $p \neq 3$; and $(\Delta_2, f) = 1$ or 3. Further $p^2 \nmid \Delta_2$ if $p \neq 2$. A prime p ramifies completely in K if and only if $p|f$.

We want to show that $F_K \in U_p$ for all p. If $p^2 \nmid \mathfrak{d}_K$, this follows at once from the definition of U_p. Hence we may assume that $\mathfrak{d}_K \equiv 0 \pmod{p^2}$.

If $p > 3$, the last congruence implies $p|f$, and p ramifies completely in K, so that by lemma 11 $(F_K, p) = (1^3)$. As $p^3 \nmid \mathfrak{d}_K$, it follows from lemma 6 that $F_K \in U_p$.

If $p = 2$, we have either $4|\Delta_2$ or $2|f$. If $4|\Delta_2$, then $\Delta_2 \equiv 8$ or $12 \pmod{16}$, $f^2 \equiv 1 \pmod 8$, hence $\mathfrak{d}_K \equiv 8$ or $12 \pmod{16}$, $F_K \in V_2 \subset U_2$. If $2|f$, 2 ramifies completely in K, hence by lemma 11 $(F_K, 2) = (1^3)$. As $\mathfrak{d}_K \equiv 4 \pmod 8$, it follows from lemma 6 that $F_K \in U_2$.

There remains only the case $p = 3, f \equiv 0 \pmod 3$. Let $\mathfrak{p}$ denote the unique prime ideal in K which divides 3. Because 3 is not a 'non-essential divisor' of the discriminant, there exists in K an integer α such that

$$3\mathfrak{d}_K \nmid \mathfrak{d}(\alpha).$$

Without loss of generality we may assume that $\alpha \equiv 0 \pmod{\mathfrak{p}}$, otherwise consider $\alpha - 1$ or $\alpha + 1$. Hence $\operatorname{tr}(\alpha) \equiv 0 \pmod 3$. It is easy to verify the identity

$$\mathfrak{d}(\alpha^2) = \mathfrak{d}(\alpha)\,\mathrm{Nm}^2(\operatorname{tr}(\alpha)-\alpha).$$

If $\alpha \not\equiv 0 \pmod{\mathfrak{p}^2}$, then

$$\mathrm{Nm}(\mathrm{tr}\,(\alpha)-\alpha) \equiv \pm 3 \pmod 9,$$

$$\mathfrak{d}(\alpha^2)\,\mathfrak{d}_K^{-1} = \mathfrak{d}(\alpha)\,\mathfrak{d}_K^{-1}\mathrm{Nm}^2(\mathrm{tr}\,(\alpha)-\alpha) \equiv \pm 9 \pmod{27}.$$

This means that $F_K(x,y)$ represents a number $\equiv \pm 3 \pmod 9$, i.e. $F_K(x,y) \in U_3$.

If $\alpha \equiv 0 \pmod{\mathfrak{p}^2}$, our identity gives

$$\mathfrak{d}(\tfrac{1}{3}\alpha^2) = 3^{-6}\,\mathfrak{d}(\alpha^2) = 3^{-6}\,\mathfrak{d}(\alpha)\,\mathrm{Nm}^2(\mathrm{tr}\,(\alpha)-\alpha),$$

$$\mathfrak{d}(\tfrac{1}{3}\alpha^2)\,\mathfrak{d}_K^{-1} = \mathfrak{d}(\alpha)\,\mathfrak{d}_K^{-1}\{3^{-3}\mathrm{Nm}(\mathrm{tr}\,(\alpha)-\alpha)\}^2$$

and, since $\frac{1}{3}\alpha^2$ is an integer in K,

$$3^3 | \mathrm{Nm}(\mathrm{tr}\,(\alpha)-\alpha).$$

This implies that $3|\alpha$, and therefore

$$\mathfrak{d}(\tfrac{1}{3}\alpha) = 3^{-6}\,\mathfrak{d}(\alpha) \equiv 0 \pmod{\mathfrak{d}_K},$$

$$\mathfrak{d}(\alpha) \equiv 0 \pmod{3^6\,\mathfrak{d}_K}$$

which is a contradiction.

Lemma 13. *Let F_1 and F_2 be two forms in U which are rationally equivalent. Then they are equivalent.*

Proof. Rational equivalence between F_1 and F_2 means explicitly that

$$F_1(x_1,y_1) = \sigma F_2(x_2,y_2),$$

$$(x_1,y_1) = M(x_2,y_2),$$

where $\sigma \neq 0$ is rational and M is a non-singular 2 by 2 matrix over $\boldsymbol{Z}$. If we replace F_1 by an equivalent form, M will be multiplied by a unimodular matrix on the left. Similarly, replacing F_2 by an equivalent form means multiplication of M with a unimodular matrix on the right.

Thus we may replace M by $M_1 M M_2$, where M_1 and M_2 are unimodular. Elementary divisor theory tells us that we can choose M_1 and M_2 in such a way that

$$M_1 M M_2 = \begin{pmatrix} m & 0 \\ 0 & 1 \end{pmatrix},$$

where $m = |\det(M)|$. If $m = 1$, our forms are equivalent.

Otherwise, there exists a prime $p|m$. Write $m = p^l m_0$, $\sigma = p^k \sigma_0$ so that $l \geqslant 1$, and m_0, σ_0 are prime to p. Then our transformation takes the form

$$F_1(p^l m_0 x, y) = p^k \sigma_0 F_2(x,y).$$

Equating coefficients we obtain

$$a_1 = p^{k-3l}\tau_a a_2,$$
$$b_1 = p^{k-2l}\tau_b b_2,$$
$$c_1 = p^{k-l}\tau_c c_2,$$
$$d_1 = p^{k}\tau_d d_2,$$

where $\tau_a, \tau_b, \ldots$ are rationals prime to p.

If $k-l>0$, we have $p|c_1$, $p^2|d_1$. If $k-l \leqslant 0$, we have $p|b_2$, $p^2|a_2$. Because of symmetry, we may restrict ourselves to the first case, $p|c_1$, $p^2|d_1$ implies $p^2|D_1$. As $F_1 \in U_p$, it follows that $(F_1, p) = (1^3)$, and therefore $p|b_1$. As $F_1 \in U_p$ and $p^2|D_1$, the congruence

$$F_1(x,y) \equiv ep \pmod{p^2}$$

has a solution for some $e \not\equiv 0 \pmod{p}$. As $b_1 \equiv c_1 \equiv d_1 \equiv 0 \pmod{p}$, it follows that $x \equiv 0 \pmod{p}$. But this implies

$$F_1(x,y) \equiv c_1 xy^2 + d_1 y^3 \equiv 0 \pmod{p^2},$$

$$e \equiv 0 \pmod{p}.$$

This contradiction completes the proof of the lemma.

LEMMA 14. *To every $F \in \Phi$ there belongs a cubic field K such that F and F_K are rationally equivalent.*

Proof. Write F in the form

$$F(x,y) = a(x-\lambda y)(x-\lambda' y)(x-\lambda'' y).$$

Then λ generates a cubic field K. We can write F_K in the form

$$F_K(x,y) = a_K(x-\mu y)(x-\mu' y)(x-\mu'' y),$$

where $\mu \in K$. If K is not cyclic, μ is unique, but if K is cyclic any of the three conjugates can be used. As λ and μ are irrationals in K, there exists a relation $k\lambda + l - m\mu\lambda - n\mu = 0$, $(k,l,m,n) = 1$, which is unique apart from a factor ± 1. Thus we have

$$\mu = (k\lambda + l)(m\lambda + n)^{-1}$$

and this also holds if we replace λ, μ by their two pairs of conjugates.

The transformation

$$x^* = kx + ly, \quad y^* = mx + ny$$

transforms the form $\quad F(x,y) = a(x-\lambda y)(x-\lambda' y)(x-\lambda'' y)$

into a form $\quad \rho(x^*-\mu y^*)(x^*-\mu' y^*)(x^*-\mu'' y^*)$,

which is a constant multiple of $F_K(x^*, y^*)$.

PROPOSITION 4. *There exists a 1–1 mapping Λ of triplets of conjugate cubic fields K onto the equivalence classes of U. And Λ preserves the discriminant.*

Proof. The map $\Lambda: K \rightarrow F_K$ maps the triplets into classes of U by lemma 12. By lemmas 14 and 13 every class in U contains an F_K. And it was stated at the beginning of this section that distinct triplets are mapped into distinct classes of U, and that $D(F_K) = \mathfrak{d}_K$.

7. Proof of theorems 1, 2 and 3

Proof of theorem 1. It follows from proposition 4 that

$$N_3(\xi, \eta) = N(\xi, \eta; U).$$

This identity in conjunction with proposition 2 gives theorem 1.

Proof of theorem 2. Let p be a fixed prime. By virtue of lemma 11 the mapping considered in the preceding proof maps the classes of forms in $U \cap T_p(111)$; $U \cap T_p(3)$ and $U \cap T_p(12)$ into cubic fields in which p factorizes as $(p) = \mathfrak{p}_1\mathfrak{p}_2\mathfrak{p}_3$, $(p) = (p)$, $(p) = \mathfrak{p}_1\mathfrak{p}_2$ respectively.

It is easily seen that the relative density of our 3 classes in U equals

$$A(T_p(111); p^2)A^{-1}(U_p; p^2), \quad \text{etc.}$$

By lemmas 1 and 5 these three relative densities are

$$\tfrac{1}{6}(1+p^{-1}+p^{-2})^{-1}, \quad \tfrac{1}{3}(1+p^{-1}+p^{-2})^{-1}, \quad \tfrac{1}{2}(1+p^{-1}+p^{-2})^{-1}$$

respectively.

As the cyclic cubic fields have relative density 0, they may be ignored. For non-cyclic cubic fields it is well known that the three types of factorization correspond to the three values I, A_3–I, S_3–A_3 of the Frobenius–Artin symbol $\{(K_6/Q)/p\}$.

Proof of theorem 3. Let K be a cubic field in which no prime ramifies completely, so that K is automatically not cyclic. This means, in the notation used in the proof of lemma 12, that $f = 1$ and that $\mathfrak{d}_K = \Delta_2$, where Δ_2 is discriminant of a quadratic field. For a given Δ_2 the number of triplets of such cubic fields K equals (Hasse 1930)

$$\tfrac{1}{2}(h_3^*(\Delta_2) - 1).$$

On the other hand, the mapping Λ maps these triplets into the classes of V. Hence

$$\tfrac{1}{2} \sum_{\xi<\Delta_2<\eta} (h_3^*(\Delta_2) - 1) = N(\xi, \eta; V).$$

An easy calculation shows that, as $X \to \infty$,

$$X^{-1} \sum_{0<\Delta_2<X} 1 \to 3\pi^{-2},$$

$$X^{-1} \sum_{-X<\Delta_2<0} 1 \to 3\pi^{-2}.$$

Hence by proposition 3

$$\lim_{X\to\infty} X^{-1} \sum_{0<\Delta_2<X} (h_3^*(\Delta_2) - 1) = \lim_{X\to\infty} 2X^{-1}N(0, X; V)$$

$$= \pi^{-2} = \lim_{X\to\infty} X^{-1} \sum_{0<\Delta_2<X} \tfrac{1}{3};$$

$$\lim_{X\to\infty} X^{-1} \sum_{-X<\Delta_2<0} (h_3^*(\Delta_2) - 1) = \lim_{X\to\infty} 2X^{-1}N(-X, 0; V)$$

$$= 3\pi^{-2} = \lim_{X\to\infty} X^{-1} \sum_{-X<\Delta_2<0} 1.$$

This completes the proof of our theorems.

References

Davenport, H. 1951 *a* On the class-number of binary cubic forms (I). *J. Lond. Math. Soc.* **26**, 183–192. (Corrigendum, *ibidem* **27**, 512.)

Davenport, H. 1951 *b* On the class-number of binary cubic forms (II). *J. Lond. Math. Soc.* **26**, 192–198.

Davenport, H. & Heilbronn, H. 1969 On the density of discriminants of cubic fields. *Bull. Lond. Math. Soc.* **1** (1969), 345–348.

Hasse, H. 1930 Arithmetische Theorie der kubischen Zahlkörper auf klassenkörpertheoretischer Grundlage. *Math. Z.* **31**, 565–582.

[Department of Mathematics, University of Toronto, Canada]

[Received 13 August, 1970]

43.

On the 2-Classgroup of Cubic Fields

Studies in Pure Mathematics (L. Mirsky, ed.), Academic Press, London–New York, 1971, 117–119.
[Commentary on p. 602]

Let k_2 be a quadratic numberfield of discriminant d_2. It is well known **(2)** that there exists a cubic numberfield k_3 of discriminant $d_3 = d_2$ if and only if the class-number of k_2 is divisible by 3. The object of this paper is to prove the following analogue.

Let k_3 be a cubic numberfield of discriminant d_3. Then there exists a quartic numberfield k_4 of discriminant $d_4 = d_3$ such that the normal closure $\bar{k}_4$ of k_4 over Q contains k_3 if and only if the class-number h_3 of k_3 (in the narrow sense) is even. More precisely, I shall prove the

THEOREM. *Let $g = g(k_3) = \frac{1}{4}$ or 1 for cyclic or non-cyclic k_3 respectively. Let h_3^* denote the number of elements in the ideal class-group of k_3 of order 2. Then there exist $g(h_3^* - 1)$ quadruplets of conjugate quartic fields k_4 such that $d_4 = d_3$, $k_3 \subset \bar{k}_4$.*

Proof. We first deal with the case when k_3 is not cyclic. Let k_6 denote the normal closure of k_3. Then it follows from class-field theory, or from the theory of relative quadratic fields (**3**, Kap. VIII) that k_3 has exactly $h_3^* - 1$ unramified quadratic extensions, say $k_{6,j} = k_3(a_j^{\frac{1}{2}})$, where $1 \leqslant j \leqslant h_3^* - 1$. Because $k_{6,j}$ is unramified over k_3, the ideal (a_j) is the square of an ideal in k_3 and it can be assumed that $a_j \equiv 1 (\mathrm{mod}\, 4)$. a_j cannot be chosen in Q because then all rational primes dividing a_j would become squares of ideals in k_3, which no rational prime can do in k_3. The case $a_j = -1$ will be excluded in a moment.

Let a_j', a_j'' denote the conjugates of a_j over Q. Then

$$a_j a_j' a_j'' = N_{k_3/Q}(a_j) = \pm N_{k_3/Q}((a_j)) = \pm A_j^2,$$

where $A_j \in Q^*$. And our congruence relation $a_j \equiv 1 (\mathrm{mod}\, 4)$ ensures that $a_j a_j' a_j'' = A_j^2$. Hence $k_{24,j} = k_6(a_j^{\frac{1}{2}}, a_j'^{\frac{1}{2}})$ is normal over k_6, and therefore over Q.

We shall now construct the Galois group $\mathrm{Gal}(k_{24,j}/Q)$ which will turn out to be isomorphic to S_4. To each permutation of $a_j^{\frac{1}{2}}, a_j'^{\frac{1}{2}}, a_j''^{\frac{1}{2}}$ in this order we assign the corresponding element in S_3 written as the group of permutations on the figures 1, 2, 3 in this order. To the automorphism

$$a_j^{\frac{1}{4}} \to a_j^{\frac{1}{4}}, \qquad a_j'^{\frac{1}{4}} \to -a_j'^{\frac{1}{4}}, \qquad a_j''^{\frac{1}{4}} \to -a_j''^{\frac{1}{4}}$$

we assign the permutation (14) (23). These assignments fix the isomorphism between $\mathrm{Gal}(k_{24,j}/Q)$ and S_4. Let V_4 be the subgroup of S_4 generated by the 2-cycles (14) and (23); it is clear that $V_4 = \mathrm{Gal}(k_{24,j}/k_{6,j})$. Further, let W_8 be the subgroup of S_4 generated by V_4 and the substitution (13) (24); it is clear that $W_8 = \mathrm{Gal}(k_{24,j}/k_3)$ as the automorphism $(13)(24) = (12)^{-1}(14)(23)(12)$ transforms a_j into itself. Finally, let $k_{4,j}$ be defined as the extension $Q(a_j^{\frac{1}{4}} + a_j'^{\frac{1}{4}} + a_j''^{\frac{1}{4}})$. Then $S_3 = \mathrm{Gal}(k_{24,j}/k_{4,j})$.

The field $k_{4,j}$ is, apart from conjugacy, the only quartic subfield of $k_{24,j}$. In $k_{24,j}$ there exist only 3 quadratic extensions of k_3, namely $k_{6,j}, k_6, k_3((da_j)^{\frac{1}{2}})$. Of these only the first is unramified over k_3. Hence, if $j \neq i, k_{24,j} \neq k_{24,i}$, and so $k_{4,j} \neq k_{4,i}$.

It remains to show that $d_3 = d_{4,j}$ for all j. Direct computation would be very tedious, but the Artin L-series provide a quick proof. We write down the character table of S_4.

C	$\sigma(C)$	$n(C)$	χ_1	χ_2	χ_3	χ_4	χ_5	$\psi(W_8)$	$\psi(V_4)$	$\psi(S_3)$
ident.	1	1	1	1	2	3	3	3	6	4
(12) (34)	2	3	1	1	2	−1	−1	3	2	0
(123)	3	8	1	1	−1	0	0	0	0	1
(12)	2	6	1	−1	0	1	−1	1	2	2
(1234)	4	6	1	−1	0	−1	1	1	0	0

Here C denotes the conjugacy class; $\sigma(C)$ the order of C; $n(C)$ the cardinality of C; $\chi_1, \ldots, \chi_5$ the simple characters of S_4; and $\psi(W_8), \psi(V_4), \psi(S_3)$ the characters of S_4 induced by the principal characters of W_8, V_4, and S_3 respectively. One verifies at once that $\psi(W_8) = \chi_1 + \chi_3$, $\psi(V_4) = \chi_1 + \chi_3 + \chi_4$, $\psi(S_3) = \chi_1 + \chi_4$. Writing $\zeta(s), \zeta_3(s), \zeta_{6,j}(s), \zeta_{4,j}(s)$ for the Dedekind ζ-functions of Q, k_3, $k_{6,j}$ and $k_{4,j}$ respectively, these relations yield (**1**, Chapter VIII, 3)

$$\zeta(s)\,\zeta_{6,j}(s) = \zeta_3(s)\,\zeta_{4,j}(s);$$

and the functional equation of these ζ-functions gives

$$d_{6,j} = d_3\, d_{4,j}.$$

Because $k_{6,j}$ is unramified over k_3, we have $d_{6,j} = d_3^{\,2}$; and it follows that $d_3 = d_{4,j}$.

Conversely, if k_3 and $k_{4,j}$ are given such that $k_3 \subset \bar{k}_{4,j}$, then $k_{6,j}$ can be determined as the field such that $\mathrm{Gal}(\bar{k}_{4,j}/k_{6,j}) = V_4$; from the previous calculations it follows that $k_{6,j}$ is unramified over k_3 as $d_3 = d_{4,j}$.

It remains to prove the theorem when k_3 is cyclic. a_j can be determined as before. a_j must again be irrational; because otherwise the field $k_{6,j}$ would be in the union of the absolutely abelian fields k_3 and $Q(a_j^{\frac{1}{2}})$ and its discriminant $d_{6,j}$ would equal $d_3^2 \mathrm{disc}^3(Q(a_j^{\frac{1}{2}})) \neq d_3^2$. But now the fields $k_3(a_j'^{\frac{1}{2}})$ and $k_3(a_j''^{\frac{1}{2}})$ are also unramified extensions of k_3 itself, which accounts for the factor $g = \frac{1}{3}$ in our theorem. To justify this argument we have, of course, to show that $k_{6,j}$ is different from its conjugates. Otherwise $k_{6,j}$ would be abelian over Q and we could choose a_j in Q, a possibility that was excluded above.

One sees as in the first case that $a_j a_j' a_j'' = A_j^2$, $A_j \in Q^*$. Calling $k_{12,j}$ the normal closure of $k_{6,j}$ over Q, it follows that $\mathrm{Gal}(k_{12,j}/Q) = A_4$; $\mathrm{Gal}(k_{12,j}/k_{6,j}) = V_2$, the subgroup of A_4 generated by the permutation (14) (23); $\mathrm{Gal}(k_{12,j}/k_3) = W_4$, the subgroup of order 4 of A_4. Calling again $k_{4,j} = Q(a_j^{\frac{1}{2}} + a_j'^{\frac{1}{2}} + a_j''^{\frac{1}{2}})$, it follows that $\mathrm{Gal}(k_{12,j}/k_{4,j}) = A_3$, the subgroup of A_4 generated by the cycle (123).

Our character table now becomes:

C	$\sigma(C)$	$n(C)$	χ_1	χ_2	χ_3	χ_4	$\psi(W_4)$	$\psi(V_2)$	$\psi(A_3)$
ident.	1	1	1	1	1	3	3	6	4
(12) (34)	2	3	1	1	1	-1	3	2	0
(123)	3	4	1	ρ	ρ^2	0	0	0	1
(132)	3	4	1	ρ^2	ρ	0	0	0	1

where ρ denotes a 3rd root of 1. One sees at once that $\psi(W_4) = \chi_1 + \chi_2 + \chi_3$, $\psi(V_2) = \chi_1 + \chi_2 + \chi_3 + \chi_4$, $\psi(A_3) = \chi_1 + \chi_4$.

Hence, again with the obvious notation,

$$\zeta(s)\zeta_{6,j}(s) = \zeta_3(s)\zeta_{4,j}(s),$$

and the relation

$$d_3 = d_{4,j}$$

follows as before. The converse argument works similarly.

References

1. CASSELS, J. W. S. and FRÖHLICH, A. (editors). *Algebraic Number Theory* (Academic Press, 1967).
2. HASSE, H. Arithmetische Theorie der kubischen Zahlkörper auf klassenkörpertheoretischer Grundlage. *Math. Zeitschrift* **31** (1930), 565–582.
3. HECKE, E. *Theorie der algebraischen Zahlen* (Akademische Verlagsgesellschaft, 1923).

44.

On Real Simple Zeros of Dedekind ζ-Functions

Proceedings of the 1972 *Number Theory Conference*, University of Colorado, Boulder, 1972, 108–110.

Theorem. Let K be a normal extension of an algebraic number field k ; let k_2 be the union of all quadratic extensions of k which are contained in K . Then every real simple zero of $\zeta_K(s)$ is a (simple) zero of $\zeta_{k_2}(s)$.

For the proof of this theorem we require the following facts.

I. Artin.
$$\zeta_K(s) = \prod_i L(s; K/k, \chi_i)^{\chi_i(1)} ,$$

where the $L(s; K/k, \chi_i)$ are L-series formed by the simple characters of $\mathrm{Gal}(K/k)$.

II. Frobenius. Let H be a subgroup of a finite group G . Let χ_i run through the simple characters of G , and let ψ_j run through the simple characters of H . Then there exist rational integers $r_{ij} \geq 0$ such that

$$\psi_j^*(\gamma) = \sum_i r_{ij}\chi(\gamma) , \quad \text{for } \gamma \in G ,$$

$$\chi_i(\eta) = \sum_j r_{ij}\psi(\eta) , \quad \text{for } \eta \in H ;$$

here $\psi^*(\gamma)$ denotes the character of G , which is induced by the character $\psi(\eta)$ of H .

III. Artin. Let H be a subgroup of $Gal(K/k)$, K_H the corresponding subfield of K, ψ_1 the principal character of H . Then

$$\zeta_{K_H}(s) = \prod_i L(s; K/k, \chi_i)^{r_{i1}} .$$

IV. R. Brauer. The $L(s; K/k, \chi)$ are meromorphic for all finite s .

V. R. Brauer, Aramata. If K is normal over k , $\zeta_k(s)^{-1}\zeta_K(s)$ is an entire function of s .

VI. Elementary. Let H be a finite group, $\rho(\eta)$ a function defined for all $\eta \in H$. If

$$\sum_{\eta \in H} \rho(\eta) \neq 0 ,$$

$$\sum_{\eta \in H^*} \rho(\eta) = 0 \quad \text{for all genuine subgroups } H^* \text{ of } H ,$$

then H is cyclic.

VII. Definition. Let s_0 be a real simple zero of $\zeta_K(s)$. Assume that $L(s; K/k, \chi_i)$ has a zero of order m_i at s_0. (m_i may be negative.) Put

$$\phi = \sum_i m_i \chi_i ,$$

so that ϕ is a 'general' character on G .

Outline of proof of the Theorem. Let k_a be the maximal abelian extension of k which is contained in K , so that $G' = \mathrm{Gal}(K/k_a)$ is the commutator subgroup of G . If s_0 is a real zero of $\zeta_{k_a}(s)$, it must be simple by V, and hence for reality reasons a zero of $\zeta_{k_2}(s)$.

Now let H be a subgroup of G' , and K_H the corresponding subfield of K . A routine calculation based on III and IV shows that $\zeta_{K_H}(s)$ has a zero at s_0 of order

$$|H|^{-1} \sum_{y \in H} \phi(\eta) ,$$

and this expression takes only the value 0 or 1 . Let H^* be the minimal subgroup of G' for which the sum vanishes. By VI, H^* is cyclic. And by reality considerations H^* is of order 2^m , $m \geq 1$. Hence, if τ is a generator of H^* ,

$$2^{-m} \sum_{n=1}^{2^m} (-1)^n \phi(\tau^n) = 1 .$$

The value of this sum indicates that -1 occurs exactly once as a characteristic root of $\phi(\tau)$. Thus the product of the characteristic roots of $\phi(\tau)$ would be -1 , which is impossible as $\tau \in G'$.

[University of Toronto]

45.

On Real Zeros of Dedekind ζ-Functions

Canadian Journal of Mathematics 25 (1973), 870–873
[Commentary on p. 571]

1. Introduction. Let K be a finite normal extension of an algebraic number field k; let k_2 be the compositum of all quadratic extensions of k which are contained in K. Let $\zeta_k(s)$, $\zeta_K(s)$ and $\zeta_{k_2}(s)$ denote the Dedekind ζ-functions of these fields. The main purpose of this paper is to prove

THEOREM 1. *Any real simple zero of $\zeta_K(s)$ is a zero of $\zeta_{k_2}(s)$.*

In particular, if k is the rational field, any real simple zero of $\zeta_K(s)$ is a zero of an L-series

$$L_\Delta(s) = \sum_{n=1}^{\infty} (\Delta/n)n^{-s}$$

where Δ is a rational integral divisor of disc (K/Q).

The motivation arises from the following well-known facts. Let C be a number field, d its absolute discriminant, κ the residue of its ζ-function $\zeta_C(s)$ at $s = 1$. Then either $\kappa^{-1} = O(\log|d|)$ or $\zeta_C(s_0) = 0$ for some $s_0 < 1$ with $\log|d| = O((1 - s_0)^{-1},)$ in which case the lower bound for κ may be very poor indeed. Moreover, the zero s_0 is simple and unique.

Now let K be the normal closure of C over $\mathbf{Q}$, of absolute discriminant D such that

$$|D| \leqq |d|^{n!}, \; n = \text{degr } C.$$

Then s_0 will also be a zero of $\zeta_K(s)$. The application of Theorem 1 to K yields

THEOREM 2. *Let C any number field of degree n and discriminant d. Then either*

$$\kappa^{-1} = O(n! \log |d|),$$

or there exists a divisor Δ of d such that

$$L_\Delta(s_0) = 0, \quad 1 - s_0 = O(\kappa).$$

Thus the task to find an effective realistic lower bound for κ is, at least in principle, reduced to the same problem for quadratic number fields. In the case where C is a totally complex quadratic extension of a totally real field, J. Sunley [**4**] and L. Goldstein [**3**] have already obtained results of this nature. I wish to record my gratitude to Prof. L. Goldstein who made me familiar with these researches, and thus provided the stimulus which led me to the present investigation.

Received April 20, 1972 and in revised form, July 10, 1972.

2. Proof of Theorem 1. The proof is based on the use of Artin L-series. We shall make use of two fundamental results of R. Brauer [**1**; **2**].

B.1. The Artin L-series are meromorphic functions of s.

B.2. If K is a normal extension of k, then

$$(\zeta_k(s))^{-1}\,\zeta_K(s)$$

is an integral function of s.

Let k_a be the maximal abelian extension of k contained in K, so that

$$k \subset k_2 \subset k_a \subset K.$$

Let $G = \mathrm{Gal}(K/k)$, so that $G' = \mathrm{Gal}\ (K/k_a)$ is the commutator group of G. Then

$$\zeta_{k_a}(s) = \zeta_{k_2}(s) \prod_{\gamma} L(s; k, \gamma)$$

where γ runs through the complex characters of G/G'.

Because the γ are abelian characters, the $L(s; k, \gamma)$ are integral functions. Because $L(s; k, \gamma) = 0 \Rightarrow L(s; k, \bar{\gamma}) = 0$ for real s, any real zero s_0 of $\zeta_{k_a}(s)$ is either a zero of $\zeta_{k_2}(s)$ or a zero of multiplicity $\geqq 2$ of $\zeta_{k_a}(s)$. By B.2 the last case is impossible, hence we assume from now on that $\zeta_{k_a}(s) \neq 0$.

Let χ_b run through all irreducible characters of G. Then

$$\zeta_K(s) = \prod_{b} L(s; k, \chi)^{\chi_b(1)},$$

where $\chi_b(1)$ denotes the dimension of the character which equals its value for the unit element of G.

It follows from B.1 that $L(s; k, \chi_b)$ has a zero of order m_b at $s = s_0$, where $m_b \in \mathbf{Z}$, and nothing may be assumed about the sign of m_b. We now define the general character

$$\phi = \sum_{b} m_b \chi_b.$$

Let k_j be any field in the range $k_a \subset k_j \subset k$, and ψ_j the character of G induced by the principal character of the subgroup $G_j = \mathrm{Gal}\ (K/k_j)$. Then it is well-known that

$$\zeta_{k_j}(s) = \prod_{b} L(s; k, \chi_b)^{r_{j,b}},$$

where the non-negative rational integers $r_{j,b}$ are determined by the decomposition

$$\psi_j = \sum_{b} r_{j,b}\chi_b.$$

By virtue of the Frobenius reciprocity the $r_{j,b}$ are explicitely given by the formula

$$r_{j,b} = |G_j|^{-1} \sum_{\gamma \in G_j} \chi_b(\gamma).$$

Thus, the order of the zero of $\zeta_{k_j}(s)$ at $s = s_0$ is given by

$$S(G_j) = \sum_{b} r_{j,b} m_b = |G_j|^{-1} \sum_{b} m_b \sum_{\gamma \in G_j} \chi_b(\gamma)$$
$$= |G_j|^{-1} \sum_{\gamma \in G_j} \phi(\gamma),$$

and we know from B.2 that $S(G_j) = 0$ or 1 for all j, $S(G') = 0$, and $S(\{1\}) = 1$, where $\{1\}$ denotes the trivial subgroup of G consisting of the unit only.

Now let H^* be a minimal subgroup of G', such that $S(H^*) = 0$ and $S(H) = 1$ for each genuine subgroup H of H^*. Then we have for every genuine subgroup H of H^*

$$\sum_{\gamma \in H} (-1 + \phi(\gamma)) = 0,$$

whereas $\sum_{\gamma \in H^*} \phi(\gamma) = 0$.

It is easy to verify that these relations are compatible only if H^* is cyclic. If H^* were not cyclic, we should have for each $\gamma \in H^*$ of order N

$$\sum_{n=1}^{N} (-1 + \phi(\gamma^n)) = 0;$$

and by virtue of the Möbius inversion formula

$$\sum_{n=1,(n,N)=1}^{N} (-1 + \phi(\gamma^n)) = 0.$$

We can find group elements $\gamma_1, \ldots, \gamma_q$ of order $N_1, \ldots, N_q$ respectively such that the elements $\gamma_i^{n_i}$, $1 \leqq i \leqq q$, $1 \leqq n_i \leqq N_i$, $(n_i, N_i) = 1$ represent all group elements uniquely. Thus

$$\sum_{\gamma \in H^*} (-1 + \phi(\gamma)) = 0,$$

which is a contradiction. Thus we have shown that H^* is cyclic.

Moreover, the order of H^* cannot be divisible by an odd prime p. Otherwise the field K^*, corresponding to the subgroup H^*, would have a cyclic extension of degree p, say K_p^*, and K_p^* would be a subfield of K. The function $\zeta_{K_p^*}(s)$ would have a simple 0 at s_0. But

$$\zeta_{K_p^*}(s) = \zeta_{K^*}(s) \prod_{i=1}^{p-1} L(s; K^*, \eta_i).$$

In this product η_i runs through non-principal abelian characters in K_p^* which occur in pairs of conjugate complex characters. Hence, if the product vanishes at s_0, it must have a zero of multiplicity $\geqq 2$; this contradicts our assumption. Hence the order of H^* is a power of 2, say 2^t.

Let τ be a generator of H^*, and let H^{**} denote the subgroup of H^* which is generated by τ^2. We have

$$1 = S(H^{**}) - S(H^*) = 2^{-t} \sum_{n=1}^{2^t} (-1)^n \phi(\tau^n).$$

The general character ϕ can be decomposed into two genuine characters ϕ_+, ϕ_- by the formula $\phi = \phi_+ - \phi_-$. This decomposition is not unique, but as ϕ is real, ϕ_+ and ϕ_- can be chosen as real characters of G. We now remember that ϕ_+ and ϕ_- are the sum of the characteristic roots of the corresponding matrix representation of G. Since the characters ϕ_+ and ϕ_- are real, conjugate roots occur with equal multiplicity. The characteristic roots forming $\phi_+(\tau)$ and $\phi_-(\tau)$ are 2^tth roots of unity. Because $\tau \in G'$, the determinants of the corresponding matrices are $+1$, and the products of the characteristic roots are $+1$. As the complex roots cancel in the product, the root -1 occurs in ϕ_+ exactly a_+ times, and in ϕ_- exactly a_- times, where a_+ and a_- are even.
As

$$2^t = \sum_{n=1}^{2^t} (-1)^n \phi(\tau^n) = \sum_{n=1}^{2^t} (-1)^n \phi_+(\tau^n) - \sum_{n=1}^{2^t} (-1)^n \phi_-(\tau^n)$$

$$= 2^t(a_+ - a_-),$$

$$1 = a_+ - a_-,$$

We obtain the desired contradiction.

Postscript. The referee has kindly pointed out to me that the result B.2 quoted above was proved originally by H. Aramata, Proc. Japan Acad. *9*(1933), 31-34.

REFERENCES

1. R. Brauer, *On Artin L-series with general group characters*, Ann. of Math. *48* (1947), 502–514.
2. ——— *On the zeta-functions of algebraic number fields.* Amer. J. Math. *69* (1947), 243–250.
3. L. Goldstein, *Relatively imaginary quadratic fields of class number* 1 *or* 2, Trans. Amer. Math. Soc. *165* (1972), 353–364.
4. J. S. Sunley, *Class numbers of totally imaginary quadratic extensions of totally real fields*, Ph.D. Thesis, University of Maryland, 1971.

University of Toronto,
Toronto, Ontario

Commentaries

The Papers on Prime Number Problems and Applications of Brun's Sieve

R. C. Vaughan

[1] Über die Verteilung der Primzahlen in Polynomen

In his first paper, Heilbronn gives a masterly and economical exposition of the "difficult" Brun sieve method, in the form worked out by Landau [1930]. It is applied to show that if f is a polynomial of degree $n \geq 2$ such that $f(\mathbb{Z}) \subset \mathbb{Z}$, then $P(\xi, f)$, the number of natural numbers $x \leq \xi$ such that $f(x)$ is prime, satisfies

$$P(\xi, f) \ll \frac{\xi}{\log \xi}.$$

There is one slight complication in that, to estimate the main term arising in the sieve, a corollary of the prime ideal theorem (Nagell [1922]) is required.

Remarkably, the sentence at the top of p. 795 still represents the state of our knowledge (or lack of it!) 54 years later.

For more recent developments in sieve theory, see Halberstam and Richert [1974].

[2] Über den Primzahlsatz von Herrn Hoheisel

This is a highly technical paper. Hoheisel had earlier shown the existence of a constant $c < 1$ such that

$$\pi(x + x^c) - \pi(x) \sim x^c/\log x \quad \text{as } x \to \infty.$$

An essentially equivalent formulation is that $\Psi(x) = \Sigma_{n \leq x}\Lambda(n)$, where $\Lambda(n) = \log p$ when n is of the form p^k and $\Lambda(n) = 0$ otherwise, satisfies

$$\Psi(x + x^c) - \Psi(x) \sim x^c \quad \text{as } x \to \infty. \tag{1}$$

In order to explain briefly the underlying ideas, we outline a slightly different line of attack, given first by Ingham [1937]. This has three ingredients.

(i) An explicit formula for $\Psi(x)$ of the kind

$$\Psi(x) = x - \sum_{\varrho}{}' \frac{x^{\varrho}}{\varrho} + O\left(\left(\frac{x}{T} + 1\right)(\log xT)^2\right) \qquad (x \geq 2), \tag{2}$$

where Σ' denotes summation over roots $\varrho = \beta + i\gamma$ of $\zeta(s)$ with $0 < \beta < 1$, $|\gamma| \leq T$ and $T \geq 2$.

(ii) A zero free region for $\zeta(s)$ of the form

$$\left\{s = \sigma + it\colon \tau = |t| + 3, \sigma > 1 - A\frac{\log\log\tau}{\log\tau}\right\}, \tag{3}$$

where A is a constant.

(iii) An estimate for the number $N(\sigma, T)$ of zeros $\varrho = \beta + i\gamma$ of $\zeta(s)$ with $\beta > \sigma$ and $|\gamma| \leq T$, of the form

$$N(\sigma, T) \ll \min\left(T^{B(1-\sigma)}(\log T)^C, T\log T\right), \tag{4}$$

where B and C are constants.

Heilbronn's method, which is an elaboration of Hoheisel's, does not use (2) and (4) explicitly, but the function theoretic techniques whereby (2) and (4) can be obtained are closely allied. For a revival of the Heilbronn-Hoheisel approach, see Heath-Brown [1982].

The estimate (2) is applied to $\Psi(x + x^c) - \Psi(x)$ with $T = x^{1-c}(\log x)^3$ (say). The general term in $\sum_{\varrho}'((x + x^c)^{\varrho} - x^{\varrho})/\varrho$ is estimated by the bound $x^{c+\beta-1}$. Then systematic use of (3) and (4) gives (1) provided that c is sufficiently close to 1. The relative weakness of (4) near $\sigma = 1$ [because of the presence of the factor $(\log T)^C$] and of (3) mean that c depends on A and C rather than on B. Heilbronn works very hard to obtain (implicitly) a relatively small value for C.

The improved zero free region for $\zeta(s)$ of the form

$$\left\{s = \sigma + it\colon \tau = |t| + 3, \sigma > 1 - A'(\log\tau)^{-D}\right\}, \tag{3'}$$

where D is a constant with $D < 1$, obtained by Tchudakoff [1936], means that the factor $(\log T)^C$ occurring in (4) is no longer of importance, and the optimal choice for c depends instead on B. Ingham showed that any value of c with $c > 1 - 1/B$ is permissible and obtained (4) with $B = 8/3$ (whence $c > 5/8$). This was later improved by Montgomery [1969] to $B = 5/2$ ($c > 3/5$) and then by Huxley [1972] to $B = 12/5$ ($c > 7/12$). Heath-Brown [1984] has refined the method so as to show that $c = 7/12$ is also permissible.

By combining these techniques with others, mostly derived from sieve methods, Iwaniec and Jutila [1979] and Heath-Brown and Iwaniec [1979] have shown that

$$\varliminf_{x \to \infty} \frac{\pi(x + x^c) - \pi(x)}{x^c} \log x > 0$$

for $c = \frac{13}{23}$ and $c > \frac{11}{20}$, respectively.

[9] (With E. Landau and P. Scherk). Alle grossen ganzen Zahlen lassen sich als Summe von höchstens 71 Primzahlen darstellen

This represents a considerable improvement over the bounds of Schnirelman and Romanov quoted, and takes Schnirelman's original method about as far as was possible at the time without a major innovation. The result was rapidly superseded by Vinogradov's celebrated result (Vinogradov [1937]) on sums of three primes. However, the method retained some independent interest, in part because its more elementary nature enables it to be readily adapted to show that *every* natural number is the sum of at most s primes with a relatively small value of s.

The method has been susceptible to two significant improvements.

(i) Here Brun's sieve method is used (Satz 12 and the calculations on p. 140) to show, essentially, that the number of representations of a number n as the sum of two primes is at most about 26 times the expected asymptotic value. Innovations in sieve techniques and the theory of the distribution of primes in arithmetic progressions have enabled the 26 to be reduced to 4 (see Halberstam and Richert [1974], Theorem 3.11) and 2 for almost all n (see Vaughan [1976]).

(ii) Schnirelman's method makes use of a version of Cauchy's inequality, $(\Sigma a_n)^2 \le \left(\sum_{a_n > 0} 1 \right) \Sigma a_n^2$ (here embedded in the proof of Satz 7) and this is rather inefficient because the a_n vary appreciably in size. Shapiro and Warga [1950] introduced a more precise method based on the weighted sum $\Sigma a_n w(n)$ with the sum taken over a short interval $[x, x(1 + \varepsilon)]$.

With these innovations the method can now be used to show that every large odd integer is the sum of at the most five primes (Vaughan [1976]) and that *every* even number is the sum of at most 18 primes (Riesel and Vaughan [1983]).

For other references and further historical background, see Vaughan [1976] and [1977].

Comment by Peter Scherk*. When Heilbronn sent his manuscript to Landau, he had an estimate of roughly ≤ 201 instead of ≤ 71. At that time he was not too interested in having it published. By refinements, mainly in the analytic parts, Landau could improve it. During that summer, I was Landau's "Privatassistent." When Landau gave me his own manuscript, I could improve both the numerical and the combinatorial parts. Conceptually, this was no great shakes, but it did more than halve Landau's upper bound.

[18] On an inequality in the elementary theory of numbers

The inequality (2) is motivated in part by a problem in sieve theory. A_n is the crucial factor in the main term of an asymptotic formula that arises from an application of the inclusion–exclusion principle similar to that occurring in the Brun sieve. See Chapter 1 of C. Hooley [1976].

For a general discussion of sets of positive integers not divisible by any a_ν, see Chapter V of H. Halberstam and K. F. Roth [1966], and in particular §5 where A_n occurs.

Correction. p. 209, line -5, "$v(1)$" should be "$o(1)$."

References

H. Halberstam and H. G. Richert

[1974] *Sieve Methods*. Academic Press, London.

H. Halberstam and K. F. Roth

[1966] *Sequences*. Clarendon Press, Oxford; 2nd ed., Springer Verlag, 1982.

D. R. Heath-Brown

[1982] Prime numbers in short intervals and a generalized Vaughan identity. *Canadian J. Math*. 34, 1365–1377.

[1984] Unpublished, but see: Finding primes by sieve methods. *Proc. Intern. Congr. Math. Warsaw 1983*, North-Holland, Amsterdam, vol. I, 487–492.

D. R. Heath-Brown and H. Iwaniec

[1979] On the difference between consecutive primes. *Invent. Math*. **55**, 45–69.

C. Hooley

[1976] *Applications of Sieve Methods to the Theory of Numbers* (*Cambridge Tracts in Mathematics*, No. 70), Cambridge University Press, Cambridge.

M. N. Huxley

[1972] On the difference between consecutive primes. *Invent. Math*. **15**, 164–170.

A. E. Ingham

[1937] On the difference between consecutive primes. *Quart. J. Math., Oxford Ser.* **8**, 255–266.

H. Iwaniec and M. Jutila

[1979] Primes in short intervals. *Arkiv för Math*. **17**, 167–176.

* Letter of July 12, 1984 to the Editor. See also P. Scherk's recollections on p. 49.

E. Landau

[1930] Die Goldbachsche Vermutung und der Schnirelmannsche Satz. *Nachr. Gesell. Wiss. Göttingen*, 1930, 255–276.

H. L. Montgomery

[1969] Zeros of L-functions. *Invent. Math.* **8**, 346–354.

T. Nagell

[1922] Zur Arithmetik der Polynome. *Abh. Math. Sem. Hamburg* **1**, 175–194.

H. Riesel and R. C. Vaughan

[1983] On sums of primes. *Arkiv för Math.* **21**, 45–74.

H. N. Shapiro and J. Warga

[1950] On the representation of large integers as sums of primes. *Comm. Pure Appl. Math.* **3**, 153–176.

N. Tchudakoff

[1936] On zeros of Dirichlet's L-functions. *Mat. Sbornik* (1) **43** (*Rec. Math.* (*N.S.*) **1**), 591–602.

R. C. Vaughan

[1976] A note on Snirel'man's approach to Goldbach's problem. *Bull. London Math. Soc.* **8**, 245–250.

[1977] On the estimation of Schnirelman's constant. *J. Reine Angew. Math.* **290**, 93–108.

I. M. Vinogradov

[1937] Some theorems concerning the theory of primes. *Mat. Sbornik* (1) **44** (*Rec. Math.* (*N.S.*) **2**), 179–195.

The Papers on Analysis

W. K. Hayman

Introduction

Heilbronn was a complex figure in several senses. Though to many he could be formidable, I always found him friendly and approachable. Our first meeting was in the early 1950s when Halberstam, Kennedy, and I came over from Exeter to Bristol to listen to a lecture by C. L. Siegel. At that stage, Heilbronn was living in a hotel where he entertained us all splendidly with dinner after the lecture. He also talked amusingly, partly about making money on the stock exchange. What struck me at the time was that he and his colleagues called each other by their Christian names, which was then rather unusual in English universities.* I met Heilbronn a few times after that and became aware of how much trouble he took to maintain his colleagues' interest in mathematics and to put problems before them. Towards the end of his time in Bristol, he confessed to me that he no longer found mathematics as exciting as he used to, but he tried not to make this obvious to anybody. Later in Toronto, he seemed to have completely regained his vigour. His marriage had a greatly rejuvenating effect on him.

In his mathematics, Heilbronn, trained by Landau, strove for elegance and simplicity. His main interest was in number theory, and papers [3]–[5] can be considered to have a number theoretic motivation but [6], [26], and [30] represent analysis for its own sake. Of this group of papers [26] is the deepest and develops a discrete analogue of harmonic functions.

In [30], a pretty question in topological function theory is raised and answered in all cases except one, which is settled by me below. But Heilbronn's greatest influence on complex analysis was probably through Problem 2.26 in my research problems [1967]. In it he suggested that Waring's problem might be extended to the idea of representing a function, for instance an entire function $f(z)$ as a sum of kth powers of entire functions $\phi_\nu(z)$. It is enough to consider $f(z) = z$. Although Heilbronn himself barely worked in this area, except perhaps in [34], this idea stimulated work by Toda [1971], Green [1975], Newman and Slater [1979], and others. All the positive theorems can be

* *Ed. note*: See P. Du Val's recollections for an explanation.

deduced from a paper of Cartan [1933]. One needs at least n kth powers of entire functions where $n(n-1) \geq k$. This gives the right answer for $k = 2, 3$, but for large k no example of a representation with less than k kth powers is known. Similar results for polynomials, rational functions, and meromorphic functions can also be obtained (see Hayman [1984]).

[3] (With E. Landau) Bemerkungen zur vorstehenden Arbeit von Herrn Bochner

[4] (With E. Landau) Ein Satz über Potenzreihen

[5] (With E. Landau) Anwendungen der N. Wienerschen Methode

In this early group of papers written with Edmund Landau, the authors refine the recently proved Ikehara–Tauberian Theorem, which led to a more elementary proof of the Prime Number Theorem than that of Hadamard and de la Vallée Poussin. The Ikehara–Wiener (I.W.) Theorem for Dirichlet series states that if

$$f(s) = \sum_{1}^{\infty} a_n e^{-\lambda_n s}$$

is a Dirichlet series with nonnegative coefficients and

$$\lim_{\varepsilon \to 0} f(1 + \varepsilon + it) - \frac{1}{\varepsilon + it} = F(1 + it) \tag{1}$$

exists uniformly for $|t| < a$ and every finite a, then

$$A(y) = \sum_{\lambda_n \leq y} a_n \sim e^y.$$

By applying this result to $f(s) = \zeta'(s)/\zeta(s)$, where $\zeta(s)$ is the zeta function, we obtain the Prime Number Theorem on the assumption that $\zeta(s)$ has no zeros on the line $\sigma = 1$.

In [3] the authors consider what happens if it is only known that (1) holds for some positive a and deduce that in this case $e^{-y}A(y)$ oscillates at most between constants which depend on a and which tend to 1 as $a \to \infty$. Thus from the regularity of $\zeta(s)$ at $s = 1$ we obtain Chebychev's result that $\pi(x)$ lies between constant multiples of $x/\log x$.

In [5] the authors show that (1) for some a leads to

$$\sum_{\lambda_n \le y} a_n e^{-\lambda_n} = y + O(1).$$

In [4] they give an astonishingly short proof of a related result for power series.

Until the method of Selberg and Erdös was discovered just after the war, the I.W. Theorem gave the most elementary proof of the Prime Number Theorem, and in many ways the approach in [3] is as simple and elementary as any proof known to this day.

[6] Zu dem Integralsatz von Cauchy

The author gives a short proof of Cauchy's Theorem for a rectifiable Jordan curve C and f regular in the interior D of C and continuous in $\overline{D}$. Naturally he assumes the fairly deep result that $\overline{D}$ is homeomorphic to the closed unit disk, but given this fact the argument is as simple as it could be.

[26] On discrete harmonic functions

In this paper the author develops a pretty theory of discrete harmonic (d.h.) functions. These are functions defined on a set D of lattice points in R^n with the property that the value of the function at any point z_0 is the average of the values of the function at all neighbouring points z_i, provided that z_0 and the z_i lie in D. Many of the properties of harmonic functions remain valid for d.h. functions, but there are some interesting differences. Thus a function d.h. in a bounded plane convex domain can always be represented by a d.h. polynomial. This is false for a certain L-shaped domain given on p. 198. There are a number of attractive questions left open by the author, and it is to be hoped that this theory may be taken up by others.

[30] On the representation of homotopic classes by regular functions

In this interesting paper the author poses the following problem (p. 181). Let A and B be two open sets in the complex sphere. Consider all homotopy classes of continuous functions which map A into B. Is it true that each such class contains a function meromorphic in A?

The author gives the answer to this question in all cases, except when B is equivalent to the punctured disk $0 < |w| < 1$ and A has infinite connectivity. In this case the answer is negative. Suppose that ∞ is the limit of points z_n, $n = 1, 2, \ldots$, on different components of the complement of A. Let γ_n be a sequence of rectifiable Jordan curves in A through a fixed point z_0, and such that γ_n surrounds such a z_n. Suppose that f is a regular map from A to B, so that

$$0 < |f(z)| < 1.$$

Write $|f(z_0)| = e^{-\alpha}$,

$$v(z) = \frac{\log|f(z)|}{-\alpha}.$$

Then the variation V_n of $\arg f(z)$ along γ_n is given by

$$|V_n| = \left|\frac{1}{2\pi i}\int_{\gamma_n}\frac{f'(z)}{f(z)}\,dz\right| \leq \frac{\alpha}{2\pi}\int_{\gamma_n}|\text{grad } v|\;|dz|.$$

The harmonic function v has positive real part in A and satisfies $v(z_0) = 1$.

Thus Harnack's Theorem (see, e.g., Hayman and Kennedy [1976], p. 37) shows that $|\text{grad } v|$ is bounded on γ_n by a constant depending on n only. In other words,

$$|V_n| \leq \alpha K_n, \tag{1}$$

where the constant K_n depends on n only.

Let now $F(z)$ be an entire function having zeros of multiplicity p_n at z_n, and no other zeros, where

$$\frac{p_n}{K_n} \to \infty \quad \text{with } n, \tag{2}$$

and consider

$$\phi(z) = \frac{F(z)}{2|F(z)|}.$$

Clearly $\phi(z)$ maps A into B. If $\phi(z)$ is homotopic to $f(z)$, where $f(z)$ is a regular map from A into B, the variation V_n of $\arg f$ along γ_n satisfies $|V_n| \geq 2\pi p_n$. This contradicts (2) and shows that ϕ cannot be homotopic to f.

References

H. Cartan

[1933] Sur les zéros des combinaisons linéaires de p fonctions holomorphes données. *Mathematica* **7**, 5–29.

M. L. Green

[1975] Some Picard Theorems for holomorphic maps to algebraic varieties. *Amer. J. Math.* **97**, 43–75.

W. K. Hayman

[1967] *Research Problems in Function Theory*. Athlone Press of the University of London, London.

[1984] Waring's Problem fűr analytische Funktionen. *Bayer. Akad. Wiss. Math.-Natur. Kl. S.-B.* 1984, 13 pp.

W. K. Hayman and P. B. Kennedy

[1976] *Subharmonic Functions*. Academic Press, London.

D. J. Newman and M. Slater

[1979] Waring's Problem for the Ring of Polynomials. *J. Number Theory* **11**, 477–487.

N. Toda

[1971] On the functional equation $\sum_{i=1}^{p} a_i f_i^{n_i} = 1$ or z. *Tohoku Math. J.* **23**, 289–299.

The Papers on Class Numbers of Imaginary Quadratic Fields

H. Stark

[7] On the class-number in imaginary quadratic fields

[8] (With E. H. Linfoot) On the imaginary quadratic corpora of class-number one

[45] On real zeros of Dedekind ζ-functions

These three papers span almost 40 years of research. Hecke turned the class-number problem into an analytic question in 1918 (cited in [7] and [8]) by showing that if d is the discriminant of a complex quadratic field of small class number compared to $|d|$, then the associated Dirichlet L-function, $L(s, \chi_d)$, has an exceptional real zero very close to $s = 1$. In 1933, Deuring (also cited in both [7] and [8]) made the remarkable discovery that if an infinite number of complex quadratic fields of class number 1 exist then the Riemann hypothesis is true. Several authors generalized Deuring's work to other class numbers, but Heilbronn had the crucial realization that it could also be generalized to Dirichlet L-functions.

In analogy to Deuring's work, Heilbronn in [7] showed that the existence of an infinite number of complex quadratic fields of fixed class number implied the generalized Riemann hypothesis for Dirichlet L-series with real characters. Thanks to Hecke's result, this proved Gauss's conjecture that the class number, $h(d)$, of a complex quadratic field of discriminant d goes to infinity with $|d|$. The method was ineffective: for a given class number, it did not give any way to calculate a bound on $|d|$ beyond which this class number does not occur. Heilbronn and Linfoot carried this analysis further and were able to show in [8] that the existence of an eleventh discriminant of class number one contradicted the existence of a tenth discriminant; namely the exceptional zero of the one L-function contradicts the existence of an exceptional zero of the other.

In a related paper, Deuring [1934] also showed that if $h(d) = 1$ for large $|d|$, then the Riemann hypothesis holds for the product $\zeta(s)L(s, \chi_d)$ up to height approximately $|d|^c$ for some $c > 0$ except for the one real exceptional zero of $L(s, \chi_d)$. Actually Deuring proved this result for Epstein zeta functions associated to binary forms with lead coefficient 1 and discriminant d. Stark later showed that $c = 1/2$ would do by the same methods without knowing of Deuring's earlier work. It may also be of interest to note that Deuring announced his result at the April 1933 meeting of the AMS in New York and the abstract was published in the May 1933 *Bulletin* of the AMS. I do not know if Heilbronn knew of this work at the time of [7] and [8].

In any event, these papers of Deuring and Heilbronn have given rise to what is called the Deuring–Heilbronn phenomenon. The phenomenon is that if an L-function has a zero very near $s = 1$, then not only that L-function, but other L-functions as well, must behave nicely in some neighborhood of $s = 1$. In this context, "nice" usually means no zeros except on the line $\sigma = 1/2$. The closer the zero of the one L-function is to $s = 1$, the bigger the neighborhood is. No classical L-function has ever been found with a zero near 1, but Goldfeld [1976] has shown that a third-order zero at the center of its critical strip of a Dirchlet series corresponding to a modular form is enough to evoke the phenomenon in a very small neighborhood. Gross and Zagier [1986] have just shown that such a series exists, and the combination of the work of Goldfeld, Gross, and Zagier combines to give the first effective proof that class numbers of complex quadratic fields go to infinity with the absolute value of the discriminant.

To return to [7] and 1934, many authors proceeded to find (ineffective) bounds on the class number in terms of d. The ultimate result was Siegel's theorem [1935] that $h(d) > c(\varepsilon)|d|^{(1/2)-\varepsilon}$ where $c(\varepsilon)$ is a constant depending upon $\varepsilon > 0$ and is ineffective if $\varepsilon < 1/2$. This is still the case today (1984); the Goldfeld–Gross–Zagier result will have a logarithmic bound at best (the exact expression with constants has not yet been worked out at this writing).[1] It was natural that many authors would attempt to generalize the Heilbronn and Linfoot paper also. And indeed, the literature is full of "at most one more" theorems. The most noteworthy of these are the Chowla and Briggs' at most one more, one class per genus theorem [1954] and Tatuzawa's at most one more version of Siegel's theorem [1951] (see also Hoffstein [1980] for an updated version).

Siegel suggested that his theorem could be carried over to number fields, and Brauer [1947] did exactly that. The resulting theorem has come to be known as the Brauer–Siegel Theorem. It too was ineffective. One of its consequences is that the class number of totally complex quadratic extensions K of a fixed totally real field k goes to infinity as K varies. This was the

[1]*Added in proof*. See T. Oeslerlé, Numbres de classes des corps quadratiques imaginaires. *Sem. Bourbaki* No. 631 (1984) for more recent information.

situation in 1966 when the first recognized effective solutions to the class number 1 problem were given in principle by Baker [1966] and by Stark [1967]. There was also a purported solution by Heegner [1952], which, however, was universally deemed incorrect. It later turned out that although Heegner's presentation is very hard to read and incorrect in other parts of the paper, the class number 1 part is correct once certain parts of Weber's *Algebra* are patched up. (Stark [1967] pointed out which section of Heegner's paper were relevant to the class number 1 problem and various patches were then supplied by Deuring [1968], Birch [1969], and Stark [1969c].)

The method of Stark was closely related to Heegner's in spirit; this method has not succeeded for any other class number, although it has done certain special cases of the class number 2 problem (Weinberger [1969], Kenku [1971] (based on a 1969 thesis), Meyer [1971] and Antoniadis [1983]). The method of Baker was independently generalized by Baker [1971] and Stark [1971] and successfully solved the class number 2 problem effectively. This was the last effective progress on complex quadratic fields until the Goldfeld and Gross and Zagier works cited above.

One more possible generalization of the effective class number 1 results was possible, however, and it was to class number 1 totally complex quadratic extensions K of a fixed totally real field k. This was investigated by Goldstein [1972]. It turned out that neither the Stark nor the Baker approaches could be carried over. However, the Deuring–Heilbronn phenomenon enabled Sunley [1973] to prove an at most one more theorem. Goldstein made the assumption that k is normal over $\mathbb{Q}$ and algebraically showed that the exceptional K would have to be of the form $k(\sqrt{d})$ where $\mathbb{Q}(\sqrt{d})$ also has class number 1 and d is negative. Thus Goldstein algebraically reduced the class number 1 problem for a fixed totally real normal extension of $\mathbb{Q}$ to the classical class number 1 problem, thereby solving it.

When Heilbronn learned of this result, he decided that there should be an analytic reason for the reduction. The result was [45]. This remarkable paper was directly responsible for the first general improvements on the Brauer-Siegel theorem. A generalization of Heilbronn's theorem enabled Stark [1974] to show effectively that for $k \neq \mathbb{Q}$ there are only finitely many K with a fixed class number and, indeed, if $\deg(k)$ is fixed greater than 2, there are only finitely many k with any such K. Using his discriminant bounds, Odlyzko [1975] was able to show that for k normal over $\mathbb{Q}$ (or slightly more generally) there are no such K if $\deg(k)$ is sufficiently large. On Artin's conjecture that nonabelian L-functions are analytic outside of $s = 1$, the normality condition could be dropped completely. Odlyzko's bound on the degree of k to get $h(K) > 1$ would be quite large; Hoffstein [1979] made the estimate in a more accurate manner for k normal over $\mathbb{Q}$ (or slightly more generally again) and found, for example, that for $h(K) = 1$, $\deg(k)$ must be less than 219. The fact (assuming Artin's conjecture if we don't want a normality condition) that there are only finitely many totally real k with any class number 1 totally complex quadratic extensions was, I believe, an entirely unexpected result.

A few words should be said about Heilbronn's expansion of the Epstein zeta functions in [7] and [8]. Heilbronn used the Euler-MacLaurin formula which gives an analytic continuation for $\sigma > 1/2$. This suffices for the application that Heilbronn made to zeros near $s = 1$. Other authors have expanded these functions in other ways. Many have used (or essentially rederived) Poisson's summation formula for this purpose. It has the advantage of giving an analytic continuation and functional equation at the same time. A representational sample of papers could be those by Mordell [1930], Bateman and Grosswald [1964], and Stark [1969a]. In the latter paper, Heilbronn's expansion of

$$\frac{1}{2} \sum_{(x, y) \neq (0,0)} \left(\frac{-p}{Q(x, y)} \right) Q(x, y)^{-s}$$

in [8] (cf. Lemma 4) is carried out for general characters with composite conductor. Had this expansion been available earlier, the class number 1 problem would have been solved by transcendence methods in 1949 by Gelfond (see Stark [1969b]).

The reader is referred to the commentaries to the related papers [11], [12], [20], [27], and [28] for further remarks on this subject.

References

J. A. Antoniadis

[1983] Über die Kennzeichnung zweiklassiger imaginär-quadratischer Zahlkörper durch Lösungen diophantischer Gleichungen. *J. Reine Angew. Math.* **330**, 27–81.

A. Baker

[1966] Linear forms in the logarithms of algebraic numbers. *Mathematika* **13**, 204–216.

[1971] Imaginary quadratic fields with class-number 2. *Ann. Math.* (2) **94**, 139–152.

P. T. Bateman and E. Grosswald

[1964] On Epstein's zeta function. *Acta Arith.* **9**, 365–373.

B. J. Birch

[1969] Diophantine analysis and modular functions. In: *Algebraic Geometry* (*Intern. Colloq., Tata Inst. Fund. Res., Bombay, 1968*). Oxford Univ. Press, Oxford.

R. Brauer

[1947] On the zeta-functions of algebraic number fields. *Am. J. Math.* **69**, 243–250.

[1950] On the zeta-functions of algebraic number fields II. *Am J. Math.* **72**, 739–746.

S. Chowla and W. E. Briggs

[1954] On discriminants of binary quadratic forms with a single class in each genus. *Can. J. Math.* **6**, 463–470.

M. Deuring

[1934] Zetafunktionen quadratischer Formen. *J. Reine Angew. Math.* **172**, 226–252.

[1968] Imaginäre quadratische Zahlkörper mit der Klassenzahl Eins. *Invent. Math.* **5**, 169–179.

D. M. Goldfeld
[1976] The class number of complex quadratic fields and the conjectures of Birch and Swinnerton-Dyer. *Ann. Scuola Norm. Sup. Pisa Cl. Sci.* (4) **3**, 624–663.

L. J. Goldstein
[1972] Relative imaginary quadratic fields of class number 1 or 2. *Trans. Am. Math. Soc.* **165**, 353–364.

B. H. Gross and D. Zagier
[1986] Heegner points and derivatives of *L*-series. *Invent. Math* **84** (1986), 225–320.

K. Heegner
[1952] Diophantische Analysis und Modulfunktionen *Math. Z.* **56**, 227–253.

J. Hoffstein
[1979] Some analytic bounds for zeta functions and class numbers. *Invent. Math.* **55**, 37–47.
[1980] On the Siegel-Tatuzawa theorem, *Acta Arith.* **38**, 167–174.

M. A. Keuku
[1971] Determination of the even discriminants of complex quadratic fields of class-number 2. *Proc. London Math. Soc.* (3) **22**, 734–746.

C. Meyer
[1971] *Imaginär-quadratische Zahlkörper mit der Klassenzahl 2 und ihre Bestimmung durch elliptische Kurven.* Lecture at Oberwolfach; some details in the paper of Antoniadis cited above.

L. J. Mordell
[1930] The zeta functions arising from quadratic forms and their functional equations. *Quart. J. Math.* **1**, 77–101.

A. M. Odlyzko
[1975] Some analytic estimates of class numbers and discriminants. *Invent. Math.* **29**, 275–286.

C. L. Siegel
[1935] Über die Classenzahl quadratischer Zahlkörper. *Acta Arith.* **1**, 83–86 (*Ges. Abh.* I, 406–409).

H. M. Stark
[1967] A complete determination of the complex quadratic fields of class-number one. *Mich. Math. J.* **14**, 1–27.
[1969a] *L*-Functions and character sums for quadratic forms (II), *Acta Arith.* **15**, 307–317.
[1969b] A historical note on complex quadratic fields with class number one. *Proc. Am. Math. Soc.* **21**, 254–255.
[1969c] On the "gap" in a theorem of Heegner. *J. Number Theory* **1**, 16–27.
[1971] A transcendence theorem for class-number problems. *Ann. Math* **94**, 153–173.
[1974] Some effective cases of the Brauer–Siegel theorem. *Invent. Math.* **23**, 135–152.

J. S. Sunley
[1973] Class-numbers of totally imaginary extensions of totally real number fields. *Trans. Am. Math. Soc.* **175**, 209–232.

B. Tatuzawa
[1971] On a theorem of Siegel. *Jap. J. Math.* **21**, 163–178.

P. J. Weinberger
[1969] *A proof of a conjecture of Gauss on class-number two.* Berkeley thesis.

The Papers on Waring's Problem

B. J. Birch and R. C. Vaughan

Introduction

In 1770 E. Waring [1770] stated without proof that every positive integer is a sum of at most 4 squares, of 9 [positive] cubes, of 19 biquadrates, "and so on." This was made more precise by J. A. Euler, a son of L. Euler, who observed that $n = 2^k[(3/2)^k] - 1$ cannot be written as a sum of fewer than $f(k) = 2^k + [(3/2)^k] - 2$ positive kth powers; cf. Euler [1772]. [Note that $f(2) = 4$, $f(3) = 9$, $f(4) = 19$.] The qualitative content of this assertion was settled by D. Hilbert (cf. Hilbert [1909]), who showed that for each k there exists a least integer $g(k)$ such that each positive integer is a sum of at most $g(k)$ positive kth powers. By now the value of $g(k)$ is known for all $k \neq 4$; in fact, $g(k) = f(k)$ (the "Ideal Waring's Theorem"), except when $k > 4$ satisfies the inequality

$$(*) \qquad 3^k - 2^k\left[(3/2)^k\right] > 2^k - \left[(3/2)^k\right] - 2,$$

and even for such exceptional k a formula for $g(k)$ is known. (For more details, cf. the notes to Ch. XXI of Hardy and Wright [1979], which contain a good up-to-date history of Waring's problem.) However, no k satisfying $(*)$ is known, and at any rate there are at most finitely many such k (Mahler [1957]) and none for $k \leq 200{,}000$ (Stemmler [1964]). Aside from this uncertainty, the only case which is not yet settled is the case $k = 4$; here we know only that $19 \leq g(4) \leq 21$ (Balasubramanian [1979]) and that every integer $n > 10^{1409}$ is a sum of at most 19 fourth powers (Thomas [1973]).[1]

A related and more interesting problem is the determination of the least number $G(k)$ such that every *sufficiently large* integer can be represented by at

[1]*Added in proof*. Recently, R. Balasubramanian, J.-M. Deshouillers and F. Dress [*Comptes Rendus de l'Academie des Sciences Paris*), *Série I-Mathematique* **303** (1986), 161–163] showed that, in fact, $g(4) = 19$.

most $G(k)$ positive kth powers, and it is this (much harder!) problem and other similar problems which Heilbronn, mostly together with H. Davenport, tackles in his papers.

[10] · Über das Waringsche Problem

Heilbronn's first paper on Waring's problem is a simplified and decidedly more readable account of Vinogradov's fundamental breakthrough (Vinogradov [1935]) that

$$G(k) \leq 6k \log k + (4 + \log 216)k.$$

In fact, by using his "exponential sums" version of the circle method, Heilbronn obtains the slightly better estimate

$$G(k) \leq 6k \log k + \left\{4 + 3\log\left(3 + \frac{2}{k}\right)\right\} + 3,$$

which for more than a decade gave the best upper bounds for $G(k)$ for general k. They were later superseded by Vinogradov [1947], Chen [1958], and Thanigasalam [1980], who proved that

$$G(k) \leq 3k \log k + ck,$$

with $c = 11$, $c = 5.2$, and $c = \log 108 \doteq 4.682\ldots$, respectively. For large k, the best bounds at present are given by Vinogradov's estimate (Vinogradov [1959]):

$$G(k) < k(2\log k + 4\log\log k + 2\log\log\log k + 13),$$

(valid for $k \geqq 170{,}000$), and for $k \leq 20$, Thanigasalam [1980], [1982], [1985] has recently obtained some further improvements. For the best upper bounds now known for $G(k)$ when $k \leq 9$ see Vaughan [1981], [1986a], [1986b].

Shortly after this paper was written, Heilbronn was elected to the Bevan fellowship at Trinity, and began to collaborate with Davenport, at first particularly on Waring's problem. Davenport recollected that many of their most fruitful discussions took place in the course of walking tours in Devon.

[13] (With H. Davenport) On Waring's problem for fourth powers

In their first joint paper, Davenport and Heilbronn prove that $G(4) \leq 17$; the same result was found at about the same time by Estermann, and his paper immediately precedes theirs in the *Proceedings of the London Mathematical*

Society. The 17 improves Hardy and Littlewood's 19 (cf. Hardy and Littlewood [1925]). Both papers follow Vinogradov's method (Vinogradov [1935]) fairly closely; there are extra difficulties (described by Davenport as "minor") which are overcome by Estermann in one way, and by Davenport and Heilbronn by appealing to a lemma on exponential sums proved in [14]. Later, Davenport [1939] succeeded in settling the issue by proving $G(4) = 16$; this, and the well-known fact that $G(2) = 4$, are the only values of k for which $G(k)$ is known.

[14] (With H. Davenport) On an exponential sum

The estimation of this sum is a prerequisite for a good estimate on the major arcs in the Hardy–Littlewood method as applied to Waring-type problems. For an account of the history of this sum and the best estimates we now have for it, see Vaughan [1984].

[15] (With H. Davenport) On Waring's problem: Two cubes and one square

In this paper Davenport and Heilbronn prove, again by Vinogradov's method, that almost all positive integers are of the form $x^2 + y^3 + z^3$. Davenport recalls,

> A difficulty in this paper is that owing to the small number of variables there is no singular series in the ordinary sense. Instead, we had to work with a finite product and prove that this is sufficiently close to the finite form of the singular series to be used in place of it, almost always.

To be more specific, one of the main difficulties here is that the series

$$\sum_{q=1}^{\infty} q^{-3} \sum_{\substack{a=1 \\ (a,q)=1}}^{q} \left| \left(S_{a,q}^{(3)}\right)^2 \cdot S_{a,q}^{(2)} \right|$$

diverges, and so there is no simple way of treating the singular series. One needs to know that the finite sum

$$\mathfrak{S}(R, n) = \sum_{q=1}^{R} A(n, q) \qquad (1)$$

is a good approximation to the infinite sum

$$\sum_{q=1}^{\infty} A(n,q)$$

or, what should be the same thing, to the Euler product

$$\mathfrak{S}(n) = \prod_p \left[\sum_{k=0}^{\infty} A(n, p^k) \right]$$

when R is a small power of n. This is similar to, but probably harder than, showing that the finite character sum

$$\sum_{n \leq Q} \frac{\chi(n)}{n}$$

is a good approximation to the infinite character sum

$$\sum_{n=1}^{\infty} \frac{\chi(n)}{n},$$

when χ is a quadratic character modulo q, and Q is a small power of q. The problem is that there might be an exceptional Siegel zero of $L(s, \chi)$ close to 1. Davenport and Heilbronn discuss the problem as it relates to their situation in the last section of [15]. They solve the difficulty in the paper by replacing $G(n)$ by a finite Euler product and showing that it is almost always a fairly good approximation to the finite sum (1). However, their method puts a severe restriction on the quality of the estimate that can be obtained for the exceptional set. Roth [1949] contains a refinement of their method, and Miech [1968] solves a similar, but easier, problem more directly. By combining the Davenport–Heilbronn technique with ideas from sieve theory, Vaughan shows how a much better estimate can be obtained for the exceptional set in this and similar problems (cf. Chap. 8 of Vaughan [1981]).

[16] (With H. Davenport) Note on a result in the additive theory of numbers

In this paper Davenport and Heilbronn show that almost all integers are the sum of a prime and a kth power (for any fixed k). The methods here are quite exciting. They use the Hardy–Littlewood technique to estimate the mean-square error

$$\sum_{n<P} (r(n) - \varrho(n))^2,$$

where $r(n)$ is the number of solutions of $n = p + x^k$, and $\varrho(n)$ is a plausible approximation to $r(n)$; the theorem follows when

$$\sum\left(r(n) - \varrho(n)\right)^2 = o\left(\sum\varrho(n)^2\right).$$

The most interesting part of the proof is their estimate for the sum over primes,

$$\sum_{p<P} e(\alpha p);$$

this depends on a formula for the number of primes in an arithmetic progression, with a good error term depending on Siegel's result, which was then very recent, about the (presumably nonexistent) real zero of an L-function (cf. Siegel [1935]). It is easy to modify the method to prove that almost all numbers have the form $p + x^k$ or $2p + x^k$ with p a prime congruent to 1 modulo 4, so that almost all numbers have the form $x^2 + y^2 + z^k$. Many papers (notably by Roth and Halberstam) in which the results of [15] and [16] are sharpened and generalized are listed in Section P12 of LeVeque's compendium *Reviews in Number Theory* (cf. LeVeque [1974]).

[22] (With H. Davenport) On indefinite quadratic forms in five variables

This is a seminal paper, though quite an easy one. It is in fact a good paper to be read by a graduate student who wishes to learn basic techniques of the circle method. They attack a conjecture of Oppenheim, that an indefinite quadratic form with real coefficients in five or more variables always takes small values, and they prove it for diagonal forms. To be precise, they prove that if $\lambda_1, \ldots, \lambda_5$ are real numbers not all of the same sign, then there are integers $x_1, \ldots, x_5$ not all zero so that $|\Sigma\lambda_1 x_1^2| < 1$. This is by no means a difficult paper—indeed it is one of the least complicated of the many applications of the circle method—but it made clear for almost the first time that Diophantine inequalities could be treated by the method as well as Diophantine equations. Papers on similar lines were soon published by Watson, Bambah, and others; particularly interesting is the problem of reducing the number of variables from five to four, but so far this has resisted all attacks. The methods of [22] are easy to apply to forms $\Sigma\lambda_i x_i^k$ of degree k in $2^k + 1$ variables; in their paper, Davenport and Roth [1955] prove the noticeably more difficult results that if $k > 12$, $Ck \log k$ variables suffice, and that if $k = 3$, eight variables are enough.

Oppenheim's conjecture for nondiagonal quadratic forms is enormously harder than the diagonal problem; there are obvious difficulties in applying the essentially additive circle method to a nonadditive problem. So far

Oppenheim's conjecture is unsolved, but Davenport (with the collaboration of B. J. Birch and D. Ridout) has shown that an indefinite real quadratic form in 21 variables assumes a value in absolute value less than one at some integer point; cf. Davenport [1956], [1958]; Birch and Davenport [1958]; Davenport and Ridout [1959]; and Ridout [1968]. The first paper listed was a significant breakthrough in the use of the circle method to test nonadditive problems; it was not the first paper treating "forms in many variables" from the viewpoint —there were earlier papers by Peck and notably by Tartakovsky (see the review article by Malyšev and Podsypanin [1974])—but I believe all subsequent work in the area derives from it.[2]

The generalization of Oppenheim's conjecture to forms of higher (especially odd) degrees has been investigated by several people, notably by Jane Pitman [1968] and Wolfgang Schmidt [1980]. The latter has proved the beautiful theorem that if k is odd then every form of degree k in enough variables takes arbitrarily small values. (For work on diagonal inequalities of odd degree, cf. Pitman and Ridout [1967] and Pitman [1971].) In all these papers an essential ingredient is the fact that if the form with integer coefficients has integer solutions, it has an integer zero whose height is bounded by an explicit function of the degree and the coefficients.

[25] On the distribution of the sequence $n^2\theta$ (mod 1)

The result proved here is the following. For fixed $\varepsilon > 0$ and $N > c(\varepsilon)$, there are integers m, n such that

$$|n^2\theta - m| < N^{1/2-\varepsilon} \qquad \text{and} \qquad 1 \le n \le N;$$

a crucial point is that the estimates are uniform in θ. Heilbronn wrote this paper in a very self-deprecating style—Vinogradov [1927] proved a similar theorem, with $N^{2/5}$ in place of $N^{1/2}$, and Heilbronn describes his result as a slight improvement of Vinogradov's, obtained by not very different methods. However, whereas Vinogradov's paper seems to have remained unstudied for 20 years, Heilbronn's eventually stimulated many other papers on related subjects; it is remarkable that though Heilbronn's result has been multiply generalized, it has not yet been sharpened. (The exponent 1/2 seems to be a natural boundary for the method.) An extension to monomials of degree k by Danicic [1958, 1959] and an extension to quadratic polynomials of shape $\alpha n^2 + \beta n$ by Davenport [1967] were followed by papers by Ming-Chit Liu [1970a, 1970b]; more recently, there have been contributions by R. J. Cook, by Wolfgang Schmidt (particularly, his exposition [1977]), and by Roger Baker [1982], who proves the most natural generalization of Heilbronn's result, that for any real polynomial $f(x)$ of degree k with $f(0) = 0$ there is an n with $1 \le n \le N$ such that

[2] *Added in proof*: The 21 variable result has recently been improved by R. Baker and H. Schlickewei [*Proceedings of the London Mathematical Society* (3) **54** (1987), 385–411].

$$\|f(n)\| < N^{1/K-\varepsilon},$$

where $K = 2^{k-1}$ and $\|..\|$ denotes the fractional part.

Remark. Although this paper (as well as [22]) is not, strictly speaking, a Waring-type problem, the hard ideas here, just like those of [13]–[16], are all centered on refinements of exponential sum techniques and are therefore included in this section.

[34] On the representation of a rational as a sum of four squares by means of regular functions*

In 1963, N. J. Fine posed the question, "Can $x^2 + y^2 + z^2 + w^2$ be expressed as the sum of three squares of rational functions with real coefficients?" Fine's question was answered (negatively) by Davenport [1963] and then, more perspicaciously, by Cassels [1964]; the question led directly to Pfister's beautiful series of papers on positive definite quadratic forms (cf. Pfister [1965, 1967]), and questions involving sums of squares were very much in the air. Every positive rational x can, of course, be expressed as the sum of four squares of rationals, $x = y_1^2 + \cdots + y_4^2$, and it is natural to ask whether $y_1, \ldots, y_4$ can be determined as nice functions of x. Heilbronn gives a beautiful answer to this question by showing that there are integral functions $f_1(x), \ldots, f_4(x)$ such that

(i) each $f_j(x)$ is real and positive whenever x is so,
(ii) $\Sigma f_j(x)^2 = 1$,
(iii) $x^{1/2} f_j(x)$ is rational whenever x is positive and rational.

His proof is exceedingly simple. He goes on to prove related but much less perfect results for $\Sigma g_j(x)^k = x$.

Note. The statement on p. 75, lines 15–16 of the paper is historically not quite correct, for the fact that every rational number is a sum of three cubes had already been proved by S. Ryley in 1825 (cf. Dickson [1920], p. 726; for a modern treatment of the problem, cf. Manin [1974]).

References

R. C. Baker
[1982] Weyl sums and diophantine approximation. *J. London Math. Soc.* (2) **25**, 25–34.
R. Balasubramanian
[1979] On Waring's problem: $g(4) \leq 21$. *Hardy–Ramanujan J.* **2**, 31 pp.

* See also O. Taussky's historical comments on pp. 33–34 [Ed.].

B. J. Birch and H. Davenport

[1958] Indefinite quadratic forms in many variables. *Mathematika* **5**, 8–12 (*Collected Works of H. Davenport*, Vol. III, 1077–1081).

J. W. S. Cassels

[1964] On the representation of rational functions as sums of squares. *Acta Arith.* **9**, 79–82.

Jun-jing Chen

[1958] On Waring's problem for nth powers. *Acta Math. Sinica* **8**, 253–257 (Chinese); translated as *Chinese Math. Acta* **8** (1966), 849–853.

I. Danicic

[1958] An extension of a theorem of Heilbronn. *Matematika* **5**, 30–37.

[1959] On the fractional parts of θx^2 and ϕx^2, *J. London Math. Soc.* **34**, 353–357.

H. Davenport

[1939] On Waring's problem for fourth powers. *Ann. Math.* (2) **40**, 731–747 (*Collected Works*, Vol. III, 946–962).

[1956] Indefinite quadratic forms in many variables. *Mathematika* **3**, 81–101 (*Collected Works*, Vol. III, 1035–1055).

[1958] Indefinite quadratic forms in many variables II. *Proc. London Math. Soc.* (3) **8**, 109–126 (*Collected Works*, Vol. III, 1058–1075).

[1963] A problematic identity. *Mathematika* **10**, 10–12 (*Collected Works*, Vol. IV, 1745–1747).

[1967] On a theorem of Heilbronn. *Quart J. Math., Oxford Ser.*, (2) **18**, 339–344 (*Collected Works*, Vol. III, 1307–1312).

H. Davenport and D. Ridout

[1959] Indefinite quadratic forms. *Proc. London Math. Soc.* (3) **9**, 544–555 (*Collected Works of H. Davenport*, Vol. III, 1105–1116).

H. Davenport and K. F. Roth

[1955] The solubility of certain Diophantine inequalities. *Mathematika* **2**, 81–96 (*Collected Works of H. Davenport*, Vol. III, 1019–1034).

L. E. Dickson

[1920] *History of the Theory of Numbers* II, Chelsea Publishing Co., New York, 1971.

L. Euler

[1772] *Opera posthuma* **1** (1862), 203–204.

G. H. Hardy and J. E. Littlewood

[1925] Some problems of "Partitio Numerorum" VI. Further researches in Waring's problem. *Math. Z.* **23**, 1–37 (*Collected Papers of G. H. Hardy* I, 469–505).

G. H. Hardy and E. M. Wright

[1979] *An Introduction to the Theory of Numbers*, 5th ed., Clarendon Press, Oxford.

D. Hilbert

[1909] Beweis für die Darstellbarkeit der ganzen Zahlen durch eine feste Anzahl n-ter Potenzen (Waringsches Problem). *Math. Ann.* **67**, 281–300 (*Gesammelte Abhandlungen* I, 510–527).

W. J. LeVeque

[1974] *Reviews in Number Theory* (6 vols.), Amer. Math. Soc., Providence, RI.

M.-C. Liu

[1970a] On a theorem of Heilbronn concerning the fractional part of θn^2. *Can. J. Math.* **22**, 784–788.

[1970b] On the fractional part θn^k and ϕn^k. *Quart. J. Math., Oxford Ser.* (2) **21**, 481–486.

K. Mahler

[1957] On the fractional parts of the powers of a rational number II. *Mathematika* **4**, 122–124.

A. V. Malyšev and V. D. Podsypanin

[1974] Analytic methods in the theory of systems of Diophantine equations and inequalities with a large number of unknowns (Russian). In *Algebra · Topology · Geometry* (R. V. Gamkrelidze, ed.), Vol. 12, Akad. Nauk SSSR Vsesujuz. Inst. Naucn. i Tehn. Informacii, Moscow, 1974, pp. 5–50.

Yu. I. Manin

[1974] *Cubic Forms—Algebra, Geometry, Arithmetic.* Translated from Russian by M. Hazewinkel. North-Holland Publishing Co., Amsterdam.

R. J. Miech

[1968] On the equation $n = p + x^2$. *Trans. Amer. Math. Soc.* **130**, 494–512.

A. Pfister

[1965] Zur Darstellung von -1 als Summe von Quadraten in einem Körper. *J. London Math. Soc.* **40**, 159–165.

[1967] Zur Darstellung definiter Funktionen als Summe von Quadraten. *Inv. Math.* **4**, 229–237.

J. Pitman

[1968] Cubic inequalities. *J. London Math. Soc.* **43**, 119–126.

[1971a] Bounds for solutions of diagonal inequalities. *Acta Arith.* **18**, 179–190.

[1971b] Bounds for solutions of diagonal equations. *Acta Arith.* **19**, 223–247.

J. Pitman and D. Ridout

[1967] Diagonal cubic equations and inequalities. *Proc. Roy. Soc. London Ser. A* **297**, 476–502.

D. Ridout

[1958] Indefinite quadratic forms. *Mathematika* **5**, 122–124.

K. F. Roth

[1949] Proof that almost all positive integers are sums of a square, a positive cube and a fourth power. *J. London Math. Soc.* **24**, 4–13.

W. Schmidt

[1977] Small fractional parts of polynomials. *Regional Conference Series in Mathematics* (Am. Math. Soc.), No. 32.

[1980] Diophantine inequalities for forms of odd degree. *Adv. Math.* **38**, 128–151.

C. L. Siegel

[1935] Über die Classenzahl quadratischer Zahlkörper. *Acta Arith.* **1**, 83–86 (*Gesammelte Abhandlungen* I, 406–409).

R. M. Stemmler

[1964] The ideal Waring theorem for exponents 401–200,000. *Math. Comp.* **18**, 144–146.

K. Thanigasalam

[1980] On Waring's problem. *Acta Arith.* **38**, 141–155.

[1982] Some new estimates for $G(k)$ in Waring's problem. *Acta Arith.* **42**, 73–78.

[1985] Improvement on Davenport's iterative method and new results in additive number theory I, II. *Acta Arith.* **46**, 1–31, 91–112.

H. Thomas

[1973] *A numerical approach to Waring's problem for fourth powers.* The University of Michigan, Ph.D. thesis.

R. C. Vaughan

[1981] *The Hardy–Littlewood Method* (*Cambridge Tracts in Math.* vol. 80), Cambridge Univ. Press, Cambridge–New York.

[1984] Some remarks on Weyl sums. In *Topics in Classical Number Theory* (G. Halász, ed.), Colloquia Mathematica Societatis János Bolyai **34**, North-Holland Publ. Co., Amsterdam, vol II, 1585–1602.

[1986a] On Waring's problem for smaller exponents, *Proc. London Math Soc.* (3) **52**, 445–463.
[1986b] On Waring's problem for sixth powers, *J. London Math Soc.* (2) **33**, 227–236.

I. M. Vinogradov
[1927] An analytic proof of the theorem on the distribution of the fractional parts of an entire polynomial. (Russian). *Izv. Akad. Nauk SSSR (Bull. Acad. Sci. URSS)*, Ser. 6, **21**, 567–578.
[1935] On Waring's problem. *Ann. Math.* (2), **36**, 395–405.
[1947] *The method of trigonometrical sums in the theory of numbers*. Translated, revised, and annotated by K. F. Roth and A. Davenport, Interscience, London and New York.
[1959] On an upper bound for $G(n)$. (Russian). *Izv. Akad. Nauk SSSR Ser. Mat.* **23**, 637–642.

E. Waring
[1770] *Meditationes algebraicae*. Cambridge, 1770, pp. 204–205; 3rd ed., 1782, p. 349.

The Papers on Euclid's Algorithm

H. Stark

[20] On Euclid's algorithm in real quadratic fields

[27] On Euclid's algorithm in cubic self-conjugate fields

[28] On Euclid's algorithm in cyclic fields

The problem of which quadratic fields are Euclidean is easier than the class number 1 problem. For complex quadratic fields it is almost trivial, but even for real quadratic fields it is possible to settle and, surprisingly enough, Heilbronn showed in [20] that there are only finitely many Euclidean real quadratic fields. Making use of methods of Davenport [1951] from the geometry of numbers, Chatland and Davenport [1950] later found them all (there are 16 of them; contrary to what was thought at the time $\mathbb{Q}(\sqrt{97})$ is not Euclidean). This is in spite of the fact that it is widely believed there are infinitely many real quadratic fields of class number 1. Heilbronn extended his methods to cyclic fields in [27] and [28], while Davenport proved similar results for those types of cubic fields [1950a] and quartic fields [1950b] with one fundamental unit. Again, for cubic fields which are not totally real, there are only finitely many Euclidean fields even though there are undoubtedly infinitely many class number 1 fields of this type.

In view of what is now known about class numbers (see the commentary to [45] on p. 571) it is not unreasonable to ask whether there are infinitely many Euclidean fields. According to Lenstra [1979], part 2, there were 334 Euclidean fields known in 1979. It is important that "Euclidean" here means Euclidean with the usual norm. There is a more general concept of Euclidean with other norms, and for this, Cooke and Weinberger [1975] proved that for any class number 1 number field which is not $\mathbb{Q}$ or complex quadratic (and if the Riemann hypothesis holds for number fields), the field is also Euclidean with respect to some norm. Heilbronn seems to suspect in [27] that there may be infinitely many totally real Euclidean cubic fields. In [28], he goes even further and names two families of quartic and sextic fields that should be investigated.

This brings up a very interesting possibility. We may artificially investigate higher rank unit groups in real quadratic fields by enlarging the ring of integers to so-called S-integers. That is, if S is a finite set of prime ideals of a field K, the ring of S-integers, $\mathfrak{O}_S$, of K is the set of all elements of K having denominators involving only the prime ideals of S. For a real quadratic field K, the unit group of $\mathfrak{O}_S$ has rank 1 plus the number of prime ideals in S. We can then ask whether or not the ring $\mathfrak{O}_S$ is Euclidean with respect to the obvious norm,

$$N_S(\alpha) = |N_{K/Q}(\alpha)| \prod_{\mathfrak{p} \in S} |\alpha|_{\mathfrak{p}},$$

where $| \ |_{\mathfrak{p}}$ is the normalized $\mathfrak{p}$-adic valuation.

When S consists of just one prime ideal, the situation here should be very analogous to that of the integers in a totally real cubic field but easier to investigate. In particular, we are entitled to ask whether or not infinitely many quadratic fields K have Euclidean rings $\mathfrak{O}_S$ with just one prime ideal in S. If so, this would be a great step forward in investigating real quadratic fields of class number 1, especially if one could choose principal ideals $\mathfrak{p}$ beforehand. If not, one would be more justified in suspecting that there are only finitely many Euclidean fields. One might begin by investigating fields K in which 2 splits and letting S consist of the two prime ideals dividing 2 (the ring $\mathfrak{O}_S$ then has a rank 3 unit group). This would seem to me to be an analogous situation to the quartic family of fields suggested by Heilbronn in [28].

References

H. Chatland and H. Davenport

[1950] Euclid's algorithm in real quadratic fields. *Can. J. Math.* **2**, 289–296.

G. Cooke and P. J. Weinberger

[1975] On the construction of division chains in algebraic number rings, with applications to SL_2, *Comm. Algebra* **3**, 481–524.

H. Davenport

[1950a] Euclid's algorithm in cubic fields of negative discriminant. *Acta Math.* **84**, 159–179 (*Collected Works*, Vol. I, 374–394).

[1950b] Euclid's algorithm in certain quartic fields. *Trans. Am. Math. Soc.* **68**, 508–532 (*Collected Works*, Vol. I, 404–428).

H. W. Lenstra, Jr.

[1979] Euclidean number fields 1, 2, 3. *Math. Intel.* **2**, 6–15, 73–77, 99–103.

The Papers on Zeta- and L-Functions

H. Stark

[11], [12] (With H. Davenport) On the zeros of certain Dirichlet series. I, II.

In these papers Heilbronn and Davenport have shown that for certain classes of Epstein zeta functions and Hurwitz zeta functions, not only is the Riemann hypothesis not true but in fact there are zeros with real parts greater than 1. For Hurwitz zeta functions, the missing classes of functions with algebraic coefficients were settled by Cassels [1961], who obtained the same results. His proof has not been carried over to Epstein zeta functions although it is not too much to hope that it could be carried over and that the following conjecture should be provable:

Conjecture. For real $c > 41$, the function

$$\zeta(s;1,1,c) = \sum_{(m,n)\neq(0,0)} (m^2 + mn + cn^2)^{-s}$$

has an infinite number of zeros with real part greater than 1.

A direct proof of this conjecture, which seemingly is a problem in analysis, would give a purely analytic resolution of the class number 1 problem, and generalizations could conceivably resolve the class number problem for other class numbers as well. Unfortunately, the only known method to prove this conjecture uses the result of Heilbronn and Davenport for integral c, which requires that the field $\mathbb{Q}(\sqrt{1-4c})$ not have class number 1. A direct proof will probably not be easy. If we pick a fixed zero ρ with real part greater than 1 for $c = c_0$ and then continuously follow ρ as c grows, almost certainly ρ will drift to the line $\sigma = 1/2$ and stay there. However, it would not be necessary to vary c over an interval of length greater than 1 if enough nice c_0 were available. Perhaps the task is not hopeless.

Reference

J. W. S. Cassels

[1961] Footnote to a note of Davenport and Heilbronn, *J. London Math. Soc.* **36**, 177–184.

The Papers on the Geometry of Numbers

J. H. H. Chalk

[23] (With H. Davenport) On the minimum of a bilinear form

Let $B(x, y, z, t) = axz + bxt + cyz + dyt$ be a real bilinear form, and let $Q(x, y) = B(x, y, x, y)$ be the associated quadratic form with discriminant $D = \theta^2 - 4\Delta$, where $\theta = |b - c|$ and $\Delta = ad - bc$ are equivalent invariants. Suppose that $D > 0$ and put $\omega = \theta/D^{1/2}$. Then B is symmetric if $\omega = 0$ and factorizable in the form $(\alpha x + \beta y)(\gamma z + \delta t)$ if $\omega = 1$. Suppose that $xt - yz = \pm 1$, where $x, y, z, t \in \mathbb{Z}$. Schur [1913] considered the case $\omega = 0$ and showed that $|B(x, y, z, t)| < D^{1/2}/2\sqrt{5}$ was solvable for every such B, apart from one special class of equivalent forms. Davenport and Heilbronn considered here the case $\omega = 1$ and proved much more. They showed that

$$|B(x, y, z, t)| < \tfrac{1}{3}(\sqrt{2} - 1)D^{1/2},$$

apart from the exclusion of *three* special classes of equivalent forms (and, indeed, that no further such exclusions was possible, cf. their Theorem 4). Barnes [1951] gave an alternative proof of this and later [1952] extended it to certain cases where $\omega \neq 0, 1$. I know no further work of this kind on bilinear forms (even with more variables), other than that of Chalk [1952], [1954], who considered a slightly generalized form

$$B(c) = (\alpha x + \beta y + c_1)(\gamma z + \delta t + c_2),$$

where $c = (c_1, c_2)$ and x, y, z, t are still controlled by the unimodular condition $xt - yz = \pm 1$. This curious hybrid problem evoked some response. For Davenport [1952] himself wrote a short article on the main lemma (cf. Chalk [1952], Lemma 5) which in geometrical language asserts the existence of a *primitive* point (x_1, x_2) of a plane lattice Λ of determinant Δ and having no points on the coordinate axes, in the region

$$|(x_1 + c)x_2| < \Delta/4, \quad (c \neq 0)$$

and he verified that this was best-possible [he also studied the analogous problem where the condition of primitivity was replaced by the restriction $(x_1, x_2) \neq (0, 0)$]. Unfortunately, Chalk's Lemma 5 was based upon a geometrical process previously used by D. B. Sawyer [1948], which was later found to be faulty, though capable of correction in his particular application of it. Abandoning this process, Chalk [1954] used the method developed by Davenport [1952] and found the correct inequality

$$|(x_1 + c)x_2| \leq (2 + \sqrt{2})^{-1}\Delta, \quad (c \neq 0),$$

which he showed was best-possible in the usual sense (there being a special lattice Λ with $\Delta = 2 + \sqrt{2}$ for which the equality sign is essential) and also "non-isolated" (by producing an infinite sequence of lattices Λ_n $(n = 1, 2, \ldots)$ for which the above inequality, with $(2 + \sqrt{2})^{-1}$ replaced by any smaller but fixed constant, was false for all large enough n.

Note. Similar problems associated with regions of this hybrid type have been considered by Kanagasabapathy, Varnavides, Blanksby, and Cole and possibly others, as recorded in Le Veque's *Reviews in Number-Theory*, Vol. III, Section J.

[24] (With H. Davenport) Asymmetric inequalities for non-homogeneous linear forms

This paper was the forerunner of a series of articles by various authors on asymmetric diophantine inequalities (the literature is too expansive to be summarized here, and the reader is referred to LeVeque's *Reviews in Number Theory*, Vol. III, Section J 08, 28, 32, 36, 40, 44, 48). Chalk [1947], [1948], extended the assertion of Theorem 1 by proving that the set S of solutions $(u_1, \ldots, u_n) \in \mathbb{Z}^n$ of the inequalities

$$\Pi(L_i + c_i) < \Delta, \qquad L_i + c_i > 0 \quad (1 \leq i \leq u),$$

where $L_i = a_{i1}u_1 + \cdots + a_{in}u_n$ $(1 \leq i \leq n)$ are n real linear forms in $u_1, \ldots, u_n$ with determinant $\Delta \neq 0$, satisfies card $S \neq \varnothing$. Macbeath gave an alternative and more constructive proof, which became the prototype for a far-reaching generalization (Macbeath [1952]; see also C. A. Rogers [1954] for a simpler version).

Chalk [1948] showed that if none of the L_i $(1 \leq i \leq n)$ had all their coefficients a_{ij} in a rational ratio, then card $S = \infty$. Rogers [1954] found that card $S = \infty$ even if one introduces further inequalities of the shape

$$\sum_{1 \leq t \leq r} (L_1 + c_1)^2 < \varepsilon^2, \quad (t \leq n - 1),$$

where $\varepsilon > 0$ is arbitrary. A variant on this problem was studied by Cole [1952] in which one of the $L_i + c_i$ was allowed to take values of either sign.

References

E. S. Barnes

[1951] The minimum of a factorizable bilinear form. *Acta Math. Acad. Sci. Hungar.* **86**, 323–336.

[1952] The minimum of a bilinear form. *Acta Math. Acad. Sci. Hungar.* **88**, 253–277.

J. H. H. Chalk

[1947] On the positive values of linear forms. *Quart. J. Math., Oxford Ser.* **18**, 215–227.

[1948] On the positive values of linear forms. II. *Quart. J. Math., Oxford Ser.* **19**, 67–80.

[1952] The minimum of a non-homogeneous bilinear form. *Quart. J. Math., Oxford Ser.* (2) **3**, 119–129.

[1954] On the primitive lattice points in the region $|(x + c)y| \leq 1$. *Quart. J. Math., Oxford Ser.* (2) **5**, 203–211.

A. J. Cole

[1952] On the product of n linear forms. *Quart. J. Math., Oxford Ser.* (2) **3**, 56–62.

H. Davenport

[1952] Note on a result of Chalk. *Quart. J. Math., Oxford Ser.* (2) **3**, 130–138 (*Collected Works of H. Davenport*, Vol. II, 592–600).

A. M. Macbeath

[1948] Non-homogeneous linear forms. *J. London Math. Soc.* **23**, 141–147.

[1952] A theorem on non-homogeneous lattices. *Ann. of Math.* (2) **56**, 269–293.

C. A. Rogers

[1954] A note on the theorem of Macbeath. *J. London Math. Soc.* (3) **4**, 50–83.

D. B. Sawyer

[1948] The product of two non-homogeneous forms. *J. London Math. Soc.* **23**, 250–251.

I. Schur

[1913] Zur Theorie der indefiniten binären quadratischen Formen. *Sitz.-ber. Preuss. Akad. Wiss.* 1913, *Phys. Math. Klasse*, 212–231 (*Gesammelte Abhandlungen* II, 24–43).

The Papers on Discriminants of Cubic Fields

A. Fröhlich

[41] (With H. Davenport) On the density of discriminants of cubic fields

Although this paper was published only in 1969, the work which it records had been done much earlier. Davenport and Heilbronn had obtained estimates for the numbers of cubic fields (for the two cases: totally real and otherwise) whose discriminants were below a given bound, but as they could not obtain results as precise as they had hoped, the work was put on one side. When Davenport was on the point of death Heilbronn must have felt that the results should be recorded and wrote [41]. This impelled him to consider the problem again, and the more precise results of [42], although published under joint names, were not in fact discovered until after Davenport's death.*

[42] (With H. Davenport) On the density of discriminants of cubic fields. II

This paper has stimulated further research in several directions. Its results raise a number of interesting and highly nontrivial questions.

There are three theorems we have to look at. The first one, and clearly to the authors the central one of the paper, deals with the density of a class of fields, measured by their discriminants. It will be useful to have some notation. For $X > 0$ let $N_m(X)$ be the number of nonconjugate algebraic number fields K of degree m over the rational field $\mathbb{Q}$, whose discriminant $d(K)$ satisfies $|d(K)| < X$. If we want only fields with positive or with negative discriminant, we shall write $N_m^+(X)$ or $N_m^-(X)$, respectively. If we restrict ourselves to fields whose Galois closure has group G over $\mathbb{Q}$, we shall write $N_m(X, G)$, etc. Throughout S_m is the symmetric group on m letters, and C_m is the cyclic group of order m. As usual, $f(X) \sim g(X)$ means $\lim_{X \to \infty} f(X)/g(X) = 1$.

The Davenport–Heilbronn density theorem says that $N_3^+(X) \sim aX$, $N_3^-(X) \sim bX$ with certain given positive constants a, b. The same limits will hold with $N_3^+(X, S_3)$ and $N_3^-(X, S_3)$ instead. In fact,

* For the convenience of the reader, we have inserted here lines 1–9 of p. 132 of the Cassels and Fröhlich obituary. (Ed.)

$$N_3(X, S_3) \sim N_3(X). \tag{1}$$

This follows from an earlier result of Harvey Cohn [1954], who proved

$$N_3(X, C_3) \sim c_3 X^{1/2} \quad (c_3 \text{ positive}). \tag{2}$$

He knew at the time of earlier, weaker results of Davenport and of Heilbronn, and he points out that indeed as a consequence

$$N_3(X, C_3)/N_3(X) \to 0. \tag{3}$$

Subsequently, and basing his work on the Davenport–Heilbronn paper, A. M. Bailey [1980] investigated $N_4(X, G)$ for all the groups G which can occur. Let A_n be the alternating group on n symbols and D_n the dihedral group of order $2n$. Then Bailey obtains upper and lower estimates as follows:

$$X \ll N_4(X, S_4) \ll X^{3/2} \log^4 X,$$

$$X^{1/2} \ll N_4(X, A_4) \ll X \log^4 X,$$

$$X \ll N_4(X, D_4) \ll X,$$

with some further detail on the constants involved. For the abelian groups involved he gets more precise results, in particular

$$N_4(X, C_4) \sim c_4 X^{1/2} \quad (c_4 > 0), \tag{4}$$

hence we observe that

$$N_4(X, C_4)/N_4(X) \to 0, \tag{5}$$

and even

$$N_4(X, C_4)/N_4(X, S_4) \to 0. \tag{6}$$

(In [1981] Bailey also deals with composites of three independent quadratic fields.)

One knows that under any reasonable (and certainly any known) definition of density of irreducible monic $\mathbb{Z}$-polynomials of given degree n, those with Galois group S_n have density 1. This, together with (1) and Bailey's estimates, raises the questions whether at least

$$N_n(X) \ll N_n(X, S_n) \tag{7}$$

or even

(8) $$N_n(X, S_n) \sim \alpha_n N_n(X) \quad \text{for some } \alpha_n > 0 \quad (\text{for } \alpha_n = 1).$$

What may be (hopefully) easier is

(9) $$\lim_{X\to\infty} N_n(X, C_n)/N_n(X) = 0 \quad \text{for } n > 2.$$

Another set of problems of this nature is connected with the property of whether or not field discriminants are squares. Let $G \subset S_n$, $G_+ = G \cap A_n$, but $G \not\subset A_n$, and of course G transitive. Examples: $G = S_n$, $G_+ = A_n$ or $G = D_n$, $G_+ = C_n$. Then the results quoted for $n = 3, 4$ together with similar considerations for polynomials raise the question

(10) $$N_n(X, G_+)X^{1/2} \sim BN_n(X, G) \quad (B > 0).$$

Further density theorems for field extensions have been proved by M. D. Steckel (cf. [1983] and also [1981]) and by M. J. Taylor [1984], both of whom we shall discuss here, as well as by D. J. Wright [1982], to whose work we shall return below.

Steckel deals with the general class of "Frobenius groups of maximal type." Such a group G is the semidirect product of a normal subgroup A, A being an elementary abelian p group for some prime p, and a subgroup U which acts transitively and without fixed points on $A \setminus \{0\}$. For any such G, the group A is uniquely determined by this structure. Steckel then gives himself a field K, Galois over $\mathbb{Q}$ with Galois group U and considers the class of fields L with $[L: Q] = |A|$ so that the Galois closure of L has group G with K as the fixed field of A. Here let $N(K, X)$ be the number of such fields L with $|d(L)| < X$. Write $e = \left(1 - \dfrac{1}{p}\right)|A|$ with p as above. Then

(11) $$N(K, X) \sim c(K)X^{1/e}(\log X)^{b(K)},$$

where $c(K) > 0$, $b(K) \geqq 0$, and where

(12) $$b(K) = 0 \quad \text{if } \left[K(\eta_p) \colon K\right] = p - 1.$$

Here η_r always denotes a primitive rth root of unity. Steckel also gives some further details about the constant $c(K)$.

In Taylor's paper the base field need not be $\mathbb{Q}$ but may be any number field K (of course of finite degree). He considers cyclic extensions L/K with Galois group

(13) $$\mathrm{Gal}(L/K) = C_n,$$

with the restriction that if 2^g is the highest power of 2 dividing n, then the primitive (2^g)th roots of unity lie in K. Taylor does not work with discriminants, but—which seems more natural in this context—with the "absolute conductor" $f_{L/K}$, which is defined as the absolute norm of the conductor of L/K in the sense of class field theory. Then let $\tilde{N}(K, n, X)$ be the number of fields K as in (13) with $f_{L/K} < X$. Then

$$\tilde{N}(K, n, X) \sim \alpha(K, n) X \log X^{a(K,n)-1}, \tag{14}$$

where $\alpha(K, n) > 0$ and where

$$a(k, n) = \sum_{1<m|n} [K(\eta_m): K]^{-1} \phi(m), \tag{15}$$

where ϕ is the Euler function. Note that conditions (12) and (15), imposed on the exponents of the logarithm in (11) and (14), respectively, are of the same type!

For the background to Theorem 2 in the Davenport–Heilbronn paper I recall the Chebotarev density theorem. Let L/K be a Galois extension of number fields with group G. Every prime ideal $\mathfrak{p}$ in K, nonramified in L, determines a conjugacy class $(L/K, \mathfrak{p})$, the class of Frobenius symbols of the prime divisors of $\mathfrak{p}$ in L. Then, given a conjugacy class Γ in G, the density (say in the Dirichlet sense) of prime ideals $\mathfrak{p}$ with $(L/K, \mathfrak{p}) = \Gamma$ is precisely $|\Gamma|/|G|$, the "density" of Γ in G. An inverse Chebotarev theorem would assert the following, for a given base field K and a given group G: Consider all pairs consisting of a Galois extension L/K and an isomorphism $\mathrm{Gal}(L/K) \simeq G$, so that $\mathfrak{p}$ is nonramified in L. Then the density of those for which $(L/K, \mathfrak{p})$ falls into Γ is again $|\Gamma|/|G|$.

There is a weaker Frobenius density theorem with a hypothetical inverse, replacing conjugacy classes by divisions. Here two elements g_1, g_2 of G lie in the same division if the cyclic subgroups generated by g_1 and by g_2, respectively, are conjugate.

To me the most exciting and challenging result in the Davenport–Heilbronn paper is their Theorem 2, which is an inverse Chebotarev–Frobenius theorem for base field $\mathbb{Q}$, and $G = S_3$. Sticking my neck out, I must confess to a belief that such a result cannot hold just for $n = 3$. What about S_n (for $n > 3$), or possibly D_n, which is perhaps fractionally more approachable but seems a less likely candidate?

The only results dealing with groups other than S_3 are in the paper of Taylor quoted earlier. K is again an arbitrary number field, and one considers the class of extensions L of K as in (13) with the same restrictions on n as before. Taylor then establishes an inverse Frobenius density theorem. For $\mathfrak{p}$ nonramified in L, denote by $[L/K, \mathfrak{p}]$ the division in C_n containing the Frobenius elements above $\mathfrak{p}$. [There is no need here to specify the particular isomorphism (13), as the divisions are fixed under all automorphisms of C_n.] Let $N'(X, \mathfrak{p})$ be the number of fields L in which $\mathfrak{p}$ is nonramified with

$f_{L/K} < X$, and among these let $N'(X, \mathfrak{p}, \Delta)$ be the number of those for which $[L/K, \mathfrak{p}] = \Delta$, a given division. Then Taylor's theorem, slightly reworded, is that

$$\lim_{X \to \infty} \frac{N'(X, \mathfrak{p}, \Delta)}{N'(X, \mathfrak{p})} = \frac{|\Delta|}{|C_n|}.$$

The third theorem of the paper is a mean value theorem for 3-class numbers of quadratic fields. This is obtained as a consequence of the method for the density theorem. The connection is the following: If $K = K_3$ is a noncyclic cubic field, then $K_2 = \mathbf{Q}(\sqrt{d(K)})$ is a quadratic field and the Galois closure K_6 of K_3 is cyclic over K_2. It is nonramified over K_2 precisely if all nontrivial inertia groups of $K_6/\mathbf{Q}$ are generated by transpositions, and this is the case precisely if no natural prime ramifies totally in K_3. Conversely, every cyclic cubic nonramified extension of a quadratic field is obtained in this way.

The question of generalization is obviously tied up with that of the density theorems. Let $l > 3$ and assume for simplicity's sake l to be a prime. The argument given above works just as well for l in place of 3, thus tying up the ramification of fields $K/\mathbf{Q}$ of degree l whose Galois closures has group D_l with l-class numbers of quadratic fields. Thus here D_l would appear as the natural generalization, and certainly the obvious one, of $S_3 = D_3$. But interpreted correctly the argument goes largely through also for S_l. Now we have to consider nonramified extensions of quadratic fields with relative Galois group A_l and absolute Galois group S_l. Such fields exist, but very little is known about them or even about their quadratic subfields.

The theorem on 3-class numbers has gained renewed interest in that it confirms in two particular cases some new, general conjectures on class groups by H. Cohen and H. W. Lenstra [1983], based on a new approach to this topic.

Last in this survey I must mention the beautiful work of D. J. Wright [1982][1], who extended the Davenport–Heilbronn theory from $\mathbf{Q}$ to an arbitrary number field K as base field, at present with the "technical" hypothesis that the class number of K be 1. He considers cubic extensions of K and generalizes all three of the Davenport–Heilbronn theorems. Here something must be said about their method of proof. This uses a discriminant preserving relation between cubic fields and irreducible primitive integral cubic binary forms, and the results are then based on theorems of Davenport [1951] on the mean value of class numbers of binary integral cubics. Davenport's proof in [1951] proceeds by what may be called a "fundamental domain" approach. Subsequently Shintani [1972] introduced certain Dirichlet series whose coefficients are given by the class numbers of integral cubic forms and proved that these admit a meromorphic continuation to the whole complex plane and a

[1] *Added in proof*. See also the recent paper of B. Datkowski and D. J. Wright [*J. Reine Angew. Math*. **367**, 27–75] and the references therein.

functional equation. This enabled him, among other things, to regain Davenport's mean value theorems. The main effort in Wright's work goes into extending Shintani's method from $\mathbb{Z}$ to the ring of integers in an arbitrary number field K. He then can use the analogue of the Davenport–Heilbronn procedure to deal with cubic extensions. It is at this stage where the class number hypothesis is presently needed.

References

A. M. Bailey

[1980] On the density of discriminants of quartic fields. *J. Reine Angew. Math* **315**, 190–210.

[1981] On octic fields of exponent 2. *J. Reine Angew. Math.* **328**, 33–38.

H. Cohn

[1954] The density of abelian cubic fields. *Proc. Am. Math. Soc.* **5**, 476–477.

H. Cohen and H. W. Lenstra, Jr.

[1983] Heuristics on classgroups of number fields. *Number Theory, Noordwijkerhous 1983* (*Proc. Journées Arithmétiques*, July 11–15, 1983) (Lecture Notes in Math. 1068), Springer-Verlag, Berlin, 1984, pp. 33–62.

H. Davenport

[1951] On the class numbers of binary cubic forms. (I). *J. London Math. Soc.* **26**, 183–192; (II). *J. London Math. Soc.* **26**, 192–198. (*Collected Works*, Vol. II, 509–518, 519–528).

T. Shintani

[1972] On Dirichlet series whose coefficients are class numbers of integral binary cubic forms. *J. Math. Soc. Japan*, **24**, 132–188.

H. D. Steckel

[1981] *Arithmetik and Asymptotik in der Theorie der Zahlkörper mit Frobenius Gruppe*. Dissertation, Cologne.

[1983] Dichte von Frobenius Körpern. *J. Reine Angew. Math.* **343**, 36–63.

M. J. Taylor

[1984] On the equidistribution of Frobenius in cyclic extensions of a number field. *J. London Math. Soc.* (2) **29**, 211–223.

D. J. Wright

[1982] *Dirichlet series associated with the space of binary cubic forms with coefficients in a number field*. Harvard thesis.

Miscellaneous Papers

[17] On real characters

Commentary by S. Chowla

Let

$$L_p(s) = \sum_{n=1}^{\infty} \left(\frac{n}{p}\right) \frac{1}{n^s} \quad (s > 0)$$

where $\left(\frac{n}{p}\right)$ is the Legendre symbol for an odd prime p. In the following we shall restrict our discussion to primes $p \equiv 3(4)$. Thus

$$L_3(s) = \left(\frac{1}{1^s} - \frac{1}{2^s}\right) + \left(\frac{1}{4^s} - \frac{1}{5^s}\right) + \left(\frac{1}{7^s} - \frac{1}{8^s}\right) + \cdots .$$

Here the bracketing shows that

$$L_3(s) > 0 \quad (0 < s < 1).$$

It is slightly trickier to show that

$$L_7(s) > 0 \quad (s > 0).$$

This depends on the fact that

$$\sum_1^x \left(\frac{n}{7}\right) \geq 0 \quad \text{for all } x \geq 1.$$

Consider the problem: For what $p \equiv 3(4)$ is

$$\sum_1^x \left(\frac{n}{p}\right) \geq 0 \quad \text{for all } x?$$

Several such p were listed by S. Chowla in the paper referred to by Heilbronn.
It is tempting to raise the question: Are there infinitely many primes $p \equiv 3(4)$ such that

$$\sum_{1}^{x}\left(\frac{n}{p}\right) \geq 0 \quad \text{for all } x \geq 1?$$

[19] On Dirichlet series which satisfy a certain functional equation

Commentary by R. C. Vaughan

The basis of Siegel's celebrated theorem (Siegel [1935]) that $\log L_d(1) = O(\log|d|)$ is an inequality of the form

$$(*) \qquad \phi(s) > \frac{1}{2} - \frac{c\lambda}{1-s}(d_1 d_2 d_3)^{4(1-s)}$$

valid for, say $\frac{7}{8} < s < 1$, where c is a positive constant. The identity (4) is the means whereby such an inequality was obtained. Later Estermann [1948] gave an even simpler proof of $(*)$ which does not use the identity (4). Other simple proofs have been given by Chowla [1950] and Goldfeld [1974].

[29] On the averages of some arithmetical functions of two variables

Commentary by A. Fröhlich

The background to the questions which I asked Heilbronn and which led him to this investigation is as follows: I had obtained results on class groups of abelian fields in terms of the mutual congruence behavior of their discriminant prime divisors. In particular, I had a necessary and sufficient criterion in such terms for an abelian field of degree a power of a prime l to have class number prime to l. The question as to the frequency of such fields then led to the problem which Heilbronn solved. The consequences of Heilbronn's theorem for the original question were never written up.

For a recent, hopefully more readable account, see Fröhlich [1983].

[35] (With P. Erdös) On the addition of residue classes mod p

Commentary by B. J. Birch

This is an attractive and typical "Erdös problem." Let p be a fixed prime and $a_1, \ldots, a_k$ be distinct residues modulo p; they show that every residue N modulo p is expressible as a combination $\Sigma e_i a_i$ with $e_i \in \{0,1\}$ provided that $k > \sqrt{54p}$, and in this case they give a correct estimate for the number of solutions. This result is self-explanatory; the condition $k > \sqrt{54p}$ has since been replaced by the best-possible condition $k > \sqrt{p}$ by J. E. Olson [1968]. No generalization to general abelian groups of the type "If $a_1, \ldots, a_k$ are distinct elements of the group G of order n then $\Pi a_i^{e_i} = g$ is certainly soluble for every $g \in G$ with $e_i \in \{0,1\}$, provided that $k > c\sqrt{n}$" can be true, as Ryavec [1968] has pointed out; but Szemeredi [1970] has proved that there is a constant c such that $\Pi a_i^{e_i} = 1$ is certainly soluble if $k > c\sqrt{n}$.

[38] (With S. Bachmuth and H. Y. Mochizuki) Burnside metabelian groups

Commentary by R. Foote

Although this article appears to be a singularity among Heilbronn's publications, being one of his few papers outside the realm of number theory and his only paper devoted purely to group theory, it possesses definite links to his investigation of ideal class groups viewed as modules over Galois groups, a topic which Cassels and Fröhlich say "interested him deeply and [which] he kept on coming back to" (cf. p. 14 of this volume). The object of this paper is to bound (above and below) the nilpotence class of the finite group $H = F/F''F^{p^e}$, where p is a prime, F is the free group on ϱ generators, and F^{p^e} is the subgroup of F generated by all p^eth powers. Any upper bound is valid for all ϱ-generator metabelian p-groups of exponent $\leq p^e$, i.e., any homomorphic image of H. Such metabelian p-groups arise in class field theory as Galois groups of extensions of the form $K_p^{(1)}/L$, where K is an abelian p-extension of the number field L and $K_p^{(1)}$ is the Hilbert p-class field of K; moreover, $\mathrm{Gal}(K_p^{(1)}/K)$ and the p-primary part of the ideal class group of K are isomorphic (via the Artin map) as modules over Gal (K/L).

During the summer of 1966 at the University of California, Santa Barbara, conversations with Seymour Bachmuth and Horace Mochizuki sparked Heilbronn's interest in bounding the nilpotence class of H from above. Their discussions reminded him of some calculations in the integral group ring of an abelian p-group he had worked on, and in September 1966 (as announced by his coauthors in [1967]) he established Theorem B of this paper. It is interesting to note that the relations between the nth power map, cyclotomic polynomials, and commutators which pervade this paper go back to Furtwängler's proof of the Principal Ideal Theorem (cf. Furtwängler [1930]) and that the "philosophy" of obtaining information (in number theoretic contexts) from these relations (and the transfer homomorphism) coupled to the action of a metabelian group on its commutator subgroup was further developed by Heilbronn's friend and colleague at the California Institute of Technology, Olga Taussky (see Scholz and Taussky [1934], for example). A more tenuous link of this paper to number theory is that the authors use the Magnus representation of a metabelian group (which Bachmuth had also utilized in some earlier papers); this representation was instrumental in providing the first short proof of the Principal Ideal Theorem (cf. Magnus [1934]).

Theorem B, which is the principal result and whose proof occupies the bulk of this paper (Section 4), gives an upper bound of $e \cdot \phi(p^e) + 1$ for the class of $H = F/F''F^{p^e}$, provided ϱ ($=$ rank H/H') $\leq p + 1$. More precisely, viewing H' as $\mathbb{Z}G$ module, where $G = H/H'$, the fact that H has exponent p^e forces H to be annihilated by the ideal J in $\mathbb{Z}G$ generated by p^e and $1 + x + \cdots + x^{p^e-1}$, $\forall\, x \in G$. Since the $(k+1)$st term of the lower central series of H equals I^kH', where I is the augmentation ideal of $\mathbb{Z}G$, the upper bound is achieved by showing $I^{e\phi(p^e)} \subset J$. The proof is a self-contained double induction argument on indices which measure what amount to be the layer, $p^iI/p^{i+1}I$, and the weights in each of the generators for elements in powers of I; Lemma 3 of Section 4 is the induction step. In his review of this paper, M. Newman [1969] asserts that the same bound but with the number of generators relaxed to at most $(e-1)\cdot(p-1)$ can be obtained by the methods of his paper [1968] written together with N. Gupta (although this is not proved explicitly). The lower bound of $e\phi(p^e)$ for the class of H is effected in Section 5 by means of the Magnus representation of F/F'' when $\varrho = 2$. Using this paper, the discrepancy of 1 between the upper and lower bounds for the class of H is resolved for the 2-generator group of exponent p^2 by Bachmuth and Mochizuki [1968] and for the 2-generator group of arbitrary exponent by Dark and Newell [1981], their exact nilpotence class being the lower bound $e\phi(p^e)$. Earlier, Meier-Wunderli [1950] had shown that when H requires more than 2 generators and has exponent p, its class always equals the upper bound p.

This paper is a nontrivial contribution to the combinatorial theory of finite groups. But in addition to showing Heilbronn's skill in this area, it demonstrates his ability to bring into this collaboration insight and expertise gained from a point of view other than the purely group theoretic. This diversity of insight is surely a hallmark of Heilbronn's career.

[39] On the average length of a class of finite continued fractions

Commentary by B. J. Birch

Theorem 1 of the paper is self-explanatory. Heilbronn gives an elementary estimate for his function $r(N)$; he could, of course, have quoted stronger classical estimates (e.g., Estermann [1930]). A recent reference using stronger methods is Iwaniec and Deshouillers [1982].

[43] On the 2-class group of cubic fields

Commentary by B. J. Birch

Let k_3 be a cubic field with discriminant d_3; then to any ideal class of k_3 of order 2 there corresponds an unramified quadratic extension of k_3. Heilbronn shows how such elements of order 2 in the ideal class group of k_3 correspond (many to many; he makes it precise) to quartic fields k_4 with discriminant d_4 such that $d_4 = d_3$ and $k_3 \subset \overline{k_4}$, the normal closure of k_4. One may in principle proceed on these lines to obtain a theorem about the density of cubic fields with elements of even order in their class group, which parallels the theorems of Davenport and Heilbronn on quadratic fields with class number divisible by 3 (cf. [41], [42]); unfortunately, there are difficulties to overcome (cf. Sheelagh Lloyd, Oxford thesis, 1986).

References

S. Bachmuth and H. Y. Mochizuki

[1967] Cyclotomic ideals in group rings. *Bull. Amer. Math. Soc.* **72**, 1018–1020.

[1968] The class of the free metabelian group with exponent p^2. *Comm. Pure Appl. Math.* **21**, 385–399.

S. Chowla

[1950] A new proof of a theorem of Siegel. *Ann. Math.* (2) **51**, 120–122.

R. Dark and M. Newell

[1981] On 2-generator metabelian groups of prime power exponent. *Arch. Math.* **37**, 385–400.

T. Estermann

[1930] On the representations of a number as the sum of two products. *Proc. London Math. Soc.* (2) **31**, 123–133.

[1948] On Dirichlet's *L*-function. *J. London Math. Soc.* **23**, 275–279.

A. Fröhlich

[1983] *Central Extensions, Galois Groups, and Ideal Class Groups of Number Fields* (*Contemp. Math.*, Vol. 24), Am. Math. Soc., Providence, RI.

P. Furtwängler

[1930] Beweis des Hauptidealsatzes für Klassenkörper algebraischer Zahlkörper. *Abh. Math. Sem. Univ. Hamburg* **7**, 14–36.

D. Goldfeld

[1974] A simple proof of a Siegel's theorem. *Proc. Nat. Acad. Sci. USA* **71**, 1055.

N. D. Gupta and M. F. Newman

[1968] Engel congruences in groups of prime power exponent. *Can. J. Math.* **20**, 1321–1323.

H. Iwaniec and J. M. Deshouillers

[1982] An additive divisor problem. *J. London Math. Soc.* (2) **26**, 1–14.

W. Magnus

[1934] Über des Beweis des Hauptidealsatzes. *J. Reine Angew. Math.* **170**, 235–240.

H. Meier-Wunderli

[1951] Metabelsche Gruppen. *Comment. Math. Helv.* **25**, 1–10.

M. F. Newman

[1969] Review of *Burnside metabelian groups. Math. Reviews* **38**, #4561.

J. E. Olson

[1968] An addition theorem modulo p. *J. Comb. Th.* **5**, 45–52.

C. Ryavec

[1968] The addition of residue classes modulo n. *Pacific J. Math.* **26**, 367–373.

A. Scholz and O. Taussky

[1934] Die Hauptideale der kubischen Klassenkörper imaginärquadratischer Zahlkörper, *J. Reine Angew. Math* **171**, 19–41.

C. L. Siegel

[1935] Über die Classenzahl quadratischer Zahlkörper. *Acta Arith.* **1**, 83–86 (*Gesammelte Abhandlungen* I, 406–409).

E. Szemeredi

[1970] On a conjecture of Erdös and Heilbronn. *Acta Arith.* **17**, 227–229.

Acknowledgments

The editors would like to thank the publishers of the papers of Hans Heilbronn for granting permission to reprint the following papers in this volume:

[1] Reprinted by permission from *Mathematische Annalen* **104**, 794–799, © 1931 by Springer Verlag.

[1t] Translated by permission from *Mathematische Annalen* **104**, 794–799, © 1931 by Springer Verlag.

[2] Reprinted by permission from *Mathematische Zeitschrift* **36**, 394–423, © 1933 by Springer Verlag.

[2t] Translated by permission from *Mathematische Zeitschrift* **36**, 394–423, © 1933 by Springer Verlag.

[3] Reprinted by permission from *Mathematische Zeitschrift* **37**, 10–16, © 1933 by Springer Verlag.

[3t] Translated by permission from *Mathematische Zeitschrift* **37**, 10–16, © 1933 by Springer Verlag.

[4] Reprinted by permission from *Mathematische Zeitschrift* **37**, 17, © 1933 by Springer Verlag.

[4t] Translated by permission from *Mathematische Zeitschrift* **37**, 17, © 1933 by Springer Verlag.

[5] Reprinted by permission from *Mathematische Zeitschrift* **37**, 18–21, © 1933 by Springer Verlag.

[5t] Translated by permission from *Mathematische Zeitschrift* **37**, 18–21, © 1933 by Springer Verlag.

[6] Reprinted by permission from *Mathematische Zeitschrift* **37**, 37–38, © 1933 by Springer Verlag.

[6t] Translated by permission from *Mathematische Zeitschrift* **37**, 37–38, © 1933 by Springer Verlag.

[7] Reprinted by permission from *The Quarterly Journal of Mathematics* **5**, 150–160, © 1934 by Oxford University Press.

[8] Reprinted by permission from *The Quarterly Journal of Mathematics* **5**, 293–301, © 1934 by Oxford University Press.

[9] Reprinted by permission from *Casopis pro pestování matematiky a fysiky* **65**, 117–141, © 1936 by Matematický Ústav Ceskoslovenské Akademie Ved.

[9t] Translated by permission from *Casopis pro pestování matematiky a fysiky* **65**, 117–141, © 1936 by Matematický Ústav Ceskoslovenské Akademie Ved.

[10] Reprinted by permission from *Acta Arithmetica* **1**, 212–221, © 1936 by Acta Arithmetica.

[10t] Translated by permission from *Acta Arithmetica* **1**, 212–221, © 1936 by Acta Arithmetica.

[11] Reprinted by permission from *Journal of the London Mathematical Society* **11**, 181–185, © 1936 by The London Mathematical Society.

[12] Reprinted by permission from *Journal of the London Mathematical Society* **11**, 307–312, © 1936 by The London Mathematical Society.

[13] Reprinted by permission from *Proceedings of the London Mathematical Society* **41**, 143–150, © 1936 by The London Mathematical Society.

[14] Reprinted by permission from *Proceedings of the London Mathematical Society* **41**, 449–453, © 1936 by The London Mathematical Society.

[15] Reprinted by permission from *Proceedings of the London Mathematical Society* **43**, 73–104, © 1937 by The London Mathematical Society.

[16] Reprinted by permission from *Proceedings of the London Mathematical Society* **43**, 142–151, © 1937 by The London Mathematical Society.

[17] Reprinted by permission from *Acta Arithmetica* **2**, 212–213, © 1937 by Acta Arithmetica.

[18] Reprinted by permission from *Proceedings of the Cambridge Philosophical Society* **33**, 207–209, © 1937 by Cambridge University Press.

[19] Reprinted by permission from *The Quarterly Journal of Mathematics* **9**, 194–195, © 1938 by Oxford University Press.

[20] Reprinted by permission from *Proceedings of the Cambridge Philosophical Society* **34**, 521–526, © 1938 by Cambridge University Press.

[21] Reprinted by permission from *Journal of the London Mathematical Society* **13**, 302–310, © 1938 by The London Mathematical Society.

[22] Reprinted by permission from *Journal of the London Mathematical Society* **21**, 185–193, © 1946 by The London Mathematical Society.

[23] Reprinted by permission from *The Quarterly Journal of Mathematics* **18**, 107–121, © 1947 by Oxford University Press.

[24] Reprinted by permission from *Journal of the London Mathematical Society* **22**, 53–61, © 1947 by The London Mathematical Society.

[25] Reprinted by permission from *The Quarterly Journal of Mathematics* **19**, 249–256, © 1948 by Oxford University Press.

[26] Reprinted by permission from *Proceedings of the Cambridge Philosophical Society* **45**, 194–206, © 1949 by Cambridge University Press.

[27] Reprinted by permission from *Proceedings of the Cambridge Philosophical Society* **46**, 377–382, © 1950 by Cambridge University Press.

[28] Reprinted by permission from *Canadian Journal of Mathematics* **3**, 257–268, © 1951 by Canadian Journal of Mathematics.

[29] Reprinted by permission from *Mathematika* **5**, 1–7, © 1958 by Mathematika.

[30] Reprinted by permission from *Bulletin de l'Académie Polonaise des Sciences* **6**, 181–184, © 1958 by Polska Akademia Nauk.

[31] Reprinted by permission from *The Mathematical Gazette* **45**, 77–78, © 1961 by The Mathematical Association.

[32] Reprinted by permission from *Journal of the London Mathematical Society* **37**, 6–9, © 1962 by The London Mathematical Society.

[33] Reprinted by permission from *Nature* **200**, 937–938, © 1963 by Macmillan Journals Limited.

[34] Reprinted by permission from *Journal of the London Mathematical Society* **39**, 72–76, © 1964 by The London Mathematical Society.

[35] Reprinted by permission from *Acta Arithmetica* **9**, 149–159, © 1964 by Acta Arithmetica.

[36] Reprinted by permission from *Collected Papers of G.H. Hardy* (edited), Vol I, pp. 263–264, 377–381, 533–534, © 1966 by Oxford University Press.

[37] Reprinted by permission from *Algebraic Number Theory* (J. W. S. Cassels and A. Fröhlich, eds.), pp. 204–230, © 1967 by Academic Press.

[38] Reprinted by permission from *Proceedings of the Royal Society, London* **A307**, 235–250, © 1968 by The Royal Society.

[39] Reprinted by permission from *Abhandlungen aus Zahlentheorie und Analysis zur Erinnerung an Edmund Landau* (P. Turán, ed.), pp. 89–96, © 1969 by VEB Deutscher Verlag der Wissenschaften.

[40] Reprinted by permission from *Canadian Mathematical Bulletin* **12**, 479–480, © 1969 by Canadian Mathematical Bulletin.

[41] Reprinted by permission from *Bulletin of the London Mathematical Society* **1**, 345–348, © 1969 by The London Mathematical Society.

[42] Reprinted by permission from *Proceedings of the Royal Society, London* **A322**, 405–420, © 1971 by The Royal Society.

[43] Reprinted by permission from *Studies in Pure Mathematics* (L. Mirsky, ed.), pp. 117–119, © 1971 by Academic Press.

[44] Reprinted by permission from *Proceedings of the 1972 Number Theory Conference*, pp. 108–110, © 1972 by Department of Mathematics, University of Colorado.

[45] Reprinted by permission from *Canadian Journal of Mathematics* **25**, 870–873, © 1973 by Canadian Journal of Mathematics.

The editors would also like to thank J. W. S. Cassels, A. Fröhlich, and The Royal Society for permission to reprint the biographical memoir of Hans Heilbronn in this volume. The original article appeared in *Biographical Memoirs of Fellows of the Royal Society* **22**, 119–135, © 1976 by The Royal Society.